# COMBUSTION FUNDAMENTALS

## McGraw-Hill Series in Energy, Combustion, and Environment

**Consulting Editor**

**Norman Chigier**

**Chigier:** *Energy, Combustion, and Environment*
**Fraas:** *Engineering Evaluation of Energy Systems*
**Lefebvre:** *Gas Turbine Combustion*
**Strehlow:** *Combustion Fundamentals*
**Toong:** *Combustion Dynamics: The Dynamics of Chemically Reacting Fluids*

# COMBUSTION FUNDAMENTALS

**Roger A. Strehlow**

*Professor of Aeronautical and Astronautical Engineering*
*University of Illinois at Urbana-Champaign*

**McGraw-Hill Book Company**

New York  St. Louis  San Francisco  Auckland  Bogotá  Hamburg
Johannesburg  London  Madrid  Mexico  Montreal  New Delhi
Panama  Paris  São Paulo  Singapore  Sydney  Tokyo  Toronto

This book was set in Times Roman.
The editor was Anne Murphy;
the production supervisor was Diane Renda.
R. R. Donnelley & Sons Company was printer and binder.

**COMBUSTION FUNDAMENTALS**

1234567890 DOCDOC 8987654

ISBN 0-07-062221-3

**Library of Congress Cataloging in Publication Data**

Strehlow, Roger A.
    Combustion fundamentals.

    (McGraw-Hill series in energy, combustion, and environment)
    Includes index.
    1. Combustion. I. Title. II. Series.
QD516.S887   1984     541.3′61     84-7938
ISBN 0-07-062221-3

To my wife Ruby

# CONTENTS

## Appendixes

## Index

# PREFACE

This book is intended to be an introductory textbook on combustion for advanced undergraduates or beginning graduate students in engineering. Although it is based on my 1968 book, entitled *Fundamentals of Combustion*, it is actually a major revision of that text, dictated first by pedagogical considerations and second by the tremendous increase in our understanding of combustion phenomena during the past sixteen years.

It is assumed that the student will have good understanding of gas dynamics and mathematics through advanced calculus. It is also assumed that he or she may need some support in physical chemistry. Thus, the first five chapters cover the necessary introductory subjects. These are: (1) Elementary kinetic theory of gases and a review of thermodynamic principles, (2) stoichiometry and thermochemistry, (3) the origin of the transport properties that are important to combustion, (4) chemical kinetics and chemical equilibrium, and (5) reactive gas dynamics, for both laminar and turbulent flows.

The next four chapters deal with the four fundamental combustion processes. These are: (6) combustion chemistry including adiabatic and vessel explosions, and kinetic rate measurements, (7) diffusion flames, (8) flames in premixed gases, and (9) gas-phase detonations. In the next two chapters the aerodynamics of steady laminar and turbulent flames are discussed. Chapter 12 contains a discussion of flame ignition and extinction and other important transient flame processes such as quenching and flammability limits. Detonation failure, transmission and direct initiation are discussed in Chapter 13. In Chapter 14 the aerodynamics of certain nonsteady flames, including vessel explosions and deflagration to detonation transition are discussed. Chapter 15 contains a discussion of condensed explosive detonation and solid and liquid propellant combustion. Finally, Chapter 16 is a brief survey of the major air pollutants generated by combustion processes and their fate in the earth's atmosphere.

This book is designed to be used as a teaching text. The first nine chapters, which treat material that is basic to the understanding of combustion processes, are arranged in a logically developed sequence which should give the student an understanding of not only the processes themselves but also some of the accepted techniques that have been used to model them. The next five chapters deal with the aerodynamics of flames and transient flame and detonation phenomena. Here only selected topics are presented. They have been chosen either because of their importance to the field or because they represent rather well understood complex phenomena. A thorough understanding of the contents of these fourteen chapters should train the student to independently evaluate the combustion literature and/or properly investigate any complex combustion process that is not well understood. The last two chapters cover material that is peripheral to the main thrust of the text. They are included primarily because of my interest in high explosives and propellants and because combustion-generated pollution represents an important problem that has always been associated with practical combustion devices.

This text has evolved from my previous text and from a course in the fundamentals of combustion that I have continued to teach during the intervening fifteen years since the appearance of that text. As such I owe a debt of gratitude to the many students who have contributed either directly or indirectly to my understanding of the subject matter. I would also like to thank Professor C. K. Law of Northwestern University for his helpful discussions of flame stretch, droplet combustion, and large activation energy asymptotics, Professor Craig T. Bowman of Stanford University and Professor N. S. Vlachos of the University of Illinois at Urbana–Champaign (UIUC) for their perceptive critiques of Chapters 6 and 5 respectively, and Professor Richard O. Buchius for his critique of the section on catalytic combustion. Professor Lester D. Savage was particularly helpful in the preparation of the appendixes and the alphabetization of the index. Additionally the technical content and organization of the text was undoubtedly improved by the many discussions I have had with my close colleagues at the UIUC, Professors Herman Krier, Lester D. Savage, and Spencer C. Sorenson. Finally I would like to acknowledge with thanks the work of Kelly Collier and Linda Harmening, who typed portions of the manuscript as course notes.

*Roger A. Strehlow*

# COMBUSTION FUNDAMENTALS

# KINETIC THEORY AND
# THERMODYNAMICS OF A DILUTE GAS

## 1-1 INTRODUCTION

Combustion processes are very diverse in nature. Nevertheless, two fundamental things always happen when combustion occurs. These are:

1. The species composition of the mixture changes with time, and these changes are caused by processes at the molecular level.
2. Weak molecular bonds are broken and replaced by stronger bonds and the excess bond energy is liberated to the system, usually causing a very large increase in temperature.

Furthermore, most combustion processes occur at sufficiently low pressures and high temperatures that the gaseous portion of the system can be considered to be a dilute gas in the kinetic theory sense. Therefore, it is appropriate that in this chapter we first briefly review basic thermodynamic principles, particularly as they apply to an ideal gas, and then discuss molecular collision processes in such a gas. Then, in the following chapter, we introduce the principles of mass balance and the thermochemical requirements which apply during processes involving changes in the species composition of a mixture. Other fundamental processes that can be important to combustion are discussed in the next three chapters, prior to the discussion of combustion processes, per se, in the rest of the text.

## 1-2 THE IDEAL GAS STATE

The gaseous state under ordinary conditions (i.e., the air we breathe) is extremely dilute. If we assume that the molecules are touching in a solid, we find, by measuring the relative densities, that in the gaseous state the molecules are separated by extremely large distances compared to their molecular diameters. Furthermore, we know from adequate experimental evidence that in the limit of very low density all gases obey the ideal gas law

$$P = \rho \mathcal{R} T$$

where $\mathcal{R}$ has the units of J/kg K and where the experimentally determined value of $\mathcal{R}$ is dependent upon the composition of the gas in question. It is also known from ample experimentation that at constant temperature and in the limit as $P$ approaches zero the ratio $P/\rho$ for various pure gases is equal to the inverse molecular weight ratio for these substances, i.e.,

$$\frac{\left(\dfrac{P}{\rho}\right)_A}{\left(\dfrac{P}{\rho}\right)_B} = \frac{\mathcal{R}_A}{\mathcal{R}_B} = \frac{M_B}{M_A}$$

We may therefore write

$$\mathcal{R}_A M_A = \mathcal{R}_B M_B = R$$

where $R$ is the universal gas constant for all substances with the value 8.31434 J/mole K. The ideal gas law then becomes

$$PV = RT \qquad (1\text{-}1)$$

where $V = M/\rho$, the volume of one mole or $6.02283 \times 10^{23}$ molecules of the gas. The volume of one mole of an ideal gas at $P = 101{,}325$ pascals (or one atmosphere) and $T = 273.15$ K (0°C) has been found to be 0.022414 m³. Equation (1-1) is a universal equation of state for all gases if the density is low enough. In addition, this universal behavior allows us to define a convenient temperature scale, the ideal gas temperature scale, by using Eq. (1-1) with the added definition that the temperature of melting ice at one atmosphere pressure is 273.15 K or 491.67°R.

## 1-3 REVIEW OF THERMODYNAMICS

In elementary treatments of thermodynamics one is introduced to eight fundamental thermodynamic variables. These are pressure $P$, temperature $T$, volume $V$, internal energy $E$, entropy $S$, enthalpy $H = E + PV$, Helmholtz free energy or work function $A = E - TS$, and Gibbs free energy $F = H - TS = E + PV - TS$. The first five of these variables are fundamental to the subject while the last three are defined for operational convenience. The first two are intensive variables, that is, are independent of the quantity of material under consideration. The numerical values of the last six are proportional to the quantity of material under consideration and are therefore called extensive variables.

Strictly speaking, one can only define these thermodynamic quantities for a system that has uniform properties and is in thermodynamic equilibrium. Furthermore, these eight fundamental thermodynamic variables are all state variables and may each be expressed uniquely as a function of any two other thermodynamic state variables, which are called the independent variables of the system in that representation. This is equivalent to the mathematical statement that the value of the line integral connecting two points in the plane of the independent variables is independent of the path used in its evaluation. However, this last statement is the definition of an exact differential and therefore the eight primary thermodynamic variables may all be expressed as exact differentials in terms of any two other state variables.

One more general statement may be made about the last five thermodynamic variables. In general, we are never interested in the absolute value of these variables but are interested only in difference values associated with a change from one thermodynamic state to another. In other words, if $P$ and $T$ are taken to be appropriate independent thermodynamic variables we would be interested only in determining $E_2(P_2, T_2) - E_1(P_1, T_1)$ and not the absolute value of either $E_1$ or $E_2$.

The first law of thermodynamics is a statement of conservation of energy. In a quiescent system, the first law may be written

$$E_2 - E_1 = Q - W \tag{1-2}$$

where $Q$ is the heat added to the system and $W$ is the work performed on the surroundings. If the process that is used to go from state 1 to state 2 is a reversible process such that a thermodynamic state can be defined for the system during every step of the process, Eq. (1-2) may be written

$$dE = \dbar Q_{rev} - \dbar W$$

where the two differentials on the right-hand side are crossed to indicate that they are path dependent differentials. When written in integral form we obtain

$$E_2 - E_1 = \int_1^2 \dbar Q_{rev} - \int_1^2 \dbar W \tag{1-3}$$

Conceptually, there are an infinite number of path processes or combinations of heat addition and work performed which will cause a transition from state 1 to state 2 through a completely reversible process.

If we further restrict ourselves to allow only compressible $PV$ work on the system, Eq. (1-3) may be written as

$$dE = \dbar Q_{rev} - P\, dV \tag{1-4}$$

where $V$ is the volume of the system and the $PV$ path as well as the $dQ$ term must be known to determine $E_2 - E_1$.

All substances have heat capacities. For example, solids and liquids show a characteristic increase in temperature when a fixed amount of heat is added from the surroundings. However, because of their innate compressibility, gases have heat capacities that are path dependent. Therefore, two limiting heat capacities are usually defined for a gas. These are the heat capacity when heat is added at constant volume $C_v = (dQ/dT)_v$ or at constant pressure $C_p = (dQ/dT)_p$. When heat is added at constant volume

$$C_v = \left( \frac{dQ_{rev}}{dT} \right)_v = \left( \frac{dE}{dT} \right)_v$$

since there is no $PV$ work performed during the process. However, at constant pressure a portion of the heat energy added to a gas is used to perform work on the surroundings as the gas increases its volume against the constant counter-pressure. Because of this, the definition of $C_p$ using the internal energy yields a rather complex equation in terms of state variables. This is the reason why enthalpy, $H = E + PV$, was defined. If we differentiate this definition and substitute the first law, Eq. (1-4), we obtain

$$dH = dQ_{rev} + V \, dP$$

Therefore, for a constant pressure process

$$C_p = \left( \frac{dQ_{rev}}{dT} \right)_p = \left( \frac{dH}{dT} \right)_p$$

Thus enthalpy is the natural variable to use when describing a constant-pressure process or when defining $C_p$.

Entropy and the second law of thermodynamics are concepts related to reversibility. The entropy change for a system undergoing a reversible process is defined as

$$S_2 - S_1 = \int_1^2 \frac{dQ_{rev}}{T}$$

Furthermore, it can be shown that for an isolated system $S_2 - S_1 = 0$ for a reversible process, while a positive value of $S_2 - S_1$ represents a spontaneous process which is not reversible and a negative value represents an unobservable process. For example it can be shown that the adiabatic (i.e., $dQ = 0$) expansion of a small volume of gas into a vacuum chamber always causes an increase in the entropy of the gas, while the reverse process, namely the spontaneous collection of all the gas in a large chamber into a small volume, would yield an entropy decrease. This latter process has never been observed to occur spontaneously.

If the differential definition of entropy is substituted into the first law, Eq. (1-4), we obtain the expression

$$dE = T \, dS - P \, dV \tag{1-5}$$

or in terms of enthalpy

$$dH = T\,dS + V\,dP \tag{1-6}$$

Since entropy is a state variable we may rearrange Eq. (1-5) to yield

$$S_2 - S_1 = \int_1^2 \frac{dE}{T} + \int_1^2 \frac{P\,dV}{T} \tag{1-7}$$

Using the definition of enthalpy we may also write

$$S_2 - S_1 = \int_1^2 \frac{dH}{T} - \int_1^2 \frac{V\,dP}{T} \tag{1-8}$$

Equations (1-7) and (1-8) allow the calculation of the entropy change of any system once an equation of state is known.

Because of the form of the first law the natural variables for expressing an equation of state are either $E(S, V)$ or $H(S, P)$. Using these independent variables and the first law as expressed in Eqs. (1-5) and (1-6) we find that

$$T = \left(\frac{\partial E}{\partial S}\right)_v \qquad P = -\left(\frac{\partial E}{\partial V}\right)_s \tag{1-9}$$

and

$$T = \left(\frac{\partial H}{\partial S}\right)_p \qquad V = \left(\frac{\partial H}{\partial P}\right)_s \tag{1-10}$$

The first and second expressions of either Eqs. (1-9) or (1-10) may be evaluated to yield what are commonly called the caloric and thermal equations of state.

Before discussing the implications of Eqs. (1-9) or (1-10) we must introduce one more relation which can be obtained from fundamental thermodynamic considerations, independent of the equations of state. This is the reciprocity relationship of thermodynamics which provides a relationship between the thermal and caloric equation of state for any substance. In terms of internal energy and volume the reciprocity relation is

$$\left(\frac{\partial E}{\partial V}\right)_T = T\left(\frac{\partial P}{\partial T}\right)_v - P \tag{1-11}$$

while in terms of enthalpy and pressure it is

$$\left(\frac{\partial H}{\partial P}\right)_T = V - T\left(\frac{\partial V}{\partial T}\right)_p \tag{1-12}$$

As an example of the application of the reciprocity relation consider an ideal gas, $PV = RT$. Differentiation of the right-hand side of Eq. (1-12) for an ideal gas yields

$$\left(\frac{\partial H}{\partial P}\right)_T = V - T\left(\frac{R}{P}\right) = 0$$

Therefore, for an ideal gas

$$H = H(T)$$

Using Eq. (1-11) it can also be shown that $E = E(T)$ and therefore from the definition of enthalpy we have

$$H(T) = E(T) + PV \tag{1-13}$$

Differentiating Eq. (1-13) with respect to temperature at constant pressure we obtain

$$C_p = C_v + P\left(\frac{\partial V}{\partial T}\right)_P = C_v + P\frac{R}{P} = C_v + R \tag{1-14}$$

where $C_p$ and $C_v$ still may be functions of temperature. For an ideal gas it is convenient to define a heat capacity ratio $\gamma(T) = C_p/C_v$, which is a function of temperature only.

Now, if we substitute the ideal gas law into Eq. (1-8) we obtain

$$S_2 - S_1 = \int_1^2 \frac{C_p\, dT}{T} - R \ln \frac{P_2}{P_1}$$

If we now assume that $C_p = $ constant or that the gas is calorically perfect, we obtain the equations

$$H_2 - H_1 = C_p(T_2 - T_1)$$

and

$$S_2 - S_1 = C_p \ln \frac{T_2}{T_1} - R \ln \frac{P_2}{P_1} \tag{1-15}$$

In this case the heat capacity ratio $\gamma$ is also a constant. An ideal gas which has constant heat capacity (i.e., with $\gamma = $ constant) is sometimes called a polytropic gas. If we now rearrange and combine these equations we obtain a single equation of state for a polytropic gas

$$H \propto (P)^{R/C_p} \cdot e^{S/C_p} \tag{1-16}$$

where the proportionality constant may be evaluated from initial conditions.

The last two of the eight thermodynamic variables are defined for operational convenience when dealing with questions of thermodynamic equilibrium. It turns out that most experimental measurements are made in systems which are held at constant temperature and therefore are not isolated in the thermodynamic sense. Thus the entropy argument, even though useful, may not be conveniently applied in a direct manner. This was first pointed out by Massieu in 1869 approximately 35 years after the concept of entropy was first introduced by Clausius. Massieu defined two new functions for discussing equilibrium in constant $(P, T)$ and $(V, T)$ systems. However, it was Gibbs who first demonstrated the utility of the newly defined functions and extensively discussed their possible applications. The Helmholtz free energy is useful for discussing equilibrium in a system held at constant volume and temperature, since we may write

$$dA = dE - T \, dS - S \, dT$$

$$= đQ - P \, dV - T \, dS - S \, dT$$

where the only work is $PV$ work. For a process occurring at constant volume and temperature this equation reduces to

$$dA = đQ - T \, dS$$

However, since

$$dS = \frac{đQ_{rev}}{T} \quad \text{and} \quad dS > \frac{đQ_{irr}}{T}$$

we obtain the statements that $dA = 0$ for a reversible process and that $A$ decreases for a spontaneous process occurring at constant volume and temperature.

In a similar manner we may write

$$dF = dH - T \, dS - S \, dT$$

$$= đQ - T \, dS \qquad \text{at constant pressure and temperature}$$

This leads to the statements that $dF = 0$ for a reversible process and that $F$ decreases for a spontaneous process occurring at constant pressure and temperature. Thus if we are able to express either the free-energy or work function for all possible states of a homogeneous system in terms of any two other state variables, the state of minimum $A$ (for $V$, $T$ held constant) or minimum $F$ (for $P$, $T$ held constant) will be the thermodynamic equilibrium state of that system within the imposed constraints. This general principle will be applied to chemical equilibrium in Chap. 4.

## 1-4 ELEMENTARY KINETIC THEORY FOR A DILUTE GAS: PRESSURE AND MOLECULAR VELOCITY

There are three macroscopic observations concerning dilute gases which form the foundation of elementary kinetic theory.

1. A dilute gas has an extremely low density when compared to the solid or liquid phase.
2. A dilute gas at equilibrium exerts an equal pressure on all walls of the container and obeys the equation of state $PV = RT$.
3. The internal energy of a dilute gas is a function of the temperature and independent of the pressure.

These observations when coupled with our concept of indivisible atoms or molecules having fixed properties yield the three basic premises of kinetic theory.

1. The pressure exerted by a dilute gas must be caused by individual molecules colliding with the walls, and therefore the gas molecules must be in constant motion in the container.
2. The molecules, being far apart, must be unaware of the existence of their neighbors during a good portion of the time. In other words, if you were to look at any one molecule of a dilute gas many times, the chances of finding it in collision with another molecule would be extremely small.
3. The internal energy of the gas must be stored in the individual molecules either as internal molecular energy or as translational energy associated with their average speed of motion relative to the container. Significant collisional energy storage is ruled out by the second premise.

If we are to use the laws of probability in deriving the properties of an ideal gas we must, in addition to these three premises, make an assumption concerning the behavior of the individual molecules or the system of molecules in equilibrium. We must assume (without a priori justification) that any allowable energy state or position of an individual molecule has an equal a priori probability of being observed if we look at a large number of samples of the gas or at one sample for a long enough period of time. This statement implies that the molecule will spend an equal amount of time in every available energy state and every available position over a long period of time. It also implies that we may externally restrict the number of available states or available positions. We restrict positions, for example, by defining the volume and we restrict states, for example, by requiring that a mixture of hydrogen and oxygen at room temperature cannot form any water molecules (even though it is thermodynamically possible for water to be formed). Thus these restrictions are based on conclusions drawn from our ordinary macroscopic observations.

This assumption that a state or position can either exist or not exist—and if it does exist that it has an equal a priori probability of being observed—is called the *ergodic hypothesis* of statistical mechanics. It may be applied equally well to an individual molecule in equilibrium with a large number of other molecules or to a macroscopic system of $N$ molecules in equilibrium with a large number of identical macroscopic systems. The only justification for this assumption is that it allows us to use the laws of probability with exactness, and further (as we will see) that it gives results which agree well with experimental observations. Notice that neither of these justifications validates the initial assumption.

Let us now consider a cubic meter of gas containing $N$ molecules. If we assume $N$ is large so that the system is macroscopic, the equal a priori assumption and our definition of macroscopic can be used to yield the principle of detailed balances. The principle of detailed balances states that for a macroscopic system at equilibrium each individual molecular process is balanced by an equal and exactly opposite process taking place at some other point in the system. If this were not true the distribution of the molecules would change with time and for a macroscopic system at equilibrium this type of change is extremely unlikely.

Now let us consider the velocity distribution of $N$ molecules contained in a unit volume. We first must assume (for the simple theory) that these molecules cannot store energy internally (i.e., we assume that the gas is monatomic). In addition, since the gas is rarefied, we assume that each molecule spends most of its time well removed from the other molecules. In the main, then, the only way the individual molecules can store the gas's internal energy is as molecular kinetic energy

$$\varepsilon = \tfrac{1}{2}mc^2$$

where $m$ is its mass and $c$ is the molecule's speed relative to the main body of the gas. We of course do not know the distribution of molecular velocity $f(u, v, w)\ du\ dv\ dw$, where $f(u, v, w) = f(\mathbf{V})$ is the number of molecules per unit volume in physical space in the velocity range

$$\mathbf{V} \quad \text{to} \quad (\mathbf{V} + d\mathbf{V})$$

with components

$$(u, v, w) \qquad \text{to} \qquad (u + du, v + dv, w + dw)$$

However, we can deduce that the value of $f(u, v, w)$ is independent of time in a quiescent macroscopic system and that it is not a simple constant. It must be independent of time because of the principle of detailed balances (i.e., because the system is macroscopic) and it cannot be a simple constant, since the limits on the integral

$$N = \int\!\!\!\int\!\!\!\int_{-\infty}^{\infty} f(u, v, w)\ du\ dv\ dw$$

must be infinite and the value of $N$ must be finite. Notice that we may define the distribution in terms of one velocity component

$$h(u) = \int\!\!\!\int_{-\infty}^{\infty} f(u, v, w)\ dv\ dw$$

Or in terms of two velocity components

$$g(u, v) = \int_{-\infty}^{\infty} f(u, v, w)\ dw$$

Furthermore, since the gas is isotropic we know that $h(u) = h(v) = h(w)$, and since the gas is quiescent we can state that the average molecular velocity

$$\overline{V} = \frac{\displaystyle\int\!\!\!\int\!\!\!\int_{-\infty}^{\infty} V(u,\ v,\ w) f(u,\ v,\ w)\ du\ dv\ dw}{\displaystyle\int\!\!\!\int\!\!\!\int_{-\infty}^{\infty} f(u,\ v,\ w)\ du\ dv\ dw}$$

is zero. However, the average square speed is not zero, but is related to the average kinetic energy of the molecules

$$\overline{V \cdot V} = \overline{c^2} = \frac{\displaystyle\int\!\!\!\int\!\!\!\int_{-\infty}^{\infty} c^2(u,\ v,\ w) f(u,\ v,\ w)\ du\ dv\ dw}{\displaystyle\int\!\!\!\int\!\!\!\int_{-\infty}^{\infty} f(u,\ v,\ w)\ du\ dv\ dw}$$

$$= \frac{3}{N} \int\!\!\!\int\!\!\!\int_{-\infty}^{\infty} u^2 f(u,\ v,\ w)\ du\ dv\ dw$$

since the gas is isotropic and

$$\overline{c^2} = \overline{u^2 + v^2 + w^2} = \overline{3u^2}$$

Therefore, the energy per unit volume may be written as

$$\mathscr{E} = N\bar{\varepsilon} = \frac{1}{2}\,mN\overline{c^2} = \frac{3}{2}\,m \int\!\!\!\int\!\!\!\int_{-\infty}^{\infty} u^2 f(u,\ v,\ w)\ du\ dv\ dw$$

for a monatomic gas.

Let us now consider the pressure exerted against one square meter of a wall in the $yz$ plane for the case of a unit volume of a monatomic gas containing $N$ molecules. We assume that the pressure is equal to the time average momentum change of the molecules which strike the wall. If we assume specular reflection, this momentum change is $2mu$ for a molecule with a velocity component $u$. However, we know that molecules do stick to the walls and in general are emitted in a direction and at a velocity which has little do do with their incident direction and velocity. Thus wall collisions in the microscopic sense are similar to intermolecular interactions, i.e., each causes changes that must be calculated for the individual interaction. However, in the macroscopic sense, the principle of detailed balances allows us to ignore the complicated individual processes. We simply note that if the gas and wall are in equilibrium on the

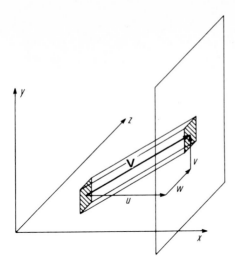

**Figure 1-1** Molecules of velocity **V** striking one square meter of a wall in the *yz* plane.

average each molecule striking the wall with a momentum *mu* is balanced by one emitted with a momentum of $-mu$ and furthermore, since there is no tangential force on a wall enclosing an equilibrium gas, the momentum terms *mv* and *mw* are conserved for a large number of collisions.

Now let us determine the integrated momentum change for all molecules which strike 1 m$^2$ of the *yz* plane in contact with an equilibrium gas. First consider molecules of the type $f(u, v, w)$. The number of these molecules hitting the wall per second are contained in a parallelepiped of base 1 m$^2$ and height *u*(m/s). Thus, since $f(u, v, w)\, du\, dv\, dw$ is the number of molecules of this type contained in a unit volume, the number of this type hitting the wall is simply

$u\, f(u, v, w)\, du\, dv\, dw$     molecules of type $(u, v, w)$ to hit the wall/m$^2$-s

(See Fig. 1-1.) Since each of these molecules carries with it an *x* component of momentum equal to *mu*, the net momentum change on reflection from the wall is

$2mu^2 f(u, v, w)\, du\, dv\, dw$     change in momentum/s (kg m/m$^2$-s$^2$)
                                                     due to molecules of type $(u, v, w)$

This expression may be integrated over *u*, *v*, and *w* to include all molecules which initially have positive values of *u*. The integrated expression gives the total momentum change per square meter per second, which is a force per square meter. We therefore write

$$P = 2m \int_{-\infty}^{\infty} \int_{-\infty}^{\infty} \int_{0}^{\infty} u^2 f(u, v, w)\, du\, dv\, dw \qquad \text{(kg m/m}^2\text{-s}^2\text{) due to all molecules}$$

for the pressure on the surface of the chamber. However the integrand in this equation is an even function of $u$, since

$$(-u)^2 f(-u, v, w) = u^2 f(u, v, w)$$

and the integral may therefore be written

$$P = m \int\int\int_{-\infty}^{\infty} u^2 f(u, v, w)\ du\ dv\ dw$$

We now note that we have already defined the mean square value of the velocity component as

$$\overline{u^2} = \frac{1}{N} \int\int\int_{-\infty}^{\infty} u^2 f(u, v, w)\ du\ dv\ dw$$

Therefore

$$P = mN\overline{u^2} \qquad \text{N/m}^2$$

and since the gas is isotropic, we find that

$$P = \frac{mN}{3}\overline{c^2} \qquad \text{Pa}$$

If we now consider a molar volume of the gas, $NV$ is Avogadro's number, $mNV$ is the molecular weight of the gas $M$, and we obtain the result that

$$PV = \frac{M}{3}\overline{c^2}$$

However

$$PV = RT$$

where $R$ is the molar gas constant. This in turn yields the equation

$$\overline{c^2} = \frac{3RT}{M}$$

for the mean-square speed of the gas molecules. We may also write the molar internal energy as

$$E = NV\bar{\varepsilon} = \tfrac{1}{2}M\overline{c^2} = \tfrac{3}{2}RT$$

which yields an expression for the molar heat capacity of a monatomic gas

$$C_v = \tfrac{3}{2}R$$

or since $C_p = C_v + R$ for an ideal monatomic gas,

$$\frac{C_p}{C_v} = \gamma = 1.6667$$

Therefore, simple kinetic theory as applied to an ideal gas tells us that

1. The molecules are moving at a root-mean-square speed which is greater than the velocity of sound in the gas. Since by definition the velocity of sound in an ideal gas is given by

$$a = \sqrt{\frac{\gamma RT}{M}}$$

$$\frac{\sqrt{\overline{c^2}}}{a} = \sqrt{\frac{3}{\gamma}} > 1$$

2. If the molecules are restricted to not storing internal energy the heat capacity ratio of the gas is $\gamma = 1.6667$.

Notice that these conclusions have been reached without knowing the exact form of the velocity distribution $f(u, v, w)$. Notice also that the first conclusion is eminently logical. A signal cannot be transmitted unless it is being carried by something (in this case the molecules). Furthermore, the second conclusion (that $\gamma = 1.6667$ for a monatomic gas) has been adequately verified by experiment.

Incidentally, the above results also verify Dalton's law of partial pressures for a dilute gas. If we consider two gases at the same pressure in separate chambers which are in thermodynamic equilibrium, the number of molecules per unit volume of each gas is the same and the average molecular kinetic energy must be the same for each gas. If we now remove the separating partition the two types of molecules in the mixture must retain the same average energy because the principle of detailed balancing applies. Therefore, in general, we may write that

$$\sum_{i=1}^{s} p_i = P \tag{1-17}$$

and

$$\bar{\varepsilon}_i = \bar{\varepsilon}_j \qquad i = 1, \ldots, s \qquad j = 1, \ldots, s$$

for a gas containing $s$ species, where the $\varepsilon$'s refer to translational energies. Since at thermodynamic equilibrium the stored rotational and vibrational energy in any diatomic or polyatomic gas is in equilibrium with the translational energy, the energy stored by each different species in a gas mixture must be independent of the dilution with other species in the mixture. Thus the internal energy and enthalpy of an ideal gas mixture can be written as

$$e = \sum_{i=1}^{s} n_i E_i \qquad \text{J/kg} \tag{1-18}$$

and

$$h = \sum_{i=1}^{s} n_i H_i \qquad \text{J/kg} \tag{1-19}$$

## 1-5 THE MAXWELL DISTRIBUTION OF VELOCITY

Let us consider the collision of two molecules in a macroscopic system of identical molecules. For a particular collision, molecules of velocity $\mathbf{V}_1$ and $\mathbf{V}_2$ will produce molecules of velocity $\mathbf{V}'_1$ and $\mathbf{V}'_2$ after the collision. The number of these particular forward collisions of type $(\mathbf{V}_1 \cdot \mathbf{V}_2) \rightarrow (\mathbf{V}'_1 \cdot \mathbf{V}'_2)$ will be proportional to the number of molecules of type $\mathbf{V}_1$ and $\mathbf{V}_2$, i.e., the forward rate $= A\,f(\mathbf{V}_1) \cdot f(\mathbf{V}_2)$. One can also write the reverse rate as $B\,f(\mathbf{V}'_1) \cdot f(\mathbf{V}'_2)$ where $B$ may be a different proportionality constant. However, the principle of detailed balances for a macroscopic system simply states that these rates must be equal at equilibrium, i.e.,

$$A\,f(\mathbf{V}_1) \cdot f(\mathbf{V}_2) = B\,f(\mathbf{V}'_1)f(\mathbf{V}'_2)$$

In addition, we must assume that energy is conserved during the collision process. Therefore, we may write

$$(\mathbf{V}_1)^2 + (\mathbf{V}_2)^2 = (\mathbf{V}'_1)^2 + (\mathbf{V}'_2)^2$$

We now consider a particular case: a collision where $V_1 = 0$. For this case

$$(\mathbf{V}_2)^2 = (\mathbf{V}'_1)^2 + (\mathbf{V}'_2)^2$$

and

$$A\,f(0) \cdot f\left(\sqrt{(\mathbf{V}'_1)^2 + (\mathbf{V}'_2)^2}\right) = B\,f(\mathbf{V}'_1) \cdot f(\mathbf{V}'_2)$$

The only nontrivial form of the function $f(\mathbf{V})$ which will satisfy this equation is

$$f(\mathbf{V}) = \alpha e^{-\beta \mathbf{V}^2} = \alpha e^{-\beta(u^2 + v^2 + w^2)}$$

where $A = B$, as may be verified by direct substitution. However,

$$f(\mathbf{V}) = f(u,\,v,\,w) = \alpha e^{-\beta(u^2 + v^2 + w^2)}$$

$$= \alpha e^{-\beta u^2} \cdot e^{-\beta v^2} \cdot e^{-\beta w^2}$$

Therefore

$$f(\mathbf{V}) = h(u) \cdot h(v) \cdot h(w)$$

where from before

$$h(u) = h(v) = h(w)$$

We may now use the fact that

$$N = \int\!\!\!\int\!\!\!\int_{-\infty}^{\infty} f(u,\,v,\,w)\,du\,dv\,dw$$

to eliminate $\alpha$ from this equation. We write

$$N = \int\!\!\!\int\!\!\!\int_{-\infty}^{\infty} \alpha e^{-\beta(u^2 + v^2 + w^2)} \, du \, dv \, dw$$

$$= \alpha \left[ \int_{-\infty}^{\infty} e^{-\beta u^2} \, du \right]^3 = \alpha \left[ \sqrt{\frac{\pi}{\beta}} \right]^3$$

Therefore

$$\alpha = N \left( \frac{\beta}{\pi} \right)^{3/2}$$

The value of $\beta$ may be determined by solving the equation for the square of one of the velocity components.

$$\overline{u^2} = \frac{1}{N} \int\!\!\!\int\!\!\!\int_{-\infty}^{\infty} u^2 f(u, v, w) \, du \, dv \, dw$$

$$= \left( \frac{\beta}{\pi} \right)^{3/2} \int_{-\infty}^{\infty} u^2 e^{-\beta u^2} \, du \left[ \int_{-\infty}^{\infty} e^{-\beta v^2} \, dv \right]^{2'}$$

$$= \left( \frac{\beta}{\pi} \right)^{1/2} \int_{-\infty}^{\infty} u^2 e^{-\beta u^2} \, du$$

By differentiating with respect to the parameter $\beta$ we obtain

$$\int_{-\infty}^{\infty} u^2 e^{-\beta u^2} \, du = -\frac{d}{d\beta} \left[ \int_{-\infty}^{\infty} e^{-\beta u^2} \, du \right] = -\frac{d}{d\beta} \left( \sqrt{\frac{\pi}{\beta}} \right)$$

$$= \frac{1}{2} \sqrt{\frac{\pi}{\beta^3}}$$

Therefore

$$\overline{u^2} = \frac{1}{2\beta}$$

However, in Sec. 1-4 we saw that

$$\overline{u^2} = \frac{1}{3} \overline{c^2} = \frac{RT}{M} = \frac{k\mathscr{A}T}{m\mathscr{A}} = \frac{kT}{m}$$

where $k$ is the gas constant per molecule (called the *Boltzmann constant*), $\mathscr{A}$ is Avogadro's number ($6.02283 \times 10^{23}$ molecules/mole) and therefore

$$\beta = \frac{m}{2kT}$$

and the Maxwell velocity distribution becomes

$$f(\mathbf{V})\ dV = f(u, v, w)\ du\ dv\ dw$$

$$= N\left(\frac{2\pi \mathscr{k} T}{m}\right)^{-3/2} e^{-(u^2 + v^2 + w^2)m/2\mathscr{k}T}\ du\ dv\ dw \tag{1-20}$$

This represents the number density of molecules of type $(\mathbf{V})$ with velocity components $(u, v, w)$. That is, if we were to plot the molecular distribution in velocity space [with coordinates $(u, v, w)$] this would represent the molecular population in a cell of size $du\ dv\ dw$ at position $(u, v, w)$. We would also like to determine the total number of molecules which have a speed $c$ irrespective of their direction in velocity space. This may be represented by a spherical shell of thickness $dc$ and radius $c$. To do this we make a transformation to spherical coordinates and integrate over all $\phi$ and $\theta$. In spherical coordinates the cell $du\ dv\ dw$ is replaced by a cell of volume $c^2 \sin \theta\ dc\ d\phi\ d\theta$ and Eq. (1-20) becomes

$$f(u, v, w)\ du\ dv\ dw = N\left(\frac{2\pi \mathscr{k} T}{m}\right)^{-3/2} c^2 e^{-(c^2 m/2\mathscr{k}T)} \sin \theta\ dc\ d\phi\ d\theta$$

If we define

$$\zeta(c, \theta) \quad \text{as} \quad \left(\frac{2\pi \mathscr{k} T}{m}\right)^{-3/2} c^2 e^{-(c^2 m/2\mathscr{k}T)} \sin \theta$$

and integrate $\zeta(c, \theta)$ for all $\phi$ and $\theta$ we obtain

$$\int_0^{2\pi} d\phi \int_0^{\pi} \zeta(c, \theta)\ d\theta = 4\pi \left(\frac{2\pi \mathscr{k} T}{m}\right)^{-3/2} c^2 e^{-(c^2 m/2\mathscr{k}T)}$$

Therefore

$$\eta(c)\ dc = 4\pi \left(\frac{2\pi \mathscr{k} T}{m}\right)^{-3/2} c^2 e^{-(c^2 m/2\mathscr{k}T)}\ dc \tag{1-21}$$

Integrating to determine the mean square speed yields

$$\overline{c^2} = \frac{N \int_0^{\infty} c^2 \eta(c)\ dc}{N} = \frac{3\mathscr{k}T}{m} = 3\mathscr{R}T \tag{1.21a}$$

which is the result obtained earlier in Sec. 1-4. In a similar manner one may obtain the mean speed (which is not zero) by evaluating

$$\bar{c} = \frac{N \int_0^{\infty} c\eta(c)\ dc}{N} = \sqrt{\frac{8\mathscr{R}T}{\pi}} \tag{1-21b}$$

by integrating by parts.

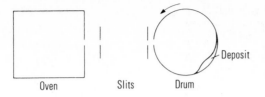

**Figure 1-2** The experiment which confirmed the Maxwell distribution in an ideal gas. *(Adapted with permission from J. F. Lee, F. W. Sears, and D. L. Turcotte, "Statistical Thermodynamics," Addison-Wesley, Reading, Mass., 1963.)*

The Maxwell distribution has been verified by experiment. Zartman and Ko in 1930–1934 measured $\eta(c)$ using a modification of a technique developed by Stern in 1920. An oven at a temperature $T$ with a slit opening is used as a source of molecules. These molecules are allowed to fall on a rotating drum which also has a narrow slit. Those molecules entering the drum travel to the opposite side at their respective velocities and deposit on the interior wall, leaving a layer whose density can be measured. (See Fig. 1-2.) Thus one can directly measure the molecular velocity distribution at the temperature of the oven. The experimental results agree very well with the theoretical distribution. The confirmation is all the more striking because there are no adjustable constants involved in plotting the experimental or theoretical curves.‡

## 1-6 MOLECULAR PROPERTIES OF A DILUTE GAS

We have determined the average speed and the root-mean-square speed for the molecules of an equilibrium dilute gas in the previous section. We may also determine some other properties using simple kinetic theory. Let us assume that all the molecules are moving at their average speed $\bar{c}$ and that $\frac{1}{6}$ of them are moving toward each wall of the container. If this were true the number of collisions with the wall per square meter per second would simply be

$$\tfrac{1}{6}N\bar{c}$$

where $N$, as before, is the number of molecules per unit volume. If we remove the restriction of constant velocity in the six normal directions and use the Maxwell velocity distribution the multiplying constant is modified to yield

$$Z_w = N\bar{c}/4 \qquad \text{molecules/m}^2\text{-s} \qquad (1\text{-}22)$$

or

$$Z_w = \frac{N\sqrt{\bar{c^2}}}{\sqrt{6\pi}} \qquad \text{molecules/m}^2\text{-s} \qquad (1\text{-}23)$$

‡ A very complete discussion of experimental methods may be found in J. Jean's *An Introduction to the Kinetic Theory of Gases*, The University Press, Cambridge, pp. 124–130 (1940).

We may also calculate the number of molecular collisions occurring in one cubic meter of gas by using a similar argument. However, in order for collisions to occur the molecules must have a finite size. Let us assume that each molecule is a hard sphere of diameter $\sigma_0$ and that one molecule is moving at the velocity $\bar{c}$. This molecule will hit another molecule when its centerline passes within $\sigma_0$ m of the other molecule's centerline. It therefore sweeps out a volume $\pi\sigma_0^2\bar{c}$ between collisions. However, there are $N$ molecules per cubic meter so it will strike $N\pi\sigma_0^2\bar{c}$ other molecules per second. At the same time though each other molecule is sweeping its own collision path. Therefore, there are $(N^2\pi\sigma_0^2\bar{c})/2$ collisions per cubic meter per second. The factor of $\frac{1}{2}$ enters because we have counted each collision twice. Thus

$$Z_{\text{molecules}} = \frac{\pi}{2}\,N^2\sigma_0^2\,\bar{c} \text{ molecules colliding per m}^3/\text{s}$$

for our simple model. If one uses the Maxwell distribution in the analysis the constant changes and the number of molecular collisions becomes

$$Z_{\text{molecules}} = \sqrt{2}\,\pi N^2\sigma_0^2\,\bar{c} \text{ collisions per m}^3/\text{s} \qquad (1\text{-}24)$$

for hard-sphere molecules.

The average transit distance between collisions is called the *mean free path*. Since we showed that a molecule moving at a velocity $\bar{c}$ suffers $N\pi\sigma_0^2\bar{c}$ collisions/s, the mean free path is

$$l = \frac{\bar{c}}{\pi\sigma_0^2 N\bar{c}} = \frac{1}{\pi\sigma_0^2 N} \text{ m/collision}$$

If one assumes hard spheres with a Maxwell distribution this becomes

$$l = \frac{1}{\sqrt{2}\,\pi\sigma_0^2 N} \qquad (1\text{-}25)$$

Representative values of $\sigma_0$ for various molecules are tabulated in App. A. These are not truly effective diameters for the hard-sphere model but are the zero-potential approach distance for the Lennard–Jones interaction potential. The properties of this potential will be discussed in Chap. 3.


## 1-7 STATISTICAL THERMODYNAMICS OF AN IDEAL GAS

We have seen that elementary kinetic theory, as applied to a dilute gas, yields results which agree well with our experience. We further saw that when the molecules are not allowed to store energy internally, the heat capacity and gamma of the gas is accurately predicted by simple kinetic theory. However, the accurate thermodynamic functions needed to handle combustion problems adequately cannot be obtained using only this simple approach. The sources of

these thermodynamic functions are experimental heats of formation and dissociation energies coupled with spectroscopic measurements interpreted by means
of a statistical thermodynamic approach as it applies to the calculation of the
thermodynamic properties of a dilute gas.

We found in our discussion of the Maxwell velocity distribution that this
distribution is of the form

$$f(\mathbf{V}) = \alpha e^{-\beta \mathbf{V}^2}$$

where $\alpha$ and $\beta$ are constants. When one considers the statistical mechanics of a
large number of "independent" particles which are in equilibrium it may be
shown that this statement can be generalized to yield the distribution law in
terms of the energy stored by each particle:

$$\Xi(\varepsilon) = \text{const} \cdot e^{-\varepsilon/kT} \tag{1-26}$$

where for a classical atomic model $\varepsilon$ is assumed to be a continuous function
while for a quantum model $\varepsilon$ is assumed to be limited to a number of discrete
values. For a dilute or ideal gas we have already discussed the observation that
molecules behave as though they are "independent" of each other even though
they are in statistical equilibrium. In short, this behavior occurs because we
only seldom observe a molecule undergoing a collision process even though
there are sufficient collisions to cause equilibrium in the thermodynamic sense.

We now assume that the storage of energy in the vibrational and rotational
degrees of freedom of the molecule are decoupled. With that assumption the
energy of an individual molecule may be written as

$$\varepsilon = \varepsilon_t + \varepsilon_v + \varepsilon_r$$

where $\varepsilon_t$ is the translational energy considered earlier, $\varepsilon_v$ is the energy stored in
the internal vibrations and $\varepsilon_r$ is the energy stored in rotation (the energy stored
in internal electronic excitation is neglected in this treatment). Notice that this
type of decoupling allows us to discuss the distribution function for each type of
energy storage separately. We have already discussed energy storage in the
translational degrees of freedom and we summarize these results by saying that
the treatment was classical and we found that the kinetic energy stored in each
degree of translational freedom contributes a value of $R/2$ to the heat capacity
of the gas.

We now turn our attention to molecular vibration. We shall discuss only
the case of the simple harmonic oscillator model of a diatomic molecule and use
this example to compare the classical and quantum approach as it is used in the
calculation of heat capacities. The classical harmonic oscillator can store energy
as both kinetic and potential energy.

$$\varepsilon = \tfrac{1}{2}m\dot{x}^2 + \tfrac{1}{2}m\omega_0^2 x^2$$

where $\omega_0 = 2\pi\nu_0$, the characteristic oscillator frequency. We note that the
energy which is stored by this degree of freedom is a quadratic function of both
the particle's velocity (momentum) and the particle's position (displacement

from its equilibrium position). The average energy storage for a weighted equilibrium distribution $\Xi$ of a number of such oscillators is simply

$$\bar{\varepsilon} = \frac{\displaystyle\int\!\!\int_{-\infty}^{\infty} \varepsilon\, \Xi\,(\varepsilon)\, d\dot{x}\, dx}{\displaystyle\int\!\!\int_{-\infty}^{\infty} \Xi\, d\dot{x}\, dx}$$

If the expression for the energy of a harmonic oscillator is substituted into this equation it reduces to the form

$$\bar{\varepsilon} = \frac{\displaystyle\int_{-\infty}^{\infty} \frac{1}{2} m\dot{x}^2 e^{-(m\dot{x}^2/2kT)}\, d\dot{x}}{\displaystyle\int_{-\infty}^{\infty} e^{-(m\dot{x}^2/2kT)}\, d\dot{x}} + \frac{\displaystyle\int_{-\infty}^{\infty} \frac{1}{2} m\omega_0^2 x^2 e^{-(m\omega_0^2 x^2/2kT)}\, dx}{\displaystyle\int_{-\infty}^{\infty} e^{-(m\omega_0^2 x^2/2kT)}\, dx}$$

Notice that these integrals are of the same type that occurred in Sec. 1-4 and therefore they may be easily evaluated to obtain the result

$$\bar{\varepsilon} = \tfrac{1}{2}kT + \tfrac{1}{2}kT = kT/\text{molecule}$$

It may be shown that this result can be generalized to yield the law of equipartition of energy for classical mechanics: Any energy term which is quadratic in either displacement or velocity (momentum) contributes $kT/2$ to the average energy stored in each of a large number of independent particles at equilibrium.

Let us now look at the quantum model for vibration. In this case the energy states are discrete and their energies are given by the relation

$$\varepsilon_n = (n + \tfrac{1}{2})h v_0 \qquad n = 0, 1, 2, \ldots$$

Since there are now an infinite number of discrete states we obtain the average energy by writing the equation

$$\bar{\varepsilon} = \frac{\displaystyle\sum_{n=1}^{\infty} (\varepsilon_n - \varepsilon_0)\Xi_n}{\displaystyle\sum_{n=1}^{\infty} \Xi_n}$$

We are now counting all the allowable states of the system and the average energy $\bar{\varepsilon}$ is measured relative to the zero-point energy of vibration. Thus we obtain the expression

$$\bar{\varepsilon} = \frac{\displaystyle\sum_{n=1}^{\infty} n h v_0\, e^{-(n h v_0/kT)}}{\displaystyle\sum_{n=1}^{\infty} e^{-(n h v_0/kT)}}$$

However, the denominator of this expression is a geometric series which may be expressed analytically while the numerator is the negative derivative of the denominator with respect to $1/\ell T$ as the argument. Therefore

$$\bar{\varepsilon} = \frac{\ell v_0}{e^{(\ell v_0/\ell T)} - 1}$$

We now evaluate the heat capacity by noting that

$$(C_v)_v = \mathscr{A}\,\frac{d\bar{\varepsilon}}{dT}$$

which yields the relationship

$$\frac{(C_v)_v}{R} = \left(\frac{\theta_v/2T}{\sinh \theta_v/2T}\right)^2 \tag{1-27}$$

where $\theta_v = \ell v_0/\ell$ K, and is called the *characteristic vibrational temperature* of the harmonic oscillator. We note that Eq. (1-27) predicts that the vibrational contribution to the heat capacity in a system of quantum harmonic oscillators will range from 0 to $R$ as $T$ ranges from 0 to $\infty$. This disagrees rather strikingly with the classical equipartition law which states that vibration simply contributes $R$ to the heat capacity irrespective of the system's temperature.

In general, if a large number of levels of a particular degree of freedom of a mechanical system are populated according to the distribution law, Eq. (1-26), where $\varepsilon$ is a continuous function, this degree of freedom is said to behave in a classical manner. Translation is a good example. It may be also shown that at room temperature and above, rotation is classically excited in most molecules. That is, each degree of rotational freedom contributes $R/2$ to the heat capacity. We may summarize these statements for a diatomic molecule by noting that since it has two degrees of rotational freedom its heat capacity as predicted by quantum-statistical mechanics may be written as:

$$C_p = C_v + R = \underbrace{\frac{3}{2}R}_{\text{translation}} + \underbrace{R}_{\text{rotation}} + \underbrace{R\left(\frac{\theta_v/2T}{\sinh (\theta_v/2T)}\right)^2}_{\text{vibration}} + R$$

This contrasts with the heat capacity predicted by classical mechanics, which is

$$C_p = C_v + R = \underbrace{\frac{3}{2}R}_{\text{translation}} + \underbrace{R}_{\text{rotation}} + \underbrace{R}_{\text{vibration}} + R = \frac{9}{2}R$$

Figure 1-3 is a plot of the dimensionless heat capacity $C_p/R$ of three diatomic gases as a function of a dimensionless temperature based on $\theta_v$ for each gas. This plot shows that the introduction of a vibrational temperature for each diatomic molecule allows one to handle the heat capacity of diatomic gases quite adequately by means of a corresponding states theory based on a charac-

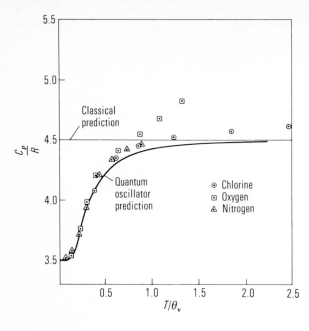

Figure 1-3 Comparison of heat capacities of three diatomic molecules with the quantum mechanical and classical mechanical predictions. The high-temperature deviations for oxygen are due to low-level electronic states in this molecule.

teristic vibrational temperature. Furthermore, note that $\theta_v = 2274$ K for oxygen, $\theta_v = 3393$ K for nitrogen, and $\theta_v = 807$ K for chlorine, and therefore on an absolute temperature scale the dimensionless heat capacity changes from 3.5 to 4.5 at markedly different temperatures for these three gases.

We have seen that quantum-statistical thermodynamics predicts reasonable heat capacities for diatomic molecules and that the energy stored in vibration in oxygen and nitrogen molecules is essentially zero at room temperature. Thus the quantum theory predicts that the heat-capacity ratio $\gamma$ of air should be in the neighborhood of 1.4. In the case of classical statistical mechanics, we obtain the prediction that the heat-capacity ratio of air should be equal to $\frac{9}{7} = 1.29$. The fact that the value of gamma for air is very close to 1.4 at room temperature further supports the quantum-mechanical approach.

The above discussion, at best, represents only a brief introduction to the use of statistical thermodynamics for the calculation of thermodynamic properties. The harmonic oscillator-rigid rotor description of molecular structure is only a first approximation to the true structure of the molecule. Nevertheless, the above exercise using the harmonic oscillator shows the utility of the quantum-mechanical approach in the calculation of thermodynamic properties. This approach has, in fact, reached such a degree of sophistication that it is now usually agreed‡ that experimentally measured heats of combustion and dissociation energies coupled with quantum-statistical mechanical calculations based on spectroscopic data yield calculated thermodynamic properties which are more accurate than directly measured thermodynamic properties.

‡ See, for example, J. S. Rowlinson, *The Perfect Gas*, Pergamon, New York, 1963, p. 57.

At present the most extensive and up-to-date compilation of thermodynamic properties to be found is the JANAF (Joint Army-Navy-Air Force) tables compiled and published by the Dow Chemical Corporation, Midland, Michigan. Appendix B contains abbreviated JANAF tables for some selected compounds of interest in the combustion field.

## PROBLEMS

**1-1** Plot the molecular velocity distribution, $\eta(c)$, for carbon dioxide at a temperature of 25°C and a pressure which is low enough to ensure that the gas is ideal. Also, calculate $\bar{c}$ and $\overline{c^2}$.

**1-2** Use the exact Maxwell relationships for hard-sphere molecules to calculate for (a) hydrogen and (b) nitrogen, at $T = 25$°C, $P = 101.325$ kPa, the following properties:
1. Wall collisions per square meter per second
2. Total molecular collisions per cubic meter per second
3. Collisions of one molecule with others per second
4. Mean free path of the molecules in meters per collision

**1-3** Hard-sphere molecules are "in collision" for an infinitely short time because they simply bounce off of each other. If we assume that they are in a "collision state" when they are closer than two diameters apart, derive the expression for a third-body collision. In other words, write an expression for the number of encounters, per cubic meter per second, where three hard-sphere molecules are simultaneously within two molecular diameters of each other. Use the "average speed" technique that was used in the text.

**1-4** Without performing a detailed calculation, show that the number of molecular collisions per second that one molecule makes with all others is proportional to the pressure—all other conditions constant—and that the number of three-body collisions per second of one molecule (as defined in Prob. 1-3) is proportional to the square of the pressure (same constraints).

**1-5** Calculate the vertical distribution of density and pressure above the earth assuming that the surface pressure is one atmosphere and gravity is constant for (a) a constant temperature atmosphere and (b) a well-stirred (isentropic) atmosphere.

**1-6** Calculate the heat capacity of water vapor and carbon dioxide at 300, 2000, and 3500 K using the harmonic oscillator approximation of Sec. 1-7. Water vapor is a triatomic molecule with a nonlinear structure (the angle between the two O—H bonds is about 105°). Thus it has three translational and three rotational degrees of freedom that can be considered classical. It thus has $3n - 6 = 3$ vibrational degrees of freedom whose characteristic temperatures are $\theta_v = 2294$, 5261, and 5403 K. Carbon dioxide is also a triatomic molecule but it has a linear structure. Therefore it has only two degrees of rotational freedom. It thus has $3n - 5 = 4$ vibrational degrees of freedom whose characteristic temperatures are $\theta_v = 960(2)$, 1932, and 3379 K. The (2) means that the low energy bending vibration of $CO_2$ has a degeneracy of two. Compare your answers to values tabulated in App. B1.

**1-7** Calculate the fraction of molecules with a velocity of at least $3\bar{c}$ and $10\bar{c}$ where $\bar{c}$ is the average speed. Use the Maxwell distribution.

**1-8** Verify Eqs. (1-21a) and (1-21b) by integrating the Maxwell distribution given in Eq. (1-21).

## BIBLIOGRAPHY

Cowling, T. G., *Molecules in Motion*, Harper and Row, New York (1960).
Glasstone, S., *Theoretical Chemistry*, Van Nostrand, New York (1944).
Golden, S., *Elements of the Theory of Gases*, Addison-Wesley, Reading, Mass. (1964).

Hill, T. L., *Introduction to Statistical Thermodynamics*, Addison-Wesley, Reading, Mass. (1960).

Jeans, J., *Introduction to the Kinetic Theory of Gases*, The University Press, Cambridge, New York (1959).

Kennard, E. H., *Kinetic Theory of Gases*, McGraw-Hill, New York (1938).

Lee, J. F., F. W. Sears, and D. L. Turcotte, *Statistical Thermodynamics*, Addison-Wesley, Reading, Mass. (1963).

Mayer, J. E., and M. G. Mayer, *Statistical Mechanics*, Wiley, New York (1940).

Penner, S. S., *Chemistry Problems in Jet Propulsion*, Pergamon, New York (1957).

Rowlinson, J. S., *The Perfect Gas*, Pergamon, New York (1963).

# TWO

# STOICHIOMETRY AND THERMOCHEMISTRY

## 2-1 INTRODUCTION

In the previous chapter we reviewed basic thermodynamic principles and showed how the properties of an ideal gas could be determined from an analysis of detailed collision processes. In this chapter we will turn our attention to two very important practical considerations relative to systems that are undergoing combustion reactions: the mass and energy balances that must be used to adequately describe such systems. Mass-balance requirements are called stoichiometric requirements by the chemist, after the Greek "stoicheion" meaning first principle or element, and the energy balances which apply when chemical reactions occur are governed by thermochemical considerations. This chapter will introduce both stoichiometric and thermochemical principles, and show by example how they apply to combustion processes.

## 2-2 PRACTICAL STOICHIOMETRY

Most practical combustion processes occur when a fossil fuel or fossil-derived fuel burns with the oxidizer, air. The majority of these fuels contain only the elements carbon, hydrogen, oxygen, nitrogen, and sulfur. The aim of practical stoichiometry is to determine exactly how much air must be used to completely oxidize the fuel to the products carbon dioxide, water vapor, nitrogen, and sulfur dioxide. This does not imply that combustion is necessarily complete in any specific practical device. Nevertheless, a stoichiometrically correct mixture of fuel with air is defined as one which would yield exactly the products listed above and have no excess oxygen *if* combustion were complete.

The common fuels one burns in practice are either solids (e.g., coal or coke), liquids (e.g., residual oils, gasoline), or gases (e.g., natural gas, liquid petroleum gas). In the case of gaseous fuels it is common practice to analyze the mixture for the component gases and to report the analysis in terms of volume (or mole) percent. It is usually feasible to do this because most gaseous fuels are mixtures of only a few chemical compounds. On the other hand, either naturally occurring or commercially available organic liquids or solids can contain thousands of compounds, many of which have very complex molecular structures. Because of this, liquid and solid fuels are analyzed for the *weight* percentages of the elements carbon, hydrogen, nitrogen, oxygen, and sulfur that are present in the fuel. This analysis is called an *ultimate analysis*. In the case of coals and some of the heavier liquid fuels the analysis also yields a percentage by weight of ash, the solid residue left after complete oxidation.

Appendix C contains some example analyses of typical gaseous fuels (by volume or mole percent) and liquid and solid fuels, on an elemental weight basis. The composition of dry air is also listed in App. C.

In a practical stoichiometric calculation, to determine the operating point for a boiler, for example, two simplifying assumptions are usually made. Firstly, one assumes that air simply contains 21 percent oxygen and 79 percent nitrogen. Secondly, one assumes that the fraction of the fuel that is ash is essentially inert to the combustion process. The first assumption is realistic in light of the application. The second assumption may yield some error because any heavy metal content of the fuel, such as iron, in all probability exists in the fuel as a compound which is not the oxide and appears in the ash as the oxide. Thus extra oxygen may be required over that calculated in a usual practical stoichiometric calculation because oxidation to form the ash is not considered.

Whenever one performs a stoichiometric calculation of any type one must always fix the mass of the system that is being considered. Furthermore, the total mass that is under consideration must remain fixed during the entire calculation. Once the total mass is conserved during a calculation the total quantity of each element present, irrespective of the species that are present, must also be conserved. Thus, the total amount of carbon that is being considered must remain constant irrespective of whether it appears as a part of the hydrocarbon component of the fuel at the start of the calculation or as part of the carbon dioxide at the end of the calculation. This is also true for all the other elements being considered in the stoichiometric calculation. Consider, for example, the combustion of one mole of a liquid petroleum gas that contains 40 mole percent propane ($C_3H_8$) and 60 mole percent butane ($C_4H_{10}$) with sufficient air to just allow complete oxidation to carbon dioxide and water. Since we assume that air contains 21 percent oxygen, one mole of oxygen carries with it 3.76 moles of nitrogen. Thus we can write a balanced stoichiometric relationship between the fuels and oxidizer as follows:

$$\begin{array}{c} \text{(C)} \quad \text{(H)} \\ \left(\begin{array}{c} 0.4C_3H_8 \\ 0.6C_4H_{10} \end{array}\right) + \left(\begin{array}{c} 1.2 + 0.8 \\ 2.4 + 1.5 \end{array}\right) \cdot (O_2 + 3.76N_2) \rightarrow 3.6CO_2 + 4.6H_2O + 22.18N_2 \end{array}$$

or

$$C_{3.6}H_{9.2} + 5.9O_2 + 22.18N_2 \rightarrow 3.6CO_2 + 4.6H_2O + 22.18N_2$$

If we now assume that the atomic weights can be rounded to the nearest gram we see that 0.052 kg of fuel will be oxidized completely to carbon dioxide and water vapor in combination with 0.810 kg of air. In other words, a stoichiometric mixture of this gaseous fuel with air will contain $100 \times 0.052/(0.052 + 0.810) = 6.1$ weight percent or $100 \times 1/(1 + 5.9 \times 4.76) = 3.4$ volume (or mole) percent of fuel. In this case the fuel has a known average molecular weight and the calculation was based on the combustion of one mole of fuel. Note that the total mass under consideration was 0.862 kg.

In the case of solid or liquid fuels the molecular weight is unknown. Therefore, one ordinarily chooses a unit mass of fuel, say 1 kg, as the basis for the calculation. Consider, for example, a rather pure gasoline as a fuel. Its ultimate analysis is 83% C and 17% H. In this case, one kilogram of fuel would have the empirical formula $C_{69.16}H_{170}$ where the subscripts represent the number of carbon atoms and hydrogen atoms respectively in one kilogram of fuel. There is no implication here that a molecular weight is known. To calculate the quantity of air required to exactly oxidize one kilogram of this fuel, one can write the balanced stoichiometric relationship

$$\underset{\text{1 kg fuel}}{C_{69.16}H_{170}} + \underset{\substack{\text{15.32 kg air} \\ \text{531.50 moles air}}}{(69.16 + 42.5)(O_2 + 3.76N_2)} \rightarrow 69.16CO_2 + 85H_2O + 419.8N_2$$

Thus we find that complete oxidation of this fuel requires $100 \times 1/(1 + 15.32) = 6.1$ weight percent of fuel in air. Notice that in this case one cannot discuss the volume (or mole) percent of fuel required for complete combustion because one does not know the average molecular weight of this mixture of liquid fuels. Notice also that one can show by calculation that the total mass under consideration is conserved because a balanced stoichiometric equation has been written. In general, one can write a balanced stoichiometric relationship for any CHONS fuel–air system as follows:

$$C_uH_vO_wN_xS_y + \left(u + \frac{v}{4} - \frac{w}{2} + y\right)(O_2 + 3.76N_2)$$

$$\rightarrow uCO_2 + \frac{v}{2}H_2O + ySO_2 + \left[3.76\left(u + \frac{v}{4} - \frac{w}{2} + y\right) + \frac{x}{2}\right]N_2 \quad (2\text{-}1)$$

where the negative sign on the oxygen in the air $(-w/2)$ indicates that less oxygen from the air is needed for complete oxidation because $w$ atoms of oxygen already exists in the fuel itself.

Even though the concept of a stoichiometrically correct mixture for combustion is a fundamental one for all combustion systems, it should be obvious that not all combustion processes occur in mixtures which are at the correct stoichiometric proportions. Thus, in most combustion studies and combustion applications it is convenient to normalize the actual mixture composition to the stoichiometric mixture composition for that fuel-oxidizer system. In all cases this normalization yields a dimensionless number whose magnitude tells one how far the mixture composition is from stoichiometric. The simplest of these dimensionless quantities is called the equivalence ratio. It is usually defined as the fuel equivalence ratio:

$$\Phi = \frac{y_{fuel}/y_{air}}{(y_{fuel}/y_{air})_{stoich}} = \frac{\eta_{fuel}/\eta_{air}}{(\eta_{fuel}/\eta_{air})_{stoich}} \qquad (2\text{-}2)$$

where the fuel/air ratio is expressed either on a mass ($y$) or mole ($\eta$) basis. The only requirement is that the denominator, which represents the fuel/air ratio of a stoichiometric mixture, be expressed on the same basis. With this definition mixtures with $\Phi < 1$ are said to be fuel-lean while mixtures in which $\Phi > 1$ are said to be fuel-rich. Two other dimensionless ratios which are commonly used to specify the composition of a combustible mixture relative to the stoichiometric composition are the percent theoretical air and percent excess air. These are defined as follows:

$$\text{Percent theoretical air} = \frac{100}{\Phi} \qquad (2\text{-}3)$$

$$\text{Percent excess air} = \frac{100}{\Phi} - 100 \qquad (2\text{-}4)$$

Thus a mixture that has an equivalence ratio of 0.8 can also be said to contain 125 percent theoretical air or 25 percent excess air.

In some cases the definition of equivalence ratio is based on the air–fuel ratio. When this is done, the term *oxidizer equivalence ratio* should be used. All oxidizer equivalence ratios are the reciprocals of *fuel equivalence ratios*.

In rare cases the fuel equivalence ratio is calculated on the basis that carbon monoxide, water, and sulfur dioxide are the final products. When this is done the full title, *fuel equivalence ratio based on carbon monoxide* should be used to avoid confusion.


## 2-3 DETAILED STOICHIOMETRY

The study of high-temperature combustion processes often requires that one follow the details of the destruction and production of species during the chemical processes that are occurring. One overriding basic requirement during such processes is that the elemental composition of the mixture remain unchanged even though the species composition is changing with time. Thus, while the basic stoichiometric principles used in the previous section still pertain, they will be amplified and generalized in this section so that we may apply these principles to more detailed discussions of chemical reaction kinetics and chemical equilibrium in later chapters.

For our discussion of detailed stoichiometry we restrict ourselves to a sample which does not lose or gain material by diffusion. Since we allow chemical reactions to occur, the mass conservation equations must require only that the total quantity of each atom in the mixture remain constant. Therefore, let us consider one kilogram of a mixture of molecules containing $c$ distinct elements as components with the $i$th element having the symbol ${}^iE$ and an atomic weight $A_i$: $i = 1, 2, \ldots, c$. Furthermore, let the mixture contain $s$ species, the general formula of the $j$th species being

$$ {}^1E_{\xi_{1j}}\,{}^2E_{\xi_{2j}} \cdots {}^cE_{\xi_{cj}} $$

where the $\xi_{ij}$ are all simple integers which may be zero. The molecular weight of the $j$th species is, therefore,

$$ M_j = \sum_{i=1}^{c} \xi_{ij} A_i \qquad \text{kg/mole} $$

where a mole is defined as $6.023 \times 10^{23}$ molecules. Thus, in a system where chemical change is allowed, there are $c$ mass-balance criteria which may be written

*MASS-BALANCE EQ.*
$$ Y_{i_E} = \text{const} = \sum_{j=1}^{s} \xi_{ij}\, y_j\, \frac{A_i}{M_j} \qquad (i = 1, 2, \ldots, c) \tag{2-5} $$

where $Y_{i_E}$ is the mass of the $i$th element and $y_j$ is the mass of the $j$th species present in one kilogram of the mixture. Notice that since each element has a distinct weight, this weight may be divided out of each mass-balance equation [Eq. (2-5)] to yield the $c$ equations

$$ \frac{Y_{i_E}}{A_i} = N_{i_E} = \text{const} = \sum_{j=1}^{s} \xi_{ij}\, \frac{y_j}{M_j} = \sum_{j=1}^{s} \xi_{ij} n_j \qquad (i = 1, 2, \ldots, c) \tag{2-6} $$

where $N_{i_E}$ is the number of atoms of the $i$th element divided by $\mathscr{A}$ present in one kilogram of the mixture and the $n_j$'s are the number of moles of each species present in one kilogram of the mixture. Equation (2-6) have mixed units (moles/kilogram). However, these units are usually found to be most convenient for representing mass balance in reactive systems.

For numerical simplicity it is quite often advantageous to write the mass-balance equation for a quantity of material different than one kilogram. For example, it is common in combustion problems to consider an air–fuel mixture as containing one mole of oxygen plus 3.76 moles of nitrogen plus $x$ moles of fuel. The total number of kilograms for this case is then $0.032 + 3.76 \times 0.028 + x \cdot M_{\text{fuel}} = \mathscr{G}$ where $\mathscr{G}$ is a constant throughout the calculation. If we define $\mathfrak{N}_i$ as the number of atoms of the $i$th element divided by $\mathscr{A}$ present in $\mathscr{G}$ kilograms of the mixture and $\eta_j$ as the number of moles of each species in $\mathscr{G}$ kilograms of the mixture, we obtain a third form for the mass-balance requirement.

$$\mathscr{G}N_{i_E} = \mathfrak{N}_{i_E} = \text{const} = \sum_{j=1}^{s} \xi_{ij}\mathscr{G}\frac{y_j}{M_j} = \sum_{j=1}^{s} \xi_{ij}\eta_j \qquad (2\text{-}7)$$

A set of general-process equations may be obtained by differentiating any of Eqs. (2-5) to (2-7). For the moment we write only the equations obtained by differentiating Eqs. (2-6):

$$0 = \sum_{j=1}^{s} \xi_{ij}\,dn_j \qquad (i = 1, 2, \ldots, c) \qquad (2\text{-}8)$$

Equations (2-8) represent a constraint on the species composition of the mixture during the course of any chemical reaction. In short, the principle of conservation of mass simply requires that these equations be satisfied. Since the Eqs. (2-8) are a set of $c$ equations in $s$ unknowns, many linear parametric equations may be written to satisfy them and there are, in fact, many sets of $s - c$ linearly independent parametric equations which may be used to completely describe the reactive system. We will find that each of these parametric equations is equivalent to a balanced stoichiometric equation as usually written by the chemist.

This may best be illustrated by means of an example. Consider a system containing the elements carbon, C, oxygen, O, and hydrogen, H ($c = 3$), and the species carbon dioxide, $CO_2$, carbon monoxide, CO, hydrogen, $H_2$, hydrogen atoms, H, oxygen, $O_2$, oxygen atoms, O, hydroxyl radicals OH, and water vapor, $H_2O$ ($s = 8$). The three independent mass-balance equations written on a mole-per-kilogram basis are

$$N_C = n_{CO} + n_{CO_2}$$
$$N_O = 2n_{O_2} + n_{H_2O} + n_{CO} + 2n_{CO_2} + n_{OH} + n_O$$
$$N_H = 2n_{H_2} + 2n_{H_2O} + n_H + n_{OH}$$

These may be differentiated to yield

$$0 = dn_{CO} + dn_{CO_2}$$
$$0 = dn_{CO} + 2dn_{CO_2} + dn_{H_2O} + dn_{OH} + dn_O + 2dn_{O_2} \qquad (2\text{-}9)$$
$$0 = 2dn_{H_2} + 2dn_{H_2O} + dn_H + dn_{OH}$$

Equations (2-9) are satisfied by many different parametric coordinates (or reaction coordinates, $\lambda_k$). Consider, for example, the set

$$dn_{H_2O} = -d\lambda_1$$
$$dn_{H_2} = d\lambda_1$$
$$dn_{CO} = -d\lambda_1 \qquad (2\text{-}10)$$
$$dn_{CO_2} = d\lambda_1$$

with the $d\lambda_1$ coefficients for the other species equal to zero. This set satisfies Eq. (2-9) as may be verified by substitution and it is equivalent to writing the balanced stoichiometric equation

$$H_2O + CO \rightarrow H_2 + CO_2 \qquad (2\text{-}11)$$

Notice that any reaction coordinate $\lambda_k$ has the units of moles of forward reaction per mass unit of the system considered. Since, in this example, the system contains one kilogram, the units on $\lambda$ are moles of reaction per kilogram of mixture. Thus, a change in $\lambda$ of $+1$ in the parametric set Eqs. (2-10) represents one unit of forward reaction per kilogram of mixture as usually expressed by the chemist in Eq. (2-11). We will find that the parametric notation used in Eqs. (2-10) is much more convenient for mathematical operations than the chemist's notation as represented by Eq. (2-11).

We may also choose the set

$$dn_{CO_2} = -2d\lambda_2$$
$$dn_{CO} = 2d\lambda_2$$
$$dn_{O_2} = d\lambda_2$$

with all other $d\lambda_2$ coefficients equal to zero. This is equivalent to the balanced stoichiometric equation

$$2CO_2 \rightarrow 2CO + O_2$$

Notice that we must make three other choices of independent balanced stoichiometric equations to yield five parameters which completely specify the system. Also notice that another choice, the stoichiometrically balanced equation

$$2H_2O \rightarrow 2H_2 + O_2$$

is not independent of the above two equations because, while we may write

$$dn_{H_2O} = -2d\lambda_3$$
$$dn_{H_2} = 2d\lambda_3 \qquad (2\text{-}12)$$
$$dn_{O_2} = d\lambda_3$$

we find that $d\lambda_3$ may also be written as a linear combination of the first two reaction coordinates. That is,

$$dn_{H_2O} = 2(-d\lambda_1) + (0d\lambda_2) = -2d\lambda_3$$
$$dn_{H_2} = 2(d\lambda_1) + (0d\lambda_2) = 2d\lambda_3$$
$$dn_{O_2} = 2(0d\lambda_1) + (d\lambda_2) = d\lambda_3$$
$$dn_{CO} = 2(-d\lambda_1) + (2d\lambda_2) = 0d\lambda_3$$
$$dn_{CO_2} = 2(d\lambda_1) + (-2d\lambda_2) = 0d\lambda_3$$

which yields

$$2d\lambda_1 + d\lambda_2 = d\lambda_3$$

The proof of dependence may also be obtained by adding the balanced chemical equations to obtain

$$2[H_2O + CO \rightarrow H_2 + CO_2]$$
$$+ [2CO_2 \rightarrow 2CO + O_2]$$
$$\overline{\phantom{+ [2CO_2 \rightarrow 2CO + O_2]}}$$
$$2H_2O \rightarrow 2H_2 + O_2$$

We have thus shown by example that balanced chemical equations written either in the usual chemist's notation or in the parametric reaction coordinate notation satisfy the general mass-balance requirements [Eqs. (2-5) and (2-6)] for the system under consideration.

Unfortunately, using these rather simple stoichiometric equations to discuss the composition of reactive systems presupposes a knowledge of (1) the species present in the system, and (2) the processes which produce the chemical change. It must be emphasized that the knowledge of both the species that are expected to be present in any situation as well as the knowledge of the specific reactions that they undergo is not in any way dependent on stoichiometric restrictions. Only a careful chemical analysis of the systems composition and possible chemical kinetic processes can ensure that *all* important species and *all* important reaction processes have been considered. The neglect of one important species or chemical kinetic process can completely obviate any results based on an otherwise correct stoichiometric analysis.

The application of a stoichiometric restriction, once the species are known and the process is defined, may best be illustrated with two simple examples. In the first example, consider one kilogram of a mixture initially containing $y_{H_2, i}$ kg of hydrogen, $y_{O_2, i}$ kg of oxygen, and $y_{H_2O, i}$ kg of water vapor capable of reacting to either produce or destroy water with the simultaneous disappearance or appearance of hydrogen and oxygen. In the chemist's notation, the reaction is

$$2H_2O \rightarrow 2H_2 + O_2 \qquad \text{(on a mole basis)}\ddagger$$

or

$$H_2O \rightarrow 0.1111H_2 + 0.8889O_2 \qquad \text{(on a kilogram basis)}\S$$

for $\Delta\lambda = +1$. Since, for this system, the number of species $s$ is equal to 3, all allowable composition changes may be represented by plotting the mass of the water molecules, the mass of the hydrogen molecules, and the mass of the oxygen molecules present in the fixed weight of reactive mixture as the three coordinate directions in a cartesian coordinate system. This is illustrated in Fig.

---

‡ It should be pointed out that this direct process is not physically realizable and is used here only as a simple example to illustrate the meaning of a stoichiometric restriction.

§ Molecular weight has been rounded to the nearest gram for convenience.

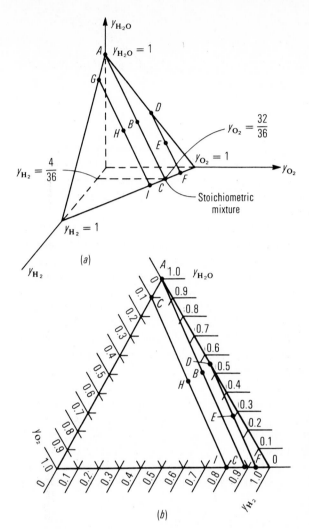

**Figure 2-1** Compositions of the $H_2$–$O_2$–$H_2O$ system on a 1-kilogram basis. (a) The plane in three dimensions. (b) The triangular coordinate representation. The line *ABC* represents the hydrogen–oxygen mixture which is stoichiometrically correct to produce water. All the straight lines are reaction coordinates (see Table 2-1).

2-1. The plane in Fig. 2-1a represents the condition $y_{H_2O} + y_{O_2} + y_{H_2} = 1$ and thus represents all allowable compositions in one kilogram of mixture. Furthermore, any point on the plane, or inside the triangular region of Fig. 2-1b, represents a fixed species composition.

To further examine the significance of a stoichiometric restriction using Fig. 2-1, the stoichiometric Eqs. (2-12) will be written more explicitly as progress equations using the reaction coordinate notation.

$$dy_{H_2O} = -2M_{H_2O}\,d\lambda \qquad\qquad dn_{H_2O} = -2d\lambda$$

$$dy_{O_2} = +M_{O_2}\,d\lambda \qquad \text{or} \qquad dn_{O_2} = +d\lambda \qquad (2\text{-}13)$$

$$dy_{H_2} = +2M_{H_2}\,d\lambda \qquad\qquad dn_{H_2} = +2d\lambda$$

where $\lambda$ is the common progress variable or reaction coordinate for this one reaction and where $y$ and $n$ are the mass or moles of the subscripted species in 1 kg of the mixture. The reaction coordinate $d\lambda$ has the same units (moles of reaction per kilogram of mixture) in both representations.

Equations (2-13), when integrated, allow one to express the composition of all stoichiometrically possible mixtures in terms of the initial composition and one reaction coordinate. For example, we may write

$$\int_{y_{H_2O,i}}^{y_{H_2O}} dy_{H_2O} = -2M_{H_2O} \int_0^\lambda d\lambda \qquad \int_{n_{H_2O,i}}^{n_{H_2O}} dn_{H_2O} = -2 \int_0^\lambda d\lambda$$

etc., which yield the relations

$$y_{H_2O} = y_{H_2O,i} - 2M_{H_2O}\lambda \qquad n_{H_2O} = n_{H_2O,i} - 2\lambda$$

$$y_{O_2} = y_{O_2,i} + M_{O_2}\lambda \qquad n_{O_2} = n_{O_2,i} + \lambda$$

$$y_{H_2} = y_{H_2,i} + 2M_{H_2}\lambda \qquad n_{H_2} = n_{H_2,i} + 2\lambda$$

Notice two things. First, the choice of $\lambda = 0$ for the initial composition, while arbitrary, must be the same for all species for any specific reaction coordinate; second, since the mass of any species in the mixture cannot be negative, $\lambda$ has an upper bound given by

$$\lambda = \frac{y_{H_2O,i}}{2M_{H_2O}} \qquad \lambda = \frac{n_{H_2O,i}}{2}$$

and a lower bound given by the most restrictive of the two conditions

$$\lambda = -\frac{y_{O_2,i}}{M_{O_2}} \qquad \lambda = -n_{O_2,i}$$

or

$$\lambda = -\frac{y_{H_2,i}}{2M_{H_2}} \qquad \lambda = -\frac{n_{H_2,i}}{2}$$

where the $i$ represents the initial composition. We have thus used the restrictions imposed by stoichiometry to define a single coordinate which, by itself, defines all possible compositions for any specific initial mixture of three species containing two components. We have also seen how the initial composition of the mixture defines the limits on the reaction coordinate. We note, in addition, that this definition of $\lambda$ is compatible with the mass-balance equations [(2-5) and (2-6)] because direct substitution yields

$$Y_O = [y_{H_2O,i} - 2M_{H_2O}\lambda]\frac{M_{O_2}}{2M_{H_2O}} + [y_{O_2,i} + M_{O_2}\lambda]$$

$$= y_{H_2O,i}\frac{M_{O_2}}{2M_{H_2O}} + y_{O_2,i}$$

and

$$Y_H = [y_{H_2O, i} - 2M_{H_2O}\lambda] \frac{M_{H_2}}{M_{H_2O}} + [y_{H_2, i} + 2M_{H_2}\lambda]$$

$$= y_{H_2O, i} \frac{M_{H_2}}{M_{H_2O}} + y_{H_2, i}$$

on a mass basis and

$$N_O = n_{H_2O} + 2n_{O_2}$$

$$= [n_{H_2O, i} - 2\lambda] + 2[n_{O_2, i} + \lambda]$$

$$= n_{H_2O, i} + 2n_{O_2, i}$$

$$N_H = 2n_{H_2O} + 2n_{H_2}$$

$$= 2[n_{H_2O, i} - 2\lambda] + 2[n_{H_2, i} + 2\lambda]$$

$$= 2n_{H_2O, i} + 2n_{H_2, i}$$

on a moles per kilogram basis.

Specific examples keyed to Fig. 2-1 are presented in Table 2-1. Three representative compositions are considered for each of three equivalence ratios; $\Phi = 1.0$, $\Phi = 0.5$, and $\Phi = 2.0$. In each case the values of the reaction coordinates are calculated assuming that $\lambda = 0$ for an initial mixture that contains (1) no water, (2) 50 percent of the maximum possible amount of water, and (3) the maximum possible amount of water. Notice, that even though the total mass of the system is fixed, as it must be, the total number of moles of species present, $n_T$, is dependent on the extent of reaction.

In the second example we use the $\mathcal{G}$ mass system and consider 0.048 kg of the mixture which initially contains $\eta_{O_3, i}$ moles of ozone, $\eta_{O_2, i}$ moles of oxygen molecules, and $\eta_{O, i}$ moles of oxygen atoms, all capable of reacting. This system again contains three species but now contains only one component. Therefore, there is only one mass-balance equation which may be written

$$3 = 3\eta_{O_3} + 2\eta_{O_2} + \eta_O$$

Since we now have only one equation in three unknowns, two reaction coordinates are necessary to completely specify the system, each corresponding to a balanced stoichiometric equation for the system. However, three stoichiometric equations are obviously available.‡

$$2O \rightarrow O_2$$

$$3O \rightarrow O_3$$

$$3O_2 \rightarrow 2O_3$$

‡ As in the previous example, these do not necessarily correspond to processes which occur in nature but are used here only to present a simple example of stoichiometric restrictions.

# Table 2-1 Compositions and reaction coordinate values for the points on Fig. 2-1

Note that while the numerical value of the reaction coordinate depends on the initial composition and extent of reaction the compositions depend only on the extent of reaction

| Reaction coordinate $\lambda$ for different initial compositions (on a mass basis) | | | Moles/kg $n_T \neq$ constant | | | | Point on Fig. 2-1 | kg/kg $y_T = 1.000$ | | | Extent of reaction | Equivalence ratio |
|---|---|---|---|---|---|---|---|---|---|---|---|---|
| 0% $H_2O$ | 50% $H_2O$ | 100% $H_2O$ | $n_{H_2O}$ | $n_{H_2}$ | $n_{O_2}$ | $n_T$ | | $y_{H_2O}$ | $y_{H_2}$ | $y_{O_2}$ | | |
| 27.78 | 13.88 | 0.0 | 55.56 | 0.0 | 0.0 | 55.56 | $A$ | 1.000 | 0.0 | 0.0 | All $H_2O$ | Stoichiometric, $\Phi = 1$ |
| 13.88 | 0.0 | -13.88 | 27.78 | 27.28 | 13.88 | 69.44 | $B$ | 0.500 | 0.056 | 0.444 | Intermediate | $N_H = 111.11$; $N_O = 55.56$ |
| 0.0 | -13.88 | -27.78 | 0.0 | 55.56 | 27.78 | 83.33 | $C$ | 0.0 | 0.111 | 0.889 | No $H_2O$ | $Y_H = 0.1111$; $Y_O = 0.8889$ |
| 14.71 | 7.35 | 0.0 | 29.41 | 0.00 | 14.71 | 44.12 | $D$ | 0.529 | 0.0 | 0.471 | Most $H_2O$ | Lean, $\Phi = 0.5$ |
| 7.35 | 0.0 | -7.35 | 14.71 | 14.71 | 22.05 | 51.47 | $E$ | 0.265 | 0.029 | 0.706 | Intermediate | $N_H = 58.82$; $N_O = 58.82$ |
| 0.0 | -7.35 | -14.71 | 0.0 | 29.41 | 29.41 | 58.82 | $F$ | 0.0 | 0.059 | 0.941 | No $H_2O$ | $Y_H = 0.0588$; $Y_O = 0.9412$ |
| 25.00 | 12.50 | 0.0 | 25.00 | 50.00 | 0.0 | 75.00 | $G$ | 0.900 | 0.100 | 0.0 | Most $H_2O$ | Rich, $\Phi = 2.0$ |
| 12.50 | 0.0 | -12.50 | 12.50 | 75.00 | 12.50 | 100.00 | $H$ | 0.450 | 0.150 | 0.400 | Intermediate | $N_H = 200.0$; $N_O = 50.0$ |
| 0.0 | -12.50 | -25.00 | 0.0 | 100.00 | 25.00 | 125.00 | $I$ | 0.0 | 0.200 | 0.800 | No $H_2O$ | $Y_H = 0.2000$; $Y_O = 0.8000$ |

Since these three equations are linearly dependent, we may choose any two. We choose the first two equations which may be written as follows using the reaction coordinates $\alpha$ and $\beta$ respectively.

| $2O \to O_2$ | $3O \to O_3$ |
|---|---|
| $d\eta_O = -2d\alpha$ | $d\eta_O = -3d\beta$ |
| $d\eta_{O_2} = d\alpha$ | $d\eta_{O_2} = 0$ |
| $d\eta_{O_3} = 0$ | $d\eta_{O_3} = d\beta$ |

The above relationships are constructed on the assumption that each reaction is occurring independent of the others. However, the net change of composition is due to the net effect of both processes occurring and one therefore obtains the following stoichiometric relations for the reactions written in terms of the reaction coordinates.

$$d\eta_O = -2d\alpha - 3d\beta$$

$$d\eta_{O_2} = d\alpha \qquad\qquad (2\text{-}14)$$

$$d\eta_{O_3} = d\beta$$

We now have an arbitrary choice for initial composition at our disposal. We choose the set $\eta_{O,i} = 3$, $\eta_{O_2,i} = 0$, $\eta_{O_3,i} = 0$ and assume that $\alpha = \beta = 0$ for this initial composition. Integration of the equations with these limits therefore yields the set

$$\eta_O = 3 - 2\alpha - 3\beta$$

$$\eta_{O_2} = \alpha$$

$$\eta_{O_3} = \beta$$

We note that the requirement that $\eta_j \geq 0$ means that in our notation

$$0 < \alpha < \tfrac{3}{2}(1 - \beta)$$

$$0 < \beta < 1 - \tfrac{2}{3}\alpha$$

Also notice that (1) since there is only one mass-balance equation, the entire triangular area is accessible to the system, (2) the limits on $\alpha$ and $\beta$ are very dependent on the assumptions used in their derivation. However, every correctly constructed set will yield correct compositions, irrespective of its internal complexity.

The two example systems we have used serve to illustrate one important consequence of a stoichiometric restriction. In a system that contains one component reactive element, all mixture compositions, based on the species that are assumed present, are available to the system. In other words, in such a system stoichiometric restrictions by themselves do not limit the available species compositions based on initial composition. However as soon as two or more component elements are in the system (i.e., $c \geq 2$), stoichiometric requirements severely limit the species compositions available from any initial mixture composition. Notice that for $c \geq 2$ one can always define $c - 1$ atomic ratios or atomic mass ratios which are the only variables needed to implement the restrictions of stoichiometry. In the water–hydrogen–oxygen example $c = 2$ and only one dimensionless number, the ratio

$$\frac{N_H}{N_O} = \overline{N}_{H/O}$$

need be specified to restrict the species composition to one specific line in Fig. 2-1 (e.g., $N_{H/O} = 2$ is the stoichiometric mixture line $ABC$ in that figure). In a general CHON hydrocarbon–air system it is convenient to specify the three ratios relative to $N_O$ (i.e., $\overline{N}_{C/O}$, $\overline{N}_{H/O}$, and $\overline{N}_{N/O}$) to completely specify the stoichiometry of the system for high-temperature equilibrium calculations. This will be discussed further in Chap. 4.

These results may now be generalized. In the case of an inert system the composition is fixed and inviolate (if we disallow second-order transport processes such as thermal diffusion). Thus an inert system may be treated as a simple gas having an average molecular weight and a set of composite thermodynamic properties. However, in the case of a nonequilibrium reactive gas we must know the number of components $c$, the number of species $s$, and the stoichiometry of the individual reaction processes which determine the overall molecular composition changes—i.e., we must know the stoichiometric coefficients in the equations

$$dy_i = \sum_{j=1}^{\not{h}} v_{ij} M_i \, d\lambda_j \qquad i = 1, 2, \ldots, s \qquad (2\text{-}15)$$

or

$$dn_i = \sum_{j=1}^{\not{h}} v_{ij} \, d\lambda_j \qquad i = 1, 2, \ldots, s \qquad (2\text{-}16)$$

where there are $\not{h}$ distinct reaction processes being considered, each with its own reaction coordinate $\lambda_j$ with units of moles of reaction per kilogram of system. Note that the $v_{ij}$'s are the coefficients that ordinarily appear in the balanced stoichiometric equations as written by the chemist, and that Eqs. (2-15) and (2-16) are a generalization of the equation set (2-14) as it was written for the ozone–oxygen system. Here, if all the $v_{ij}$'s for an $i$th species are identically zero the species is assumed to be inert in the system under consideration and if one or more of the $v_{ij}$'s are not zero the equation represents the net production of that species in terms of the net changes of the individual reaction coordinates $\lambda_j$. In addition we should point out that a nonequilibrium reactive gas always exhibits a net change in the molecular composition with time and we may therefore write Eq. (2-16), for example, as

$$\frac{dn_i}{dt} = \sum_{j=1}^{\not{h}} v_{ij} \frac{d\lambda_j}{dt} \qquad (2\text{-}17)$$

where the $d\lambda_j/dt$ are determined by the reaction rates. Under the most advantageous conditions these may be directly evaluated for the individual reactive collision processes that are occurring in the system. If this is possible, direct integration of the $i$ equations will allow one to determine the composition of the mixture as a function of time. Also note that in a system containing two or more independent reaction coordinates it is generally true that their limit values are coupled in a complex manner. The evaluation of elementary $d\lambda_j/dt$'s for reacting gases is discussed in Chaps. 4 and 6.

## 2-4 THERMOCHEMISTRY

The second most important fundamental change that must be followed during a combustion process is the change in the thermodynamic state of the system associated with any species composition changes that occur. We must of course always work within the constraints of stoichiometry, but we are now extending our discussion to include changes in the fundamental thermodynamic properties during a chemical change. For the basic discussion of the thermodynamics of change we are not interested in the details of the actual process but only in changes from state 1 to state 2. Thus, to perform our thermodynamic calculations we can conceptually use rather simple paths to formalize our approach. Specifically, we allow (1) chemical reactions to occur at either constant pressure and temperature or constant volume and temperature and (2) allow temperature changes (at either constant pressure or volume) to occur only in the absence of chemical reaction. If we do this we can restrict ourselves to discussing (1) the chemical internal energy, enthalpy, or entropy change for a specific reaction occurring at constant temperature, and (2) the physical (sensible) internal energy, enthalpy, or entropy change when the temperature of some reactant or product mixture is changed without the occurrence of chemical reaction. In order to simplify the treatment we will further assume that the ideal gas law applies.

Consider some specific chemical reactions occurring at constant temperature and pressure in a mixture of ideal gases. We can write, for example,

$$CO + H_2O(g) \rightarrow CO_2 + H_2 \qquad \Delta H_r = -41.16 \text{ kJ} \qquad (T = 298.15 \text{ K})$$

which means that 41.16 kJ of enthalpy will be released to the surroundings if one mole of carbon monoxide reacts completely with one mole of water vapor at constant pressure to produce one mole of carbon dioxide and one mole of hydrogen. Thus this reaction is exothermic, and if this enthalpy were added to the products their temperature would be increased because of the occurrence of the reaction. We can also write the reactions

$$H_2 + \tfrac{1}{2}O_2 \rightarrow H_2O(g) \qquad \Delta H_r = -241.83 \text{ kJ} \qquad (T = 298.15 \text{ K})$$

and

$$CO + \tfrac{1}{2}O_2 \rightarrow CO_2 \qquad \Delta H_r = -282.99 \text{ kJ} \qquad (T = 298.15 \text{ K})$$

Recall from stoichiometry that the last two of these three equations can be subtracted from each other to yield the first. This is also true of the heats of reaction that are associated with the chemical changes. Thus we have just presented a redundant set of heats of reaction. This redundancy proves to be very useful because it means that we really need to tabulate only one basic heat of reaction for each species of interest.

This heat of reaction has been formalized to be the heat of formation of any desired species ($\Delta H_f^\circ$), in the desired state (solid, liquid, or ideal gas) from the standard state elements at the specified temperature and a pressure of one atmosphere (101,325 Pa). Here the standard state elements are taken to be the most stable state of that element at the temperature of interest and a pressure of one atmosphere. For the common elements of combustion interest at 298.15 K these states are carbon (C) as graphite, molecular hydrogen ($H_2$), oxygen ($O_2$) and nitrogen ($N_2$) as ideal gases, and atomic sulfur (S) as a solid.

Therefore, we have effectively required that the standard state elements at 298.15 K be the primary reference state for all other states of a system. This means that we can write the enthalpy of any pure substance as

$$H(T) = (\Delta H_f^\circ)_{298.15} + \int_{298.15}^{T} C_p \, dT$$

$$H(T) = (\Delta H_f^\circ)_{298.15} + (H_T - H_{298.15})$$

and for a mixture of ideal gases

$$h(T) = \sum_{i=1}^{s} n_i H_i(T) = \sum_{i=1}^{s} n_i[(\Delta H_f^\circ)_{298.15} + (H_T - H_{298.15})]_i \qquad (2\text{-}18)$$

when the $n_i$'s are moles per kilogram and $h(T)$ is the enthalpy of the mixture in joules per kilogram of mixture.

For an ideal gas the same statement may be made for the internal energy

$$e(T) = \sum_{i=1}^{s} n_i E_i(T) = \sum_{i=1}^{s} n_i[(\Delta E_f^\circ)_{298.15} + (E_T - E_{298.15})]_i \qquad (2\text{-}19)$$

where, because of the definition of enthalpy, for an ideal gas

$$(E_T - E_{298.15}) = (H_T - H_{298.15}) - R(T - 298.15) \qquad (2\text{-}20)$$

and

$$(\Delta E_f^\circ)_{298.15} = (\Delta H_f^\circ)_{298.15} - 298.15 \, \Delta n \, R \qquad (2\text{-}21)$$

where $\Delta n$ = the mole change *in the gas phase* for one mole of formation reaction. For example, for the following formation reactions at 298.15 K we obtain

| Formation reaction | $\Delta H_f^\circ$ (kJ) | $\Delta n$ | $\Delta E_f^\circ$ (kJ) |
|---|---|---|---|
| $H_2 + \frac{1}{2}O_2 \rightarrow H_2O(g)$ | $-241.83$ | $-\frac{1}{2}$ | $-240.59$ |
| $C(s) + \frac{1}{2}O_2 \rightarrow CO$ | $-110.53$ | $+\frac{1}{2}$ | $-111.77$ |
| $C(s) + O_2 \rightarrow CO_2$ | $-393.52$ | $0$ | $-393.52$ |
| $C(s) + 2H_2 \rightarrow CH_4$ | $-74.87$ | $-1$ | $-72.39$ |

For an ideal gas the same type of statement can be made for entropy. The formalism here is that each species has an entropy of formation at a specified temperature and one atmosphere.

However, in this case since entropy is a function of temperature and pressure we write for a single ideal gas species

$$S(T, P) = S^\circ(T_0, P_0) + \int_{T_0}^{T} \frac{C_p}{T} \, dT - R \ln \frac{P}{P_0} \qquad (2\text{-}22)$$

where $P_0 = 101{,}325$ Pa.

For an ideal gas mixture this yields the equation

$$s(T, P) = \sum_{i=1}^{s} n_i S_i(T, P) = \sum_{i=1}^{s} n_i \left[ S_i^\circ(T_0, P_0) + \int_{T_0}^{T} \frac{C_{pi}}{T} \, dT - R \ln \frac{p_i}{P_0} \right] \qquad (2\text{-}23)$$

where $p_i$ is the partial pressure of the $i$th species.

The thermochemical properties of many species of interest to the combustion field have been tabulated in standardized form in two major compilations. These are the JANAF (Joint Army-Navy-Air Force) compilation of simple species from 0 to 6000 K and the Stull et al. compilation of more complex organic molecules, usually from 0 to 1500 K. Portions of these compilations are given in tabular form in App. B.‡

These tables have been compiled primarily from spectroscopic data which allows one to calculate the heat capacity of a species, using sophisticated statistical thermodynamic principles similar to that used in the very simple example presented in Sec. 1-6. The only additional data needed to complete the table are the entropy and enthalpy of formation of the species at one reference temperature. These data are obtained from very careful thermochemical measurements, usually of the heat of combustion of the substance. The relation between heat of formation and heat of combustion will be discussed in the next section.

‡ The heats of formation and reaction used in the examples in this section were obtained from the JANAF tables in App. B.

## 2-5 THERMOCHEMICAL CALCULATIONS

The thermochemistry of a reactive system becomes particularly easy to calculate if all the species involved have properties that are tabulated either in the JANAF tables or Stull et al.‡ Consider, for example, the stoichiometric oxidation of a 40 volume percent propane 60 volume percent butane mixture with air that contains 21% $O_2$ and 79% $N_2$ and assume that the process is adiabatic and occurs at 1 atmosphere pressure. This means that $\Delta h = 0$ for the process. Let us further assume that the initial temperature of the fuel and air are different and that we wish to calculate the final temperature when combustion is complete. We will simplify the approach by assuming that the reaction occurs at 298.15 K. Note that this assumption is arbitrary and will not affect the final temperature that is calculated because enthalpy is a state variable.

To perform the calculation we assume as our base that the enthalpy of the standard state elements that make up the mixture are all zero at 298.15 K and calculate the enthalpy of the reactants relative to that standard state enthalpy. For a system that contains 0.862 kg of mass we obtain

$$h_{reactants} = 0.4[(\Delta H_f^\circ)_{298.15} + (H_{T_{fu}} - H_{298.15})]_{C_3H_8}$$
$$+ 0.6[(\Delta H_f^\circ)_{298.15} + (H_{T_{fu}} - H_{298.15})]_{C_4H_{10}}$$
$$+ 5.9[(\Delta H_f^\circ)_{298.15} + (H_{T_{air}} - H_{298.15})]_{O_2}$$
$$+ 22.18[(\Delta H_f^\circ)_{298.15} + (H_{T_{air}} - H_{298.15})]_{N_2}$$

where $T_{fu}$ and $T_{air}$ are the initial fuel and air temperatures respectively. See Fig. 2-2a for a graphical representation.

For the products we obtain

$$h_{prod} = 3.6[(\Delta H_f^\circ)_{298.15} + (H_{T_{prod}} - H_{298.15})]_{CO_2}$$
$$+ 4.6[(\Delta H_f^\circ)_{298.15} + (H_{T_{prod}} - H_{298.15})]_{H_2O}$$
$$+ 22.18[(\Delta H_f^\circ)_{298.15} + (H_{T_{prod}} - H_{298.15})]_{N_2}$$

where $T_{prod}$ is the final product temperature. In these equations the condition $h_{reactants} = h_{prod}$ determines the product temperature. Note that for any set of initial conditions all terms except the three product terms involving $(H_{T_{prod}} - H_{298.15})$ are constants. The solution is iterative but convergence is usually relatively rapid because enthalpies vary almost linearly with temperature, and thus linear interpolation usually yields a good next guess for the product temperature.

In many cases the "fuel" in question is not a pure substance and has only an ultimate analysis. Also, the heat of formation has not been determined for all pure substances. When one deals with either of these cases one must have the

‡ Referenced in App. B.

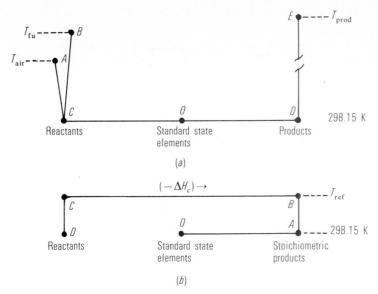

**Figure 2-2** Formalized paths used to calculate enthalpy changes (*a*) for a general reaction, (*b*) to convert a heat of combustion to a heat of formation.

composition of the fuel and a heat of combustion, if one is to perform a thermochemical calculation. The heat of combustion of any general CHONS fuel is the heat released by the complete oxidation of that fuel at a temperature near room temperature in accordance with the stoichiometry of Eq. (2-1). Either or both of two values may be reported. The high value is for the product water in the liquid state at the temperature of interest while the low value is for the hypothetical state when water is in the vapor state and is an ideal gas. The difference between the two is just the heat of vaporization of $v/2$ moles of $H_2O$ per kilogram of fuel [as per Eq. (2-1)]. Additionally, it must be pointed out that even though all heats of combustion are large negative numbers (because heat must be extracted to keep the temperature from rising), they are always reported in the literature as positive numbers. Heats of combustion for some common fuels are tabulated in App. C.

In any case the heat of formation of the fuel at 298.15 K must be determined before a thermochemical calculation can be performed using the JANAF tables. Since the heat of combustion may have been determined at some temperature other than the JANAF standard state reference temperature of 298.15 K (25°C) (77°F) reference temperature corrections must be included in the calculation. Consider Fig. 2-2*a* where heats of formation are known.

We calculated the enthalpy change from state *O* through *C* to *B* and *A* to determine the relative enthalpy content of the fuel and air at temperatures $T_{fu}$ and $T_{air}$ respectively. Here the line *OC* represents the heat of formation of the fuel and air at 298.15 K. We also calculated the enthalpy of the products at point *E* from point *O* by going through point *D*.

For the calculation of the effective enthalpy of formation we use a similar diagram, Fig. 2-2b. In this figure $T_{ref}$ is the reference temperature at which the heat of combustion was reported and $\Delta H_c$ is the heat of combustion for the process at the temperature $T_{ref}$. The positive value reported in the literature must be multiplied by $-1$ to yield a thermodynamic heat of reaction which represents the transition from point $C$ to point $B$ in Fig. 2-2b.

We note that the effective heat of formation of the fuel is given by the transition $O$ to $D$ in Fig. 2-2b and may be calculated by following the path $O–A–B–C–D$. When this is done the net change of the sensible enthalpy of the nitrogen in the air cancels out. The expression for the heat of formation per kilogram of fuel becomes

$$
\begin{aligned}
(\Delta h_f^\circ)_{298.15} = \; & u[(\Delta H_f^\circ) + (H_{T_{ref}} - H_{298.15})]_{CO_2} \\
& + \frac{v}{2}[(\Delta H_f^\circ) + (H_{T_{ref}} - H_{298.15})]_{H_2O} \\
& + \frac{x}{2}[(\Delta H_f^\circ) + (H_{T_{ref}} - H_{298.15})]_{N_2} \\
& + y[(\Delta H_f^\circ) + (H_{T_{ref}} - H_{298.15})]_{SO_2} \\
& - (-\Delta H_c) \\
& - c(T_{ref} - T_{298.15})_{fu} \\
& - \left(u + \frac{v}{4} - \frac{w}{2} + y\right)[(\Delta H_f^\circ) + (H_{T_{ref}} - H_{298.15})]_{O_2}
\end{aligned}
\qquad (2\text{-}24)
$$

where $c$ is the heat capacity of the fuel in joules per kilogram (if the fuel is a gas, $c_p$ should be used).

The $\Delta h_f^\circ$ calculated using this approach can be used in the enthalpy balance of Fig. 2-2a to determine an enthalpy balance for any fuel-oxidizer mixture when coupled with a proper mass balance or stoichiometric relationship. We note that Fig. 2-2b was drawn on the basis of one kilogram of material. It should be obvious that for a pure substance the mass basis for Fig. 2-2b could be one mole of the substance.

## 2-6 THERMOCHEMICAL MODELING

The previous section of this chapter describes the proper technique for performing thermochemical calculations for combustion processes. Unfortunately, the enthalpy-temperature relationship for virtually all species are complex functions which cannot be expressed analytically. For exact work the tabulated heat capacities are fitted to a high-order polynomial to yield an accurracy of about

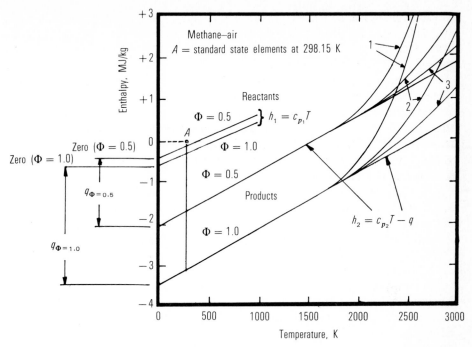

**Figure 2-3** Reactant and product enthalpies for the methane–air system. Curves 1, $P = 1$ kPa; curves 2, $P = 100$ kPa; curves 3, $P = 10,000$ kPa.

0.01% in enthalpy. JANAF curve-fit data for a large number of compounds are given as a part of App. B.

If one wants to use a fully analytical technique to describe a particular combustion process, one cannot use the full enthalpy relationships that are presented in the previous section. One must instead approximate these relationships by some type of simplified analytical expression which will facilitate the construction of closed form solutions. Fortunately, for most fuel–air combustion systems the enthalpy-temperature relationship of the product gas mixture is nearly linear over a large temperature range. Furthermore for such a system, if the fuel is gaseous, the average molecular weight of the products is quite close to the molecular weight of the reactants (because approximately 80 percent of the mixture is nitrogen). Under these circumstances one can closely approximate the enthalpy-temperature relationships and the mass-balance requirements for the system by using a working fluid–heat addition model to represent the chemistry.

Figure 2-3 is a plot of the actual enthalpy-temperature relationship for the reactants and products of a $\Phi = 0.5$ and $\Phi = 1.0$ methane–air mixture for the case of full chemical equilibrium of the products at pressures of 1, 100, and 10,000 kPa. The straight lines in Fig. 2-3 are plotted through the enthalpy-temperature relationship for the reactants and for the products at low temperature. These can be written analytically as

$$h_1 = c_{p1} T \tag{2-25a}$$

$$h_2 = c_{p2} T - q \tag{2-25b}$$

where $q$ is an effective heat of reaction for one kilogram of mixture.

Notice that when one uses the working fluid heat addition model the reference zero has been shifted. Specifically, the point $A$ in Fig. 2-3 is at $h = 0$ and $T = 298.15$ K and represents the enthalpy of the standard state elements in the system that is to be modeled. In the heat addition model Eqs. (2-25), as written, shift the reference zero to the enthalpy of the ideal gas reactants at $T = 0$ K. This is a hypothetical state which does not exist in nature. However, since only differences in enthalpy are used in any analysis for initial conditions near room temperature, the model is quite accurate. Notice that the product enthalpy can be quite adequately modeled up to about 2200 K in this system when $P > 100$ kPa. This is typical of most hydrocarbon–air combustion systems.

Note that the slope of the reactant and product $(h, T)$ curves are not parallel. This means, of course, that in general $c_{p1} \neq c_{p2}$. However, because only small changes in initial conditions usually occur, it is common practice to set $c_{p1} = c_{p2}$ because this greatly reduces the complexity of the analytical expressions that are derived.

When using this thermochemical model one must be fully aware of the fact that the chemical system is being replaced by an inert working fluid with one or two constant heat capacities and a constant molecular weight in which the combustion process is represented by heat addition from some "magic" external source. Thus even though this type of model can be used to quite adequately represent the effects of "heat" release during the combustion process, there are certain aspects such as reversibility and the second law of thermodynamics which cannot be discussed using such a model.

In modeling the details of some combustion processes this heat-addition–working-fluid model can be extended by replacing the extent of combustion with an extent of heat addition to the working fluid. This approach will be used in this text and this type of application of the model will be discussed more fully after Chap. 4. Furthermore the relationship of this type of modeling to more sophisticated models is discussed in Sec. 6-11.

## PROBLEMS

**2-1** Calculate the weight of air (in kilograms) required to exactly oxidize a gaseous fuel that contains 40% $C_3H_8$, 25% $CH_4$, and 35% $N_2$ (i.e., the stoichiometric requirement). Assume air is 21% $O_2$ and 79% $N_2$. How many volumes of air per volume of fuel (at the same initial temperature and pressure) are required? What is the final volume of the mixture at 25°C if $H_2O$ is in the vapor phase?

**2-2** Subbituminous coal (Colo.) is to be burned at 30% excess air. Calculate the volume of air (21% $O_2$, 79% $N_2$) at 25°C needed per kilogram of fuel.

**2-3** What would be the CO equivalence ratio for stoichiometric mixtures of the fuels hydrogen, acetylene, and methane with air?

**2-4** Set up a general relationship that relates volume percent fuel to equivalence ratio for a pure gaseous $C_uH_v$ fuel and for a pure gaseous $C_uH_vO_wN_xS_y$ fuel.

**2-5** Assume that for a fuel-rich mixture of a $C_uH_v$ fuel and 21% $O_2$ air the oxygen deficient products contain only the species CO, $CO_2$, $H_2O$ and $N_2$. Write a general equation for the $CO/CO_2$ ratio as a function of the equivalence ratio. Use this relationship to calculate the fuel-equivalence ratio at which the $CO/CO_2$ ratio is 1 and at which only CO and $H_2O$ are present for the fuels acetylene, methane, and ethane. Assume that $H_2$ is absent in the product gas and there is no dissociation.

**2-6** Consider air at room temperature to be composed of 21% $O_2$ and 79% $N_2$. Also assume that at high temperatures it dissociates to form a mixture of $N_2$, $O_2$, NO, and O atoms. How many reaction coordinates are needed to specify the composition of high-temperature air if only these species are present? Based on your choice of reaction coordinates, the initial composition of air and the mass that you are considering, what are the limits on your reaction coordinates?

**2-7** Assume that the methane–air system with an equivalence ratio of 1.2 contains, after combustion, only the species $CO_2$, $H_2O$, $N_2$, $O_2$, CO, $H_2$, H, O, and OH. How many reaction coordinates are needed to specify this system? For 21% $O_2$ air, set up the equations which determine the molar composition of the final mixture in terms of the reaction coordinates that you have used.

**2-8** Use heats of formation tabulated in the JANAF tables (App. B) to calculate the heat of reaction for the following reactions:

(a) $C_2H_2 + 2.5O_2 \rightarrow 2CO_2 + H_2O$ at 25°C;
(b) $H_2 + O \rightarrow OH + H$ at 1500 K;
(c) $O_2 \rightarrow 2O$ at 3000 K;
(d) $CO + H_2O \rightarrow CO_2 + H_2$ at 1100 K.

**2-9** Calculate the enthalpy change for the following processes:

(a) $CH_4 + 2O_2 + 4N_2$ at 25°C yields $CO_2 + 2H_2O + 4N_2$ at 2200 K;
(b) $C_2H_5OH(g) + 2O_2 + 7N_2$ at 25°C yields $2CO + 3H_2O + 7N_2$ at 1700 K;
(c) $O_2$ at 25°C yields 2O at 3000 K.

**2-10** Calculate the heat of formation at 25°C for:

(a) bituminous coal, high volatiles, F, KY;
(b) a natural gas that contains 6.5% $CO_2$, 77.5% $CH_4$, 10% $C_2H_6$, and 6% $N_2$.

**2-11** Calculate the high and low heat of combustion at 25°C of a liquid petroleum gas (LPG) that contains 40% propane and 60% butane.

**2-12** The Federal Register, Vol. 36, #159, Tuesday, August 17, 1971, page 15712 defines the percent excess air used in a furnace in terms of the %$CO_2$, %$O_2$, %$N_2$, and %CO measured in the dry stack gases as

$$\%EA = \frac{(\%O_2) - 0.5(\%CO)}{0.264(\%N_2) - (\%O_2) + 0.5(\%CO)} \times 100$$

Using the general CHONS fuel of Eq. (2-1), derive the correct expression for %EA to determine the assumptions used in the derivation of the Federal Register equation.

**2-13** The heat of combustion of isopropyl alcohol (l) (formula $C_3H_8O$) is 1.986 MJ/mole at 20°C (products $CO_2$ and $H_2O(l)$). Calculate the standard heat of formation of this substance at 25°C and also calculate the constant pressure explosion temperature for the combustion of this compound with air at $\Phi = 1$. (Assume $H_2O(g)$ $CO_2$ and $N_2$ are products and that air contains 21% $O_2$ and 79% $N_2$.)

**2-14** Calculate the final temperatures which would be reached at constant pressure if:

(a) a mixture of one mole of CO plus one mole of $O_2$ at 300 K yielded a mixture which contained one mole of $CO_2$ plus one-half mole of $O_2$;

(b) the mixture in (a) above yielded a mixture of one-half mole of CO plus one-half mole of $CO_2$ plus three-quarters mole of $O_2$.

**2-15** Acetone $CH_3COCH_3$ and propylene oxide $CH_2OCHCH_3$ are isomers, i.e., have the same atomic composition ($C_3H_6O$). Calculate the heat of combustion of these two compounds at 25°C if $H_2O(g)$ and $CO_2$ are the only products.

**2-16** For the above two compounds (the isomers acetone and propylene oxide) calculate the flame temperature (constant pressure) for the four following cases:

(a) stoichiometric $CH_3COCH_3$ + air (21% $O_2$) at 25°C yields $H_2O(g)$ + $CO_2$ + $N_2$ at $T_f$;

(b) stoichiometric $CH_2OCH_2CH_3$ + air (21% $O_2$) at 25°C yields $H_2O(g)$ + $CO_2$ + $N_2$ at $T_f$;

(c) stoichiometric $CH_3COCH_3$ + pure oxygen ($O_2$) at 25°C yields $H_2O(g)$ + $CO_2$ at $T_f$;

(d) stoichiometric $CH_2OCH_2CH_3$ + pure oxygen ($O_2$) at 25°C yields $H_2O(g)$ + $CO_2$ at $T_f$.

**2-17** Calculate and plot the enthalpy-temperature relationship for two systems on the same absolute scale. The systems are:

(a) one mole $H_2$ plus one-half mole of $O_2$

(b) one mole of $H_2O$ vapor.

Determine average heat capacities and an effective value of $Q$ for a working fluid–heat addition model for this system.

# BIBLIOGRAPHY

**Practical stoichiometry**

Reed, R. J., *North American Combustion Handbook*, 2d ed., North American Mfg. Co., Cleveland, Ohio (1978).

Singer, J. G., ed., *Combustion, Fossil Power Systems*, Combustion Engineering Inc., Windsor, Conn. (1981).

Smith, M. L., and K. W. Stinson, *Fuels and Combustion*, McGraw-Hill, New York (1952).

**Detailed stoichiometry and thermochemistry**

Fitts, D. D., *Nonequilibrium Thermodynamics*, McGraw-Hill, New York (1962).

Stull, D. R., and H. Prophet, *JANAF Thermochemical Tables*, 2d ed., National Bureau of Standards NSRDS-NBS 37, Washington (June 1971).

———, E. F. Westrum, and G. C. Sinke, *The Chemical Thermodynamics of Organic Compounds*, Wiley, New York (1969).

# THREE

## THE PHYSICAL PROPERTIES OF REAL GASES

### 3-1 INTRODUCTION

We saw in the first chapter that the ideal gas law and ideal gas thermodynamic properties can be obtained without considering the details of collision processes. In this chapter we will consider the more important properties of real gases which are collision determined. These include static or equilibrium nonideality (i.e., $PV \neq RT$), the primary transport properties (diffusion, viscosity, and thermal conductivity) and one important secondary transport process (thermal diffusion). We are in essence restricting ourselves to those properties that are determined by the details of collisions which do not cause chemical change or affect, in a nonequilibrium manner, the relative energy storage in internal degrees of freedom. We defer until the next chapter the discussion of collisions which produce such changes in the gas.

### 3-2 NONIDEAL GASES

The ideal gas law is well supported by experimental evidence under conditions where the mean free path is very much larger than the diameter of a molecule. However, deviations do occur at high densities and at low temperatures, and in fact all real gases are observed to condense to the liquid state if the temperature is low enough and the density sufficiently high. Experimentally, this condensation behavior is observed to be similar for all pure substances (see Fig. 3-1). In this figure the double cross-hatched area represents the region where only

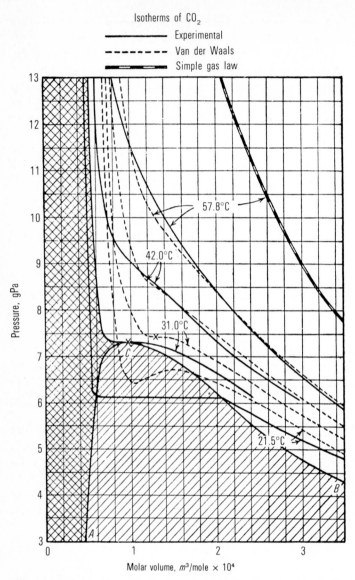

**Figure 3-1** Isotherms of carbon dioxide. *(With permission from O. A. Hougen and K. M. Watson, "Chemical Process Principles," Part II, "Thermodynamics," Wiley, New York, 1959.)*

pure liquid is present while the single cross-hatched area under curve *ACB* represents the coexistence region where liquid and gas are in equilibrium. It has been observed that there is a critical temperature $T_c$ (31.0°C in Fig. 3-1) for each gas above which condensation (i.e., the formation of a liquid phase in equilibrium with the gas phase) does not occur, even though the application of sufficient pressure will produce densities well above ordinary liquid densities. Furthermore, on the critical isotherm ($T = T_c$) one observes an inflection point at which $(\partial P/\partial V)_T = 0$ and $(\partial^2 P/\partial V^2)_T = 0$. This inflection point is called the *critical point* for the substance. The volume and pressure at this point may be measured and are unique constants for each pure substance. At temperatures below the critical temperature the gas is observed to condense when the pressure reaches its vapor pressure, and to continue to condense at this pressure as the volume is decreased until only the pure liquid phase is present. Further decreases in the volume then cause extremely large increases in the pressure of the pure liquid phase.

Many modifications of the ideal gas law have been proposed to account for these experimentally observable effects. For example, the most useful equation of state for exact work is a simple geometric series expansion of the ideal gas law in terms of either the pressure or the volume:

$$\frac{PV}{RT} = 1 + b(T)P + c(T)P^2 + \cdots \tag{3-1}$$

or

$$\frac{PV}{RT} = 1 + \frac{B(T)}{V} + \frac{C(T)}{V^2} + \cdots \tag{3-2}$$

where the coefficients are ordinarily taken to be temperature-dependent. These are commonly called virial equations of state. Experimental PVT data may be fitted quite accurately to either of these equations and the first and second coefficients for these equations may be calculated with reasonable accuracy from a theoretical consideration of the collision process for an equilibrium gas. This approach will be discussed in more detail in Sec. 3-4.

Among the other empirical equations of state there is one, the Van der Waals equation, which has a rather sound theoretical basis, is a closed form (analytical) equation, and qualitatively predicts the observed behavior of gases through the condensation region. This equation was first derived in the latter part of the nineteenth century. Clausius first proposed that the volume term in an equation of state should be the net volume available to the molecules and therefore modified the ideal gas law to read:

$$P(V - b) = RT$$

where $b$ is an effective volume of one gram-mole of molecules. In 1873 Johannes Van der Waals further modified this equation to allow for the slight attractive force between neighboring molecules. This attractive force is very weak and for ordinary molecules it decreases rapidly with the separation distance (for non-polar molecules the attractive force varies as $1/r^6$). Thus only the molecule's nearest neighbors affect its motion appreciably, and those molecules near the surface of the vessel are in an unbalanced potential field. The pressure at any surface is proportional to the number of molecules per unit volume in the gas near the surface, and the net attractive force (on the molecules away from the surface) is also proportional to the number density of the molecules in the layer near the surface. Thus a weak attractive force will tend to make the measured pressure less than the actual pressure by an amount proportional to $(\mathscr{A}/V)^2$ where $\mathscr{A}/V$ equals the number of molecules per unit volume in the gas ($\mathscr{A}$ is Avogadro's number). Therefore, the true gas pressure (i.e., on an imaginary internal surface) will be higher than the measured pressure and the Van der Waals equation is written as

$$\left(P + \frac{a}{V^2}\right)(V - b) = RT \tag{3-3}$$

where $P$ is the externally measured pressure. The units on $a$ and $b$ are (pressure volume$^2$ per mole) and (volume per mole). Appendix D contains a tabulation of critical constants and Van der Waals constants for various gases.

The Van der Waals equation is a cubic in the volume and predicts the phenomena of condensation and the existence of a critical point where the gaseous and liquid state are indistinguishable. To demonstrate this, let us define the critical properties as $P = P_c$, $V = V_c$, and $T = T_c$. At the critical point $(\partial P/\partial V)_T = 0$ and $(\partial^2 P/\partial V^2)_T = 0$ and we can take these derivatives and obtain the equations

$$\frac{RT_c}{(V_c - b)^2} = \frac{2a}{V_c^3}$$

$$\frac{2RT_c}{(V_c - b)^3} = \frac{6a}{V_c^4}$$

which, with Eq. (3-3) written at the critical point, yields

$$a = 3V_c^2 P_c \qquad b = \frac{V_c}{3}$$

and

$$R = \frac{8}{3}\frac{P_c V_c}{T_c} \qquad \text{or} \qquad \frac{P_c V_c}{RT_c} = 0.375$$

Values of $P_c V_c/RT_c$ for typical gases are tabulated in Table 3-1. Thus the Van der Waals constants are simply related to the critical constants of the gas and the behavior of all Van der Waals gases is similar when related to critical properties. This may be shown explicitly by defining the dimensionless variables $P_r = P/P_c$, $V_r = V/V_c$, and $T_r = T/T_c$ which, when substituted into Eq. (3-3), yield the dimensionless Van der Waals equation of state

$$\left(P_r + \frac{3}{V_r^2}\right)(3V_r - 1) = 8T_r \tag{3-4}$$

In practice the Van der Waals equation is most useful in applications where the deviations from ideality are small since it does not conform quantitatively to the behavior of real gases. It is a useful equation of state, however, because it

**Table 3-1 Critical constants ratios for gases‡**

(Theoretical value = 0.375)

| Gas | $P_c V_c / R T_c$ |
|---|---|
| Ar | 0.290 |
| $N_2$ | 0.291 |
| $O_2$ | 0.292 |
| $H_2$ | 0.304 |
| $Cl_2$ | 0.276 |
| NO | 0.251 |
| CO | 0.294 |
| HCl | 0.266 |
| $CO_2$ | 0.274 |
| $H_2O$ | 0.230 |
| $CH_4$ | 0.290 |
| $C_2H_2$ | 0.274 |
| $C_2H_4$ | 0.270 |
| $NH_3$ | 0.242 |

‡ Data taken from R. C. Reid and Thomas K. Sherwood, *The Properties of Gases*, McGraw-Hill, New York (1958) (*with permission*).

indicates the source of the primary corrections to the ideal gas law and because it neatly illustrates the principle of corresponding states or similitude for the PVT behavior of gases. Equation (3-4) is a dimensionless equation of state in terms of critical properties of the gas and is generally applicable to all gases with reasonable accuracy. Thus, in the same manner that the ideal gas law shows that equal volumes of different gases under the same conditions contain equal numbers of molecules, the Van der Waals equation shows that the first-order correction terms to the ideal gas law have the same origin in all gases.

The principle of corresponding states has also led to the development of graphical methods of determining the properties of real gases. In this approach one uses the equation $PV = ZRT$, where $Z$ is called the *compressibility factor* and is determined graphically as a function of reduced pressure $P_r$ and reduced temperature $T_r$. The accuracy of such an empirical all-inclusive fit for pure gases is estimated to be about $\pm 5\%$ over most of the range from condensation to $T_r = 15$ and $P_r = 30$.

The review of thermodynamics in Chap. 1 showed that there is a unique relationship between the thermal equation of state and the caloric equation of state. Specifically, the reciprocity relationship of thermodynamics gives either the pressure dependence of enthalpy or the volume dependence of internal energy once a thermal equation of state is known. Thus the thermal and caloric equations of state are not freestanding but are coupled. Therefore we note that $h = h(T)$ and $e = e(T)$ only when there are very specific constraints on the thermal equation of state; one example of which is that it be the ideal gas law.[1]

1. See, for example, R. C. Reid, J. M. Prausnitz, and T. K. Sherwood, *The Properties of Gases*, McGraw-Hill, N.Y. (1977), for a complete discussion of the structure of a caloric equation of state based on the virial thermal equation of state [Eq. (3-2)].

Any further discussion of nonideal gas behavior for pure gases and for gas mixtures is outside the scope of this text. The interested reader is referred to the bibliography for further information.

## 3-3 TRANSPORT PHENOMENA—ELEMENTARY THEORY

There are three fundamental transport properties, one for each of the properties conserved in the three conservation equations of fluid dynamics. The three conservation equations may be derived with the assumption that transport is zero or small enough to be neglected. In fact, transport phenomena are only important when the solution to the conservation equations predicts that the fluid must support a large gradient in either concentration, velocity, or temperature. Under these conditions, mass transport, momentum transport, or energy transport (respectively) will occur and the proportionality constants which relate the quantity transported to the gradient (or the driving force which causes transport) are called the *diffusion coefficient*, the *viscosity coefficient*, and the *coefficient of thermal conductivity*.

Let us consider viscosity. If two large plates are placed a short distance $y$ apart and are moved relative to each other in the $x$ direction (keeping $y$ constant) one finds that a force is required to keep their relative velocity in the $x$ direction, $u$, at a constant value. Furthermore, experimental evidence has shown that when gases or simple liquids are trapped between the plates, the steady velocity that is maintained is proportional to the force per unit area applied in the direction of slippage times the distance between the plates, i.e.,

$$y \frac{\mathscr{F}}{A} \propto u$$

or

$$\frac{\mathscr{F}}{A} = -\mu \frac{du}{dy} \tag{3-5}$$

This is called *Newton's law of viscosity*. Fluids with a constant proportionality constant (independent of $du/dy$) are called newtonian fluids. In SI units

$$\mu = \left| \frac{\mathscr{F}}{A} \frac{dy}{du} \right| = \text{kg/m-s}$$

Kinematic viscosity is defined as

$$v = \frac{\mu}{\rho}$$

where $\rho$ is the fluid density. Therefore kinematic viscosity has units of square meters per second.

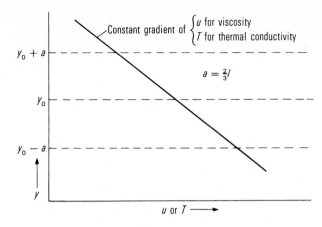

**Figure 3-2** Constant gradient assumed in the kinetic theory derivation of the transport properties viscosity and thermal conductivity.

Using the kinetic theory of gases from Chap. 1 we may relate the viscosity coefficient to the molecular properties of an ideal gas consisting of hard spheres. Recall that the average speed of a molecule is

$$\bar{c} = \sqrt{\frac{8}{\pi}\frac{RT}{M}}$$

and its mean free path is

$$l = \frac{1}{\sqrt{2}\,\pi\sigma_0^2\,N}$$

Now consider an imaginary plane in a gas at a position $y_0$ normal to a velocity gradient (see Fig. 3-2). Molecules are continually crossing this plane from above and below. On the average those that cross the plane will have had their last collision at a distance $a \cong \frac{2}{3}l$ away from the plane. From Chap. 1 we know that the number of molecules crossing the plane per second from one side is $Z_w = N\bar{c}/4$. We assume that each molecule is carrying an $x$ momentum characteristic of the $x$ momentum in the plane of its last collision. The momentum flux across the plane is therefore

$$\frac{\mathscr{F}}{A} = Z_w(mu)|_{y-a} - Z_w(mu)|_{y+a}$$

But since we assume a constant gradient

$$u|_{y-a} = u|_y - \frac{2}{3}\,l\,\frac{du}{dy}$$

and

$$u|_{y+a} = u|_y + \frac{2}{3}\,l\,\frac{du}{dy}$$

Therefore

$$\frac{\mathscr{F}}{A} = -Z_w m \frac{4}{3} l \frac{du}{dy} \cong -\frac{1}{3} N m \bar{c} l \frac{du}{dy}$$

or

$$\mu = \tfrac{1}{3} N m \bar{c} l = \tfrac{1}{3} \rho \bar{c} l \qquad (3\text{-}6)$$

Maxwell first obtained this expression in 1860. Substituting for $l$ and $\bar{c}$, we find that

$$\mu = \frac{2}{3\pi^{3/2}} \frac{\sqrt{m \& T}}{\sigma_0^2} \qquad (3\text{-}7)$$

Viscosity measurements in gases can therefore be used to determine molecular diameters. Equation (3-7) correctly predicts that the viscosity of dilute gases is independent of the pressure and the equation also yields approximately correct molecular diameters. However, the predicted temperature dependence has been found to be too weak.

Thermal conductivity without convection is quite similar to viscosity, since the heat conduction per unit area through any simple substance has been found to be simply proportional to the temperature gradient

$$\frac{Q}{A} = -\kappa \frac{dT}{dy} \qquad (3\text{-}8)$$

This is called *Fourier's law of heat conduction*, where $\kappa$ is the thermal conductivity. In SI units

$$\kappa = \left| \frac{Q}{A} \frac{dy}{dT} \right| = \text{J/m-s-K}$$

Thermal diffusivity is defined as

$$\alpha = \frac{\kappa}{\rho c_p} = \text{m}^2/\text{s}$$

where $c_p = \text{J/kg-K}$.

Kinetic theory can also be used to relate the thermal conductivity $\kappa$ to the molecular parameters of the gas. When this is done, we obtain

$$\kappa_{\text{mono}} = \tfrac{1}{2} N \& \bar{c} l = \tfrac{1}{3} \rho c_v \bar{c} l$$

for the thermal conductivity of a monatomic gas. This reduces to the expression

$$\kappa_{\text{mono}} = \frac{1}{\sigma_0^2} \sqrt{\frac{\&^3 T}{\pi^3 m}} \qquad (3\text{-}9)$$

Equation (3-9) predicts the correct pressure sensitivity for thermal conductivity (none) and the correct relationship with molecular diameter and molecular

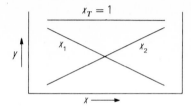

**Figure 3-3** Mole fractions in a binary gas mixture containing a concentration gradient.

weight. However, once again the predicted temperature sensitivity has been found to be too weak.

The diffusion coefficient is defined as a proportionality constant which relates the net transport of a species to the concentration gradient under conditions where the total net transport of molecules is zero. Because of the requirement that there be no net transport in the mixture as a whole, binary-diffusion coefficients are dependent on the properties of both gases. There are, therefore, self-diffusion coefficients, i.e., for the gas into itself; and binary-diffusion coefficients, for the gas into another gas (which must be specified).

It is important to note there that when binary diffusion occurs between two gases of different molecular weight the requirement that there be no net transport of molecules means that the mass center of the system moves in the direction that the heavier molecules are diffusing. The situation is even more complex when multicomponent diffusion occurs. This problem will be discussed again in Chap. 5 on Reactive Gas Dynamics.

We will consider only self- and binary diffusion here. Assume the mixture has a constant concentration gradient as shown in Fig. 3-3. *Fick's law of diffusion* states that the current density of flow (or net molecular flow of one species) past a plane normal to the gradient is proportional to the gradient, i.e.,

$$\Gamma_1 = -D \frac{dN_1}{dy} \quad \text{and} \quad \Gamma_2 = -D \frac{dN_2}{dy} \tag{3-10}$$

Notice that both $D$'s are the same. Here $\Gamma$ has units of molecules per square meter-second, and $dN/dy$ has units of molecules per cubic meter. Therefore $D$ must have units of square meters per second. These units are identical with the units for kinematic viscosity and thermal diffusivity as defined above. Using our earlier kinetic theory approach to calculate a value of the diffusion coefficient, we obtain

$$D_{11} = \tfrac{1}{3}\bar{c}l = \text{m}^2/\text{s}$$

where the subscript 11 refers to the self-diffusion coefficient. Substituting in the ordinary state variables yields the equation

$$D_{11} = \frac{2}{3}\left(\frac{k^3}{\pi^3 m}\right)^{1/2} \frac{T^{3/2}}{P\sigma_0^2}$$

This simple theory may be extended, by using the proper assumptions, to derive an expression for binary diffusion coefficients. The derivation yields the approximation

$$D_{12} \simeq \frac{2}{3} \left( \frac{k^3}{\pi^3} \right)^{1/2} \left( \frac{1}{2m_1} + \frac{1}{2m_2} \right)^{1/2} \frac{T^{3/2}}{P[(\sigma_{0_1} + \sigma_{0_2})/2]^2} \tag{3-11}$$

Once again the simple theory yields a result which is adequate except that the predicted temperature dependence is found to be too weak.

The dimensional similarity of these three transport properties suggests that their ratios may be used to define a set of convenient dimensionless parameters for discussing the properties of gases. This is indeed the case, and the three resulting parameters have had sufficient utility to each be identified with a scientist's name. They are; (1) the ratio of the kinematic viscosity and the thermal diffusivity, or Prandtl number,

$$\Pr = \frac{c_p \mu}{\kappa} = 1.667 \tag{3-12}$$

(2) the ratio of the kinematic viscosity and the self-diffusion coefficient, or Schmidt number,

$$\text{Sc} = \frac{\mu}{\rho D} = 1.0 \tag{3-13}$$

and (3) the ratio of the thermal diffusivity and the self-diffusion coefficient, or Lewis number

$$\text{Le} = \frac{\kappa}{c_p \rho D} = 0.60 \tag{3-14}$$

Some texts define Le as the reciprocal of Eq. (3-14). The numerical values quoted above are for the interaction of hard-sphere monatomic molecules using the simple kinetic theory approach. These values do not agree with the experimentally measured ratios for dilute gases.

Thus, in summary, we find that elementary kinetic theory yields the correct pressure dependence for the dilute-gas transport properties but does not yield either a correct temperature dependence or a correct set of values for the dimensionless ratios of these properties. The reason for these discrepancies will be discussed in the next section.

## 3-4 TRANSPORT PROCESSES AND THE COLLISION INTERACTION POTENTIAL

The elementary treatment of transport, developed in Sec. 3-3, adequately explains the gross behavior of the transport coefficients of dilute gases and has been found to yield order-of-magnitude agreement with experimentally determined values. However, the results are not really useful for estimating the transport coefficients of real gases with reasonable accuracy. The more rigorous Chapman–Enskog theory of transport, however, yields reasonably accurate transport coefficients. It is based on the same hypothesis as the elementary theory (i.e., dilute gas/only binary collisions contribute) but it includes the effect of an assumed interaction potential during the molecular collision process. In addition it assumes that each interaction is weighted in relation to all other possible elastic collisions using a Maxwell distribution of molecular velocities. Thus the more rigorous kinetic theory requires that an interaction potential be specified for each molecular encounter, and this theory is therefore only as realistic as the assumed potential. Using this theory, the ultimate result of a rather lengthy calculation is a set of correction factors to the elementary formulas which are dependent on the chosen interaction potential and the gas temperature. Because of the complexity of the theory only the interaction of spherically symmetric molecules has been calculated in any detail at this time.

The true molecular interaction potential (even between truly spherical molecules such as argon) cannot be represented exactly by a simple mathematical expression. However, many simple equations come quite close to the correct potential. This is usually assumed to consist of a long-distance attractive force (for a simple spherical molecule with no permanent dipole moment this is caused by an induced-dipole–induced-dipole interaction and is proportional to $r^{-6}$) and a close-approach repulsive force which is thought to be exponential. The net result is an interaction potential similar to that shown in Fig. 3-4. Notice that the potential-energy minimum (i.e., the "well") formed by this type of interaction will markedly change the character of low-velocity collision processes. In general one can describe this "well" by determining its depth $\varepsilon$ and a

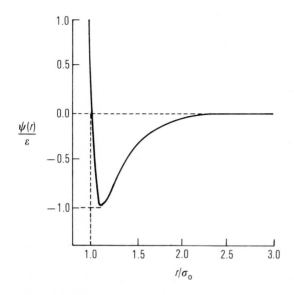

**Figure 3-4** Lennard–Jones interaction potential for the collision of spherical molecules [Eq. (3-15)].

zero-energy collision diameter $\sigma_0$ within the framework of a particular mathematical model.

The most successful and best known two-parameter relationship for the potential energy for which the Chapman–Enskog theory has been solved is the Lennard–Jones potential:

$$\Psi(r) = 4\varepsilon\left[\left(\frac{\sigma_0}{r}\right)^{12} - \left(\frac{\sigma_0}{r}\right)^{6}\right] \tag{3-15}$$

Notice in Fig. 3-4 that the depth of the well is important to a collision only as it relates to the kinetic energy of the collision. Since in a Maxwell gas the average kinetic energy is proportional to the temperature, $\varepsilon$ is made dimensionless by writing it as $kT/\varepsilon$ where $k$ is the Boltzmann constant. The theory is, therefore, a corresponding states theory, since it assumes a universal potential energy function for all molecules. Correction factors calculated from the Chapman–Enskog theory using the Lennard–Jones potential are also tabulated in App. A. Since in the theory the viscosity and thermal conductivity corrections are determined by the same averaging process, these correction factors (called the *collision integrals* $\Omega_\mu$ and $\Omega_\kappa$) are identical. The collision integral for diffusion, $\Omega_D$, is obtained by a slightly different averaging process and therefore is tabulated separately. Values of $\varepsilon/k$ in kelvins and $\sigma_0$ in nanometers for various molecules are tabulated in App. A. The transport coefficients may be calculated from the values of $\varepsilon/k$, kelvins and $\sigma_0$, nanometers, found in these tables using the following formulas:

Viscosity

$$\mu = 8.4411 \times 10^{-7} \frac{\sqrt{MT}}{\sigma_0^2 \Omega_\mu} \qquad \text{kg/m-s} \tag{3-16}$$

where $M$ is in kilograms per mole and $T$ is in kelvins.

Thermal conductivity for monatomic gases

$$\kappa_{\text{mono}} = 2.6330 \times 10^{-5} \frac{\sqrt{T/M}}{\sigma_0^2 \Omega_\kappa} \qquad \text{J/m-s-K} \tag{3-17}$$

and diffusion coefficients (binary)

$$D_{12} = 5.9543 \times 10^{-6} \frac{\sqrt{T^3[(1/M_1) + (1/M_2)]}}{P(\sigma_0)_{12}^2 \Omega_{D_{12}}} \qquad \text{m}^2\text{/s} \tag{3-18}$$

where $P$ is in pascals and $\Omega_{D_{12}}$ is a dimensionless collision integral calculated from a value of $(kT/\varepsilon)_{12}$. Here $(\varepsilon/k)_{12}$ and $(\sigma_0)_{12}$ are calculated from the individual parameters using the approximate equations

$$(\sigma_0)_{12} = \tfrac{1}{2}(\sigma_{0_1} + \sigma_{0_2})$$

and

$$\left(\frac{\varepsilon}{k}\right)_{12} = \sqrt{\left(\frac{\varepsilon}{k}\right)_1 \cdot \left(\frac{\varepsilon}{k}\right)_2}$$

Self-diffusion coefficients may be calculated using the above formulas by assuming $M_1$ equals $M_2$, etc.

Another approach to determining binary diffusion coefficients has been presented by Marrore and Mason.[2] They have curve-fitted the product $PD_{12}$ to a theoretically justified function of temperature for over 60 gas pairs. By using molecular beam data they claim that their equations are reasonably accurate to 10,000 K. They point out that their equations should not be used at lower temperatures, i.e., where $kT/\varepsilon < 1$.

In the case of thermal conductivity, the Chapman–Enskog theory allows the correct calculation of a coefficient only for the case of a monatomic gas. This is true because even the rigorous theory ignores the transport of energy stored in the internal degrees of freedom. Eucken first proposed an empirical formula to calculate the contribution of this internal-energy transport. His formula may be written as

$$\kappa = \kappa_{\text{mono}}\left[\frac{1}{3} + \frac{4}{15}\frac{\gamma}{\gamma - 1}\right] \tag{3-19}$$

where $\gamma$ equals the heat capacity ratio $c_p/c_v$ of the real gas. Hirschfelder[3] has derived a similar expression on a sound theoretical basis. The Hirschfelder formula for thermal conductivity is

$$\kappa = \kappa_{\text{mono}}\left[1 + \delta_f\left(\frac{c_p}{c_{pf}} - 1\right)\right]$$

where

$$\delta_f = \frac{\rho D c_{pf}}{\kappa_{\text{mono}}} = \frac{1}{\text{Le}_{\text{mono}}}$$

and

$$c_{pf} = \tfrac{5}{2}\mathscr{R}$$

Hirschfelder evaluated $\delta_f$ for a number of realistic potential functions and found that it is relatively constant at $\delta_f \cong 0.885$ over a large range of $kT/\varepsilon$. With this value of $\delta_f$ the Hirschfelder formula reduces to

$$\kappa = \kappa_{\text{mono}}\left[0.115 + 0.354\frac{\gamma}{\gamma - 1}\right] \tag{3-20}$$

The more rigorous kinetic-theory approach yields quite reasonable values of the dimensionless ratios of the transport properties. A few typical values for the Prandtl number are listed in Table 3-2. In the case of self-diffusion the data are much more meager, so that extensive comparisons cannot be made. However, the rigorous theory is quite successful in fitting experimental diffusion data and therefore the same type of fit for the dimensionless Schmidt and Lewis numbers should be obtained using the rigorous theory.

2. T. R. Marrore and E. A. Mason, *J. Phys. Chem., Ref. Data*, 1:3–118 (1972).
3. J. O. Hirschfelder, *J. Chem. Phys*, **26**:282 (1957).

**Table 3-2 Predicted and observed values of $(c_p \mu / \kappa)$ for gases at atmospheric pressure‡**

| Gas | $T$, K | $(c_p \mu / \kappa)$ From Eqs. (3-16), (3-17), and (3-19) | $(c_p \mu / \kappa)$ from observed values of $c_p$, $\mu$, and $\kappa$ |
|---|---|---|---|
| Ne | 273.2 | 0.667 | 0.66 |
| Ar | 273.2 | 0.667 | 0.67 |
| $H_2$ | 90.6 | 0.68 | 0.68 |
| | 273.2 | 0.73 | 0.70 |
| | 673.2 | 0.74 | 0.65 |
| $N_2$ | 273.2 | 0.74 | 0.73 |
| $O_2$ | 273.2 | 0.74 | 0.74 |
| Air | 273.2 | 0.74 | 0.73 |
| CO | 273.2 | 0.74 | 0.76 |
| NO | 273.2 | 0.74 | 0.77 |
| $Cl_2$ | 273.2 | 0.76 | 0.76 |
| $H_2O$ | 373.2 | 0.77 | 0.94 |
| | 673.2 | 0.78 | 0.90 |
| $CO_2$ | 273.2 | 0.78 | 0.78 |
| $SO_2$ | 273.2 | 0.79 | 0.86 |
| $NH_3$ | 273.2 | 0.77 | 0.85 |
| $C_2H_4$ | 273.2 | 0.80 | 0.80 |
| $C_2H_6$ | 273.2 | 0.83 | 0.77 |
| $CHCl_3$ | 273.2 | 0.86 | 0.78 |
| $CCl_4$ | 273.2 | 0.89 | 0.81 |

‡ With permission from R. B. Bird, W. E. Stewart, and E. N. Lightfoot, *Transport Phenomena*, Wiley, New York, 1960.

There is one second-order transport property, thermal diffusion, which is important in certain combustion processes. Dufour, in 1873 first showed that the ordinary diffusion of one gas into another causes a transient local temperature change. The reciprocal effect, thermal diffusion, was first discovered by Soret in 1879 when he measured steady concentration gradients in salt solutions that contained a thermal gradient. Then in the early 1900s the general kinetic theory of gases developed by Chapman and Enskog predicted that diffusion should occur in the presence of a thermal gradient as well as a concentration gradient. Chapman and Doatson showed the effect experimentally in 1917 by filling two bulbs connected with a tube with a uniform binary gas mixture and analyzing the gas in the two bulbs after they had been held at different temperatures for a long period of time. In this experiment, thermal diffusion is offset by ordinary diffusion when steady state is reached and one can define a thermal diffusion ratio for species 1, $K_{T_1}$, for a binary system by writing the equation

$$\frac{dx_1}{dx} = -K_{T_1} \frac{1}{T} \frac{dT}{dx} \qquad (3\text{-}21)$$

where the logarithmic temperature dependence arises from theoretical considerations. Here if $K_{T_1}$ is positive that species moves toward the colder bulb. Note that $K_{T_1}$ is dimensionless as defined. For a binary system one can easily define a thermal diffusion coefficient[4] $D_1^T$ and a thermal diffusion factor $\alpha_1$

$$D_1^T = \frac{\mathcal{N}^2}{\rho} M_1 M_2 D_{12} K_{T_1} = \text{kg/m-s} \tag{3-22}$$

$$\alpha_1 = \frac{K_{T_1}}{x_1 x_2} \tag{3-23}$$

Based on these definitions it should be obvious that $D_2^T = -D_1^T$, $K_{T_2} = -K_{T_1}$ and $\alpha_2 = -\alpha_1$ in any binary system.

As it turns out, in the gas phase the binary thermal diffusion ratio $\alpha_i$ is almost independent of concentration. However, in multicomponent mixtures it is a quite complex function of the local mixture composition. This problem can be handled (to a certain extent) when the species of interest is a trace species, by assuming that the background gas has a uniform composition and that thermal diffusion of the trace species occurs independently of others in that background gas. We defer further discussion of this problem to the next section on transport in multicomponent mixtures.

The Chapman–Enskog theory may also be used to determine the second and third virial coefficients in the equation of state, Eq. (3-2):

$$\frac{PV}{RT} = 1 + \frac{B(T)}{V} + \frac{C(T)}{V^2} + \cdots$$

The formulas for this estimation are

$$B(T) = B^* \left( \frac{\ell T}{\varepsilon} \right) \cdot b_0$$

$$C(T) = C^* \left( \frac{\ell T}{\varepsilon} \right) \cdot b_0^2$$

where

$$b_0 = \tfrac{2}{3}\pi \mathcal{A} \sigma_0^3 \times 10^{-27} \ \text{m}^3/\text{mole}$$

and $\mathcal{A}$ represents Avagadro's number. Notice that $V$ must have the same units as $b_0$. Values of $B^*$, $C^*$, and $b_0$ are tabulated in App. A.

The Chapman–Enskog theory of transport, while based on a sound interaction model, suffers some loss of generality in its application. This is caused by the lack of flexibility that the theory allows in the choice of potential models. Once a model is fixed—say a two-parameter Lennard–Jones potential—there is a great deal of calculation needed to evaluate the correction factors. After this

---

4. There are different definitions of $D_1^T$ in the literature (with different units). This text will follow the approach used by J. O. Hirschfelder, C. F. Curtis, and R. B. Bird, *The Molecular Theory of Gases and Liquids*, Wiley, N.Y. (1954) as adopted by R. B. Bird, W. E. Stewart, and E. N. Lightfoot in *Transport Phenomena*, Wiley, N.Y. (1960).

effort has been made, the predicted temperature dependence is compared to experimental data to determine the model constants for this particular substance, model, and transport property. Thus the theory, because of its inherent complexity, does not allow one to determine the best molecular model but only the best parameters in an a priori model. This is why the Lennard–Jones parameters as evaluated from experimental viscosity and equation of state data are different for most substances (see App. A). Furthermore, it has been found that the evaluation of "best fit" parameters for other "realistic" potential models yields interaction potentials which are markedly different for the same substance, even though the different models yield transport coefficients which fit the experimental data with similar standard deviations. This is why only Lennard–Jones parameters are presented in this text. For ordinary work they should be adequate to estimate transport coefficients and equation of state nonideality. The reader is referred to references in the bibliography and to the current literature on transport properties for a more detailed discussion of this problem.

## 3-5 MULTICOMPONENT MIXTURES TRANSPORT PROPERTIES

It has been observed experimentally that the transport coefficient of a binary mixture of dilute gases does not vary linearly with the composition. This is particularly true for those cases where the transport coefficients of the pure substances are markedly different. In the case of viscosity the properties of a mixture may be estimated with reasonable accuracy by using the semiempirical formula of Wilke[5]

$$\mu_{mix} = \sum_{i=1}^{s} \frac{\mu_i}{1 + \frac{1}{x_i} \sum_{\substack{j=1 \\ j \neq i}}^{s} x_j \chi_{ij}} \tag{3-24}$$

where

$$\chi_{ij} = \frac{1}{\sqrt{8}} \left(1 + \frac{M_i}{M_j}\right)^{-1/2} \left[1 + \left(\frac{\mu_i}{\mu_j}\right)^{1/2} \left(\frac{M_j}{M_i}\right)^{1/4}\right]^2 \tag{3-25}$$

In this equation $x$ is the mole fraction and $M$ is the molecular weight of the species involved. The $i$ and $j$ refer to the specific species in the mixture.

The thermal conductivity of mixtures may be determined using an equation given by Mason and Saxena[6]

$$\kappa = \sum_{i=1}^{s} \frac{\kappa_i}{1 + \frac{1.065}{x_i} \sum_{\substack{j=1 \\ j \neq i}}^{s} x_j \chi_{ij}} \tag{3-26}$$

5. C. R. Wilke, *J. Chem. Phys.*, **18**: 517 (1950).
6. E. A. Mason and C. S. Saxena, *Phys. Fluids*, **1**: 361–369 (1958).

where the $\chi_{ij}$ term is calculated from Eq. (3-25) using viscosities (or monatomic thermal conductivities) and the $\kappa_i$'s are the polyatomic thermal conductivities calculated using the Hirschfelder correction. The 1.065 multiplier is an approximation to a complex coefficient whose value remains close to 1 at all times.

In the case of diffusion the problem is more complicated. The reader will recall that the diffusion coefficient of a binary mixture is dependent on the properties of both species because the coefficient is defined for no net transport of molecules in the mixture as a whole. In the same manner the diffusion coefficient of the $i$th species in a multicomponent mixture is dependent on the local composition of the mixture. The problem is compounded, however, by the fact that each individual component in a multicomponent mixture diffuses at its own rate relatively independent of the other species. Thus, observation of the diffusion of carbon dioxide into a helium–nitrogen mixture[7] shows clearly that the helium concentration gradient disappears very rapidly whereas the nitrogen–carbon dioxide gradient remains. In other words, the rapid diffusion of helium into the pure carbon dioxide will initially cause a negative net mass flux of the nitrogen in the direction of the predominate diffusion gradient.

Furthermore, it is well known that the diffusion coefficient in a multicomponent mixture, $D'_{ij}$, does not have the same numerical value as the diffusion coefficient, $D_{ij}$, for a simple binary mixture.[2] The relationship between these two has been investigated for a trace species in a multicomponent mixture and it has been found that deviations are usually less than 10 percent. This, coupled with the fact that exact multicomponent diffusion coefficients are very difficult to calculate, has led to the use of the approximation $D'_{ij} \cong D_{ij}$. In other words, in practice it is reasonable to simply use binary diffusion coefficients as calculated from Eq. (3-18) for calculations in multicomponent mixtures.

If a single trace species is diffusing in a background gas mixture which is otherwise uniform (air, for example) one can calculate an effective diffusion coefficient using Blanc's law[2]

$$\frac{1}{D_1} = \sum_{i=2}^{s} \frac{x_i}{D_{1i}} \tag{3-27}$$

Thermal diffusion factors for two species in a multicomponent mixture, $\alpha_{ij}$, can be calculated using the formula[8]

$$\alpha_{ij} = \frac{M_i M_j}{M_i + M_j} \cdot \frac{(6C_{ij}^\dagger - 5)}{5RD_{ij}} \left( \frac{\kappa'_j}{\rho_j} - \frac{\kappa'_i}{\rho_i} \right) \tag{3-28}\ddagger$$

---

7. R. E. Walker, N. de Haas, and A. A. Westenberg, *J. Chem. Phys.*, **32**:1314 (1960).

8. I. Monchick, R. J. Munn, and E. A. Mason, *J. Chem. Phys.*, **45**:3051 (1966).

‡ The collision integral ratio in Eq. (3-28) is usually tabulated as $C^*$; since we use $C^*$ for the dimensionless third viral coefficient in App. A we use $C^\dagger$ for the collision integral ratio.

where

$$\kappa_i' = \frac{(\kappa_{\text{mono}})_i}{1 + \dfrac{1.065}{x_i} \sum_{\substack{j=1 \\ j \neq i}}^{s} x_i \chi_{ij}}$$ (3-29)

and the $C_{ij}^\dagger$'s are ratios of appropriate collision integrals evaluated using $\varepsilon_{ij}/k$ as before. These are tabulated in App. A. Using this approach, the thermal diffusion ratios for a multicomponent mixture can be determined using the general form of Eq. (3-23)

$$K_{T_1} = x_i \alpha_i = x_i \sum_{\substack{j=1 \\ j \neq i}}^{s} x_j \alpha_{ij}$$ (3-30)

In practice, these are usually used with the assumption that the species of interest, $i$, is dilute in a background gaseous mixture. In this case the $\alpha_i$'s have been observed to have a simple universal temperature dependence

$$\alpha_i = (\alpha_\infty)_i \left[ 1 + \frac{(\alpha_T)_i}{T} \right]$$

where $(\alpha_\infty)_i$ is the infinite temperature value of $\alpha_i$ and $(\alpha_T)_i$ yields the temperature dependence. The values of $(\alpha_\infty)_i$ and $(\alpha_T)_i$ for thermal diffusion into air are tabulated for a few species of interest in the combustion field in Table 3-3.

These thermal diffusion factors can be directly equated to the thermal diffusion coefficient using Eq. (3-22) only when the thermal diffusion of a trace species is being considered. The more general case is much more complex. It must be pointed out, however, that thermal diffusion is dependent on a second-order collision effect (i.e., a ratio of first-order corrections). Thus the accuracy of the simple expressions for the thermal diffusion coefficient is less than that of the first-order transport coefficients.

The reader is referred to the bibliography for further information on this subject.

9. D. E. Rosner, *Physicochemical Hydrodynamics*, **1**:159 (1980), with permission.

**Table 3-3**[9]

| Species | $\alpha_\infty$ | $\alpha_T$ |
|---------|---------|---------|
| H | $-0.2901$ | 17.94 |
| $H_2$ | $-0.2881$ | 30.37 |
| $H_2O$ | $-0.1424$ | 290.6 |
| CO | $-0.00888$ | 32.02 |
| $CO_2$ | 0.1480 | 49.56 |
| He | $-0.2707$ | $-29.95$ |

## 3-6 TRANSPORT MODELING

We learned earlier that simple kinetic theory yields transport coefficients which have the correct pressure dependence but too weak a temperature dependence. We also learned that Chapman–Enskog theory not only yields much more accurate transport coefficients but also predicts a relatively accurate Prandtl number for many gases. However, Chapman–Enskog theory does not yield a simple analytic expression for these coefficients. For modeling purposes we need an analytic expression which rather faithfully approximates the real implicit relationships for each transport coefficient. To a good first approximation we can use the expressions

$$\mu \propto P^0 T^{0.7}$$

$$\kappa \propto P^0 T^{0.7}$$

and

$$D \propto P^{-1} T^{1.7}$$

for the pressure and temperature dependences of the primary transport coefficients. It turns out that in many cases it is extremely convenient to use the approximations

$$\mu \propto P^0 T^1$$

$$\kappa \propto P^0 T^1$$

and

$$D \propto P^{-1} T^2$$

This is because, with this approximation, the quantities $\mu\rho$ and $\kappa\rho$ are simple constants in a constant pressure system.

Additionally, as we will see later, many combustion processes involve simultaneous transport such as the transport of mass and momentum in a boundary layer over a catalytic surface, the transport of energy and momentum in the boundary layer of a cold or hot wall, or the transport of energy and mass in a premixed flame. In all these cases, a significant simplification of the equations of motion can be made if we assume that the dimensionless quantities that couple these processes, namely the Schmidt, Prandtl, or Lewis number, respectively, are equal to unity. Since in the gas phase these numbers are reasonably close to unity [see Eqs. (3-12) to (3-14) and Table 3-2], this approximation yields reasonably accurate models for many combustion processes.

## PROBLEMS

**3-1** Use the virial equation of state to calculate the pressure at which the density of air at ambient temperature deviates from the ideal gas density by (a) 1% and (b) 10%.

**3-2** Calculate the Prandtl, Schmidt, and Lewis numbers for pure oxygen (based on the self-diffusion coefficient) at 300 and 1000 K.

**3-3** DiPippo, Kestin, and Oguchi, *J. Chem. Phys.* **46**:4758 (1967) report on a measurement of the viscosity of three binary mixtures. Use the molecular parameters given in App. A to calculate pure component viscosities for comparison to their values. Also use the mixture formula to calculate the effect of mixture composition on viscosity for He–$CO_2$ mixtures for comparison to their curve.

**3-4** Calculate the thermal conductivity of NO and $N_2O$ at three temperatures (60.6, 253.7, and 434°C) using the following assumptions:

   (*a*) $\kappa = \kappa_{mono}$;
   (*b*) Eucken's correction;
   (*c*) Hirschfelder's correction.

Compare your answers with the experimental results tabulated by Choy and Raw, *J. Chem. Phys.* **45**:1413 (1966).

**3-5** Calculate the thermal diffusion coefficient of hydrogen in nitrogen at 300 and 1000 K. Compare your answers to the ordinary diffusion coefficient and estimate the separation between two bulbs held at 250 and 350 K and 950 and 1050 K.

**3-6** A vertical cylinder 0.1 m high and 0.01 m in diameter is closed at the bottom and open at the top and contains 0.01 m of pure ethyl alcohol. Assume that the diffusion coefficient of ethyl alcohol in air is $10^{-5}$ m²/s and that its vapor pressure is 10 kPa. If there are no air currents in the room, calculate the length of time it will take for the alcohol to evaporate completely.

**3-7** The Loschmidt cell is an apparatus that has been used to measure the diffusion coefficient of gases. It is a vertical tube with a horizontal thin sliding valve at the midline and the top and bottom are filled with a light and heavy gas respectively. At time $t = 0$, the valve is opened and diffusion proceeds until the composition is uniform throughout the tube. Solve the diffusion equation

$$\frac{\partial x_i}{\partial t} = D \frac{\partial^2 x_i}{\partial x^2}$$

for constant $D$ and show that at late time the concentration profile reduces to a simple sine or cosine function. Also use the solution to determine the motion of the center of mass of the system as diffusion proceeds.

**3-8** Assume that a liquid hydrogen–liquid oxygen rocket engine produces pure water vapor in the combustion chamber at a temperature of 2900 K. At what operating pressure would this gas deviate from ideal gas behavior by (*a*) 1% and (*b*) 10%? Use the virial equation of state.

**3-9** A classic experiment to determine the nonideality of a gas at low pressure involves the use of two thermostated vessels that are held at the same temperature and connected with a stopcock. The procedure is to initially pressurize the two bulbs to some relatively high pressure and then successively isolate one bulb, evacuate the other, and then allow the remaining gas to expand to fill both bulbs. The pressure is read very accurately after each expansion, when the bulbs have again reached equilibrium with the thermostat bath. It is well known that at low thermostat temperatures the pure gas will exhibit a negative deviation from ideal behavior and that as the temperature is raised it will eventually show no deviation and then positive deviation. The no-deviation temperature is called the Boyle temperature. Devise a mathematical construct that would allow you to determine the second virial coefficient using this technique and show how the Boyle temperature, if measured, can be used to get the well depth energy $\varepsilon/\ell$ for the gas.

   *Hint:* The third virial coefficient is not important under the conditions of this experiment.

**3-10** Calculate the temperature change if nitrogen initially at 10 MPa and 500 K is expanded reversibly and isentropically to a pressure of 100 kPa. Assume:

   (*a*) ideal gas, $\gamma = 1.4$;
   (*b*) ideal gas, real gas enthalpies from App. B.

**3-11** Do the calculation of Prob. 3-10 for the water vapor of Prob. 3-8 at initial conditions of 10 MPa and 2900 K.

**3-12** Calculate the kinematic viscosity $v$ and thermal diffusivity $\alpha$ of pure nitrogen at 300 and 2300 K and 1 atmosphere pressure. The large changes that occur with this type of temperature change has important implications in many combustion problems.

## BIBLIOGRAPHY

Bird, R. B., W. E. Stewart, and E. N. Lightfoot, *Transport Phenomena*, Wiley, New York (1960).

Chapman, S., and T. G. Cowling, *Mathematical Theory of Non Uniform Gases*, Cambridge (1970).

Cunningham, R. E., and R. J. J. Williams, *Diffusion in Gases and Porous Media*, Plenum, New York (1980).

Hirschfelder, J. O., C. F. Curtiss, and R. B. Bird, *Molecular Theory of Gases and Liquids*, Wiley, New York (1954).

Kreith, F., *Principles of Heat Transfer*, 3d ed., Donnelly, New York (1976).

Reid, R. C., J. M. Prausnitz, and T. K. Sherwood, *The Properties of Gases and Liquids*, McGraw-Hill, New York (1977).

# FOUR

## CHEMICAL KINETICS AND EQUILIBRIUM

### 4-1 INTRODUCTION

There are three limit types of macroscopic behavior relative to reactive systems which can be rather sharply defined. These are (1) inert systems—all collisions are nonreactive; (2) nonequilibrium reactive systems—all collisions are not inert, and local composition changes occur which are not balanced macroscopically by reverse processes, causing the species composition of the system to change with time; and (3) equilibrium systems—reactive collisions occur but are balanced on the macroscopic scale by equal and opposite reverse processes so that, for constant external conditions, the composition of the system is observed to remain unchanged with time. In case (3) there are two types of equilibrium which will be defined and discussed later, either restrictive or complete equilibrium.

Because of the microscopic and dynamic nature of the processes occurring in the last two categories, they admit to the existence of reactive species which can only be observed in situ. For example, one can imagine the possible reactive collision

$$M + H_2O \rightarrow M + H + OH \qquad (4-1)$$

where $M$ is any inert collision partner. As written this microscopic process is producing a neutral hydrogen atom and a neutral hydroxyl radical. These species are both highly reactive and therefore cannot be extracted from the mixture and stored in any ordinary sense. However, their occurrence in the reacting mixture contributes markedly to the systems macroscopic behavior and their presence cannot be ignored if we wish to obtain an accurate description of reactive processes.

Notice that Eq. (4-1) does not describe the only reactive process available for an $M + H_2O$ collision where $M$ is completely inert. If we limit ourselves to products which are neutral species, we can also imagine the reactions

$$M + H_2O \rightarrow M + H_2 + O$$

and

$$M + H_2O \rightarrow M + H + H + O$$

However, there is no reason why a collision of this type should produce only neutral species, and we may consider the possibility that the products are charged particles, i.e.,

$$M + H_2O \rightarrow M + H_2^+ + O^-$$

$$M + H_2O \rightarrow M + H^+ + OH^-$$

$$M + H_2O \rightarrow M + H^- + OH^+$$

$$M + H_2O \rightarrow M + H^+ + OH + e, \text{ etc.}$$

(4-2)

where $e$ is a free electron. In general, we observe that these ionization processes (Eq. 4-2) are much less likely to occur because the formation of ionized species requires more energy than processes involving only the separation of neutral fragments. Thus in most cases the first departure from inert collision behavior is a "chemical reaction" occurring through the formation and interaction of neutral fragments. Ionization processes can be important at extremely high temperatures but will not be discussed further in this text. The interested reader is referred to the bibliography at the end of this chapter.

## 4-2 REACTION KINETICS

The quantitative study of chemical reactions, i.e., the evaluation of the $d\lambda_j/dt$ terms in Eq. (2-17), is called chemical reaction kinetics or chemical kinetics. It involves the experimental determination and theoretical calculation of the rates of production or disappearance of chemical species during a chemical reaction and the effect of temperature, pressure, species concentrations, etc. on these rates. Reaction kinetics was first placed on a modern quantitative basis in 1884 when Van't Hoff stated the simple laws of uni-, bi-, and trimolecular reactions and predicted the exponential temperature dependence of reaction rate which was verified by Arrhenius in 1889. Van't Hoff's statements concerning the rate of formation or disappearance of a species during a chemical reaction are still valid (if slightly modified) today. Today we apply his definition of uni-, bi-, or trimolecular reactions not to the overall reaction, but to the individual kinetic steps of a reaction process. In the following sections we will relate the occurrence of these steps to individual collision processes through the use of kinetic theory. However, we will first compare the chemist's notation for reaction processes to the reaction coordinate notation and define a few terms.

The general form of the rate law preferred by the chemist is

$$\left(\frac{d[I]}{dt}\right)_j = {}^nk_j(T)\,\frac{v_{ij}}{|v_{ij}|}\,\prod_{k=1}^{n}[B_k] \tag{4-3}$$

where $n$ is the reaction order‡ ($n = 1, 2,$ or $3$), ${}^nk_j$ is the rate constant of the $j$th reaction (always a positive number), the symbol [ ] represents a concentration, moles per cubic meter, $I$ and $i$ represent the $i$th species in the $j$th reaction (the one whose concentration change is being monitored), the $v_{ij}$ is the stoichiometric coefficient of the $i$th species in the $j$th balanced chemical reaction whose reaction coordinate is $\lambda_j$ [see Eqs. (2-15) to (2-17)], and the $B_k$'s are the reactants in the forward reaction (i.e., are written on the left side of the balanced chemical equation that represents the reaction that is occurring).

Strictly speaking this expression is valid only when applied to a rate measured in a constant-volume, constant-temperature environment. This is because the term on the left-hand side of this equation is expressed in a restrictive form in terms of the modern interpretation of reaction kinetic processes. The early chemists could only observe overall processes and their rate equations were usually empirical equations determined by fitting experimental data obtained in a specific environment. Thus, the convention exists that the reaction rate is expressed in terms of the rate of change of concentration of a specific species. However, as we will see, the kinetic theory of gases yields a rather fundamental relation between the rate of change of a *reaction coordinate* and the *concentrations* of the reacting species.

In general (i.e., $V$, $T \neq$ constant) the rate of change of concentration of a single species may be written as

$$\frac{d[I]}{dt} = n_i\,\frac{d\rho}{dt} + \rho\,\frac{dn_i}{dt}$$

by noting that $[I] = \rho n_i$ where $\rho$ is the mixture density. However, substituting Eq. (2-17) we obtain

$$\frac{d[I]}{dt} = n_i\,\frac{d\rho}{dt} + \rho\sum_{j=1}^{p} v_{ij}\,\frac{d\lambda_j}{dt} \tag{4-4}$$

where $v_{ij}$ is the stoichiometric coefficient of the species $I$ in the $j$th chemical reaction and $\lambda_j$ is the reaction coordinate of the $j$th reaction. The first term on the right-hand side of this equation is a contribution to concentration change due to density changes in the system. The reaction coordinates that occur in the

---

‡ The superscript on the left will be used in the next few pages and in Chap. 8 to identify the order of the reaction. Usually the stoichiometry of the reaction itself identifies order and $n$ is not used.

second term are related to the concentration changes caused by the individual reactions as ordinarily reported by the chemist.

By rearranging Eq. (4-4), holding the density constant (i.e., constant volume), and substituting Eq. (4-3) we obtain

$$\frac{d\lambda_j}{dt} = \frac{1}{\rho v_{ij}} \left[ \frac{d[I]}{dt} \right]_{V, T, j} = \frac{1}{\rho |v_{ij}|} \left[ {}^n k_j(T) \prod_{k=1}^{n} [B_k] \right]_{V, T, j} \tag{4-5}$$

The subscript $V, T, j$ in Eq. (4-5) implies (or at least should imply) that for the $j$th reaction the rate coefficient ${}^n k_j$ was measured at constant volume and temperature. We will use the chemist's notation on our discussion primarily because it is the conventional notation. We now turn our attention to the definition of the elementary rate laws.

A *unimolecular* or first-order reaction is one where the rate of disappearance of a species is proportional to the instantaneous concentration of that species. The reaction is written as

$$A \to \text{products}$$

and the rate is given by

$$\frac{dA}{dt} = - {}^1 k[A]$$

At constant temperature and volume we may integrate this expression

$$^1 k = \frac{1}{t} \ln \frac{[A]_0}{[A]} \qquad \text{s}^{-1}$$

or

$$[A] = [A]_0 e^{-{}^1 kt}$$

Thus, the decay of a radioactive nucleus is a special case of a unimolecular reaction. The half-life (the time when half of the original material is gone) is independent of the initial concentration for a unimolecular reaction.

$$t_{1/2} = \frac{1}{{}^1 k} \ln 2$$

In a similar fashion a *bimolecular* or second-order reaction is written as

$$A + B \to \text{products}$$

and has a rate

$$\frac{d[A]}{dt} = \frac{d[B]}{dt} = - {}^2 k[A] \cdot [B]$$

Since the concentrations of $A$ and $B$ are not necessarily equal, this equation may be integrated for a constant-temperature, constant-volume process after writing $[A] = a - x$ and $[B] = b - x$, where $a$ and $b$ are constants, to obtain

$$\frac{dx}{dt} = {}^2k(a - x)(b - x)$$

This yields

$$^2kt = \frac{1}{a - b} \ln \frac{b(a - x)}{a(b - x)} \qquad cm^3/mole$$

or if $a = b$

$$^2k = \frac{x}{ta(a - x)} \qquad cm^3/mole\text{-}s$$

The half-life of this reaction is given by expression

$$t_{1/2} = \frac{1}{{}^2ka}$$

A *trimolecular*, or third-order, reaction is written as

$$A + B + C \rightarrow products$$

and the rate is given by

$$\frac{d[A]}{dt} = \frac{d[B]}{dt} = \frac{d[C]}{dt} = -{}^3k[A] \cdot [B] \cdot [C]$$

Once again, in general, $[A] \neq [B] \neq [C]$. For the case where they are equal, we can write $A = B = C = (a - x)$ and the equation becomes

$$\frac{dx}{dt} = {}^3k(a - x)^3$$

This may be integrated to yield

$$^3k = \frac{1}{2t}\left[\frac{1}{(a - x)^2} - \frac{1}{a^2}\right] \qquad cm^6/mole^2\text{-}s$$

and the equation for half-life becomes

$$t_{1/2} = \frac{\frac{3}{2}}{{}^3ka^2}$$

Therefore, in general, the half-life of a reaction occurring at constant temperature and volume is

$$t_{1/2} \propto \frac{1}{a^{(n-1)}}$$

where $n$ is the order of the reaction.

The effect of temperature on the rate constant $k$ of simple first- and second-order reactions is given quite accurately by the Arrhenius equation

$$^i k = A_i e^{-E_i^*/RT} \qquad (4\text{-}6)$$

where $E_i^*$ is called the activation energy and $A_i$ is the frequency factor and has the same units as $^i k$. For most third-order reactions the temperature dependence is found to be very small. It may usually be expressed by the relation

$$^3 k = BT^\zeta \qquad (4\text{-}7)$$

where $\zeta$ has a value which typically lies in the range $\pm 2$.

In the next sections we will relate these elementary macroscopic equations to the collision processes which must be responsible for the macroscopic behavior of the system.

## 4-3 REACTIVE COLLISION POTENTIALS— ACTIVATION ENERGY

Our kinetic theory concept of a gas as an aggregate of colliding molecules separated on the average by large distances leads us immediately to the observation that any measurable reaction rate must be the result of relatively infrequently occurring "reactive collisions." For example, consider a mixture at room temperature and pressure containing an equal number of molecules of gas $A$ and $B$ which are reactive ($A + B \rightarrow$ products). The kinetic theory which we developed in Chap. 1 states that one molecule of $A$ will collide with a molecule of $B$ approximately $10^9$ times per second. Thus, if each collision led to reaction, the process would be completed in an extremely short time. Therefore we can assume that even rapid reactions (half-lives of about $10^{-4}$ s) must be occurring only on extremely infrequent collisions between the reactive species. This fact allows us to consider a reacting gas mixture as being slightly perturbed from thermodynamic (but not chemical) equilibrium at all times during the course of the reaction. There are some limitations to this approach but they will not concern us in the present section.

In order to discuss the reasons for the apparent inertness of most collisions between inherently reactive molecules, we must look at some of the details of the collision process for an encounter of reactive species. We do this by considering the collision potential for an encounter. A typical collision potential for unreactive spherical molecules is given by the Lennard–Jones potential discussed in Chap. 3. In the case of reactive collisions we are interested mainly in the close approach portion of the potential because the slowness of reactions indicates that highly energetic collisions are necessary before chemical changes will occur. However, reactive collision potentials cannot be constructed easily for a general collision of complicated molecules. This is true because orientation and molecular distortion are important in describing a reactive collision. Nevertheless, reactive collision potentials have been constructed for a few simple

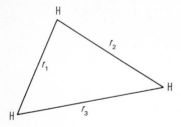

**Figure 4-1** Typical configuration for calculating an interaction potential.

collision processes, and the behavior of these potential curves leads to a very useful generalization concerning reaction processes.

If we consider the elementary reaction

$$H + H_2 \rightarrow H_2 + H$$

as a three-body problem (3H atoms) we can calculate the potential energy of this system for any three separation distances $r_1$, $r_2$, $r_3$ (see Fig. 4-1). Unfortunately, this extremely simple process is not sufficiently simple to allow a convenient graphical representation of the potential energy as a function of the configuration. However, the potential energy for the two-body problem consisting of a fixed hydrogen molecule (i.e., $r_1$ fixed) and a neighboring hydrogen atom has been calculated and is shown in Fig. 4-2. The most interesting feature of this potential energy diagram is the potential well which is observed at the end of the fixed hydrogen molecule. This peculiarity in the potential field leads

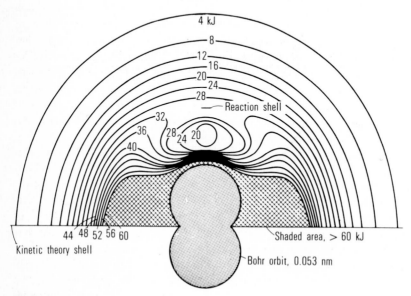

**Figure 4-2** Potential-energy contours for approach of a hydrogen atom to a rigid hydrogen molecule. *(Adapted from J. O. Hirschfelder, H. Eyring, and B. Topley, J. Chem. Phys., vol. 4, p. 170 (1936), with permission.)*

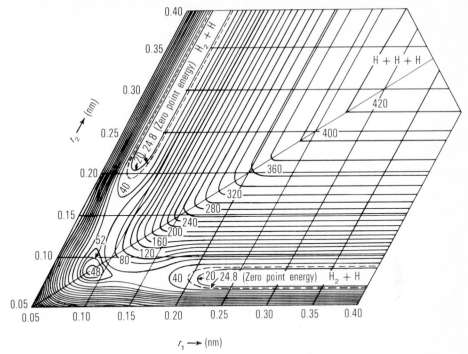

**Figure 4-3** Potential-energy surface for the system of three hydrogen atoms based on 14 percent coulombic energy. *(Adapted from H. Eyring, H. Gershinowitz, and C. E. Sun, J. Chem. Phys., vol. 3, p. 786 (1935), with permission.)*

one to the conclusion that a chemical reaction is most likely to occur during an in-line collision of $H_2$ and H. Fortunately, the potential energy for such an in-line collision can be represented graphically because it is a two-parameter problem. A plot of the potential energy for the case of an in-line collision is shown in Fig. 4-3. Notice the existence of a saddle point on this plot at $r_1 = r_2$. The restricted (in-line) collision behavior of a hydrogen molecule and hydrogen atom may be discussed using the potential energy surface (Fig. 4-3) by following the motion, on this surface, of a mass point in a gravitational field with an initial potential energy and arbitrarily directed velocity. Approaching the saddle point from either $r_1$ or $r_2 \cong 0.05$ nm represents an $H_2$ + H collision process and leaving the potential well in the $r_2$ or $r_1$ direction respectively represents the occurrence of a reactive collision. Thus Fig. 4-3 predicts that only energetic collisions will have any probability of leading to reaction and that even energetic collisions will have only a finite probability of producing chemical change. This behavior is illustrated schematically in Fig. 4-4.

This concept of a reactive collision allows us to place the Arrhenius equation on a relatively sound semiquantitative foundation. All we need require is that the number of reactive collisions be proportional to the total number of collisions with an intermolecular kinetic energy above the critical collision

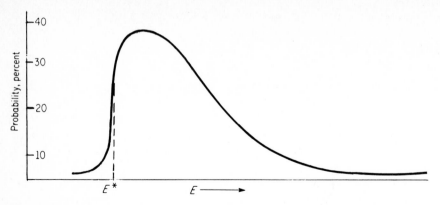

**Figure 4-4** Probability of reactive collision vs. collision energy (schematic).

energy for a particular reactive process. We also must assume that this critical energy for reaction is high when compared to the average kinetic energy of the molecules so that only a few collisions will yield products. If these assumptions are true, we note that we can replace the relative collisional kinetic energy by the kinetic energy of one of the molecules relative to the gas sample as a whole without greatly altering the results. Let us therefore calculate the fraction of the total number of molecules in an equilibrium gas at temperature $T$ having a translational kinetic energy above a certain value $E^*$, where $E^* = \frac{1}{2}\mathscr{A}mc^{*2}$. This fraction may be determined by integrating Eq. (1-21) from $E^*$ to $\infty$. Replacing $\eta(c)\,dc$ by $dn/n_T$ where $n$ = number of molecules with velocity $c$ to $c + dc$ and $n_T$ = total number of molecules, we obtain

$$\frac{dn}{n_T} = 4\pi \left(\frac{M}{2\pi RT}\right)^{3/2} \exp\left(-\frac{Mc^2}{2RT}\right) c^2 \, dc$$

or

$$\frac{n^*}{n_T} = \int_{E^*}^{\infty} \frac{2}{\sqrt{\pi}} \frac{1}{RT^{3/2}} \exp\left(-\frac{E}{RT}\right) E^{1/2} \, dE$$

$$= \frac{2}{\sqrt{\pi}} \left(\frac{E^*}{RT}\right)^{1/2} e^{-E^*/RT} \left[1 + \left(\frac{RT}{2E^*}\right) - \left(\frac{RT}{2E^*}\right)^2 + \cdots\right]$$

which yields

$$\frac{n^*}{n_T} \simeq \frac{2}{\sqrt{\pi}} \left(\frac{E^*}{RT}\right)^{1/2} e^{-E^*/RT}$$

However, since the exponential factor varies extremely rapidly with temperature and the $T^{-1/2}$ factor varies only slowly, as a first approximation the number of molecules above an energy $E^*$ may be written as

$$\frac{n^*}{n_T} \propto e^{-E^*/RT}$$

Therefore this simple concept yields a functional temperature dependence which is in good agreement with the experimentally observable temperature dependence exhibited by the Arrhenius equation. It also presents the concept of a unique activation energy for each collision pair and the concept of an activated complex as a loose molecular aggregate which has some identity during the reactive-collision process.

A further consideration of the energetics of a reactive collision allows us to discuss limits on the magnitude of the activation energy in some detail. Consider the reactive-collision process illustrated in Fig. 4-5:

$$XY + Z \rightarrow X + YZ \qquad (4\text{-}8)$$

which destroys the molecule $XY$ and produces in its place the molecule $YZ$. This is a typical "exchange" reaction. Each of these molecules has a dissociation energy or heat of formation and in general the process represented by Eq. (4-8) is not thermally neutral. Referring again to Fig. 4-5 we will assume, for discussion, that the lowest possible potential energy of the system $XY + Z$ is five arbitrary units above the lowest possible potential energy of the system $X + YZ$. In other words, if the dissociation energy of molecule $XY = 65$ units, the dissociation energy of molecule $YZ = 70$ units and the plateau at large $r_1$ and $r_2$ is at a height of 70 units on the diagram. This means that occurrence of this reaction will cause the liberation or absorption of heat energy from the system in which the reaction is occurring. On the microscopic scale, each reactive collision will either produce or absorb some kinetic energy from the collision partners, since the collision is conservative overall and the product molecules will have a new potential energy associated with their bond structure. We have already discussed the manifestation of this process at the macroscopic level in Chap. 2. Specifically, if one mole of the reaction $XY + Z \rightarrow X + YZ$ were to be carried out at constant temperature, $5\mathscr{A}$ arbitrary units of bond energy per mole would have to be rejected to the surroundings. Thus, this reaction is exothermic and the reverse reaction $X + YZ \rightarrow XY + Z$ is endothermic.

Referring to Fig. 4-5, we see that this exothermic reaction has an activation energy of $E^* = 21$, while the reverse endothermic reaction will have an activation energy of $E^* = 26$ arbitrary units. This is summarized in Fig. 4-6.

There are many reactions which are either highly exothermic or highly endothermic and in these cases $\Delta E \cong \mathscr{A} \Delta \Psi$ where $\Delta \Psi$ is the potential energy change in a single reactive encounter. Thus we note from Fig. 4-6 that every endothermic reaction must have an activation energy at least as great as its endothermicity, whereas highly exothermic reactions can have extremely low activation energies and their behavior is not predictable from the $\Delta H$ of the reaction.

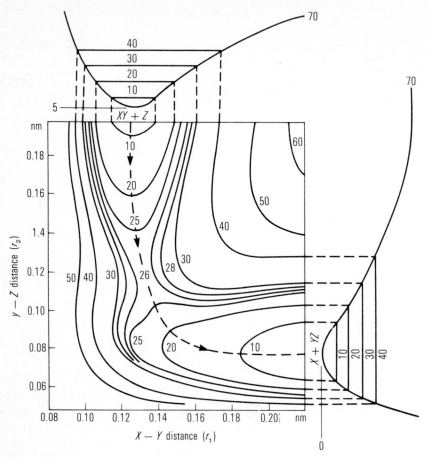

**Figure 4-5** Typical potential-energy surface for a three-atom reaction. Zero point energies are neglected in this example. *(Adapted with permission from S. Glasstone, T. Laidler, and H. Eyring, "The Theory of Rate Processes," McGraw-Hill, New York (1941).)*

## 4-4 THE COLLISION PROCESS AND KINETIC ORDER

It is quite easy to see how collision-determined reactions can yield a second-order rate dependence at constant temperature and volume. Let us consider molecules $A$ and $B$ which are capable of reacting to produce products

$$A + B \rightarrow \text{products}$$

The reactive collision concept yields the conclusion that the rate of this reaction is proportional (at any one temperature) to the total number of collisions of molecules $A$ with molecules $B$. However, the total number of available collisions is simply proportional to the product of the number density of molecules $A$

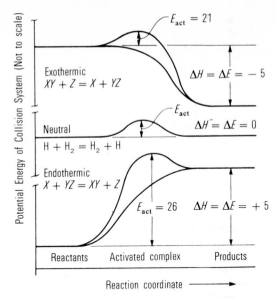

Potential Energy of Collision System (Not to scale)

$E_{act} = 21$

Exothermic
$XY + Z = X + YZ$

$\Delta H = \Delta E = -5$

$E_{act}$

Neutral
$H + H_2 = H_2 + H$

$\Delta H = \Delta E = 0$

Endothermic
$X + YZ = XY + Z$

$E_{act} = 26$    $\Delta H = \Delta E = +5$

Reactants    Activated complex    Products

Reaction coordinate

**Figure 4-6** Activitation energy and the heat of reaction for the reactions of Figs. 4-3 and 4-5 (not to scale).

multiplied by the number density of molecules $B$, and these number densities are proportional to concentrations. Thus, second-order rate constants arise as a direct consequence of the collision-determined nature of a reaction process.

Apparently-first-order rate constants are observed experimentally only for a very specific type of decomposition—one where the molecule is capable of storing a considerable quantity of energy internally as vibrational energy and then decomposing when this vibrational energy resides in a weak bond.‡ In essence, this reaction becomes first-order because sufficient energy for dissociation can be stored for an appreciable length of time in the molecule before spontaneously causing decomposition. The stoichiometric rate equation is written as

$$A + A \underset{k_2}{\overset{k_1}{\rightleftharpoons}} A + A^* \downarrow^{k_3}$$

products

where three separate and distinct processes are occurring. Two of these are collision determined. These are (1) activation by collision

$$\left(\frac{d[A^*]}{dt}\right)_1 = k_1[A]^2$$

‡ This mechanism was first proposed by Lindeman in a discussion meeting of the Faraday Society (F. A. Lindeman, *Trans Faraday Soc.* **17**: 598 (1922).)

and (2) deactivation by collision

$$\left(\frac{d[A^*]}{dt}\right)_2 = -k_2[A] \cdot [A^*]$$

The third process is the spontaneous decomposition of the activated species and is therefore first-order:

$$\left(\frac{d[A^*]}{dt}\right)_3 = -k_3[A^*]$$

In these simultaneous processes we cannot, of course, measure either $[A^*]$ or $d[A^*]/dt$. However, we see that

$$\frac{d[\text{prod}]}{dt} = k_3[A^*]$$

and, therefore knowledge of $[A^*]$ will allow us to calculate the rate of production of products. We note that three reactions are competing for either the production or destruction of the activated species $A^*$ and we may therefore write that

$$\frac{d[A^*]}{dt} = k_1[A]^2 - k_2[A] \cdot [A^*] - k_3[A^*]$$

As a first approximation this equation may be solved by assuming that the composite rate of production of $A^*$ during the course of the reaction is small compared to the rate of the three competing steps. This is called the *steady state approximation* of chemical kinetics and is reasonably valid for most simple decomposition reactions. With this assumption we obtain

$$[A^*] = \frac{k_1[A]^2}{k_2[A] + k_3}$$

or

$$\frac{d[\text{prod}]}{dt} = \frac{k_1[A]^2}{(k_2/k_3)[A] + 1}$$

Let us examine the denominator of this equation more closely. Since the $[A]$ term depends on the concentration of $A$, this term should be pressure dependent. Thus at high pressures $(k_2/k_3)[A] > 1$ and

$$\frac{d[\text{prod}]}{dt} = \frac{k_3 k_1}{k_2}[A]$$

which is a first-order rate equation. However, at low pressure $(k_2/k_3)[A] < 1$ and the kinetic equation becomes second-order in $A$

$$\frac{d[\text{prod}]}{dt} = k_1[A]^2$$

Thus the collision theory of first-order processes predicts that the reaction should behave as a second-order process at low pressures where the collision deactivation process is slower than the spontaneous decomposition process and should be first-order at high pressures where the reverse is true. This is observed experimentally. The gas-phase decomposition of a complicated molecule is second-order at low pressures and slowly becomes first-order as the pressure increases to the point where collisional deactivation is important.

Third-order rates are observed for recombination reactions where the species which is formed cannot conveniently store the heat of formation internally. Consider the collision of two hydrogen atoms. The potential curve for this process is given in Fig. 4-3 if either $r_1$ or $r_2$ is assumed to be large, for example, 0.4 nm. On this figure the approach kinetic energy of a pair of hydrogen atoms (relative to their center of mass) can be represented as a horizontal plane well above the energy level of 420 which represents the potential energy of two (or three) well-separated hydrogen atoms. Since, for this collision process, the velocity of the center of mass of the collision pair cannot change, it is impossible to store the energy that must be released to the surroundings to form a stable H–H bond. Emission of radiation is one possible process but photo recombination does not occur often. However, if a third body (which is sometimes called a "chaperon" molecule) is present at the moment of collision we may write the equation

$$H + H + M \rightarrow M + H_2$$

where the heat of reaction appears as recoil velocity of the $M$ and $H_2$ species. Thus recombinations occur mainly through three-body collision processes where the third body is any available species in the gas mixture. Characteristically, these processes are slow because three-body collisions are infrequent. Notice that our collision concept shows that the possibility of forward reaction for this process should not be greatly dependent on the relative collision energy because it is a highly exothermic reaction. Thus we expect to find only a weak temperature dependence for atom recombination, which is confirmed by experiment.

In addition to the gas-phase collisions that yield chemical reaction, there can be wall collisions which lead to composition changes in the gas. This is particularly true because wall collisions are usually "sticky" in the sense that the molecule resides in the neighborhood of the wall for a relatively long time and therefore has an increased probability of contacting another reactive molecule. The occurrence of wall reactions greatly complicated the work of the earlier kineticists, since they were restricted to work on relatively slow reactions in small chambers. Thus almost every molecule in the system had an opportunity to contact the wall during the course of the reaction and the investigator was required to study wall reactions as an important part of the overall process. Wall reactions, if relatively rapid, can lead to a diffusion controlled rate, but if sufficiently slow simply yield a reaction rate which is proportional to the internal surface area of the container. Recombination processes, which are by nature three-body processes in the gas, can conveniently occur on the wall of the vessel because the wall can act as the requisite heat sink for the reaction. Since the number of wall collisions (per square centimeter) increases as the square of the pressure, wall collisions become more important to the recombination process as the pressure is lowered.

The above "theoretical" treatment illustrates the fundamental processes that are occurring at the microscopic level during a chemical reaction. Furthermore they show, in a semiquantitative way, how first-, second-, and third-order reactions occur and why their temperature dependence is either approximately exponential in $1/T$ or relatively independent of temperature. Even though extensions of the kinetic-theory approach can be used to make estimates of the actual rates of elementary reaction steps, it has been found that, except for the simplest reactions (such as $H + H_2 \rightarrow H_2 + H$), these rates are not sufficiently accurate to be used for predictive purposes. In other words the determination of the rate constants for elementary reaction steps is still predominately an experimental science.

## 4-5 FAST REACTIONS AND RELAXATION PROCESSES

It is well known that exchange and recombination reactions usually produce product molecules that have vibrational and rotational states that are not in equilibrium with the translational temperature of the system. In most cases the molecules that are produced leave the reactive collision in a state in which both vibration and rotation are highly excited when compared with their equilibrium states at the translational temperature of the system. Consider, for example, the oxygen atom–ozone reaction in which the reactive O atoms are produced by flash photolysis of the ozone

$$O_3 + h\nu \rightarrow O_2 + O$$

and the very reactive oxygen atom reacts with the remaining ozone to produce excited oxygen

$$O + O_3 \rightarrow 2O_2^*$$

Observations at room temperature have shown that the $O_2^*$ that is formed is almost entirely in the 14th, 15th, and 16th vibrational levels and only slowly decays to the ground state, which is its thermal equilibrium state. In a similar manner the hydroxyl radical produced by the rapid reaction

$$H + O_3 \rightarrow OH^* + O_2$$

has been found to have an effective rotational temperature of 3200 K in a system where the translational gas temperature is 300 K. The fact that chemical reactions can produce inverted population distributions (in which higher energy-level states are more populated than lower) has led to the development of the chemical laser, in which the return of the molecules to a lower state is accompanied by the emission of an intense coherent beam of light.

Another deviation from the simple kinetic behavior presented in the previous section is the slowing of the decomposition of a molecule that is very rapidly heated to a very high temperature. In this case the molecule is initially "vibrationally cold" and dissociation can only take place after some vibrational relaxation occurs.

Both the vibrational and rotational relaxation processes are important in many applications and occur because energy storage is quantized for these internal degrees of freedom. Translational equilibration is very rapid, taking only about 5 to 10 collisions for an individual molecule. Vibrational relaxation is particularly slow, because the quantized energy storage levels are large by comparison with the thermal energy of the average collision process. Vibrational and rotational relaxation processes are not only associated with the occurrence of a chemical reaction but are also observed when a gas is either compressed or expanded very rapidly. This causes a rapid change in the translational temperature and delays in vibrational and rotational equilibration have been observed. It should be pointed out that during compression the occurrence of relaxation means that the translational temperature will temporarily overshoot its final equilibrium value and then drop as the energy stored in vibration increases to the equilibrium value. In a similar manner, for rapid expansion the translational temperature will initially be lower than the final equilibrium temperature because the energy stored in vibration is released slowly to the translational degrees of freedom. Relaxation during shock compression is discussed in Sec. 5-5.

## 4-6 CHEMICAL EQUILIBRIA

We indicated earlier that the occurrence of chemical equilibrium could be considered as a possible restrictive behavior for a reactive system (case 3, Sec. 4-1). In this case many detailed composition changes are occurring on the local or molecular level but the overall composition of the gas remains invariant with time. We infer that the system contains competing reactions which do not allow any one reaction to go to completion. If this is true, the principle of detailed balancing, discussed in Chap. 1, allows us to define equilibrium concentrations in terms of the rate constants for the reversible reactions responsible for equilibrium. Consider as an example the forward and reverse reactions

$$A + B \xrightarrow{\ k_f\ } C + D$$

and

$$C + D \xrightarrow{\ k_r\ } A + B$$

Local concentration changes due to these reactions may be written as

$$-\frac{d[A]}{dt} = -\frac{d[B]}{dt} = \frac{d[C]}{dt} = \frac{d[D]}{dt} = k_f[A][B]$$

$$\frac{d[A]}{dt} = \frac{d[B]}{dt} = -\frac{d[C]}{dt} = -\frac{d[D]}{dt} = k_r[C][D] \qquad (4\text{-}9)$$

Therefore at equilibrium Eq. (4-9) predicts that

$$\left(\frac{d[I]}{dt}\right)_{net} = 0 = k_f[A][B] - k_r[C][D]$$

where

$$I = A, B, C, \text{ or } D$$

This equation may be arranged to yield

$$\frac{k_f}{k_r} = \frac{[C][D]}{[A][B]} = K_c \text{ for the reaction}$$

$$A + B \rightleftharpoons C + D$$

This definition of $K_c$ involves a convention in that the concentration of the products in the forward stoichiometric equation are always placed in the numerator of the defining equation while the reactant concentrations are placed in the denominator.

Even though this is a useful definition of equilibrium it does not ordinarily allow one to determine an equilibrium constant experimentally (our macroscopic measuring instruments do not allow a measurement of the individual rates at equilibrium). However, from stoichiometry, we know that in general in an equilibrium system we can define $s$–$c$ reaction coordinates whose possible range of numerical values defines all the mixture compositions available to the system. We will now see how thermodynamics may be used to determine the composition that the system actually reaches within these constraints. This is possible because chemical equilibrium represents a more general case of thermodynamic equilibrium.

Consider a closed but reactive system. The conditions for equilibrium in such a system at constant temperature are either that the Gibbs free energy, $f$, be a minimum at constant pressure or that the Helmholtz free energy, $a$, be a minimum at constant volume. We will use the Gibbs free energy and therefore discuss a closed system at constant pressure and temperature. In such a closed reactive system conservation of mass requires that changes in the species composition be restricted to changes in the $n_i$'s through the definition of any set of $r = s$–$c$ linearly independent reaction coordinates. When these are specified, their values, plus the $c$ independent mass-balance equations [Eqs. (2-6)], are sufficient to completely specify the species composition of the mixture. For simplicity, let us consider a single general reaction whose stoichiometry is given by the equation

$$\nu_1 I + \nu_2 II + \cdots \rightleftharpoons \nu_{n+1}(N+1) + \nu_{n+2}(N+2) + \cdots \qquad (4\text{-}10)$$

which may also be written as the equation set

$$dn_i = \nu_i \, d\lambda \qquad (4\text{-}11)$$

Here, by convention, if any $\nu_i$ is negative it means that the species appears on the left in Eq. (4-10). Since the reactive mixture is an ideal mixture of gases which obeys Dalton's law, we can write simple expressions for the thermodynamic properties of this mixture. The enthalpy is given by the expression

$$h = \sum_{i=1}^{s} n_i H_i$$

where $H_i$ is a molar enthalpy of the $i$th species. Similarly, the free energy may be written as

$$f = \sum_{i=1}^{s} n_i F_i = \sum_{i=1}^{s} n_i (H_i - TS_i)$$

However, we must allow for the appearance or disappearance of materials in determining the thermodynamic properties. For example, the heat release accompanying chemical reaction cannot be neglected in determining the enthalpy. Thus, the enthalpy of any species must be defined as

$$H_i(T) = \int_{T_0}^{T} C_{p_i} \, dT + [\Delta H_f^{\circ}(T_0)]_i$$

where $[\Delta H_f^{\circ}(T)]_i$ is the standard heat of formation of the $i$th species at temperature $T_0$. Similarly, the entropy for the species $S_i$ at a pressure $p_i$ and a temperature $T$ is given by the expression

$$S_i(T, p_i) = \int_{T_0}^{T} \frac{C_{p_i} \, dT}{T} - R \ln p_i + [\Delta S_f^{\circ}(T_0, P_0)]_i + R \ln P_0 \qquad (4\text{-}12)$$

where $[\Delta S_f^{\circ}(T_0, P_0)]_i$ is evaluated at some standard temperature $T_0$ and a reference pressure of one atmosphere. Thus in Eq. (4-12), $P_0 = 101,325$ Pa. In this text we will define a dimensionless partial pressure $\not{p}_i = p_i/P_0$ which will be used throughout the treatment of chemical equilibrium. With this definition Eq. (4-12) becomes

$$S_i(T, p_i) = \int_{T_0}^{T} \frac{C_{p_i} \, dT}{T} - R \ln \not{p}_i + [\Delta S_f^{\circ}(T_0, P_0)]_i$$

The free energy of the species, $F_i$, is therefore

$$F_i = H_i - TS_i = \int_{T_0}^{T} C_{p_i}\, dT$$

$$+ [\Delta H_f^\circ(T_0)]_i - T \int_{T_0}^{T} C_{p_i} \frac{dT}{T}$$

$$+ RT \ln p_i - T[\Delta S_f^\circ(T_0, P_0)]_i$$

or

$$F_i(T, p_i) = \omega_i(T) + RT \ln p_i \tag{4-13}$$

We are now ready to discuss equilibrium conditions for the general reaction expressed in Eqs. (4-10) or (4-11). For Eq. (4-11) we may write

$$f = f(T, P, \lambda)$$

and since we are working at constant pressure and temperature the equilibrium condition $df = 0$ implies that

$$\frac{df}{d\lambda} = 0$$

or that

$$\sum_{i=1}^{s} \frac{dn_i}{d\lambda} [\omega_i(T) + RT \ln p_i] + \sum_{i=1}^{s} n_i RT \frac{d}{d\lambda} (\ln p_i) = 0$$

However, the last term vanishes because the total pressure is constant

$$\sum_{i=1}^{s} n_i RT \frac{d}{d\lambda} (\ln p_i) = \sum_{i=1}^{s} \frac{n_i RT}{p_i} \frac{dp_i}{d\lambda} = V P_0 \sum_{i=1}^{s} \frac{dp_i}{d\lambda} = 0$$

Thus we obtain

$$\sum_{i=1}^{s} \frac{dn_i}{d\lambda} [\omega_i(T) + RT \ln p_i] = 0 \tag{4-14}$$

However, since

$$\frac{dn_i}{d\lambda} = v_i$$

we obtain

$$\sum_{i=1}^{s} v_i [\omega_i(T) + RT \ln p_i] = 0 \tag{4-15}$$

which may be rearranged to yield the equation

$$-\frac{1}{RT} \sum_{i=1}^{s} v_i \omega_i(T) = \sum_{i=1}^{s} \ln p_i^{v_i} = \ln K_p \tag{4-16}$$

Equation (4-16) relates a thermodynamic quantity (on the left) to a ratio of the partial pressures of the reacting species at equilibrium (in the middle) and this in turn is used to define an equilibrium constant (on the right). Thus, through the use of Eq. (4-16), a knowledge of the thermodynamic properties of the individual species in the mixture allows a determination of the equilibrium composition of the mixture. Equation (4-16) has been verified experimentally for many different systems over a wide range of conditions and its use in making theoretical performance calculations is quite well justified.

The use of Eq. (4-16) for the calculation of chemical equilibrium requires a self-consistent set of rules in each calculation. These formalities have been standardized by the chemists and, at present, the standardized definitions are those used in preparing the JANAF tables, some of which are reproduced in App. B. Unfortunately, if other sources of thermodynamic data are used they may not be consistent with these tables. Because of this, one should always be careful to check the standards when data from two different sources are used in an equilibrium calculation. In the SI system, as used in this text, the units on the $\omega_i$'s are joules per mole. Thus if $p_i$ is taken to be unity, i.e., the species' partial pressure is taken to be the reference pressure of 101,325 Pa, Eq. (4-13) reduces to

$$\omega_i(T) = F_i^\circ(T) = [\Delta F_j^\circ(T)]_i$$

where those symbols with a superscript $^\circ$ represent a standard state value of the thermodynamic function for the compound relative to the thermodynamic properties of the constituent elements in their standard states. Thus, Eq. (4-16) may be written

$$\ln K_p = \frac{-1}{RT} \sum_{i=1}^{s} v_i (\Delta F_f^\circ)_i \qquad (4\text{-}17)$$

where the $(\Delta F_f^\circ)_i$'s can be obtained from the JANAF tables. Actually, the equilibrium constant for any reaction can be more easily obtained by defining an equilibrium constant of formation, $K_{p_f}$, for each species relative to the elements in their standard states. With this definition the $v_i$'s in Eq. (4-17) become the stoichiometric coefficients of the formation reaction and we can now sum $\ln K_{p_f}$'s to get the $\ln K_p$ of any desired reaction just as heats of formation are summed to get a heat of reaction. If we use log instead of ln, Eq. (4-17) becomes

$$(\log K_p)_j = \sum_{i=1}^{s} v_{ij} (\log K_{p_f})_i \qquad (4\text{-}18)$$

The JANAF values of log $K_{p_f}$ are tabulated in App. B.

We make three more observations pertinent to the use of Eq. (4-17). First, the numerical value of $K_p$ is dependent on the stoichiometric relationship, and this relationship must be specified before a stated value of $K_p$ will have meaning. For example, the statement, "The equilibrium constant for the formation of water at 2500 K is 167.5," is meaningless. One should say, instead, that for the reaction

$$H_2 + \tfrac{1}{2}O_2 \rightleftharpoons H_2O$$

$$K_p = 167.5 \text{ at } 2500 \text{ K}$$

This means that at the indicated temperature

$$167.5 = \frac{p_{H_2O}}{p_{H_2} \cdot p_{O_2}^{1/2}}$$

Notice that this is equivalent to the statement that

$$K_p = (167.5)^2 = \frac{p_{H_2O}^2}{p_{H_2}^2 \cdot p_{O_2}}$$

for the reaction

$$2H_2 + O_2 \rightleftharpoons 2H_2O$$

Furthermore, in these two relations the $p_i$'s must be expressed in the proper dimensionless units, at least for the convention adopted in the JANAF tables and this text. Second, we observe that our derivation of Eq. (4-17) places no restriction on the occurrence of simultaneous equilibria. For complex systems the original thermodynamic equations may be written as a set of $s{-}c$ linearly independent equations each with a definite reaction coordinate. Therefore, once a set of linearly independent reaction coordinates are discovered, $s{-}c$ independent equilibrium relations may be deduced, each governing the extent of reaction along one reaction coordinate. Unfortunately the resulting $s$ equations in $s$ unknowns are a coupled set of $c$ linear and $s{-}c$ nonlinear equations and this set is usually very difficult to solve for cases where $s{-}c > 3$.

Thirdly, we have seen that for each of $s{-}c$ independent reactions occurring in a system containing $s$ species we may write

$$dn_i = v_{ij} \, d\lambda_j \qquad [j = 1, 2, \ldots, (s{-}c)]$$

and we may define $j$ equilibrium constants for these reactions as

$$(K_p)_j = \prod_{i=1}^{s} (p_i)^{v_{ij}} \qquad [j = 1, 2, \ldots, (s{-}c)] \tag{4-19}$$

However, for an ideal gas Dalton's law states that

$$\mathcal{P} = \sum_{i=1}^{s} p_i$$

and since

$$n_T = \sum_{i=1}^{s} n_i$$

the partial pressure of any specie may be expressed as

$$\rlap{/}p_i = n_i \frac{\mathscr{P}}{n_T} = [I] \frac{RT}{P_0}$$

where $[I]$ represents the concentration of the $i$th species.

Therefore we may rewrite Eq. (4-19) in terms of molar quantities to yield

$$(K_p)_j = \left(\frac{\mathscr{P}}{n_T}\right)^{\sum\limits_{i=1}^{s} v_{ij}} \cdot \prod_{i=1}^{s} (n_i)^{v_{ij}} = \left(\frac{RT}{P_0}\right)^{\sum\limits_{i=1}^{s} v_{ij}} \cdot \prod_{i=1}^{s} [I]^{v_{ij}} \qquad (4\text{-}20)$$

where

$$\prod_{i=1}^{s} (n_i)^{v_{ij}} = (K_n)_j$$

and

$$\prod_{i=1}^{s} [I]^{v_{ij}} = (K_c)_j$$

The quantity $K_c$ was defined at the beginning of this section.

Equation (4-20) explicitly illustrates the effect of pressure changes on the composition of an equilibrium mixture. It states that if the sum of the $v_{ij}$'s is zero for a specific reaction, pressure changes will not affect the composition and that, if the sum is not zero, the equilibrium composition will be shifted by a change in pressure.

## 4-7 DISSOCIATIVE EQUILIBRIA AND THE TEMPERATURE DEPENDENCE OF $K_p$

One of the more important types of simple equilibrium is the dissociation equilibrium which occurs when a simple homonuclear diatomic molecule decomposes at high temperature. Consider, for example, the dissociation of pure hydrogen:

$$H_2 \rightleftharpoons 2H$$

This system contains two species and one component. We can analyze this system on either the mole/kilogram basis, the kilogram/kilogram basis or the mole/$\mathscr{G}$ kg basis. We choose to use the mole/$\mathscr{G}$ kg basis. If we consider a system which initially contains one mole of $H_2$ and no H atoms and set the reaction coordinate $\alpha = 0$ at $\eta_{H_2} = 1$, $\eta_H = 0$ we may integrate the two progress equations to obtain the moles/$\mathscr{G}$ kg of the species in terms of $\alpha$ (called the *degree of dissociation* in this case)

$$\eta_{H_2} = 1 + \int_0^\alpha - d\alpha = 1 - \alpha$$

$$\eta_H = \int_0^\alpha + 2 \, d\alpha = 2\alpha$$

Thus we find that one mole of hydrogen dissociates to produce $2\alpha$ moles of hydrogen atoms and $1 - \alpha$ moles of hydrogen molecules at equilibrium. The system, therefore, must contain $1 + \alpha$ moles of gas at equilibrium. Dalton's law allows us to write

$$\wp_H = \frac{2\alpha}{1 + \alpha}\,\mathscr{P} \quad \text{and} \quad \wp_{H_2} = \left(\frac{1 - \alpha}{1 + \alpha}\right)\mathscr{P}$$

or

$$K_p = \frac{\wp_H^2}{\wp_{H_2}} = \frac{4\alpha^2 \mathscr{P}}{(1 + \alpha)(1 - \alpha)} = \frac{4\alpha^2}{1 - \alpha^2}\,\mathscr{P}$$

which is the equilibrium equation in terms of the degree of dissociation $\alpha$ and the total pressure of the system. Thus at a specified temperature we can find a value of $K_p$ from the tables in App. B and, with the total pressure, we can directly determine the value of $\alpha$ and thereby all the thermodynamic properties of the system. This dissociative-equilibrium equation will now be used to derive the temperature dependence of the equilibrium constant. To do this we write the reciprocity relation from thermodynamics for the equilibrium system

$$\left(\frac{\partial H}{\partial P}\right)_T = V - T\left(\frac{\partial V}{\partial T}\right)_P$$

However, the enthalpy of this system is governed by the relation

$$H = 2\alpha H_A + (1 - \alpha)H_m$$

$$= H_m + \alpha(2H_A - H_m) = H_m + \alpha H_D$$

Here the subscripts $A$ and $m$ refer to the atomic and molecular species respectively. $H_D$ is the dissociation enthalpy of the molecule and is, in general, a function of the temperature. We also note that the equations

$$P = \frac{RT}{V}(1 + \alpha) \tag{4-21}$$

and

$$K_p = \frac{4\alpha^2}{1 - \alpha^2}\,\mathscr{P} \qquad \mathscr{P} = \frac{P}{P_0} \tag{4-22}$$

are governing equations for the system. Substituting these into the reciprocity relation yields

$$H_D\left(\frac{\partial \alpha}{\partial P}\right)_T = \frac{RT}{P}(1 + \alpha) - T\left[\frac{R}{P}(1 + \alpha) + \frac{RT}{P}\left(\frac{\partial \alpha}{\partial T}\right)_P\right]$$

or

$$\frac{H_D}{RT^2} = -\frac{(\partial\alpha/\partial T)_P}{(\partial\alpha/\partial P)_T} \cdot \left(\frac{1}{P}\right)$$

The two partial derivatives may be evaluated using the equilibrium equation to yield

$$\frac{\partial (\ln K_p)}{\partial T} = \frac{H_D}{RT^2}$$

This is called Van't Hoff's equation and may be integrated if $H_D$ is independent of temperature to yield

$$K_p = \mathscr{P}_D \exp\left(\frac{-\theta_D}{T}\right) \tag{4-23}$$

where $\mathscr{P}_D$ is a characteristic dissociation pressure (dimensionless) and $\theta_D = (H_D/R)$, $(K)$ is a characteristic dissociation temperature.

Substituting into Eq. (4-22) gives the degree of dissociation of a homonuclear molecule in terms of the dimensionless parameters $\mathscr{P}/\mathscr{P}_D$ and $\theta_D/T$.

$$\alpha = \sqrt{\frac{1}{1 + (4\mathscr{P}/\mathscr{P}_D)\exp(\theta_D/T)}} \tag{4-24}$$

The enthalpy of a dissociating gas may be written as

$$H = H_m + \alpha H_D$$

where

$$H_m = \frac{7}{2} RT + RT\left[\frac{\theta_v/2T}{\sinh \theta_v/2T}\right]^2 \cong 4RT$$

Thus

$$\frac{H}{H_D} = \frac{(4 + \alpha)T}{\theta_D} + \alpha \cong \frac{4T}{\theta_D} + \alpha \tag{4-25}$$

This approximation was first introduced by Lighthill[1] and is satisfactory in light of the earlier assumption that $H_D$ is constant.

Table 4-1 lists dissociation temperatures and pressures for a number of homonuclear diatomic species. The multiple values listed for oxygen give an indication of the validity of this approach. Figures 4-7 and 4-8 are plots of the degree of dissociation and the dimensionless enthalpy of an ideal dissociating gas in terms of the dimensionless pressure $\mathscr{P}/\mathscr{P}_D$ and temperature $(T/\theta_D)$.

## 4-8 THE EFFECT OF INERTS OR CONDENSATION ON EQUILIBRIA

The value of an equilibrium constant is not affected by the presence of an inert gas. Equation (4-16) is still valid. However, in this case the partial pressures of the reactive components do not add to give the total pressure of the system

1. M. J. Lighthill, *J. Fluid Mech.*, **2** (1957).

**Table 4-1**

| Molecule | Dissociation energy* $\theta_D$, K at 298.15 K | Best fit constants to Eq. (4-23) using a least-squares fit to $K_p$'s taken from App. B2 | | |
|---|---|---|---|---|
| | | $\theta_D$, K | log $\mathscr{P}_D$ | Temperature range, K |
| $H_2$ | 52,440 | 53,915 | 6.17 | 600–4000 |
| $O_2$ | 59,937 | 61,030 | 6.92 | 600–4000 |
| $O_2$ | 59,937 | 60,938 | 6.88 | 600–3000† |
| $O_2$ | 59,937 | 61,176 | 7.02 | 1600–4000† |
| $N_2$ | 113,705 | 114,962 | 6.91 | 600–4000 |
| $F_2$ | 18,982 | 19,885 | 6.68 | 600–4000 |
| $Cl_2$ | 29,176 | 30,171 | 6.31 | 600–4000 |
| $Br_2$ | 23,209 | 23,947 | 5.93 | 600–4000 |

* Taken from Appendix B1. Here $\theta_D = (\Delta H_f^\circ)_{298.15} * 2/R$. The $Br_2$ value has been corrected for the heat of vaporization at 298.15 K.

† For comparison to indicate the sensitivity of the fit constants, $\mathscr{P}_D$ and $\theta_D$, to the temperature range.

because the inert gas exerts a partial pressure in the mixture. Therefore, $n_T$ in Eq. (4-20) now becomes

$$n_T = n_1 + n_2 + \cdots + n_{s-1} + n_s + n_I$$

where $n_I$ represents the number of moles of inert gas present in one kilogram of mixture. Obviously, if $\sum v_{ij}$ in Eq. (4-20) is zero the addition of an inert will not

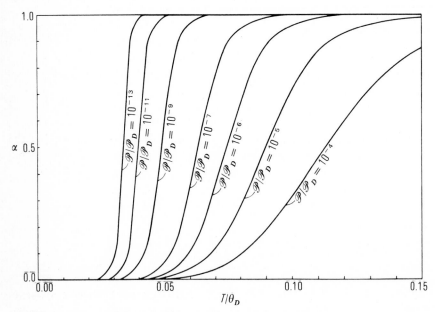

**Figure 4-7** Degree of dissociation for an ideal dissociating diatomic gas in terms of the reduced pressure and temperature [Eq. (4-24)].

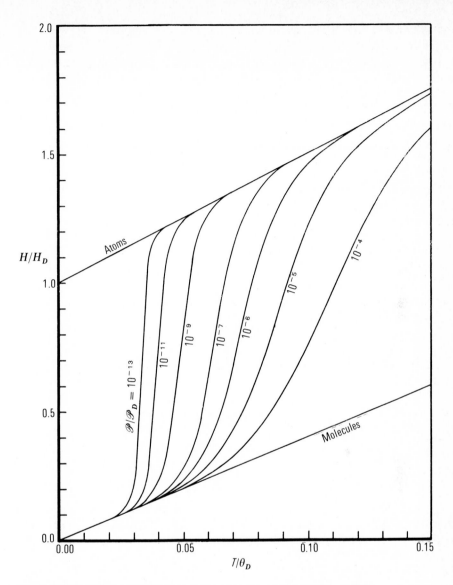

**Figure 4-8** The dimensionless enthalpy of an ideal dissociating gas [Eq. (4-25)].

shift the equilibrium composition. However, for a constant total pressure, if this exponent is not zero the equilibrium composition will shift with the addition of an inert.

The presence of a solid or liquid at equilibrium greatly affects the composition of the equilibrium mixture. If the solid or liquid is *pure*, one can assume that it exerts its equilibrium vapor pressure in the gaseous portion of the mixture independent of the quantity of solid or liquid present. To solve an equilibrium problem in this case one must first determine if a solid or liquid can be present at equilibrium. An equilibrium calculation is made assuming no condensed phases are present. The equilibrium partial pressures of all the species are then calculated and if these are all below the vapor pressures of the respective liquid or solid phases the calculation is valid. However, if any one of these partial pressures is above the vapor pressure of that component, it will condense and a new equilibrium calculation must be made.

Chemists have established a convenient convention in defining the standard state free energies of formation of solids or liquids. Since a solid or liquid is in equilibrium with its vapor at its vapor pressure, the free energy must be the same in the two phases. Furthermore, this free energy is independent of the total pressure in an ideal gaseous system. Therefore, by definition,

$$[\Delta F_f^{\circ}(T)]_{jth \text{ solid or liquid}} = \omega_j(T) + RT \ln p_{jv}$$

where $p_{jv}$ is the dimensionless equilibrium vapor pressure of the solid or liquid. Now consider a general system of $s$ species, $g$ of which are ideal gases and $s-g$ of which are solids or liquids, each exerting its own vapor pressure in the system. Substitution into Eq. (4-25) yields the relation

$$RT \sum_{i=1}^{g} \ln (p_i)^{v_i} + RT \sum_{i=g+1}^{s} \ln (p_{iv})^{v_i} = - \sum_{i=1}^{s} v_i[\Delta F_f^{\circ}(T)]_i + RT \sum_{i=g+1}^{s} \ln (p_{iv})^{v_i}$$

where the second terms on each side cancel and we may therefore write

$$RT \sum_{i=1}^{g} \ln (p_i)^{v_i} = - \sum_{i=1}^{s} v_i[\Delta F_f^{\circ}(T)]_i$$

or

$$RT \ln K_p' = - \sum_{i=1}^{s} v_i[\Delta F_f^{\circ}(T)]_i \qquad (4\text{-}26)$$

where the $\Delta F_f^{\circ}(T)$ are now for solids, liquids, or gases as appropriate. The new equilibrium constant $K_p'$ contains only terms for the partial pressure of the gaseous components. When these partial pressures satisfy this value of $K_p'$ the solid or liquid formed will be in exact equilibrium with its vapor and complete chemical equilibrium will exist. The convention, therefore, replaces the extra step of calculating the vapor pressures of solid or liquid components and their substitution into Eq. (4-16). Notice that the numerical value of $K_p$ in Eq. (4-16) is not equal to the value of $K_p'$ in Eq. (4-26). In Eq. (4-16), $K_p$ is equal to the sum of the logarithms of all the partial pressures and they are all variable. In Eq. (4-26) the $K_p'$ is equal to the sum of the logarithms of the partial pressures of only the completely gaseous components. The remaining partial pressures are fixed at the equilibrium vapor pressure of that component and divided out of the equation. The above discussion is valid only for the formation of pure solids or liquids in equilibrium with the gas phase. If an intimate mixture (a solution) of solids and/or liquids is formed, another difficulty arises because the vapor pressure of a component in a solution is not equal to the vapor pressure of the pure component. In this case allowances can be made by defining an activity, $a_i$, of the component where

$$a_i = p_i(\text{mixture})/p_i(\text{pure solid or liquid})$$

and Eq. (4-26) must contain activities $a_i$ raised to the proper stoichiometric exponents $(a_i^{v_i})$ before equilibrium concentrations may be calculated.

## 4-9 ACCURACY AND RESTRICTIONS IN EQUILIBRIUM CALCULATIONS

In the previous sections we discussed chemical kinetics and chemical equilibria as limit phenomena with the implication that rate processes in some way yield equilibrium systems after the passage of sufficient time. Furthermore, we found that reaction kinetics is a rather experimental science, in which a directly measured elementary rate is generally more useful than a theoretically calculated rate. We now observe that this contrasts rather sharply with the sophistication of our theoretical understanding of chemical equilibria, which in many cases appears to allow a more accurate calculation of the equilibrium state than can be measured experimentally. There are four major limitations to our ability to calculate the equilibrium behavior of chemical systems.

1. Thermodynamic functions must be known with extremely high accuracy to obtain realistic values for equilibrium constants. There are two main sources for this data. Spectroscopic information allows one to calculate the ordinary thermodynamic functions of a compound, and calorimetry is required to determine heats and entropies of formation. However, at high temperatures the molecules cannot be described as containing a series of "normal" independent energy storage modes, and cross-term contributions become important. This constitutes a challenge to the spectroscopist; high-temperature thermodynamic functions are only as good as the interpretation of complex spectra. In addition, the heat of reaction and the entropy of reaction cannot always be determined unambiguously using spectroscopic evidence. Careful calorimetric work must be performed to establish heats and entropies of formation. The importance of obtaining correct thermodynamic values can be shown using Eq. (4-17). At $T = 300$ K, an uncertainty of 240 J in the value of $\sum v_i(\Delta F_f^\circ)_i$ will cause $K_p$ to have an uncertainty of 10 percent, while at 3000 K an uncertainty of 2400 J will cause the same uncertainty in $K_p$. Since most standard heats of formation have values which lie between 100 and 600 kJ, they must be measured with extreme accuracy to estimate equilibrium constants reasonably well. This problem is continually under attack and the thermodynamic values tabulated in App. B represent only the best current values of the thermodynamic functions.

2. All species must be included in the calculation. We mentioned earlier that the presence of a species is in no way determined by the writing of a stoichiometric equation. These equations can only be written after the participating species are chosen. Furthermore, one can determine those species which are present at equilibrium only from experimental experience. This problem is particularly difficult if we are attempting to determine the thermodynamic behavior of a system which contains an atom whose properties and whose high-temperature compounds have not been studied extensively. This problem is not trivial because the formation of even a few percent of a compound with a high heat of formation can drop an equilibrium flame temperature hundreds of degrees and thus greatly modify performance calculations.

3. Nonideal effects can be large. This problem becomes important at high pressures and it can be treated with fair accuracy by introducing fugacities and fugacity coefficients. The fugacity of the $i$th species, $f_i$, is a fictitious pressure which the species must exert to satisfy Eq. (4-16). It is related to the mole fraction by the fugacity coefficients, $\xi_i$

$$f_i = \xi_i \frac{n_i}{n_T} \mathscr{P}$$

Therefore in an ideal gas mixture all $\xi_i = 1$ and $f_i = p_i$. In a nonideal system the ideal value of $K_p$ [from Eq. (4-16)] is given in terms of fugacities

$$K_p = \prod_{i=1}^{s} (f_i)^{\nu_i} = \prod_{i=1}^{s} (n_i)^{\nu_i} \cdot \prod_{i=1}^{s} (\xi_i)^{\nu_i} \cdot \left(\frac{\mathscr{P}}{n_T}\right)^{\sum_{i=1}^{s} \nu_i}$$

where

$$K_\xi = \prod_{i=1}^{s} (\xi_i)^{\nu_i}$$

The mole fraction at equilibrium may, therefore, be determined once a value for $K_p$ and $K_\xi$ are known. $K_\xi$ may be determined with reasonable accuracy in most cases by using nonideal thermodynamic functions derived from an assumed equation of state.[2] However, $K_\xi$ is a function of both temperature and pressure and the calculations become considerably more laborious.

4. Equilibrium might not be reached. This will occur when the number of available independent reactions is less than the $s$–$c$ required reaction paths. In certain simple systems it is relatively easy to construct a set of restricted equilibrium equations to account for this behavior. However, in a general case, the specification of the needed additional equations is not obvious and a generalized technique is required.

---

2. See, for example, O. A. Hougen and K. M. Watson, *Chemical Process Principles*, Part II, *Thermodynamics*, Wiley, New York, 1947.

Let us first consider a case of full equilibrium in a system containing $s$ species, $c$ components, and, therefore, $s-c = r$ independent reactions. Assume that there are $p > r$ possible reversible reaction paths and construct a $p \times s$ matrix of the $r_i$ available stoichiometric coefficients.

$$
\begin{array}{c c c c c c}
 & s_1 & s_2 & s_3 & & s_s \\
r_1 & v_{11} & v_{12} & v_{13} & \cdots & v_{1s} \\
r_2 & v_{21} & v_{22} & v_{23} & \cdots & v_{2s} \\
r_3 & v_{31} & v_{32} & v_{33} & \cdots & v_{3s} \\
 & \vdots & \vdots & \vdots & & \vdots \\
r_p & v_{p1} & v_{p2} & v_{p3} & & v_{ps}
\end{array}
$$

Since this matrix represents a set of linear equations containing only $r$ variables, it must have a rank no greater than $r$. If inspection shows that this matrix has a rank $r$, we are dealing with a case of nonrestricted or full equilibrium. If this matrix has a rank $r^* < r$ we have a case of restricted equilibrium.

When the rank is $r$ we can add or subtract rows to reduce this matrix to an $r \times s$ matrix. We may then diagonalize any $r$ independent columns, writing the matrix so that a $(-1)$ appears at each position on the main diagonal of the $r$ columns. When this is done we will have, in effect, written the stoichiometric equations for the formation of the $r$ species using the remaining $s-r = c$ species as components.

$$
\begin{array}{c c c c c c c c c}
 & c_1 & c_2 & \cdots & c_c & s_1 & s_2 & \cdots & s_r \\
r_1 & v'_{11} & v'_{12} & \cdots & v'_{1c} & -1 & 0 & \cdots & 0 \\
r_2 & v'_{21} & v'_{22} & \cdots & v'_{2c} & 0 & -1 & \cdots & 0 \\
\vdots & \vdots & \vdots & & \vdots & \vdots & \vdots & & \vdots \\
r_r & v'_{r1} & v'_{r2} & \cdots & v'_{rc} & 0 & 0 & & -1
\end{array}
$$

This matrix may be used to construct the mass-balance equations for these components in terms of the species present at equilibrium by constructing an $s \times c$ matrix consisting of a $c \times c$ unit matrix written over the component coefficient matrix.

$$
\begin{array}{c c c c c}
 & c_1 & c_2 & & c_c \\
c_1 & 1 & 0 & \cdots & 0 \\
c_2 & 0 & 1 & \cdots & 0 \\
\vdots & \vdots & \vdots & & \vdots \\
c_c & 0 & 0 & \cdots & 1 \\
r_1 & v'_{11} & v'_{12} & \cdots & v'_{1c} \\
r_2 & v'_{21} & v'_{22} & \cdots & v'_{2c} \\
\vdots & \vdots & \vdots & & \vdots \\
r_r & v'_{r1} & v'_{r2} & & v'_{rc}
\end{array}
$$

An example will be useful. Consider the system containing $H_2$, $O_2$, $H_2O$, H, O, and OH. Here $s = 6$, $c = 2$, and $r = 4$. Consider the equilibrium reactions

$$H_2 + O_2 \rightleftharpoons 2OH$$

$$H_2 + \tfrac{1}{2}O_2 \rightleftharpoons H_2O$$

$$H_2 \rightleftharpoons 2H$$

$$O_2 \rightleftharpoons 2O \tag{4-27}$$

$$H + OH \rightleftharpoons H_2 + O$$

$$O + H_2 \rightleftharpoons OH + H$$

These reactions yield the matrix

| $H_2$ | $O_2$ | $H_2O$ | H | O | OH |
|---|---|---|---|---|---|
| $-1$ | $-1$ | $0$ | $0$ | $0$ | $+2$ |
| $-1$ | $-\tfrac{1}{2}$ | $+1$ | $0$ | $0$ | $0$ |
| $-1$ | $0$ | $0$ | $+2$ | $0$ | $0$ |
| $0$ | $-1$ | $0$ | $0$ | $+2$ | $0$ |
| $+1$ | $0$ | $0$ | $-1$ | $+1$ | $-1$ |
| $-1$ | $0$ | $0$ | $+1$ | $-1$ | $+1$ |

which is of rank 4 and may therefore be altered by adding and subtracting rows to yield the matrix

| O | H | OH | $H_2$ | $O_2$ | $H_2O$ |
|---|---|---|---|---|---|
| $+1$ | $+1$ | $-1$ | $0$ | $0$ | $0$ |
| $0$ | $+2$ | $0$ | $-1$ | $0$ | $0$ |
| $+2$ | $0$ | $0$ | $0$ | $-1$ | $0$ |
| $+1$ | $+2$ | $0$ | $0$ | $0$ | $-1$ |

This matrix is equivalent to the reactions

$$OH \rightleftharpoons H + O$$

$$H_2 \rightleftharpoons 2H$$

$$O_2 \rightleftharpoons 2O$$

$$H_2O \rightleftharpoons 2H + O$$

which represent the independent formation of the species OH, $H_2$, $O_2$, and $H_2O$ from the components H and O. Rewriting to express the mass-balance equations, we obtain

$$
\begin{array}{c|cc}
 & \text{O} & \text{H} \\
\hline
\text{O} & 1 & 0 \\
\text{H} & 0 & 1 \\
\text{OH} & 1 & 1 \\
\text{H}_2 & 0 & 2 \\
\text{O}_2 & 2 & 0 \\
\text{H}_2\text{O} & 1 & 2
\end{array}
$$

which represents the equations

$$N_\text{O} = n_\text{O} + n_\text{OH} + 2n_{\text{O}_2} + n_{\text{H}_2\text{O}}$$

$$N_\text{H} = n_\text{H} + n_\text{OH} + 2n_{\text{H}_2} + 2n_{\text{H}_2\text{O}}$$

where the $N$'s represent component moles and are therefore constants of the system. Notice that while this choice of components is not unique, we are restricted to choose independent species for the diagonalization procedure and therefore do not have a completely unrestricted choice of components. For example, the determinant represented by the columns $\text{O}_2$, H, O, and OH has a value of zero and therefore H and $\text{H}_2$ cannot be components in this system. However, we may rewrite the matrix to yield the new matrix,

$$
\begin{array}{cccccc}
\text{OH} & \text{O} & \text{H}_2\text{O} & \text{H} & \text{O}_2 & \text{H}_2 \\
\end{array}
$$
$$
\begin{vmatrix}
+2 & -1 & -1 & 0 & 0 & 0 \\
+1 & -1 & 0 & -1 & 0 & 0 \\
0 & +2 & 0 & 0 & -1 & 0 \\
+2 & -2 & 0 & 0 & 0 & -1
\end{vmatrix}
$$

The equilibrium equations for this case are

$$\text{O} + \text{H}_2\text{O} \rightleftharpoons 2\text{OH}$$

$$\text{O} + \text{H} \rightleftharpoons \text{OH}$$

$$\text{O}_2 \rightleftharpoons 2\text{O}$$

$$2\text{O} + \text{H}_2 \rightleftharpoons 2\text{OH}$$

and the mass-balance matrix becomes

$$
\begin{array}{c|cc}
 & \text{OH} & \text{O} \\
\hline
\text{OH} & 1 & 0 \\
\text{O} & 0 & 1 \\
\text{H}_2\text{O} & 2 & -1 \\
\text{H} & 1 & -1 \\
\text{O}_2 & 0 & 2 \\
\text{H}_2 & 2 & -2
\end{array}
$$

This matrix represents the equations

$$N_{OH} = n_{OH} + 2n_{H_2O} + n_H + 2n_{H_2}$$

$$N_O = n_O - n_{H_2O} - n_H + 2n_{O_2} - 2n_{H_2}$$

This is a valid set of mass-balance equations for the six-specie system in terms of the components OH and O. Notice that for $\Phi > 0.5$, $N_O$ will be a negative number in this representation.

Restricted equilibrium occurs if the reduced matrix has a rank $r^* < r$. In the hydrogen–oxygen system the early reactions yield a state of restricted equilibrium because complete equilibrium requires a net reduction in the number of moles of gas present in the mixture. Since, in this system, mole reduction can only occur via three-body collisions, the rapid reactions result in the momentary appearance of a restricted equilibrium state. Let us examine a representative set of available binary exchange reactions:

$$H + O_2 \rightleftharpoons OH + O$$

$$O + H_2 \rightleftharpoons OH + H$$

$$OH + H_2 \rightleftharpoons H_2O + H \qquad (4\text{-}28)$$

$$H_2 + O_2 \rightleftharpoons 2OH$$

$$\text{etc.}$$

These may be reduced to the $3 \times 6$ matrix

| | $O_2$ | $H_2$ | $H_2O$ | H | O | OH |
|---|---|---|---|---|---|---|
| | $+\frac{1}{2}$ | $+1\frac{1}{2}$ | $-1$ | $-1$ | $0$ | $0$ |
| | $+1$ | $+1$ | $-1$ | $0$ | $-1$ | $0$ |
| | $+\frac{1}{2}$ | $+\frac{1}{2}$ | $0$ | $0$ | $0$ | $-1$ |

representing the reactions

$$H + H_2O \rightleftharpoons \tfrac{1}{2}O_2 + 1\tfrac{1}{2}H_2$$

$$O + H_2O \rightleftharpoons H_2 + O_2$$

$$OH \rightleftharpoons \tfrac{1}{2}O_2 + \tfrac{1}{2}H_2$$

The mass-balance matrix becomes

| | $O_2$ | $H_2$ | $H_2O$ |
|---|---|---|---|
| $O_2$ | $1$ | $0$ | $0$ |
| $H_2$ | $0$ | $1$ | $0$ |
| $H_2O$ | $0$ | $0$ | $1$ |
| H | $\frac{1}{2}$ | $1\frac{1}{2}$ | $-1$ |
| O | $1$ | $1$ | $-1$ |
| OH | $\frac{1}{2}$ | $\frac{1}{2}$ | $0$ |

which yields the equations

$$N_{O_2} = n_{O_2} + \tfrac{1}{2}n_H + n_O + \tfrac{1}{2}n_{OH}$$

$$N_{H_2} = n_{H_2} + 1\tfrac{1}{2}n_H + n_O + \tfrac{1}{2}n_{OH} \qquad (4\text{-}29)$$

$$N_{H_2O} = n_{H_2O} - n_H - n_O$$

In this case one must specify three components to discuss the restricted equilibrium behavior. The *initial* species composition of this mixture is therefore important in the determination of a restricted equilibrium composition. Once again, negative component mole numbers are possible.

Notice that the third of these three mass-balance requirements is new and represents the requirement that the number of moles of gas remain constant during the reaction. This is because this equation states that the production of an atomic species must be accompanied by the production of the triatomic molecule $H_2O$. It should be obvious that the occurrence of the exchange reactions, Eqs. (4-28), cannot change the number of moles in the system. Notice also that Eqs. (4-27) were sufficient to produce a state of full equilibrium and they do contain reactions in which mole number changes occur.

## 4-10 A CALCULATION OF AN EXPLOSION TEMPERATURE

In this section we present two explosion-temperature calculations based on the assumptions that the products of the explosion process are at equilibrium and that the explosion process is either (1) a constant-pressure process or (2) a constant-volume process. We choose as the system one mole of nitric oxide at $T_i = 298.15$ K and $P_i = 101,325$ Pa. To simplify the calculation we limit the equilibrium mixture to one containing only nitrogen molecules, $N_2$, oxygen molecules, $O_2$, and nitric oxide, NO.‡

We use the stoichiometric equation

$$\tfrac{1}{2}N_2 + \tfrac{1}{2}O_2 \rightleftharpoons NO$$

and call $\alpha$ the number of moles of NO remaining at equilibrium. Thus

$$K_p = \frac{2\alpha}{1 - \alpha}$$

and $\alpha$ is a function of temperature only

$$\alpha = \frac{K_p}{2 + K_p}$$

‡ The final temperature calculated in these examples will be too high because oxygen atoms, O, should be included to obtain a realistic final state. However, if oxygen atoms are included the enthalpy and internal energy of the system become pressure-dependent and this complicates the calculation.

**Table 4-2 Constant-pressure explosion-temperature calculation‡**

| | | Method $a$ $(T_r = T_f)$ | | | Method $b$ $(T_r = T_i)$ | |
|---|---|---|---|---|---|---|
| $T, K$ | $\alpha$ | $(\alpha - 1)[(\Delta H_f^\circ)_{T_f}]_{NO}$ | $(H_{T_f} - H_{298.15})_{NO}$ | $\Delta H_{process}$ | $(\alpha - 1)[(\Delta H_f^\circ)_{T_i}]_{NO}$ | $\sum$§ |
| 2600 | 0.03388 | $-87,184.3$ | 80,074.0 | $-7,110.3$ | $-87,273.5$ | 80,163.2 |
| 2700 | 0.03932 | $-86,620.9$ | 83,803.7 | $-2,817.2$ | $-86,782.1$ | 83,964.9 |
| 2800 | 0.04508 | $-86,021.9$ | 87,533.4 | 1,511.5 | $-86,261.8$ | 87,773.3 |
| 2900 | 0.05119 | $-85,383.9$ | 91,275.7 | 5,891.8 | $-85,709.6$ | 91,601.4 |
| 3000 | 0.05758 | $-84,714.2$ | 95,022.2 | 10,308.0 | $-85,132.8$ | 95,440.8 |

‡ Interpolation yields $T_{eq} = 2765$ K; $\alpha_{eq} = 0.0430$ (see Fig. 4-9).
§ $\sum = \frac{1}{2}(1 - \alpha)(H_{T_f} - H_{298.15})_{N_2} + \frac{1}{2}(1 - \alpha)(H_{T_f} - H_{298.15})_{O_2} + \alpha(H_{T_f} - H_{298.15})_{NO}$.

In this notation, at any temperature, the extent of reaction in the forward direction is

$$\tfrac{1}{2}(\alpha - 1) \text{ moles of reaction (note that this is negative)}$$

and the composition is given by

$$\eta_{NO} = \alpha$$

and

$$\eta_{O_2} = \eta_{N_2} = \tfrac{1}{2}(1 - \alpha)$$

Notice that because of our specific choice of a stoichiometric equation the $K_p$ which we are using is the $K_p$ tabulated as log $K_p$ in the nitric oxide table in App. B. The quantity $\alpha(T)$ is tabulated in Table 4-2 and graphed in Fig. 4-9.

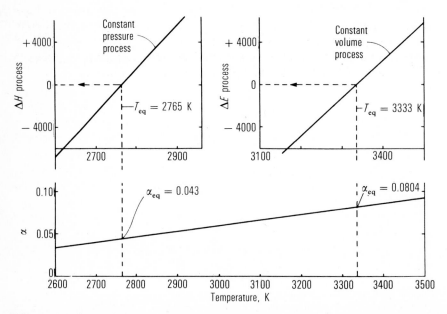

**Figure 4-9** Degree of reaction $\alpha$, $\Delta H$ process, $\Delta E$ process, plotted for the calculation of Sec. 4-10.

For a constant-presssure process the enthalpy of the system is conserved and the equilibrium explosion temperature may be determined by requiring that $\Delta H_{\text{process}} = 0$. There are a number of possible ways of performing this calculation. We choose to use the technique outlined in Sec. 2-5. First, for purposes of calculating an enthalpy or energy balance for the system we write an equation to summarize the chemical changes that occur.

$$(1)\text{NO} \quad \rightarrow \underbrace{\alpha(T_f)\text{NO} + \tfrac{1}{2}[1 - \alpha(T_f)]\text{O}_2 + \tfrac{1}{2}[1 - \alpha(T_f)]\text{N}_2}_{} \qquad (4\text{-}30)$$

$$\underbrace{\phantom{(1)\text{NO}}}_{\substack{\text{initial} \\ \text{system} \\ \text{(mole basis)}}} \qquad \underbrace{\phantom{\alpha(T_f)\text{NO} + \tfrac{1}{2}[1 - \alpha(T_f)]\text{O}_2}}_{\substack{\text{equilibrium mixture composition at temperature } T_f \\ \text{final system} \\ \text{(mole basis)}}}$$

These coefficients are not stoichiometric coefficients in the usual sense but simply represent the net quantity of each species present in the initial mixture and at an assumed final temperature $T_f$. In general, a final pressure would also be needed to fully evaluate the right-hand side of this equation.

We now note, as we did in Sec. 2-5, that in terms of tabulated values available to evaluate the condition $\Delta H = 0$, there are two contributions to the enthalpy change in this system. These are the chemical enthalpy change caused by composition changes at constant temperature and the sensible enthalpy change caused by temperature changes at constant composition. Therefore, for calculation purposes one should evaluate the change of reactants and products relative to the standard state elements at 298.15 K. If $T_r = 298.15$ K, the equations of Sec. 2-5 reduce to

$$(H_{T_r} - H_{T_i})_{\text{reactants}} + (\Delta H_{\text{chem}})_{T_r} + (H_{T_f} - H_{T_r})_{\text{prod}} = \Delta H_{\text{process}}(T_f) \quad (4\text{-}31)$$

Where the choice of the temperature $T_r$ is either arbitrary (if sufficient thermodynamic data is available) or is limited to a short range (by the lack of thermodynamic data for either the sensible enthalpy or the enthalpy of formation of the reactants). As written, $\Delta H_{\text{process}}$ is a function of the assumed final temperature; in general it is also a function of pressure. The equilibrium temperature $T_{\text{eq}}$ may now be determined by using this equation and adjusting $T_f$ until $\Delta H_{\text{process}} = 0$.

In our example we evaluated Eq. (4-31) for two assumed values of $T_r$: (a) $T_r = T_f$ and (b) $T_r = T_i$. Method (b) is identical to the procedure proposed in Sec. 2-5 because $\Delta H_f^\circ$ of $\text{O}_2$ and $\text{N}_2$ are zero by definition. These results are tabulated in Table 4-2 and summarized in Fig. 4-9. Note that both techniques of evaluation yield the same value of $T_{\text{eq}}$ and in fact yield identical values of $\Delta H_{\text{process}}$. In general, if thermodynamic data are available, the evaluation of $T_{\text{eq}}$ by process $a$ is simpler.

If we want to calculate a constant-volume explosion temperature, internal energy is conserved and we write

$$(E_{T_r} - E_{T_i})_{\text{reactants}} + (\Delta E_{\text{chem}})_{T_r} + (E_{T_f} - E_{T_r})_{\text{products}} = \Delta E_{\text{process}}(T_f) \quad (4\text{-}32)$$

**Table 4-3 Constant-volume explosion calculation‡**
Method *a* only

| $T$, K | $\alpha$ | $(\alpha - 1)[(\Delta E_f^\circ)_{T_f}]_{NO}$ | $(E_{T_f} - E_{298.15})_{NO}$ | $\Delta E_{process}$ |
|---|---|---|---|---|
| 3100 | 0.06415 | −84,025.6 | 75,468.6 | −8,557.0 |
| 3200 | 0.07111 | −83,292.2 | 78,391.6 | −4,900.6 |
| 3300 | 0.07815 | −82,548.8 | 81,322.7 | −1,226.1 |
| 3400 | 0.08534 | −81,786.5 | 84,254.6 | 2,468.1 |
| 3500 | 0.09264 | −81,012.1 | 87,194.4 | 6,182.3 |

‡ Interpolation yields $T_{eq} = 3333$ K; $\alpha_{eq} = 0.0804$ (see Fig. 4-9).

Values of $E$ for Eq. (4-32) may be determined from values of $H$ tabulated in App. B by using Eq. (2-21). Notice that in general $\Delta E_{chem} \neq \Delta H_{chem}$. However, in the case described above, $\Delta n = 0$ and therefore $\Delta E_{chem} = \Delta H_{chem}$.

Sensible internal energies may also be evaluated on a molar basis by using the definition of enthalpy from Chap. 2

$$E_T = H_T - RT$$

Table 4-3 contains a tabulation of the internal energy terms from Eq. (4-32) as used to determine the constant-volume explosion temperature of this system. The results are graphed in Fig. 4-9.

The value of

$$(T_{eq})_{V = V_i}$$

is higher than

$$(T_{eq})_{P = P_i}$$

because no external work was performed during the process in the former case. A calculation of the entropy change of the system for these two processes would show that entropy increases, indicating that the processes are thermodynamically spontaneous.

## 4-11 MULTISPECIES EQUILIBRIUM CALCULATIONS[3]

In a system that contains $s$ species and $c$ component elements the calculation of an equilibrium composition usually becomes difficult when $s-c > 3$. A good practical example is the calculation of the high-temperature equilibrium composition of the products of combustion of a hydrocarbon–air mixture at some specified temperature and pressure. Let us assume that over most of the possible composition ranges solids (such as soot) are not formed. The problem then

3. F. Van Zeggeren and S. H. Storey, *The Computation of Chemical Equilibria*, Cambridge University Press (1970).

reduces to the calculation of the gas phase equilibrium composition of a general
C–H–O–N–$I$ system where $I$ is an inert such as helium or argon.

The species composition of the initial mixture allows one to calculate the
ratios $\overline{N}_{C/O}$, $\overline{N}_{H/O}$, $\overline{N}_{N/O}$, and $\overline{N}_{I/O}$ (see Sec. 2-3). This information, the specified
equilibrium pressure and temperature, and the list of the species present at equi-
librium plus the assumption that the gas is an ideal gas, are all that is required
to perform an equilibrium calculation. Let us assume a very general case in
which the 14 species, oxygen, $O_2$, nitrogen, $N_2$, hydrogen, $H_2$, carbon dioxide,
$CO_2$, carbon monoxide, CO, water vapor, $H_2O$, hydroxyl radical, OH, nitric
oxide, NO, nitrogen dioxide, $NO_2$, oxygen atoms, O, nitrogen atoms, N, hydrogen atoms, H, methane, $CH_4$, and an inert, $I$, are present at equilibrium.

To solve for the species composition we will follow a technique described
by Weinberg.[4] The atomic ratios which specify the fixed atomic composition
can be written in terms of partial pressures as follows:

$$\overline{N}_{C/O} = \frac{\mathfrak{P}_C}{\mathfrak{P}_O} \qquad \overline{N}_{H/O} = \frac{\mathfrak{P}_H}{\mathfrak{P}_O} \qquad \overline{N}_{N/O} = \frac{\mathfrak{P}_N}{\mathfrak{P}_O} \qquad \text{and} \qquad \overline{N}_{I/O} = \frac{\mathfrak{P}_I}{\mathfrak{P}_O} \quad (4\text{-}33)$$

where

$$\mathfrak{P}_C = \not{p}_{CO_2} + \not{p}_{CO} + \not{p}_{CH_4}$$

$$\mathfrak{P}_O = 2\not{p}_{O_2} + 2\not{p}_{CO_2} + \not{p}_{CO} + \not{p}_{H_2O} + \not{p}_{OH} + \not{p}_{NO} + 2\not{p}_{NO_2} + \not{p}_O$$

$$\mathfrak{P}_H = 2\not{p}_{H_2} + 2\not{p}_{H_2O} + \not{p}_{OH} + \not{p}_H + 4\not{p}_{CH_4} \qquad\qquad (4\text{-}34)$$

$$\mathfrak{P}_N = 2\not{p}_{N_2} + \not{p}_{NO} + \not{p}_{NO_2} + \not{p}_N$$

$$\mathfrak{P}_I = \not{p}_I$$

with the added equation that the specified pressure of the system is the sum of
the partial pressures

$$\mathscr{P} = \sum_{i=1}^{s} \not{p}_i \qquad\qquad (4\text{-}35)$$

At this point we have 14 unknowns (the $\not{p}_i$'s) and only five equations [Eqs.
(4-33) and (4-35)]. Thus we need nine independent equilibrium equations to
solve for the equilibrium composition. The choice is arbitrary and we will use
the following independent equations.

1. Carbon dioxide dissociation: $CO_2 \rightleftarrows CO + \frac{1}{2}O_2$

$$K_{p1} = \frac{\not{p}_{CO} \cdot \not{p}_{O_2}^{1/2}}{\not{p}_{CO_2}}$$

2. The dissociation of water: $H_2O \rightleftarrows H_2 + \frac{1}{2}O_2$

$$K_{p2} = \frac{\not{p}_{H_2} \cdot \not{p}_{O_2}^{1/2}}{\not{p}_{H_2O}}$$

4. F. J. Weinberg, *Proc. R. Soc.*, London, **241A**:132 (1957).

3. The formation of hydroxyl radical: $H_2O \rightleftarrows \frac{1}{2}H_2 + OH$

$$K_{p3} = \frac{p_{H_2}^{1/2} \cdot p_{OH}}{p_{H_2O}}$$

4. The dissociation of hydrogen: $\frac{1}{2}H_2 \rightleftarrows H$

$$K_{p4} = \frac{p_H}{p_{H_2}^{1/2}}$$

5. The dissociation of oxygen: $\frac{1}{2}O_2 \rightleftarrows O$

$$K_{p5} = \frac{p_O}{p_{O_2}^{1/2}}$$

6. The formation of methane: $4H_2 + CO_2 \rightleftarrows CH_4 + 2H_2O$

$$K_{p6} = \frac{p_{CH_4} \cdot p_{H_2O}^2}{p_{H_2}^4 \cdot p_{CO_2}}$$

7. The formation of nitrogen atoms: $\frac{1}{2}N_2 \rightleftarrows N$

$$K_{p7} = \frac{p_N}{p_{N_2}^{1/2}}$$

8. The formation of nitric oxide: $\frac{1}{2}N_2 + \frac{1}{2}O_2 \rightleftarrows NO$

$$K_{p8} = \frac{p_{NO}}{p_{N_2}^{1/2} \cdot p_{O_2}^{1/2}}$$

9. The formation of nitrogen dioxide: $\frac{1}{2}N_2 + O_2 \rightleftarrows NO_2$

$$K_{p9} = \frac{p_{NO_2}}{p_{N_2}^{1/2} \cdot p_{O_2}}$$

The solution of this formidable set of equations can be reduced to a tractable process by noting that the partial pressures of the species N, NO, and $NO_2$ will always be rather small and that they can be calculated by an "add-on technique." This reduces the main calculation to the determination of the composition of a system containing only the three active components C, H, and O.

To do this one only need assume the partial pressures of three species that independently contain the component elements. Any set that satisfies this requirement may be chosen, such as $p'_{CO_2}$, $p'_{O_2}$, $p'_{H_2}$ or $p'_{CO_2}$, $p'_{H_2O}$, $p'_{O_2}$, etc., where the primes indicate that the partial pressure is a guess. Using this guess the (incorrect) partial pressures of the remaining species, $p'_i$, may be calculated using the equilibrium constants. The three mass-conservation equations [from sets (4-33) to (4-35)] are then written by substituting $p_i = p'_i + \delta_i$ where $\delta_i$ is the correction to the partial pressure which is an unknown. However, the equilibrium relations written in terms of $p_i = p'_i + \delta_i$ can be differentiated to yield an expression for each of the remaining $\delta_i$'s in terms of the guessed species $\delta_i$'s.

When these are substituted into the mass-balance set one obtains three equations for the three unknown correction factors, one for each of the three guessed species. Since these equations are linear they are easily solved in determinant form.

The calculation converges rapidly if the initial guesses of partial pressures are reasonable and if the guessed species have a relatively large partial pressure at equilibrium.

The nitrogen-containing species and the inerts are included in the calculation as "add-on" species at each step by writing

$$\mathscr{P}_{\text{calc}} = \mathscr{P} - \not{p}_N - \not{p}_{NO} - \not{p}_{NO_2} - \not{p}_I$$

where $\not{p}_I$ is calculated from $\overline{N}_{I/O}$ and the properly weighted partial pressures of the oxygen containing species [Eq. (4-34)] and the $\not{p}_N$, $\not{p}_{NO}$, $\not{p}_{NO_2}$ are calculated from the values of $\not{p}_{N_2}$ and $\not{p}_{O_2}$ and the appropriate equilibrium constants. The main calculation is then performed using $\mathscr{P}_{\text{calc}}$ as a fictitious total pressure for the C–H–O system. Appendix E contains a listing of a FORTRAN program to perform this calculation. This program, with the condition $\Delta H = 0$, has been used to calculate constant-pressure explosion temperatures for a number of hydrocarbon–air mixtures. Typical explosion-temperature curves are shown in Fig. 4-10. Figure 4-11 is a plot of the equilibrium composition for a constant pressure explosion of methane–air.

Once an equilibrium composition has been calculated on the basis of any

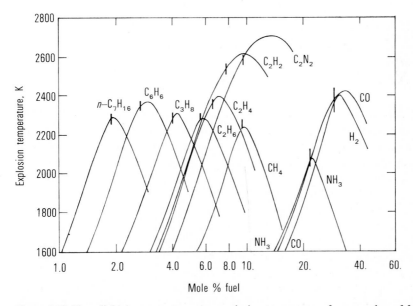

**Figure 4-10** The adiabatic constant-pressure explosion temperatures for a number of fuel–air mixtures. Initial conditions: $T = 298.15$ K, $P = 101.325$ kPa. The vertical bars represent stoichiometric mixtures. The species $H_2$ and CO share the same bar.

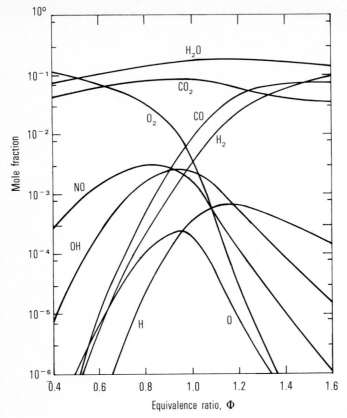

**Figure 4-11** Equilibrium composition of a methane–air constant-pressure explosion. Initial conditions: $T = 298.15$ K, $P = 101.325$ kPa. See Fig. 4-10 for the explosion temperature.

assumed species composition, it is easy to check and determine if any appreciable error was made by neglecting a specific species. Take, for example, the species $NH_2$ which was not included in the above scheme. To estimate its concentration under equilibrium conditions for any calculation, find the equilibrium constant of formation of that species at that temperature. In this case,

$$\tfrac{1}{2}N_2 + H_2 \rightleftarrows NH_2$$

$$K_{p,\,NH_2} = \frac{p_{NH_2}}{p_{N_2}^{1/2} \cdot p_{H_2}}$$

This equation may be solved to determine a first estimate of the partial pressure of that species

$$p_{NH_2} = K_{p,\,NH_2} \cdot p_{H_2} \cdot p_{N_2}^{1/2}$$

An estimate of the effect of neglecting this species on either a constant-pressure or constant-volume explosion temperature may be made by determining its heat of formation and calculating the temperature change by performing either an enthalpy or energy-balance calculation.

## PROBLEMS

**4-1** Calculate the equilibrium composition of air (composition, 21% $O_2$ + 79% $N_2$ by volume) at 2500 and 5000 K and $P = 10$ kPa. Assume that O, N, $N_2$, $O_2$, and NO are present at equilibrium.

**4-2** In Prob. 4-1 calculate

    (a) Volume ratio $V_{equilibrium}/V_{unreacted}$;

    (b) Enthalpy of the equilibrium mixture in joules per mole of unreacted mixture relative to the unreacted mixture at 298.15 K;

    (c) Heat capacity ratio (frozen composition).

**4-3** Calculate a value for the degree of dissociation $\alpha$ for $Br_2$ at 1000 K, for $O_2$ at 5000 K, and for $F_2$ and 5000 K and 100 kPa, using Table 4-1 and Fig. 4-7. Compare your values to values calculated from the thermodynamic functions given in App. B.

**4-4** Using the system presented in Sec. 4-10, show that isentropic expansion of the constant-volume explosion products to $P = P_i$ does not yield the constant-pressure explosion temperature for either an equilibrium composition or frozen composition expansion. Discuss the reason for this behavior.

**4-5** Perform the calculation of Sec. 4-10 at $P = P_i = 100$ kPa assuming that oxygen atoms are present at equilibrium. Should one also include nitrogen atoms? Why?

**4-6** Consider a mixture of diatomic oxygen, $O_2$, oxygen atoms, O, and inert argon, Ar, at 3000 K and 100 kPa total pressure. If the mass fraction of argon is 0.55555 find the equilibrium composition using two techniques:

    (a) Use the tabulated equilibrium constant (App. B) and solve for the composition;

    (b) Plot the free energy of all possible $O_2$–O–Ar mixtures at fixed argon mass fraction as a function of either the $O_2$ or O concentration to show that it is indeed a minimum at the equilibrium composition.

    *Note:* Assume $MW$ Ar = 0.040, $MW$ $O_2$ = 0.032.

**4-7** Discuss the approximate rate of the following reactions, i.e., fast, slow, and if appropriate give the approximate value of the activation energy. Base your discussion on the heats of reaction at 1800 K.

$$CO + OH \rightarrow CO_2 + H$$
$$H + O_2 \rightarrow OH + O$$
$$O + H_2 \rightarrow OH + H$$
$$H + H + M \rightarrow H_2 + M$$
$$O_2 + M \rightarrow O + O + M$$
$$CH_4 + OH \rightarrow CH_3 + H_2O$$

**4-8** Consider ideal gas water vapor. At 100 kPa pressure calculate the composition and enthalpy of the mixture as a function of temperature (range 300–6000 K) assuming that the mixture at equilibrium contains the species

    (a) $H_2O$, $H_2$, $O_2$

    (b) $H_2O$, $H_2$, $O_2$, OH, and

    (c) determine the effect of the formation of O and H atoms on your answers.

Plot your results for parts *a* and *b*.

**4-9** Calculate the final isobaric-explosion temperature of a mixture which initially contains 40% $N_2$, 40% CO, and 20% $O_2$ at 298.15 K and 100 kPa if

    (a) only $CO_2$ and $N_2$ are present at $T_f$,
    (b) an equilibrium mixture of $N_2$, CO, $CO_2$ and $O_2$ are present at $T'_f$.

Estimate the effect of the presence of O atoms on $T'_f$.

**4-10** Calculate the constant-pressure explosion temperature and composition of a rich ($\Phi = 1.2$) methane–air mixture initially at 25°C and 100 kPa. Assume that only $N_2$, $CO_2$, $H_2O$, $H_2$, and CO are present at equilibrium and that air contains 79% $N_2$ and 21% $O_2$. Determine the effect of neglecting NO, OH, O, H, and N on your answers and compare your results with values obtained from Figs. 4-10 and 4-11.

**4-11** Calculate the explosion temperature for a constant-pressure and constant-volume explosion of pure ozone at initial conditions of 300 K and 100 kPa. Also determine the respective volume and pressure increases associated with such an explosion. Assume $O_2$ and O are present at equilibrium. Compare your constant-volume explosion temperature with the equilibrium composition line on Fig. 6-16.

**4-12** Use App. B to verify that Van't Hoff's equation is valid for the equilibrium $2H_2 + O_2 \leftrightarrows 2H_2O$ over the temperature range of 1000 to 3000 K.

**4-13** Calculate the fraction of the molecules that are potentially chemically reactive at 1500 K if the activation energy is 80 or 240 kJ. What would be the initial rate of disappearance of the species $A$ if a mixture of $A$ and $B$ were equimolar, the pressure was 100 kPa, and the reaction was second-order with a preexponential constant of $10^{10}$.

**4-14** Work Prob. 2-9a. Now determine the amount of nitrogen that would be necessary to make $\Delta H_r = 0$ for this reaction. Compare your answer with the flame temperature from Fig. 4-10 where the coefficient for $N_2$ is 7.52.

# BIBLIOGRAPHY

Benson, S. W., *Thermochemical Kinetics*, 2d ed. Wiley, New York (1976).

Denbigh, K., *The Principles of Chemical Equilibria*. Cambridge, London (1961).

King, E. L., *How Chemical Reactions Occur*. Benjamin, New York (1964).

Kondrat'ev, V. N., *Chemical Kinetics of Gas Reactions*. Pergamon, Oxford (1964).

Skinner, G. B., *Introduction to Chemical Kinetics*. Academic Press, New York (1974).

Smith, W. R., and R. W. Missen, *Chemical Reaction Equilibrium Analysis: Theory and Algorithms*. Wiley, New York (1982).

Van Zeggeren, F., and S. H. Storey, *The Computation of Chemical Equilibria*. Cambridge, London (1970).

# REACTIVE GAS DYNAMICS

## 5-1 INTRODUCTION

Now that we have introduced the basic physical properties that are most important in combustion processes, we turn our attention to the conservation equations which describe flow processes. In the previous chapters we first introduced a physically accurate and usually quite complex description of the different physical properties and their dependence on pressure, temperature, etc. We then discussed, in each case, approximations which, while not exactly correct physically, simplified the description of the property to yield an analytic expression which has the major important features of the real property. In many cases a hierarchy of simplifications was presented so that the most appropriate simplification can be selected for some specific application.

We will take the same approach in this chapter. In the next section a relatively general set of conservation equations are presented, without derivation. This presentation follows, with little change, the treatment of Chap. 18 of Bird et al.[1] and the reader is referred to that text for the derivation of the equations and other more detailed comments than are presented here.

In the following sections we will discuss, within the framework of these equations, the two major subsets of flow that are important to combustion processes—laminar and turbulent flows. We make this distinction because our understanding of reactive turbulent flows is still very incomplete when compared with our understanding of reactive laminar flows. In the case of laminar

1. R. B. Bird, W. E. Stewart, and E. N. Lightfoot, *Transport Phenomena*, Wiley (1960).

**113**

flows we will discuss in a general manner the way that approximations to the equations of motion are usually made. Some general conclusions concerning inviscid flows will then be presented followed by examples which yield important specific conclusions or insights relative to the behavior of such flows. Examples of solutions for flows with transport as well as inviscid laminar flows will be presented in later chapters.

In the case of turbulent flows, simple turbulence concepts will be introduced and the complications caused by the presence of mixing and thermally neutral or highly exothermic reactions in the flow will be discussed. In addition, recent experimental observations of the nature of relatively simple nonreactive and reactive mixing turbulent flows will be described to lay the foundation for later descriptions of turbulent flows with combustion. None of the current mathematical (i.e., theoretical) descriptions of turbulent reactive flows will be presented in this text because they are complex and, in the main, still very controversial. Hopefully, the treatment that is presented will help the interested reader to understand the extensive current bibliography on this subject.

## 5-2 THE GENERAL EQUATIONS FOR REACTIVE GAS DYNAMICS

The general equations for reactive gas dynamics may be written as follows:

*Overall continuity (conservation of mass)*

$$\frac{D\rho}{Dt} + \rho(\nabla \cdot \mathbf{V}) = 0 \tag{5-1}$$

*Species continuity (or species conservation)*

$$\frac{D\rho_i}{Dt} = -\rho_i(\nabla \cdot \mathbf{V}) - (\nabla \cdot \boldsymbol{j}_i) + \rho M_i \sum_{j=1}^{p} v_{ij} \frac{\partial \lambda_j}{\partial t} \qquad (i = 1, 2, \ldots, s) \tag{5-2}$$

*Conservation of momentum‡*

$$\rho \frac{D\mathbf{V}}{Dt} = \left[ \nabla \cdot \boldsymbol{\pi} \right] + \left\{ \sum_{i=1}^{s} \rho_i \boldsymbol{f}_i \right\} \tag{5-3}$$

*Conservation of energy*

$$\rho \frac{D}{Dt} \{ e + \tfrac{1}{2} \mathbf{V} \cdot \mathbf{V} \} = -(\nabla \cdot \boldsymbol{q}) - \nabla \cdot \left[ \boldsymbol{\pi} \cdot \mathbf{V} \right] + \left\{ \sum_{i=1}^{s} \mathbf{N}_i \cdot \boldsymbol{f}_i \right\} - \left\{ (\mathscr{E} - \mathscr{A}) \right\} \tag{5-4}$$

---

‡ The terms in the large brackets, {}, in this section will never be included in any mathematical treatments in this text even though the results of calculations in which some of them were retained will be discussed.

In order to solve these one must have in addition a thermal equation of state

$$P = P(\rho, T, x_i) \tag{5-5}$$

a caloric equation of state

$$e = e(\rho, T, x_i) \tag{5-6}$$

and the rate equations for the $p$ chemical reactions, $d\lambda_j/dt$ ($j = 1, 2, ..., p$).

The terms used in Eqs. (5-1) to (5-4) are defined as follows: $\mathbf{V}$ is the vector mass average velocity, in meters per second, of the fluid with components $(u, v, w)$ in cartesian coordinates.‡ The operator $\nabla$, del, is $(\partial/\partial x, \partial/\partial y, \partial/\partial z)$ in cartesian coordinates. The total derivative, $D/Dt = \partial/\partial t + u(\partial/\partial x) + v(\partial/\partial y) + w(\partial/\partial z)$ in cartesian coordinates. The $j_i$ are mass fluxes relative to the mass velocity $\mathbf{V}$ with units of kilograms per square meter-second. The stress tensor is $\pi = \tau - P\delta_{ij}$ where $\tau$ is the viscous stress tensor and $\delta_{ij}$ is the unit tensor (or Kronecker delta). The $f_i$ are the body forces acting on the $i$th species. The quantity $q$ represents the energy fluxes relative to the mass average velocity. The quantity $\mathbf{N}_i = \rho_i \mathbf{V}_i + j_i$ and thus represents the net flux of species $i$ in a fixed coordinate system. Finally, $\mathscr{E}$ and $\mathscr{A}$ are the local rates of photon emission and adsorption per unit volume of the gas with units of joules per cubic meter-second. These last two quantities must be included in the energy equation when radiative transport becomes important.

For simplicity we will assume that the thermal equation of state is the ideal gas law

$$P = \frac{\rho RT}{\sum\limits_{i=1}^{s} x_i M_i} = [S]RT \tag{5-7}$$

where $[S]$ is the total species concentration in moles per cubic meter. There are very few combustion problems where this is not true and this assumption greatly reduces the complexity of the conservation equations. The nonideal gas treatment is presented in Bird et al.

The fluxes $j_i$ and $q$ are due to transport in the system and can be written as follows

$$j_i = j_i^{(D)} + \left\{ j_i^{(p)} + j_i^{(f)} \right\} + j_i^{(T)} \tag{5-8}$$

where

$$j_i^{(D)} = \frac{[S]^2}{\rho} \sum\limits_{\substack{j=1 \\ j \neq i}}^{s} M_i M_j D'_{ij} \nabla x_j \quad \text{(ordinary diffusion)} \tag{5-9}$$

---

‡ Cylindrical, spherical, or the curvilinear coordinates $(s, n)$ which are orthogonal along and perpendicular to a stream line are also used in combustion studies. They are not introduced here but may be used in later chapters.

$$\left\{ j_i^{(p)} = \frac{[S]^3}{\rho} \sum_{\substack{j=1 \\ j \neq i}}^{s} M_i M_j D'_{ij} \left[ x_j M_j \left( \frac{1}{\rho_j} - \frac{1}{\rho_i} \right) \frac{\nabla \cdot P}{P} \right] \right\} \quad \text{(pressure diffusion)}$$

$$\left\{ j_i^{(f)} = \frac{[S]^3}{P\rho} \sum_{\substack{j=1 \\ j \neq i}}^{s} M_i M_j D'_{ij} \left[ x_j M_j \left( f_j - \sum_{k=1}^{s} \frac{\rho_k}{\rho} f_k \right) \right] \right\} \quad \text{(forced diffusion)}$$

and

$$j_i^{(T)} = -D_i^T \nabla \ln T \quad \text{(thermal diffusion)} \tag{5-10}$$

where the $D'_{ij}$ and $D_i^T$ in Eqs. (5-9) to (5-12) have the following properties

$$\sum_{i=1}^{s} D'_{ij} = 0 \qquad \sum_{i=1}^{s} D_i^T = 0$$

$$\sum_{i=1}^{s} (M_i M_h D'_{ih} - M_i M_k D'_{ik}) = 0$$

For $s > 2$, $D'_{ij} \neq D'_{ji}$ in general. However, as was indicated in Chap. 3, in practical calculations it is usual to replace $D'_{ij}$ by $D_{ij}$, the ordinary binary diffusion coefficient.

In the same spirit

$$q = q^{(c)} + q^{(D)} + \left\{ q^{(R)} + q^{(T)} \right\} \tag{5-11}$$

where

$$q^{(c)} = -\kappa \nabla T \quad \text{(conduction)} \tag{5-12}$$

$$q^{(D)} = \sum_{i=1}^{s} h_i j_i \quad \text{(diffusion of species)} \tag{5-13}$$

$$\left\{ q^{(R)} \propto \sigma T^4 \right\} \quad \text{(radiative transport)}$$

and $q^{(T)}$ is of such minor importance that it is not even included in Ref. 1. We note, however, that $j_i^{(T)}$ can be important because reactive species will exhibit different concentration profiles in the presence of a thermal gradient and this can affect the local rates of production or destruction of species.

One can also write detailed equations for two-phase flow. The mathematics of this subject is outside the scope of this text and the reader is referred to the bibliography for further information.

## 5-3 APPROXIMATIONS TO THE GENERAL EQUATIONS— LAMINAR FLOW

Even with properly specified boundary and initial conditions it is usually impossible to solve the general equations of motion for laminar flow without making physical and/or mathematical approximations that allow truncation of the equations until they become tractable. Unfortunately, the combustion literature contains an overabundance of analytical and/or numerical treatments in which either or both of two fatal errors were made in the formulation of the equations that were solved. The first of these modeling errors is an incorrect assumption about the driving force or the nature of the phenomena itself. For example, the assumption that heat loss from a strictly one-dimensional premixed flame can lead to a flammability limit or, in a completely different problem, the assumption that all properties of a detonation wave including its structure can be modeled using one-dimensional steady gas dynamic concepts are modeling errors of this type. The second class of errors is caused by either the neglect of one or more terms in one or more of the equations or the use of a physically implausible approximation for the physical properties in one or more terms. This is normally done because it would be impossible to obtain an analytic solution if the term or terms were retained, or left in their more complex and realistic form. This ad hoc neglect or simplification of terms can yield good approximations to the correct answer, but only if the term is indeed found to be small or the simplification appropriate for a good physical reason. On the other hand, such neglect or simplification of terms can and has led to conclusions which violate the fundamental laws of physics.

These problems are similar to the species problem discussed in Chap. 2. There it was pointed out that if a species is not included because of a basic lack of understanding of chemistry, even the application of correct stoichiometric principles will yield incorrect answers. The same principle applies to the application of a truncated set of equations of motion or the use of a physically implausible assumption about properties in a particular combustion problem. In this case the full equations of motion should not be truncated nor should physical property approximations be made until it is obvious from experimental observations of the phenomena that removing that term or making that assumption is indeed physically viable. Also, just as with species assumptions one can solve the equations once with the offending term in place or with the assumption relaxed to see if the truncation or approximation is adequate.

In order to indicate how one should approach the problem of truncating the equations of motion to obtain tractable equations we follow a simple procedure which has evolved since Reynolds time. Brodkey[2] calls this technique "inspection analysis." For our treatment of this technique we first truncate the equations of motion by removing terms for second-order transport and radiation, and for chemical simplicity we assume that the system contains only one reaction, $A \to B$, of order $v$ and that the two species have the same molecular weight and same heat capacity, which may be a function of temperature.

2. R. S. Brodkey, *The Phenomena of Fluid Motions*, Addison-Wesley, Reading, Mass., p. 162 (1967).

With these assumptions Eqs. (5-1) and (5-2) yield two complementary species conservation equations

$$\underbrace{\rho\,\frac{\partial y_i}{\partial t} + \nabla(\rho y_i\,\mathbf{V})}_{1} + \underbrace{\nabla(\rho y_i\,\mathbf{v}_i)}_{2} = \underbrace{\rho M_i v_i\,\frac{\partial \lambda}{\partial t}}_{3} \qquad (i = 1, 2) \qquad (5\text{-}14)$$

where $\lambda$ has units of moles of reaction per kilogram of mixture and the diffusion velocity $\mathbf{v}_i$ is obtained by simplifying Eq. (5-9)

$$\mathbf{v}_i = -D\nabla \ln y_i$$

The momentum equation, (5-3), with these assumptions may be written as

$$\underbrace{\frac{\partial \mathbf{V}}{\partial t} + (\mathbf{V} \cdot \nabla)\mathbf{V}}_{4} = \underbrace{-\frac{1}{\rho}\,\nabla P}_{5} + \underbrace{\frac{1}{3}\frac{\mu}{\rho}\,\nabla(\nabla \cdot \mathbf{V}) + \frac{\mu}{\rho}\,\nabla^2\mathbf{V}}_{6} + \underbrace{\mathbf{g}}_{7} \qquad (5\text{-}15)$$

where $g$ is now simply considered to be the force of gravity. The energy equation, (5-4), with these assumptions plus the definitions $h = e + P/\rho$ and

$$h = \sum_{i=1}^{2} y_i \left[ \Delta h_i^{\circ} + \int_{T_0}^{T} c_{pi}\,dT \right]$$

becomes

$$\underbrace{\frac{\partial}{\partial t}\left(\rho \int_{T_0}^{T} c_p\,dT\right) + \nabla\left(\rho\mathbf{V}\int_{T_0}^{T} c_p\,dT\right)}_{8} + \underbrace{\nabla\left(\rho\mathbf{v}\int_{T_0}^{T} c_p\,dT\right)}_{9}$$

$$\underbrace{-\nabla(\kappa\,\nabla T)}_{10} = \underbrace{\rho \sum_{i=1}^{2} \Delta h_i^{\circ} M_i v_i\,\frac{\partial \lambda}{\partial t}}_{11} \qquad (5\text{-}16)$$

To illustrate the principle of inspection analysis we will derive two familiar ratios and one new dimensionless ratio which are important in discussing fluid-flow problems. In this type of analysis of the equations one asks if one can "nondimensionalize" the equations such that any two terms of one of the equations will be of equal magnitude under grossly different conditions. Consider, for example, terms 4 and 6 in Eq. (5-15). If we pick two bodies of a fixed geometrical shape or some fixed boundary shape we can define a length scaling ratio such that $L_1 = c_1 L_2$ and a scaled time ratio $t_1 = c_2 t_2$. This means that a characteristic reference velocity in the flow will scale as $\mathbf{V}_1 = (c_1/c_2)\mathbf{V}_2$. We must also scale a reference density and viscosity, i.e., $\rho_1 = c_3 \rho_2$ and $\mu_1 = c_4 \mu_2$.

When this is done one finds that the ratio of terms 4 and 6 will remain the same in the (1) and (2) systems only if

$$\frac{c_1}{c_2^2} = \frac{c_4}{c_1 c_2 c_3} \qquad \text{or if} \qquad \frac{c_1^2 c_3}{c_2 c_4} = 1$$

Substituting in the definitions of these ratios yields the relationship

$$\frac{L_1 \mathbf{V}_1 \rho_1}{\mu_1} = \frac{L_2 \mathbf{V}_2 \rho_2}{\mu_2}$$

Thus if this quantity, based on some reference conditions, is a constant in two-flow situations, terms 4 and 6 will have exactly the same ratio in those two flows.

This dimensionless ratio is the Reynolds number, Re, and is based on a characteristic length and a reference velocity, density, and viscosity of the fluid. Since the two terms that were compared represent an inertial force (term 4) and a viscous force (term 6) we have found a parameter which can be used to characterize flow behavior if body forces and pressure forces are negligible. Also, in the absence of viscous forces where $\text{Re} = \infty$ (because $\mu = 0$) we obtain the inviscid equations of motion in which inertial forces interact with pressure forces to determine the nature of the flow. The nature of these "inviscid" flows will be discussed extensively in later sections of this chapter. In the other limit as $\text{Re} \to 0$ the flow becomes a "viscous" flow in which the actual momentum of the flow makes only a minor contribution to the structure of the flow.

In another example we compare terms 4 and 5 of Eq. (5-15). Now we must also define a scaling ratio for pressure, $P_1 = c_5 P_2$. Substituting and again requiring equivalence yields the relation

$$\frac{c_1}{c_2^2} = \frac{c_5}{c_1 c_3} \qquad \text{or} \qquad \frac{c_1^2 c_3}{c_2^2 c_5} = 1$$

This reduces to the expression

$$\frac{\mathbf{V}_1^2}{P_1/\rho_1} = \frac{\mathbf{V}_2^2}{P_2/\rho_2}$$

Recall that the velocity of sound for an ideal gas is given by the expression $a^2 = \gamma P/\rho$. Thus the comparison of terms 4 and 5 in Eq. (5-15) shows that flow Mach number is an important dimensionless ratio when inertial forces cause the fluid to become compressible.

In 1936, Damköhler[3] generalized some earlier observations of Föster and Geib and presented four dimensionless ratios involving the chemical source terms in Eqs. (5-14) and (5-16). For example, in Eq. (5-14), he compared the

3. G. Damköhler, Z. *Electrochem.* **42**:846 (1936).

convective mass flow, term 1, to the chemical source strength, term 3, to define the first Damköhler number

$$Da_I = \frac{(M_i \nu_i / y_i) \frac{\partial \lambda}{\partial t} \cdot L}{V} = \frac{L/V}{\rho y_i / \omega}$$

where $\omega = \rho M_i \nu_i (\partial \lambda / \partial t)$ with units of kilograms per cubic meter-second. This can be considered to be the ratio of a characteristic flow time and a characteristic reaction time, $\tau_f / \tau_r$. Obviously if $Da_I \propto \tau_f / \tau_r \to 0$ the flow is inert and if $\tau_f / \tau_r \to \infty$ the reaction occurs quickly relative to the specified length scale of the problem.

Any two terms in each of the Eqs. (5-14) to (5-16) may be analyzed in a similar manner to obtain dimensionless ratios which are important for scaling flows, both reactive and nonreactive, and for determining limit behaviors. This type of consideration, which is based on the physical observations of the behavior of the flow for various initial and boundary conditions, when properly used will allow proper truncation of the equations before analysis begins. A number of ratios which have been identified as being important in reactive fluid flow are listed in Table 5-1.

In addition to these dimensionless ratios we have already defined the transport ratios in Chap. 3. These are the Prandtl number, the Schmidt number, and the Lewis number. It should be noted that certain of the dimensionless numbers

**Table 5-1 Dimensionless ratios used in reactive flow studies**

| Name | Symbol | Grouping | Eq. | Term | Ratio of |
|------|--------|----------|-----|------|----------|
| Reynolds | Re | $\rho V L / \mu$ | (5-15) | 4 and 6 | Inertial to viscous forces |
| Mach | M | $V/a$ | (5-15) | 4 and 5 | Inertial to pressure forces |
| Froude | Fr | $V^2 / Lg$ | (5-15) | 4 and 7 | Inertial to gravity forces |
| Peclet | Pe | $V L c_p \rho / \kappa$ | (5-16) | 8 and 10 | Convective to conductive heat transfer |
| Peclet, mass transfer | Pe, s | $V L / D$ | (5-14) | 1 and 2 | Convective to conductive mass transport |
| Damköhler 1 | $Da_I$ | $\omega_i L / \rho y_i V$ | (5-14) | 1 and 3 | Mass source from chemistry to convective transport |
| Damköhler 2 | $Da_{II}$ | $\omega_i L^2 / y_i D$ | (5-14) | 2 and 3 | Mass source from chemistry to diffusive transport |
| Damköhler 3 | $Da_{III}$ | $q_{Da} L / h_s V$ | (5-16) | 10 and 11 | Heat source from chemistry to convective heat transport |
| Damköhler 4 | $Da_{IV}$ | $q_{Da} L^2 / \kappa T$ | (5-16) | 8 and 11 | Heat source from chemistry to conductive heat transport |
| Damköhler 5‡ | $Da_V$ | $q_{Da} L^2 / D h_s$ | (5-16) | 9 and 11 | Heat source from chemistry to diffusive heat transport |

Notation: $h_s = \rho \int_{T_0}^{T} c_p \, dT$ J/m³    $\omega_i = \rho M_i \nu_i \frac{\partial \lambda}{\partial t}$ kg/m³-s    $q_{Da} = \sum \Delta h_{f_i}^\circ \omega_i$ J/m³-s.

‡ Not defined by Damköhler.

of Table 5-1 which are derived from flow considerations can be used to yield the transport ratios. For example one can write

$$\text{Pe} = \text{Re} \cdot \text{Pr}, \qquad \text{Pe}, s = \text{Re} \cdot \text{Sc}$$

and

$$\text{Le} = \frac{\text{Pe}, s}{\text{Pe}}$$

It turns out that a significant simplification of the equations of motion can be made if the transport ratios can be assumed to be unity. As was mentioned in Chap. 3, this is a reasonable approximation for most gaseous systems. In the field of nonreactive homogeneous composition gas dynamics the assumption that $\text{Pr} = 1$ leads to the Reynolds analogy for boundary layer flows with heat transfer. In this approximation the equations for momentum transport, Eq. (5-15), and for heat transport, Eq. (5-16), become identical in form and the solution to either also satisfies the other.

As Williams[4] points out in his text, an even more powerful simplification can be made if the flow is reactive and steady and the Lewis number is equal to one. In this case, the left-hand sides of the energy and species conservation equations reduce to the same differential operator and the highly nonlinear source term stands alone on the right-hand sides of the equations. Penner and Williams[4] first called this the Shvab-Zel'dovich formulation of the equations of motion in honor of their pioneering work in the late 1940s. Williams points out that the formulation becomes particularly simple if one assumes, as we have, a single step $A \to B$ irreversible reaction with constant fluid properties and a fixed heat of reaction $q$. In this case, one can subtract either of the properly non-dimensionalized species and temperature or two species equations from each other to remove the highly nonlinear source term. In other words if

$$\mathscr{L}(\beta_1) = \omega$$

$$\mathscr{L}(\beta_T) = \omega$$

we obtain

$$\mathscr{L}(\beta_T - \beta_1) = 0$$

throughout the entire flow field even though the nonlinear reactive terms are still present. This approach is particularly useful in situations where diffusion plays an important role. The application of this type of approximation is discussed extensively by Williams[4] and will be applied in simplified form within the framework of specific reactive flows in later chapters of this text.

Another very powerful technique, called large activation energy asymptotics, has been extensively applied to the analysis of a wide variety of flows

4. F. A. Williams, *Combustion Theory*, Addison-Wesley, Reading, Mass. (1965).

with combustion since the late 1960s. This technique takes advantage of two basic properties of combustion reactions; they are highly exothermic and their overall rate has a very high temperature-dependence. When taken in combination these two properties mean that in many cases the chemistry can be assumed to occur in a very thin region of the flow separated by two mathematical surfaces from the rest of the flow, where only diffusion, thermal conduction, and convection occur. Using this technique, relatively simple but different differential equations can be derived for the "inner" or chemically reactive region and the "outer" or chemically inert region. Then the variable or variables of interest are written as a power series of a natural quantity that is small, such as the inverse dimensionless activation energy, and properties and slopes are matched to determine a solution. The technique has been used to solve many previously intractable problems and results that have been obtained using the technique will be presented in later chapters as appropriate. A detailed example of how the technique is applied is presented in Sec. 8-3 on laminar flame theory.

There are, of course, other approximations which can be used to reduce the equations to a more tractable form. Inviscid flow approximations will be discussed extensively in the next sections because they represent a class of flows that are important in many combustion applications. Certain other simplified approaches will also be used in later sections, as appropriate. In each case that is treated, it will be seen that the approximations that are used are based primarily on physical observations of the phenomenon itself. In many cases the consequences of these assumptions will be discussed. This approach is taken because a complete discussion of mathematical solution techniques is well outside the scope of this text.

This section on approximate techniques would not be complete without mentioning the impact of large digital computers on reactive flow modeling. At the present time large digital computers can be used to solve the general equations with a minimum number of simplifications or assumptions, and these calculations have yielded insights into certain aspects of combustion behaviors that could not be gained using the simplified or truncated equations discussed throughout this text. While results of such detailed calculations will be presented, as appropriate, the treatment of complex computational techniques is outside the scope of this text. The interested reader is referred to reviews of this subject by McDonald[5] and Oran and Boris.[6]

5. H. McDonald, "Combustion Modeling in Two and Three Dimensions—Some Numerical Considerations," *Prog. Energy Comb. Sci.*, **5**: 97–122 (1979).

6. E. S. Oran and J. P. Boris, "Detailed Modeling of Combustion Systems," *Prog. Energy Comb. Sci.*, **7**: 1–72 (1981).

## 5-4 COMPRESSIBLE INVISCID REACTIVE GAS DYNAMICS

There are many situations in combustion where the gradients of velocity, temperature and concentrations of species are so low in major regions of the flow and so concentrated in other regions that the flow can be treated as though it contained large regions where transport is absent, divided, in some cases, by narrow regions which may be considered as surface or wave discontinuities, such as shock waves, heat addition waves, boundary layers, or contact discontinuities. In this and the next five sections we develop some general consequences of the equations of motion for such "inviscid" bulk flows and their associated discontinuities.

With the restriction that transport be absent the equations of Sec. 5-1 reduce to the following conservation equations

$$\frac{D\rho}{Dt} + \rho(\nabla \cdot \mathbf{V}) = 0 \qquad \text{(mass)} \tag{5-17}$$

$$\rho \frac{D\mathbf{V}}{Dt} = -\nabla P \qquad \text{(momentum)} \tag{5-18}$$

$$\rho \frac{De}{Dt} = -P(\nabla \cdot \mathbf{V}) \qquad \text{(energy)} \tag{5-19}$$

$$\rho \frac{Dy_i}{Dt} = \omega_i \qquad (i = 1, 2, \ldots, s) \qquad \text{(species mass balance)} \tag{5-20}$$

where

$$\sum_{i=1}^{s} \frac{Dy_i}{Dt} = 0 \qquad \text{everywhere}$$

and

$$\omega_i = \rho M_i \sum_{i=1}^{p} v_{ij} \frac{\partial \lambda_j}{\partial t} \qquad (i = 1, 2, \ldots, s)$$

the composite rate of production of each species due to $p$ chemical reactions with units of kilograms per cubic meter-second.

Since these equations represent inviscid flow we know that they also represent equilibrium flow for an unreactive or fixed-composition system. Thus Eqs. (5-17) and (5-19) yield the first law of thermodynamics for a conservative system written as a rate equation

$$\frac{De}{Dt} = \frac{P}{\rho^2} \frac{D\rho}{Dt} \tag{5-21}$$

This verifies the statement that

$$\frac{Ds}{Dt} = 0$$

in the unreactive flow case. Implicit in this statement is the assumption that we can define all the thermodynamic properties at every point in the flow. This concept will now be extended to include the reactive case as one which is every-where definable as a thermodynamic system, but which may or may not be in full chemical equilibrium.

The equation for the internal energy of an equilibrium open system which is not in full chemical equilibrium can be written using the chemical potentials $\mu_i$ of the $s$ species

$$de = T \, ds - P d\left(\frac{1}{\rho}\right) + \sum_{i=1}^{s} \left(\frac{\mu_i}{M_i}\right) dy_i \qquad (5\text{-}22)$$

where by definition[7]

$$\mu_i = \left(\frac{\partial E}{\partial n_i}\right)_{S, V, n_j; \, i \neq j} = \left(\frac{\partial H}{\partial n_i}\right)_{S, P, n_j; \, i \neq j} = \left(\frac{\partial A}{\partial n_i}\right)_{T, V, n_j; \, i \neq j} = \left(\frac{\partial F}{\partial n_i}\right)_{T, P, n_j; \, i \neq j}$$

Equation (5-22) is correct either for an open system or for the particular closed system where the $dy_i$'s can vary only along the reaction coordinates given by Eq. (2-17). Combining Eqs. (5-21) and (5-22) we obtain

$$\frac{Ds}{Dt} = -\frac{1}{T} \sum_{i=1}^{N} \left(\frac{\mu_i}{M_i}\right) \frac{Dy_i}{Dt} \qquad (5\text{-}23)$$

which states the entropy will be increased in this thermodynamic system by the occurrence of irreversible chemical reactions. Note that Eq. (5-23) reduces to an isentropic law for either fixed composition (all $dy_i = 0$) or for the occurrence of chemical equilibria throughout the flow [i.e., $\sum \mu_i \, dn_i = 0$ from Eqs. (4-14) and (5-23)].

Based on the above concepts the specifying equation for reactive flow may be derived if we assume that every point in the flow has a thermodynamically definable state and that the flow is isentropic except for the contribution of irreversible chemical reactions. We can then write

$$\rho = \rho(P, s, y_i) \qquad (i = 1, 2, \ldots, N)$$

which allows us to write the equation

$$\frac{D\rho}{Dt} = \left(\frac{\partial \rho}{\partial P}\right)_{s, \, y_i} \frac{DP}{Dt} + \left(\frac{\partial \rho}{\partial s}\right)_{P, \, y_i} \frac{Ds}{Dt} + \sum_{i=1}^{N} \left(\frac{\partial \rho}{\partial y_i}\right)_{P, s, \, y_j} \frac{Dy_i}{Dt}$$

where $(\partial/\partial y_i)_{y_j}$ means that all the $y_j$'s are held constant for $j \neq i$. Therefore

$$\frac{D\rho}{Dt} = \frac{1}{a_f^2} \frac{DP}{Dt} + \left(\frac{\partial \rho}{\partial s}\right)_{P, \, y_i} \left[ \frac{Ds}{Dt} - \sum_{i=1}^{N} \left(\frac{\partial s}{\partial y_i}\right)_{P, \rho, \, y_j} \frac{Dy_i}{Dt} \right]$$

---

7. See for example: K. Denbigh, *The Principles of Chemical Equilibrium*, Cambridge Press (1961).

where $a_f$ is the frozen-composition velocity of sound. This may be written

$$\frac{D\rho}{Dt} = \frac{1}{a_f^2}\frac{DP}{Dt} - T\rho\frac{\beta_f}{c_{pf}}\left[\frac{Ds}{Dt} - \sum_{i=1}^{N}\left(\frac{\partial s}{\partial y_i}\right)_{P,\rho,y_j}\frac{Dy_i}{Dt}\right]$$

where

$$\beta_f = \frac{1}{V}\left(\frac{\partial V}{\partial T}\right)_{P,y_i}$$

is the frozen-composition coefficient of thermal expansion and $c_{pf}$ is the frozen-composition heat capacity. Each term of the sum may be expanded by noting that

$$s = s(P, T, y_i)$$

or

$$ds = \left(\frac{\partial s}{\partial P}\right)_{T,y_i}dP + \left(\frac{\partial s}{\partial T}\right)_{P,y_i}dT + \sum_{i=1}^{N}\left(\frac{\partial s}{\partial y_i}\right)_{P,T,y_j}dy_i$$

which yields

$$\left(\frac{\partial s}{\partial y_i}\right)_{P,\rho,y_j} = \left(\frac{\partial s}{\partial y_i}\right)_{T,P,y_j} + \left(\frac{\partial s}{\partial T}\right)_{P,y_i}\left(\frac{\partial T}{\partial y_i}\right)_{P,\rho,y_j} + \left(\frac{\partial s}{\partial P}\right)_{T,y_i}\left(\frac{\partial P}{\partial y_i}\right)_{P,\rho,y_j}$$

where the last term is zero and the second term becomes

$$\frac{c_{pf}}{\beta_f T\rho}\left(\frac{\partial \rho}{\partial y_i}\right)_{P,T,y_j}$$

Therefore

$$\frac{D\rho}{Dt} = \frac{1}{a_f^2}\frac{DP}{Dt} - \frac{T\rho\beta_f}{c_{pf}}\left[\frac{Ds}{Dt} - \sum_{i=1}^{N}\left\{\left(\frac{\partial s}{\partial y_i}\right)_{T,P,y_j}\frac{Dy_i}{Dt} + \frac{c_{pf}}{\beta_f T\rho}\left(\frac{\partial \rho}{\partial y_i}\right)_{P,T,y_j}\frac{Dy_i}{Dt}\right\}\right]$$

Substituting Eq. (5-23) for $Ds/Dt$ we obtain

$$\frac{D\rho}{Dt} = \frac{1}{a_f^2}\frac{DP}{Dt} + \frac{T\rho\beta_f}{c_{pf}}\sum_{i=1}^{N}\left[\frac{c_{pf}}{\beta_f T\rho}\left(\frac{\partial \rho}{\partial y_i}\right)_{P,T,y_j} + \frac{1}{T}\frac{\mu_i}{M_i} + \left(\frac{\partial s}{\partial y_i}\right)_{P,T,y_j}\right]\frac{Dy_i}{Dt}$$

However, from Eq. (5-22)

$$\left(\frac{\mu_i}{M_i}\right) = \left(\frac{\partial f}{\partial y_i}\right)_{P,T,y_j} = \left(\frac{\partial h}{\partial y_i}\right)_{P,T,y_j} - T\left(\frac{\partial s}{\partial y_i}\right)_{P,T,y_j}$$

Therefore, in its final form the specifying equation becomes

$$\frac{D\rho}{Dt} = \frac{1}{a_f^2}\frac{DP}{Dt} + \rho\sum_{i=1}^{N}\left[\frac{1}{\rho}\left(\frac{\partial \rho}{\partial y_i}\right)_{P,T,y_j} + \frac{\beta_f}{c_{pf}}\left(\frac{\partial h}{\partial y_i}\right)_{P,T,y_j}\right]\frac{Dy_i}{Dt}$$

For an ideal gas mixture $\beta_f = 1/T$ and

$$\frac{D\rho}{Dt} = \frac{1}{a_f^2}\frac{DP}{Dt} + \sum_{i=1}^{N}\left[\left(\frac{\partial \rho}{\partial y_i}\right)_{P,T,y_j} + \frac{\rho}{c_{pf}T}\left(\frac{\partial h}{\partial y_i}\right)_{P,T,y_i}\right]\frac{Dy_i}{Dt} \qquad (5\text{-}24)$$

For the case of no reaction Eq. (5-24) reduces to the well-known isentropic relation for conservative flow

$$\frac{D\rho}{Dt} = \frac{1}{a^2} \frac{DP}{Dt}$$

The extra terms represent respectively a contribution arising from a change in the density of the gas at constant pressure and temperature due to the net production or destruction of species during reaction and a contribution representing the enthalpy change during chemical reaction. If we assume that there is no molecular weight change during chemical reaction, we can represent the chemical reaction by a programmed heat addition along a fluid particle path

$$\frac{Dq}{Dt} = -\sum_{i=1}^{N} \left(\frac{\partial h}{\partial y_i}\right)_{P, T, y_j} \frac{Dy_i}{Dt} \tag{5-25}$$

Equation (5-25) can be written in the form

$$\frac{Dq}{Dt} = -\left(\frac{Dh_{\text{chem}}}{Dt}\right)_{P, T} \tag{5-26}$$

where $h_{\text{chem}}$ is the net change in enthalpy of a fluid element due to a chemical reaction occurring at constant pressure and temperature. Thus $Dq/Dt$ is positive for exothermic and negative for endothermic chemical reactions. The right-hand sides of Eqs. (5-25) and (5-26) represent the real chemical system while the left sides represent the model. Substitution of Eq. (5-25) into (5-24) yields the expression

$$\frac{D\rho}{Dt} = \frac{1}{a^2} \frac{DP}{Dt} - \frac{\rho}{c_p T} \frac{Dq}{Dt}$$

if $c_p$ is assumed to be constant. This assumption causes the gas to be polytropic and we can replace $1/T$ by $\gamma \mathcal{R}/a^2$ to obtain the usual specifying equation for the heat-addition–working-fluid model

$$\frac{D\rho}{Dt} = \frac{1}{a^2} \frac{DP}{Dt} - \rho \frac{(\gamma - 1)}{a^2} \frac{Dq}{Dt} \tag{5-27}$$

We note that Eq. (5-27) states that if heat is added to a fluid element at constant pressure the density will decrease (i.e., the temperature of the working fluid must increase). Thus the more general development of the heat-addition model in this section agrees with the simpler form of the concept that was introduced in Sec. 2-6.

Equations (5-27), (5-17), (5-18), and (5-19) constitute an equation set which has been used extensively for studying combustion systems. Using these equations, thermal energy or heat is assumed to appear spontaneously in the flow from some external source, thus representing the effect of chemical reaction. Accordingly, the apparent entropy of the flow increases markedly when a heat-addition model is used. This artifice of the heat-addition model has no physical significance because the entropy change in the universe must be discussed to determine reversibility or spontaneity, and the model, by its simplification, allows no such discussion. However, the heat-addition model can still be considered a useful approximation for discussing many combustion processes.

We now wish to examine a number of relatively simple reactive compressible inviscid flows in some detail, and include techniques for solving the simplified equations.

## 5-5 STEADY AND STRICTLY ONE-DIMENSIONAL INVISCID FLOW

For the steady flow of an unreactive gas in a constant area stream tube the equations of motion are known to yield a nontrivial solution: the jump conditions for a steady shock wave. Furthermore, for this adiabatic flow a calculation of the entropy difference between stations located on opposite sides of the shock wave adequately justifies our experimental observation of compression shocks, representing transitions from supersonic to subsonic flow, and our inability to observe steady expansion discontinuities.

For the more general case of reactive flow the jump equations may be obtained by integrating Eqs. (5-17), (5-18), and (5-19) for steady one-dimensional flow with the additional stipulation that $h = e + P/\rho$. This operation yields the following equations:

$$\rho_1 u_1 = \rho_2 u_2 \quad \text{(mass)} \tag{5-28}$$

$$P_1 + \rho_1 u_1^2 = P_2 + \rho_2 u_2^2 \quad \text{(momentum)} \tag{5-29}$$

$$h_1 + \tfrac{1}{2}u_1^2 = h_2 + \tfrac{1}{2}u_2^2 \quad \text{(energy)} \tag{5-30}$$

where $h$ is the enthalpy per kilogram and includes chemical enthalpy.

In our analysis of this system of equations let us first determine the final states (2) allowable from any initial state (1) in the pressure-specific volume $(P, v)$ plane. The flow velocity may be written in terms of the initial and final pressures and volumes by rearranging the mass and momentum equations

$$P_1 + \rho_1 u_1^2 = P_2 + \frac{\rho_1^2}{\rho_2} u_1^2$$

to yield

$$u_1^2 = v_1^2 \left( \frac{P_2 - P_1}{v_1 - v_2} \right) \tag{5-31}$$

and similarly

$$u_2^2 = v_2^2 \left( \frac{P_2 - P_1}{v_1 - v_2} \right) \tag{5-32}$$

Notice that these equations state that the mass-flow velocity ($\rho u$) is equal to the square root of the negative slope of a line in the ($P, v$) plane connecting the final state and the initial state for the particular process under consideration. Thus a truly steady transformation from state 1 to state 2 with no transport is restricted to occur through states which lie on a straight line, usually called the *Rayleigh line* in the English literature, and the *Mikhel'son line* in the Russian literature. At this point, without imposing any further limiting assumptions we can additionally see one more limitation to steady-flow processes implicit in Eqs. (5-31) and (5-32). In the $P, v$ plane a steady-flow process cannot simultaneously lower or raise both the pressure and volume; thus there are two restricted areas in this plane which cannot contain end states for any steady-wave process, see Fig. 5-1.

The energy and momentum equations may be rearranged to yield the expressions

$$h_2 - h_1 = \tfrac{1}{2}(u_1^2 - u_2^2)$$

and

$$(P_2 - P_1)\left(\frac{1}{\rho_2} + \frac{1}{\rho_1}\right) = u_1^2 - u_2^2 + \frac{\rho_1}{\rho_2}\,u_1^2 - \frac{\rho_2}{\rho_1}\,u_2^2$$

Substituting, the mass-balance equation yields

$$(P_2 - P_1)\left(\frac{1}{\rho_2} + \frac{1}{\rho_1}\right) = u_1^2 - u_2^2$$

or

$$h_2 - h_1 = \tfrac{1}{2}(P_2 - P_1)(v_1 + v_2) \tag{5-33}$$

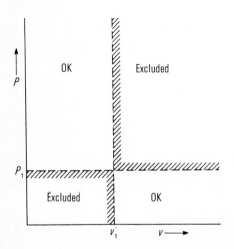

Figure 5-1 Excluded areas for steady flow.

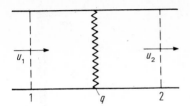

**Figure 5-2** Steady flow with heat addition.

This is called the Rankine–Hugoniot equation or simply the Hugoniot equation for steady flow. Notice that for steady, constant area flow the Hugoniot and Rayleigh line equations are completely independent of the choice of an equation of state or of the process occurring between stations 1 and 2.

To discuss the behavior of these equations in more detail we will now limit ourselves to a constant molecular weight polytropic gas as the working fluid and replace the chemical change by simple heat addition to the working fluid. [See Eq. (2-25) and Fig. 5-2.] With these assumptions,

$$\frac{P_1}{\rho_1 T_1} = \frac{P_2}{\rho_2 T_2} \qquad \text{(state)}$$

$$h_1 = h_1' = c_P T_1$$

and

$$h_2 = h_2' - q = c_P T_2 - q$$

where a positive value of $q$ represents heat addition to the flow. Therefore

$$\frac{\gamma}{\gamma - 1}(P_2 v_2 - P_1 v_1) - q = \frac{1}{2}(P_2 - P_1)(v_2 + v_1) \qquad (5\text{-}34)$$

Equation (5-34) represents the locus of final states $(P_2, v_2)$ for any initial state $(P_1, v_1)$ and heat addition $q$.

Asymptotic values of both $P_2$ and $v_2$ exist for this equation for any value of $P_1$, $v_1$, and $q$. Assume that $P_2$ approaches infinity. Then

$$\frac{\gamma}{\gamma - 1} P_2 v_2 - q = \frac{1}{2}(P_2 v_2 + P_2 v_1)$$

or

$$\frac{2q}{P_2 v_1} = \frac{\gamma + 1}{\gamma - 1} \cdot \frac{v_2}{v_1} - 1$$

Therefore

$$\frac{v_2}{v_1} \to \frac{\gamma - 1}{\gamma + 1} \qquad \text{as} \qquad P_2 \to \infty$$

Similarly, if $v_2$ approaches infinity we see that

$$\frac{\gamma}{\gamma - 1} P_2 v_2 - q = \frac{1}{2} (P_2 v_2 - P_1 v_2)$$

or

$$\frac{2q}{P_1 v_2} = \frac{\gamma + 1}{\gamma - 1} \cdot \frac{P_2}{P_1} + 1$$

Therefore

$$\frac{P_2}{P_1} \rightarrow -\frac{\gamma - 1}{\gamma + 1} \qquad \text{as} \qquad v_2 \rightarrow \infty$$

Since the asymptotic value of $P_2/P_1$ as $v_2$ approaches infinity is negative, the solution curve must cross the $P = 0$ line at some point. The value of $v_2/v_1$ at this intersection may be easily determined by substituting $P_2 = 0$ in the original equation and introducing the velocity of sound $a_1^2 = \gamma P_1 v_1$

$$\frac{\gamma}{\gamma - 1} P_1 v_1 + q = \frac{1}{2} (P_1 v_2 + P_1 v_1)$$

or

$$\frac{2q}{P_1 v_1} = -\frac{\gamma + 1}{\gamma - 1} + \frac{v_2}{v_1}$$

Therefore

$$\frac{v_2}{v_1} = \frac{\gamma + 1}{\gamma - 1} + \frac{2\gamma q}{a_1^2} \qquad \text{at} \qquad P_2 = 0$$

The values of $P_2$ at $v_2 = v_1$ and of $v_2$ at $P_2 = P_1$ may also be determined quite easily. At $v_2 = v_1$

$$\frac{\gamma}{\gamma - 1} v_1 (P_2 - P_1) - q = \frac{1}{2} (P_2 - P_1) 2 v_1$$

Therefore

$$\frac{P_2}{P_1} = 1 + \frac{\gamma(\gamma - 1)q}{a_1^2} \qquad \text{at} \qquad v_2 = v_1$$

At $P_2 = P_1$,

$$\frac{\gamma}{\gamma - 1} P_1 (v_2 - v_1) - q = 0$$

Therefore

$$\frac{v_2}{v_1} = 1 + \frac{(\gamma - 1)q}{a_1^2} \qquad \text{at} \qquad P_2 = P_1$$

We have now determined the value of $P_2$ and $v_2$ on the Rankine–Hugoniot curve at five points in the pressure-volume plane for the working-fluid–heat-addition model. To show the functional form of the Rankine–Hugoniot equation for the case of a constant heat capacity gas with heat addition we substitute the variables

$$\eta = \frac{v_2}{v_1} - \frac{\gamma - 1}{\gamma + 1}$$

$$\xi = \frac{P_2}{P_1} + \frac{\gamma - 1}{\gamma + 1}$$

which transposes the coordinates such that the equation will asymptote the axes in the new coordinate system. After some algebraic manipulation the Rankine–Hugoniot relation is found to be a regular hyperbola in the $(\eta, \xi)$ plane.

$$\eta \xi = \frac{4\gamma}{(\gamma + 1)^2} + \frac{2\gamma q}{a_1^2} \left( \frac{\gamma - 1}{\gamma + 1} \right) \tag{5-35}$$

The results of this analysis are illustrated in Fig. 5-3. An inspection of Eq. (5-31) and the Hugoniot equation (5-35) shows, as before, that there are two excluded

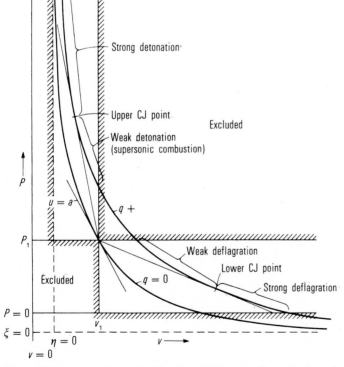

**Figure 5-3** Pressure-volume plot of end states for a one-dimensional steady process, with heat addition indicating excluded regions and upper and lower Chapman–Jouguet (CJ) states.

quadrants in the $(P, v)$ plane corresponding to imaginary wave propagation velocities, and of course, since negative pressures are impossible there is a lower solution limit imposed by the $P_2 = 0$ intersection. Since the isotherms for this gas are represented by hyperbolas which are asymptotic to the axes of the $P, v$ diagram, the Hugoniot curves always cross the isotherms such that increasing $P_2$ along a Hugoniot increases temperature. The $P_2 = 0$, $v_2 = $ constant intersection therefore represents a value of $T_2 = 0$. This particular solution to the equation represents the hypothetical and impossible case where all the random kinetic energy of the molecules has been converted to an ordered velocity in the flow direction.

Notice that if no heat is added the equations reduce to the simple steady-shock equations and there is exactly one solution (i.e., $(P_2, v_2)$ value) for each supersonic incident-flow velocity. However, if heat is added the situation changes markedly. In the velocity region above a limiting subsonic velocity and below a limiting supersonic velocity, there are no possible solutions to the steady equations, while below and above these two critical velocities there are two possible end states for each velocity of propagation. Furthermore, on the subsonic branch, due to the intersection of the Hugoniot with a $P = 0$ curve, very low subsonic-flow velocities yield only one solution. From the earlier relation for the velocity of the flow we see that we can write the expression for the Mach number of the Rayleigh process at stations 1 and 2 as

$$M_1^2 = \frac{v_1}{\gamma P_1}\left(\frac{P_2 - P_1}{v_1 - v_2}\right) = \frac{1}{\gamma}\frac{(y-1)}{(1-\varepsilon)} \tag{5-36}$$

and

$$M_2^2 = \frac{v_2}{\gamma P_2}\left(\frac{P_2 - P_1}{v_1 - v_2}\right) = \frac{\varepsilon}{\gamma y}\frac{(y-1)}{(1-\varepsilon)} \tag{5-37}$$

where

$$\varepsilon = \frac{v_2}{v_1} \quad \text{and} \quad y = \frac{P_2}{P_1}$$

To show that the points of tangency of the Hugoniot and Rayleigh lines in Fig. 5-3 correspond to exactly sonic flow at station 2 we differentiate the Hugoniot equation, (5-34), to determine its slope. This yields

$$\left.\frac{dy}{d\varepsilon}\right|_{\text{Hugoniot}} = -\left\{\frac{[(\gamma + 1)/(\gamma - 1)]y + 1}{[(\gamma + 1)/(\gamma - 1)]\varepsilon - 1}\right\}$$

Since the slope of the Rayleigh line is

$$\left.\frac{dy}{d\varepsilon}\right|_{\text{Rayleigh}} = \frac{y-1}{\varepsilon - 1}$$

we obtain the equation

$$y = \frac{\varepsilon}{(\gamma + 1)\varepsilon - \gamma}$$

as the condition for tangency. However, this may be rearranged to yield

$$1 = \frac{\varepsilon}{\gamma y} \frac{(y - 1)}{(1 - \varepsilon)}$$

which is identical to Eq. (5-37) and therefore proves that $M_2 = 1$ at the upper and lower tangency points. The existence of these unique points for steady flow was discovered around the turn of the century, and this type of flow is called *Chapman–Jouguet* flow, abbreviated CJ.

For any other possible steady flows Fig. 5-3 shows that

$$\left.\frac{dy}{d\varepsilon}\right|_{\text{Hugoniot}} > \left.\frac{dy}{d\varepsilon}\right|_{\text{Rayleigh}}$$

for a strong detonation or weak deflagration, while

$$\left.\frac{dy}{d\varepsilon}\right|_{\text{Hugoniot}} < \left.\frac{dy}{d\varepsilon}\right|_{\text{Rayleigh}}$$

for a weak detonation or strong deflagration.

Thus the equal sign is replaced by an inequality in Eq. (5-37) and we have proved the general statement that

$$M_2 < 1$$

for a strong detonation or weak deflagration, and

$$M_2 > 1$$

for a weak detonation or strong deflagration.

Also note that for a weak shock wave the slope of the Hugoniot is $-\gamma$ and the equation for $M_1$, Eq. (5-36), reduces to $M_1 = 1$ at $y = \varepsilon = 1$. Thus the slope of the Rankine–Hugoniot curve for an infinitesimal adiabatic shock wave is the same as that of an acoustic wave.

We have thus developed a set of equations which show multiple solutions for steady one-dimensional flow with heat addition. Only certain of these steady-flow solutions are observable experimentally and represent truly physical processes. While it is true that any steady one-dimensional flow phenomena which is observed must satisfy these equations, the justification for the physical existence of these particular solutions cannot be made wholly on the basis of an examination of these equations. Indeed, the stability of any of the above solutions can be answered only by investigating their true transient behavior in the neighborhood of these steady solutions. This we will attempt to do when we individually discuss the theory of any specific heat release wave.

For purposes of continuing our discussion of one-dimensional heat addition waves we will assume in the spirit of Fig. 5-2 that weak deflagrations and weak, strong, and CJ detonations are physically observable phenomena and that a constant-volume explosion represents the limit of an infinite wave velocity, and a constant-pressure explosion is the limit of zero wave velocity.

Solving the Hugoniot and Rayleigh equations [(5-34) and (5-36) respectively] to determine the volume and pressure ratios in terms of the approach flow Mach number yields

$$\varepsilon = \frac{u_2}{u_1} = \frac{1 + \gamma M_1^2 \pm \sqrt{(M_1^2 - 1)^2 - 2(\gamma + 1)(\gamma - 1)M_1^2 q/a_1^2}}{(\gamma + 1)M_1^2} \tag{5-38}$$

and

$$y = \frac{1 + \gamma M_1^2 \mp \gamma\sqrt{(M_1^2 - 1)^2 - 2(\gamma + 1)(\gamma - 1)M_1^2 q/a_1^2}}{(\gamma + 1)} \tag{5-39}$$

We note that depending on the value of $q$, the heat-addition parameter, and $M_1$, the approach flow Mach number, these equations have either two, one, or no solutions. For any approach-flow Mach number $M_1$ the value of $q$, which yields only one solution (the CJ solution), is found by setting the quantity under the radical equal to zero. This yields

$$\frac{\gamma - 1}{a_1^2} q_{CJ} = \frac{(M_{CJ}^2 - 1)^2}{2(\gamma + 1)M_{CJ}^2} \tag{5-40}$$

Since this quartic equation is a quadratic in $M_1^2$ it can be solved to yield

$$M_{CJ} = \left[ \left( \frac{\gamma^2 - 1}{a_1^2} q_{CJ} + 1 \right) \pm \sqrt{\left( \frac{\gamma^2 - 1}{a_1^2} q_{CJ} + 1 \right)^2 - 1} \right]^{1/2} \tag{5-41}$$

where the positive sign yields the upper or detonative Chapman–Jouguet point and the minus sign the lower or deflagrative CJ point.

Based on this observation, if $q$ is less than the $q_{CJ}$ of Eq. (5-40) for any approach-flow Mach number the upper sign on the radical in Eqs. (5-38) and (5-39) yields the weak deflagration or detonation solutions while the lower signs represent strong deflagrations or detonations. Also note that if the value of $q$ is greater than dictated by Eq. (5-40) steady flow is physically impossible.

For common fuels mixed with air the equilibrium Hugoniot, Eq. (5-33), when solved using the full equilibrium products obtained by the techniques of Sec. 4-11, yields a remarkably good curve fit to the heat-addition Hugoniot. Values of these curve-fit constants for a number of stoichiometric fuel–air mixtures are listed in Table 5-2. The values of Table 5-2 have been obtained by fitting the real Hugoniot to Eq. (5-35) at the points $P_2 = P_1$ and $V_2 = V_1$. The calculated temperatures for constant-pressure and constant-volume combustion are also listed for both the real Hugoniot and the heat-addition curve fit. Differences in these temperatures are primarily caused by mole number changes during combustion.

## Table 5-2 Hugoniot curve-fit data for various stoichiometric fuel–air mixtures

$T_1 = 298.15$ K $\quad P_1 = 101,325$ Pa $\quad q = q_{CJ} \quad (y + \beta)(\varepsilon - \beta) = c \quad \beta = \dfrac{\gamma - 1}{\gamma + 1} \quad c = \dfrac{4\gamma}{(\gamma + 1)^2} + \dfrac{2\beta\gamma q}{a_1^2}$

| Fuel | Mole % | $a_1$‡ m/s | $\dfrac{q}{MJ/kg}$ fuel | $\dfrac{q}{MJ/kg}$ mix | $\dfrac{q}{a_1^2}$ | $\dfrac{q}{c_v T_1}$ | $\gamma$ | $P = $ const T, K curve-fit | T, K Hugoniot | $V = $ const T, K curve-fit | T, K Hugoniot |
|---|---|---|---|---|---|---|---|---|---|---|---|
| $H_2$ | 29.52 | 375.1 | 152.34 | 4.316 | 30.68 | 7.030 | 1.192 | 2056 | 2385 | 2394 | 2755 |
| $CH_4$ | 9.48 | 327.0 | 64.73 | 3.549 | 33.18 | 7.819 | 1.197 | 2246 | 2229 | 2629 | 2592 |
| $C_2H_6$ | 5.65 | 319.3 | 62.30 | 3.644 | 35.75 | 8.265 | 1.194 | 2362 | 2276 | 2762 | 2641 |
| $C_3H_8$ | 4.02 | 316.3 | 60.86 | 3.649 | 36.48 | 8.407 | 1.193 | 2399 | 2283 | 2805 | 2649 |
| $nC_7H_{16}$ | 1.87 | 312.5 | 58.42 | 3.610 | 36.98 | 8.529 | 1.193 | 2429 | 2278 | 2841 | 2643 |
| $C_2H_4$ | 6.53 | 319.2 | 60.53 | 3.834 | 37.63 | 8.403 | 1.188 | 2407 | 2372 | 2803 | 2738 |
| $C_2H_2$ | 7.73 | 319.5 | 58.93 | 4.127 | 40.43 | 8.779 | 1.183 | 2510 | 2541 | 2916 | 2921 |
| $C_6H_6$ | 2.72 | 311.8 | 52.44 | 3.673 | 37.79 | 8.443 | 1.188 | 2417 | 2352 | 2816 | 2715 |
| $CO$ | 29.52 | 317.7 | 12.46 | 3.592 | 35.59 | 6.975 | 1.168 | 2104 | 2420 | 2411 | 2747 |
| $NH_3$ | 21.83 | 342.4 | 23.49 | 3.314 | 28.26 | 7.777 | 1.221 | 2198 | 2079 | 2617 | 2465 |

‡ Based on the curve-fit value of gamma.

The quantity $q/c_v T_1$ is tabulated because it has special significance in explosion research. It is called the *energy density* because it is simply related to the pressure rise for a constant-volume explosion

$$y = \frac{q}{c_v T_1} + 1 \tag{5-42}$$

and the volume increase for a constant-pressure explosion

$$\varepsilon = \frac{1}{\gamma}\left(\frac{q}{c_v T_1}\right) + 1 \tag{5-43}$$

Finally, we note that the Chapman–Jouguet pressure and volume are obtained when the brackets of Eqs. (5-38) and (5-39) become zero

$$y_{CJ} = \frac{1 + \gamma M_{CJ}^2}{\gamma + 1} \tag{5-44}$$

and

$$\varepsilon_{CJ} = \frac{u_2}{u_{CJ}} = \frac{1 + \gamma M_{CJ}^2}{(\gamma + 1)M_{CJ}^2} \tag{5-45}$$

In all cases, using this model, the temperature rise and velocity of sound change across the wave may be calculated by observing that

$$\frac{a_2}{a_1} = \sqrt{\frac{T_2}{T_1}} = \sqrt{y\varepsilon} \tag{5-46}$$

Individual wave-propagation problems in exothermic systems will be discussed in Secs. 5-7 to 5-9 and in later chapters in considerable detail. There remains one problem, however—the propagation of compression waves in endothermic systems—which does not classify as a combustion process but should be discussed because of its importance. The most common endothermic reactive system is an equilibrium gas mixture which is capable of reacting with a finite rate, or a nonreactive gas which has a finite relaxation time. In either of these cases the system admits to the definition of two distinct and limiting velocities of sound—the frozen and equilibrium velocities of sound

$$a_f^2 = \left(\frac{\partial P}{\partial \rho}\right)_{s,\, \lambda_1, \lambda_2, \lambda_3, \dots, \lambda_N}$$

$$a_{eq}^2 = \left(\frac{\partial P}{\partial \rho}\right)_{s\,\text{(along an equilibrium isentrope)}}$$

For a system with a thermodynamically stable equilibrium state, $a_{eq} < a_f$ because the reaction is always endothermic as the temperature is raised by adiabatic compression. Both of these equations represent isentropic wave-propagation processes because the adiabatic compression process in the wave is assumed to follow the isentropic conditions discussed in Sec. 5-4. The equilibrium velocity of sound may be calculated using the thermodynamic relationship

$$a_{eq}^2 = \frac{-v^2}{(\partial v/\partial P)_T + \dfrac{T}{c_p}(\partial v/\partial T)_P^2}$$

once the state equations for the reacting or relaxing system are known. For the case of a simple dissociating gas obeying Eqs. (4-21), (4-22), and (4-23) we obtain

$$a_{eq} = \left[ \frac{c_{eq} P/\rho}{c_{P_{eq}}\left(1 + \dfrac{\alpha(1-\alpha)}{2}\right) - (1+\alpha)\mathscr{R}\left[1 + \dfrac{\alpha(1-\alpha)}{2(T/\theta_D)}\right]^2} \right]^{1/2}$$

where

$$c_{P_{eq}} = \left[ (4+\alpha) + \left(\frac{\theta_D}{T}\right)^2 \frac{\alpha(1-\alpha^2)}{2} \right]\mathscr{R}$$

Therefore we may evaluate the velocity decrement $(a_f - a_{eq})/a_f$ for a constant-gamma gas using the parameters $T/\theta_D$ and $\mathscr{P}/\mathscr{P}_D$ as dimensionless parameters.

$$\text{Percent decrement} = 100 \times \frac{[P(4+\alpha)/3\rho]^{1/2} - a_{eq}}{[P(4+\alpha)/3\rho]^{1/2}}$$

Figure 5-4 is a plot of the velocity decrement $(a_f - a_{eq})/a_f$ and shows that for a dissociating gas there can be quite a difference between the equilibrium and frozen-sound velocities. If we now modify our steady-flow concepts to discuss the case of an equilibrium gas we find that there are two Rankine–Hugoniot curves through the point $(P_1, V_1)$ on a $P$, $V$ plot; the equilibrium Hugoniot always lying to the left and below the frozen-composition Hugoniot (see Fig. 5-5). Furthermore, we find that a Rayleigh line crosses both Hugoniots at high flow velocities but crosses only the equilibrium Hugoniot for flow velocities which lie between $a_f$ and $a_{eq}$. Experimental behavior representing both these possibilities has been observed (see Fig. 5-6). In the strong shock case vibrational relaxation or finite rate endothermic reactions yield shock profiles which may be adequately represented as a discontinuous shock jump from $O$ to $E$, followed by a continuous relaxation along the Rayleigh line $ED$ to the end state $D$ on the equilibrium Hugoniot. Steady diffuse shock waves are represented by Rayleigh lines which lie between slopes $CO$ and $AO$ in Fig. 5-5. These also represent an observable process. The transition from $O$ to $B$, for example, cannot be abrupt because of the finite rate of the relaxation or reaction process, nor can it be preceded by a high-temperature frozen-composition shock wave because signals can propagate locally at a higher velocity than the overall compression process. However, these signals (traveling at the frozen-sound velocity)

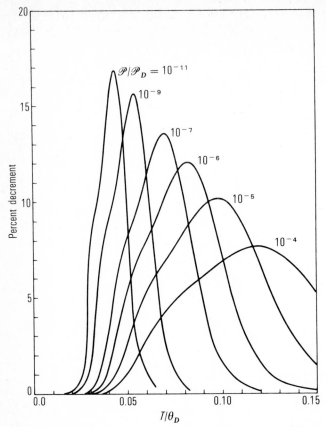

**Figure 5-4** The percent velocity of sound decrement $(a_f - a_{eq}) \cdot 100/a_f$ for an ideal dissociating gas.

attenuate rapidly in a reactive system, and the net effect is the eventual appearance of a diffuse compression wave traveling as a steady phenomenon at an intermediate velocity.

Thus, steady wave propagation through an endothermic reactive system yields two distinct compression processes—a shock followed by a distinguishable reaction zone at high Mach numbers, and at lower Mach numbers a diffuse shock wave with no preceding discontinuity.

## 5-6 QUASI ONE-DIMENSIONAL STEADY INVISCID FLOW

There is one steady inviscid reactive flow which has a considerable interest to the combustion field. This is the rocket-nozzle flow which occurs when a high-temperature equilibrium gas is exhausted through a sonic nozzle and subsequently expanded supersonically to a high velocity and low pressure. For purposes of this section we will treat this flow as a quasi–one-dimensional (variable area) steady flow of a reactive gas containing a single recombination reaction, the recombination of a dissociated Lighthill gas (see Sec. 4-7). For this case Eqs. (5-17) to (5-20) reduce to the following equations:

$$\frac{d\rho}{\rho} + \frac{du}{u} + \frac{dA}{A} = 0 \tag{5-47}$$

$$\rho u \, du + dP = 0 \tag{5-48}$$

$$u \, du + dh = 0 \tag{5-49}$$

and

$$u \frac{d\beta}{dx} = f(P, \, T, \, \beta) \tag{5-50}$$

Equations (5-48) and (5-49) are directly obtainable from Eqs. (5-18) and (5-19). However, the mass-balance equation for quasi–one-dimensional flow must be obtained from the integral form of Eq. (5-17). Equation (5-50) is a general

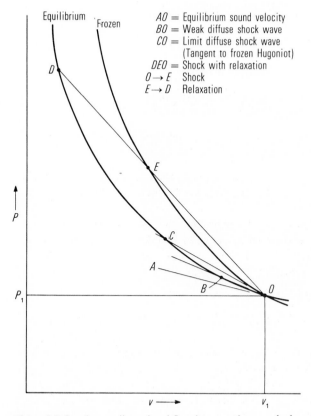

Figure 5-5 Steady one-dimensional flow in a reacting or relaxing gas.

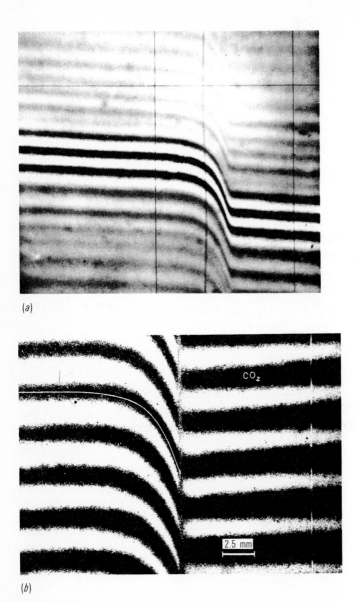

(a)

(b)

**Figure 5-6** Optical interferogram of diffuse and strong shock waves, showing the density profile through the waves. The shock wave is propagating to the right. The white curved line in (b) is the theoretical fringe shift using vibrational relaxation theory. *(Courtesy of Professor Walker Bleakney, Princeton University.)*

expression for the rate of change of the single reaction coordinate with distance. The integration of these equations requires a set of throat conditions, the rate function $f(P, T, \beta)$ and an equation of state $h(P, T, \beta)$. Here $\beta$ is a reaction coordinate based on mass fraction of atoms present in a homonuclear dissociating diatomic gas. Numerically $\beta$ has the same value as the $\alpha$ of Sec. 4-7.

These equations reduce to a set of algebraic equations for the two limiting cases of inert and equilibrium flow. In these cases the flow may be represented by a constant-entropy line on an appropriate Mollier (or $h$ vs. $s$) diagram, and the properties are uniquely determined as a function of initial conditions and the local value of any one of the state or flow variables. In the general case (that is, when the flow is fully reactive) the chemical lags will preclude the occurrence of either full equilibrium or inert flow at any point in the nozzle, and Eqs. (5-47) to (5-50) must be numerically integrated to determine the behavior for any particular nozzle shape. The flow at any point now depends in a complex manner on the previous history, and general conclusions are not easily drawn.

In 1959 Bray[8] pointed out that for most nozzle-flow problems a very simple approximation yields results which agree quite adequately with the result obtained from a full numerical integration of the equations. We summarize his model and conclusions here.

Bray assumed that downstream of the throat the nozzle has an area given by the equation

$$A = A_{\text{throat}} + \pi(\tan^2 \theta)x^2$$

and that the rate law for the recombination of a Lighthill gas could be written in a simplified form (due to Freeman)[9]

$$\frac{d\beta}{dt} = C\rho \left[ (1 - \beta) \exp\left(\frac{-\theta_E}{T}\right) - \frac{\rho}{\rho_D} \beta^2 \right] \tag{5-51}$$

where $C\rho$ has units of $s^{-1}$, $\theta_E$ is the effective characteristic dissociation temperature based on energy and related to $\theta_D$ by the equation

$$\theta_E = \theta_D - T(1 + \alpha)$$

and $\rho_D$ is a characteristic dissociation density which is equivalent to the dissociation pressure introduced in Sec. 4-7. It may be shown in fact that Lighthill's $\rho_D$ and the $\mathscr{P}_D$ of Sec. 4-7 are related by the equation

$$\rho_D = \mathscr{P}_D \left[ \frac{M}{32.82T \exp (1 + \alpha)} \right] \tag{5-52}$$

where $M$ is the molecular weight of the diatomic species and $T$ is the temperature in kelvins. We have already seen that $\mathscr{P}_D$ as determined by a best fit to

8. K. N. C. Bray, *J. Fluid Mech.*, **6**:1 (1959).
9. N. C. Freeman, *J. Fluid Mech.*, **4**:407 (1958).

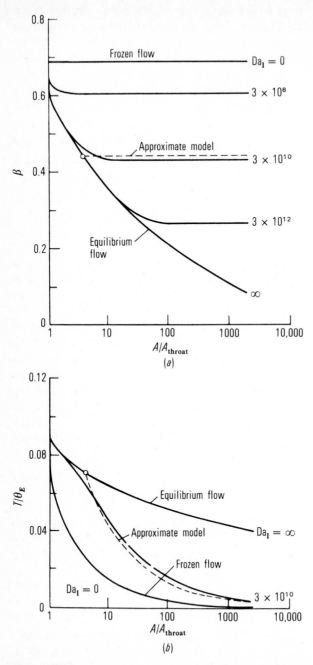

**Figure 5-7** Flow variables in a hyperbolic nozzle. $T_0/\theta_E = 0.1$, $P_0 M/R_{\rho D}\theta_E = 5.0 \times 10^{-6}$. (a) Degree of dissociation, (b) temperature. (*Adapted with permission from W. G. Vincenti and C. H. Kruger, Jr. "Introduction to Physical Gas Dynamics," John Wiley and Sons, New York (1965).*)

the equilibrium constant is not strictly a constant. Lighthill's $\rho_D$ has also been found to be a function of temperature. Thus the temperature-dependence noted in Eq. (5-52) is not too surprising.

Notice that the quantity in the brackets of Eq. (5-51) reduces to the equation

$$\frac{\beta^2}{1 - \beta} = \frac{\rho_D}{\rho} \exp\left(\frac{-\theta_E}{T}\right) \tag{5-53}$$

for $d\beta/dt = 0$ (i.e., for equilibrium). Eq. (5-53) may be obtained from Eq. (4-22) by substituting the equation of state and Eq. (5-52). For this fixed-nozzle geometry and rate equation, Bray found that all nozzle behaviors could be expressed in terms of a single dimensionless factor, a Damköhler number which contains both nozzle-size and rate constant terms

$$\mathrm{Da_I} = \frac{C\rho_D}{2\sqrt{\pi}\,\tan\theta} \sqrt{\frac{A_{\mathrm{throat}}\,M}{\theta_E}}$$

The results of numerical integration through the nozzle for a series of fixed values of $\mathrm{Da_I}$ are shown in Fig. 5-7. Notice that for values of $\mathrm{Da_I}$ which lie between $\mathrm{Da_I} = 0$ (frozen flow) and $\mathrm{Da_I} = \infty$ (equilibrium flow) the flow behaves as though the chemical reactions were in equilibrium near the throat and then suddenly changes to a flow which acts as though the chemical reactions are frozen. Also note that the transition point moves away from the nozzle as the effective reaction rate $\mathrm{Da_I}$ becomes larger. Bray pointed out that to a first approximation the nozzle expansion of a reactive flow with a single reaction coordinate could be handled as equilibrium flow to a point and thereafter as a frozen flow. He states that the reaction freezes primarily because the recombination rate drops very rapidly during the expansion process. Bray postulated that the position of a "freezing line" in the nozzle can be estimated by performing an equilibrium calculation until the condition

$$-\left(u\,\frac{d\beta}{dx}\right)_{\mathrm{eq}} = KC\left[\rho(1 - \beta)\exp\left(\frac{-\theta_E}{T}\right)\right]_{\mathrm{eq}}$$

is satisfied. Here $K$ is an adjustable fitting constant whose value has been found to be near unity. The flow that results from the application of this criterion (with $K = 1$) is shown in Fig. 5-7 as the dashed line for comparison to the case when $\mathrm{Da_I} = 3 \times 10^{10}$.

This criterion allows one to construct a composite Mollier diagram which will allow the prediction of the behavior of the flow of a specific reactive gas through a specific nozzle. In this diagram one constructs a Mollier diagram for equilibrium flow in the high enthalpy region. This region is then bounded by a line representing Bray's criterion for that specific nozzle and chemical system. The portions of the diagram below this line are constructed for the frozen compositions found at the freezing line. This composition varies with the entropy, but the construction is performed along lines of constant entropy for each specific composition.

## 5-7 NONSTEADY ONE-DIMENSIONAL INVISCID FLOW— THE METHOD OF CHARACTERISTICS

Equations (5-18), (5-20), (5-23), and Eqs. (5-17) and (5-20) substituted into Eq. (5-24) reduce to the following set for the case of one-dimensional nonsteady flow

$$\rho \frac{\partial u}{\partial t} + \rho u \frac{\partial u}{\partial x} + \frac{\partial P}{\partial x} = 0 \tag{5-54}$$

$$\frac{\partial y_i}{\partial t} + u \frac{\partial y_i}{\partial x} = \frac{\omega_i}{\rho} \qquad (i = 1, 2, \ldots, N) \tag{5-55}$$

$$\frac{\partial s}{\partial t} + u \frac{\partial s}{\partial x} = -\frac{1}{T} \sum_{i=1}^{N} \left(\frac{\mu_i}{M_i}\right) \frac{Dy_i}{Dt} \tag{5-56}$$

and

$$\rho \frac{\partial u}{\partial x} + \frac{1}{a_f^2} \frac{\partial P}{\partial t} + \frac{u}{a_f^2} \frac{\partial P}{\partial x} = -\sum_{i=1}^{N} \sigma_i \omega_i \tag{5-57}$$

where

$$\sigma_i = \left[ \left(\frac{\partial \rho}{\partial y_i}\right)_{P, T, y_j} + \frac{\rho}{c_{pf} T} \left(\frac{\partial h}{\partial y_i}\right)_{P, T, y_j} \right]$$

Multiplying Eq. (5-44) by $a_f$ and Eq. (5-49) by $a_f^2$ and adding and subtracting yields the equation

$$\frac{\partial P}{\partial t} + (u \pm a_f) \frac{\partial P}{\partial x} \pm \rho a_f \left[ \frac{\partial u}{\partial t} + (u \pm a_f) \frac{\partial u}{\partial x} \right] = -\frac{a_f^2}{\rho} \sum_{i=1}^{N} \sigma_i \omega_i \tag{5-58}$$

The set of Eqs. (5-55), (5-56), and (5-58) constitute the characteristic equations for the analysis of one-dimensional nonsteady flow situations with reaction. Equations (5-55) and (5-56) are $(s + 1)$-fold degenerate equations along a particle path, while Eqs. (5-58) are single characteristic equations along the two directions of sound propagation. The important result of this derivation is the observation that the characteristic directions at any point in the fluid are given by the expression $(u \pm a_f)$ and therefore are determined by the frozen-sound velocity for conservative flow. These equations may be simplified to yield the characteristic set for the heat-addition model, using the approximations described in Sec. 5-4 above. A discussion of the detailed numerical techniques used in solving these equations is beyond the scope of this text but has been described adequately in references listed in the bibliography.

Note that if one removes the chemical source term in the characteristic equations (5-58), the left-hand side of these equations states that a positive pressure gradient across a right-hand propagating characteristic [in which $(u + a_f) = $ constant in $x$-$t$ space] is just matched by the quantity $\rho a_f$ multiplied by a negative velocity gradient. Additionally for a left-hand propagating wave [in which $(u - a_f) = $ constant in $x$-$t$ space] a positive pressure gradient is just matched by $\rho a_f$ times a negative velocity gradient. Also note that the presence of a chemical source term in the flow alters these $P$-$u$ gradient requirements locally. Even in more general cases, this is the primary way in which finite rate combustion chemistry influences flow. It causes a volumetric dilatation which leads to pressure waves propagating away from the region of chemical activity. The quantity $\rho a_f$ is sometimes called the characteristic impedence of the fluid.

When Eqs. (5-17) to (5-23) are written for two-dimensional steady *supersonic* flow they take the same characteristic form as the equations discussed in this section. In this case the Mach angle, defined as $\mu = \sin^{-1}(1/M)$, defines the characteristic direction for wave propagation. Thus, steady supersonic flow may be analyzed in a manner that is completely analogous to the nonsteady one-dimensional analysis discussed above.

## 5-8 NONSTEADY ONE-DIMENSIONAL INVISCID FLOW— FLOW-PATCHING

In certain combustion applications a nonsteady one-dimensional flow can be approximated by assuming that very large, uniform flow regions are separated by constant-velocity discontinuities in which steep gradients occur. These gradients can be contact discontinuities between two dissimilar fluids or propagating deflagration waves, detonation waves, shock waves, or rarefaction waves. Strictly speaking, a rarefaction wave or fan is not a discontinuity because it spreads with time. However, a number of interactions of other waves and/or discontinuities can produce a centered rarefaction fan, and this centered fan can be treated as a discontinuity at its inception.

Usually, within the framework of these assumptions one also assumes that the combustion processes can be modeled using the working-fluid–heat-addition model and that simple velocity transformations (without changes of the state properties of the fluid) can be used to find the solution for any specified initial and boundary conditions.

Given the above constraints on the flow, the momentum equation yields the requirement that the pressure across a contact surface be constant while mass conservation requires that the flow velocity be constant across the same contact surface. Thus the natural plane for determining nonsteady one-dimensional flow-patching requirements is the pressure-flow velocity plane, i.e., the $(P, u)$ plane.

Since we have removed the restriction of steady flow, propagating waves must be considered and, as in the previous section, we find that waves can propagate either to the right $(R)$ or to the left $(L)$. We note that while the pressure change will depend on the type of wave and its velocity, the flow-velocity change will be either positive or negative depending on the direction that the wave is propagating. Furthermore each of the individual waves yields a different unique relationship between the pressure change and the flow-velocity change across the wave in terms of the wave velocity and its direction of propagation relative to the fluid. In the heat-addition–working-fluid model these relations also contain $\gamma$, the effective heat-capacity ratio of the gas, and $q/a_1^2$, the effective heat release in the wave. In this regard the Chapman–Jouguet, or thermally choked waves for heat addition are very special waves. In both these cases the wave-front velocity is unique, being either the subsonic or supersonic thermally choked approach velocity given by the symbol $u_{CJ} = M_{CJ} a_1$.

Furthermore, for a CJ wave, the pressure ratio and velocity change are fixed once $q/a_1^2$ and $\gamma$ are specified. The wave velocity relative to the flow ahead of it is given by Eq. (5-41), the pressure ratio by Eq. (5-44), and the dimensionless velocity change of the fluid across the wave is given by rewriting Eq. (5-45) to yield

$$\frac{\Delta u_{CJ}}{a_1} = \frac{u_2 - u_{CJ}}{(u_{CJ}/M_{CJ})} = \pm \frac{M_{CJ}^2 - 1}{(\gamma + 1)M_{CJ}} \left(\frac{R}{L}\right) \tag{5-59}$$

where the symbol $(R/L)$ refers to the use of the $(+/-)$ value for a right- or left-propagating wave respectively.

A rarefaction fan is also unique because it can propagate through the fluid only at sonic velocity. Thus for any assumed $\gamma$ there is a unique $y$, $\Delta u/a_1$ relationship given by the expression[10]

$$\frac{\Delta u}{a_1} = \mp \frac{2}{\gamma - 1} (1 - y^{(\gamma - 1)/2\gamma}) \left(\frac{R}{L}\right) \tag{5-60}$$

We now consider all possible heat-addition discontinuities which have physical significance, i.e., weak deflagrations and strong or weak detonations. These have a wave-propagation rate which either must be specified or determined from the boundary and initial conditions. This is also true for the case where the discontinuity is a nonreactive shock wave. In all these cases the values of $y$ and $\Delta u/a_1$ are functions of $M_1/a_1$, $q/a_1^2$, and $\gamma$, with $q/a_1^2 = 0$ for compressive shock waves. For those cases where heat is added at the wave discontinuity the expression for $y$ is given by Eq. (5-39) with the condition that $q/a_1^2$ be less than the value obtained from Eq. (5-40). The equation for the dimensionless flow-velocity increment across such a wave is obtained by rewriting Eq. (5-38) to yield

$$\frac{\Delta u}{a_1} = \pm \left[\frac{M_1^2 - 1 \pm \sqrt{(M_1^2 - 1)^2 - 2(\gamma + 1)(\gamma - 1)M_1^2 q/a_1^2}}{(\gamma + 1)M_1}\right]\left(\frac{R}{L}\right) \tag{5-61}$$

where the $\pm$ sign on the square root term has the same meaning as it does in Eq. (5-38) while the $\pm$ sign outside the bracket yields the sign on $\Delta u/a_1$ for the appropriate $(R/L)$ wave type.

10. H. W. Liepman and A. Roshko, *Elements of Gas Dynamics*, Wiley, N.Y. (1957).

Nonreactive shock waves represent a special case in which $M_1 > 1$ and $q = 0$. Note that in this case we obtain the limit behavior for the strong detonation solution, and Eqs. (5-39) and (5-61) reduce to

$$y = \frac{2\gamma M^2 - (\gamma - 1)}{\gamma + 1} \tag{5-62}$$

and

$$\frac{\Delta u}{a_1} = \pm \frac{2(M_1^2 - 1)}{(\gamma + 1)M_1}\left(\frac{R}{L}\right) \tag{5-63}$$

These equations also yield an analytic expression for the dependence of $\Delta u/a_1$ on $y$

$$\frac{\Delta u}{a_1} = \pm \frac{1}{\gamma}(y - 1)\left[\frac{2\gamma/(\gamma + 1)}{y + (\gamma - 1)/(\gamma + 1)}\right]^{1/2}\left(\frac{R}{L}\right) \tag{5-64}$$

Curves for $y$ vs. $\Delta u/a_1$ for all of these cases for a right-hand propagating wave are plotted in Fig. 5-8.

We consider two examples to illustrate how flow-patching can be used to solve various transient-flow problems. In the first case we wish to determine the effective deflagration wave velocity propagating away from the end wall of a tube which would just generate a compressive shock of Mach number 1.4 in the methane–air mixture of Fig. 5-8. See Fig. 5-9a and b for the flow representation in the $(x, t)$ and $(P, u)$ planes. From Eqs. (5-62) and (5-63) we find that $_1y_2 = 2.046$ and $\Delta u/a_1 = 0.624$. Therefore from Table 5-2, $_1\Delta u_2 = 0.624 \times 327 = 204$ m/s. Since $u_1 = 0$, $u_2 = 204$ on Fig. 5.9b. The velocity of sound ratio between regions 1 and 2 is obtained by solving (5-38) and (5-46) to yield $a_2/a_1 = 1.164$. Thus in region 2 the dimensionless flow velocity relative to the deflagration wave is $\Delta u/a_2 = (\Delta u/a_1)/(a_2/a_1) = 0.624/1.164 = 0.536$. Since $u = 0$ at the back wall Fig. 5-8 yields a pressure change across the deflagration wave of $_2y_3 = 0.950$ and Eq. (5-39) may be solved iteratively to yield $M_1 = 0.077$. Since $a_2 = 281$ m/s the deflagration velocity must be 21.6 m/s. The actual observed speed of the wave in the coordinate system of Fig. 5-9a will be $204 + 21.6 = 225.6$ m/s.

For the second example we consider a more complex case in which a contact surface discontinuity is generated by the intersection of two waves. This is illustrated in Fig. 5-9c and d. Consider a deflagration of constant Mach number $M_1 = 0.05$ propagating toward the right from the open end of a tube and assume that this deflagration does not disturb the fluid that is ahead of it. At the same time assume that the right-hand wall of the tube acts as an impulsive piston and generates a shock wave propagating to the left with a Mach number $M = 1.40$. Assume the same methane–air mixture properties as before. For these conditions the properties behind the two waves before their intersection are $y = 0.98$ and $\Delta u = -0.333 \times 327 = -109$ m/s for the deflagration wave and $y = 2.046$ and $\Delta u = -0.624 \times 327 = -204$ m/s for the shock wave. These are shown as regions 2 and 3 and points 2 and 3 respectively in Fig. 5-9c

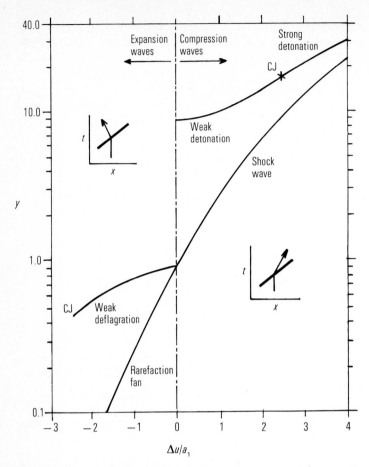

**Figure 5-8** The $y - \Delta u/a_1$ behavior of discontinuous waves. Stoichiometric methane–air modeled using the heat-addition model with constants from Table 5-2. The rarefaction fan and shock wave were also calculated for $\gamma = 1.197$.

and $d$. After the intersection of these two waves the primary shock changes its strength and continues to propagate to the left and the deflagration wave continues to propagate to the right relative to the fluid ahead of it. The intersection also generates a right-propagating wave 3–6 which may be either a shock wave or rarefaction fan depending on the conditions at the intersection point. It is labeled $R$ and is drawn as a rarefaction wave in Fig. 5-9$c$. The possible strengths of this wave are shown in Fig. 5-9$d$ as the line 6′–6–3–6″ where the 6′–6–3 branch is a rarefaction fan and the 3–6″ branch is a shock wave. For any possible strength of this wave, the flame will cause a transformation to the states 5′–5–5″ and the only valid solution occurs at the point 5 where the 5′–5–5″ curve intersects the 2–5–4 curve in Fig. 5-9$d$. This means that the fluid to the right of the contact surface has passed through the states 1–3–6–5 and the fluid to the left through the states 1–2–4 where $y_4 = y_5$ and $u_4 = u_5$. In order to obtain the solution one must calculate for an assumed range of states $6'$–$6''$ and determine the locus of the $5'$–$5''$ line to find the intersection point iteratively. The solution shown in Fig. 5-9c and d, i.e., the one with a right-propagating rarefaction fan, is the one that is observed.

The examples shown in Fig. 5-9 are intended to be illustrative. It should be noted that flow-patching is really a simplification of the method of characteristics technique described in the previous section, and that the local boundary conditions at the discontinuities can be applied in conjunction with the method of characteristics technique if the bulk flow regions are not uniform but can still be approximated as being inviscid.

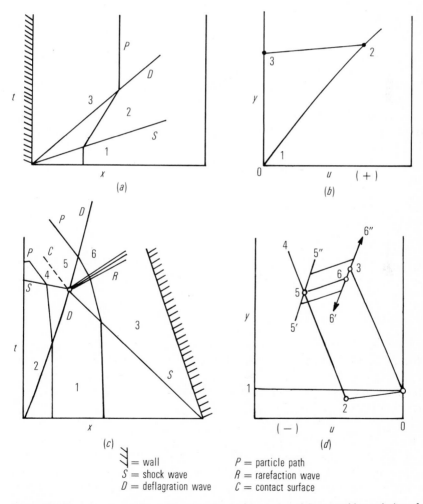

$\quad$ = wall $\qquad$ $P$ = particle path
$S$ = shock wave $\qquad$ $R$ = rarefaction wave
$D$ = deflagration wave $\qquad$ $C$ = contact surface

**Figure 5-9** Wave $(x, t)$ and $(P, u)$ diagrams for two examples of flow-patching solutions for simple one-dimensional nonsteady flows. Not to scale.

As with the method of characteristics, flow-patching techniques can also be applied to a number of supersonic and transonic two-dimensional steady flows that contain large inviscid flow regimes. In this case the contact surface must be a boundary in which the pressure $P$ and the flow-deflection angle $\theta$ are conserved. Thus the natural plane for analysis of 2-D steady flows by flow-patching is the $(P, \theta)$ plane. The application of a $(P, \theta)$ flow-patching analysis for steady 2-D flow is presented in Chap. 9.

## 5-9 NONSTEADY ONE-DIMENSIONAL INVISCID FLOW— ACOUSTICS

In many combustion applications we are interested in studying the growth of instabilities. Under these circumstances we can usually *linearize* the equations of motion by assuming that initially the deviations from some known background state are very small. One typical example of such linearization is the classical field of acoustics. Here, for the simplest case, the background "flow" is assumed to be a quiescent gas with constant state properties in which all fluctuation quantities are small. If the flow is restricted to be one-dimensional nonsteady the basic equations that apply are obtained by rewriting Eqs. (5-17) and (5-18)

$$\frac{\partial \rho}{\partial t} + \rho \frac{\partial u}{\partial x} + u \frac{\partial \rho}{\partial x} = 0 \tag{5-65}$$

$$\frac{\partial u}{\partial t} + u \frac{\partial u}{\partial x} + \frac{1}{\rho} \frac{\partial P}{\partial x} = 0 \tag{5-66}$$

Furthermore, using the heat-addition–working-fluid (constant gamma) model to determine the relationship between pressure and density we obtain

$$a^2 \left[ \frac{\partial \rho}{\partial t} + u \frac{\partial \rho}{\partial x} \right] = \frac{\partial P}{\partial t} + u \frac{\partial P}{\partial x} - \rho(\gamma - 1) \left( \frac{\partial q}{\partial t} + u \frac{\partial q}{\partial x} \right) \tag{5-67}$$

Within the framework of the acoustic approximation we assume that

$$\rho = \rho_0 + \delta\rho, \quad P = P_0 + \delta P$$

where

$$\frac{\delta P}{\gamma P_0} = s \ll 1 \qquad \text{and} \qquad \frac{u}{a_0} \ll 1$$

with

$$a_0^2 = \frac{\gamma P_0}{\rho_0} = \text{const}$$

We note here that the acoustic condensation, $s$, is defined in terms of pressure rather than density because the pressure and flow velocity are the important balance parameters along a particle path in the flow. This yields the equations

$$\frac{\partial(\delta\rho)}{\partial t} + \rho_0 \frac{\partial u}{\partial x} = 0 \tag{5-68}$$

$$\frac{\partial u}{\partial t} + \frac{1}{\rho_0} \frac{\partial(\delta P)}{\partial x} = 0 \tag{5-69}$$

$$\frac{\partial(\delta\rho)}{\partial t} = \frac{1}{a_0^2} \frac{\partial(\delta P)}{\partial t} - \frac{\rho_0}{a_0^2} (\gamma - 1) \frac{\partial q}{\partial t} \tag{5-70}$$

Cross-differentiation of Eqs. (5-68) and (5-69) and substitution of (5-70) then yields the acoustic equation in terms of the dimensionless pressure

$$\frac{\partial^2 s}{\partial t^2} - a_0^2 \frac{\partial^2 s}{\partial x^2} = \frac{\partial^2}{\partial t^2} \left( \frac{q}{c_p T_0} \right) \tag{5-71}$$

Let us now assume that $q \equiv 0$, that is, that the fluid is unreactive. With this assumption we can also cross-differentiate Eqs. (5-68) and (5-69) in the other direction to obtain

$$\frac{\partial^2 u}{\partial t^2} - a_0^2 \frac{\partial^2 u}{\partial x^2} = 0$$

This equation has a solution of the form

$$s(x, t) = f(x - a_0 t) + g(x + a_0 t)$$

which represents waves of *permanent* shape and amplitude propagating to the right (if $g = 0$) and to the left (if $f = 0$). For either of these cases (i.e., when only a right- or left-hand propagating wave is present) the waves are said to be *simple* and either of Eqs. (5-68) or (5-69) yields the relationship

$$s(x, t) = \pm \frac{u(x, t)}{a_0} \left( \frac{R}{L} \right) \tag{5-72}$$

The waves that are predicted by these equations have a permanent amplitude and shape because the linearized treatment does not include dissipative terms and assumes that $a \equiv a_0$. If dissipation were included the amplitude of the waves would decrease with time and if the effect of isentropic compression on the velocity of sound were included the wave shape would distort with time.

If wave propagation occurs in a tube of length $L$, we also have the possibility of observing standing waves at the longitudinal resonant frequencies of the gas in the tube. The simplest boundary conditions for this case are

(*a*) at an open end $s = 0$ and $\partial s / \partial t = 0$ for all time
(*b*) at a closed end $u = 0$ and $\partial u / \partial t = 0$ for all time

Using Eqs. (5-68) and (5-69) respectively, these boundary conditions also mean that $\partial u/\partial x = 0$ at an open end and $\partial s/\partial x = 0$ at a closed end of the tube. Note also that the closed end requirements are quite rigorous while the open end requirements may not occur exactly at the end of the tube. For these boundary conditions Eq. (5-71) or (5-72) are separable and yield a coupled $(u, s)$ field in the tube that represents a standing acoustic resonant mode of the tube. If the condensation at $t = 0$ is taken to be zero and both ends are open at $x = 0$ and $L$ the coupled $(u, s)$ field is given by the equations

$$s = \sum_{n=1}^{\infty} C_n \sin \frac{n\pi a_0 t}{L} \sin \frac{n\pi x}{L} \tag{5-73}$$

and

$$\frac{u}{a_0} = \sum_{n=1}^{\infty} C_n \cos \frac{n\pi a_0 t}{L} \cos \frac{n\pi x}{L} \tag{5-74}$$

where the $C_n$ are positive or negative numbers that represent the initial amplitudes of the dimensionless flow velocity and condensation for each resonant frequency. As with traveling acoustic waves there is no dissipation or distortion in this model. The coupled nature of the $(u, s)$ field is shown in Fig. 5-10. The resonant frequencies are given by the expression

$$f = \frac{n a_0}{2L}$$

for a tube which has either both ends open or both ends closed.

Returning now to the case of heat addition we note that if we assume

$$\int_{\text{cycle}} \frac{\partial q}{\partial t} \, dt = 0$$

the background conditions become constant. This assumption greatly simplifies the problem. Toong et al.[11] has experimentally observed and analyzed the case when the gas is undergoing an exothermic reaction while acoustic waves are present. We also note that if $\partial q/\partial t$ is cyclic at a resonant frequency of the tube the amplitude will increase linearly with time. This is the case for a forced oscillation where the forcing function is external to the system, and it produces a linear amplification of the mode that is being driven.

We now assume the simplest case relative to combustion-driven instability. Assume that homogeneously throughout the tube the rate of heat addition is proportional to the perturbation quantity $s$, that is, that

$$\frac{1}{c_p T_0} \left( \frac{\partial q}{\partial t} \right) = ks$$

11. T. Y. Toong, P. Arbeau, C. A. Garris, and J. P. Paturean. "Acoustic kinetic interactions in an irreversibly reacting medium," Fifteenth Symposium (International) on Combustion, The Combustion Institute, Pittsburgh, Pa, pp. 87–100 (1975).

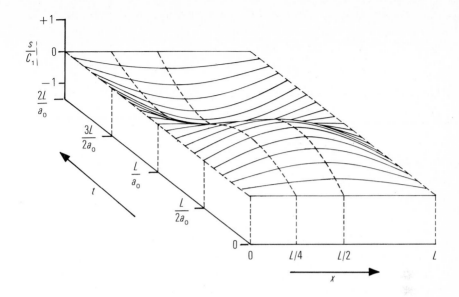

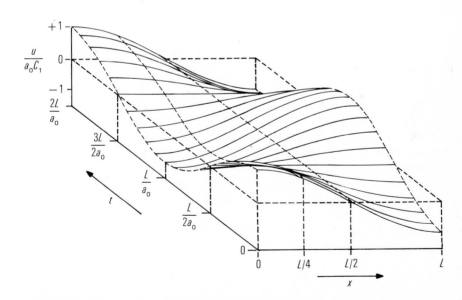

**Figure 5-10** A plot of the coupled pressure-flow velocity field for the first organ pipe mode of a pipe ($C_2 = C_3 = \cdots = C_\infty = 0$) with both ends open. Also at time $t = 0$ the acoustic pressure is arbitrarily set equal to zero. Note that initially the flow is inward and the pressure is rising. It reaches its maximum value at $t = L/2a_0$, at which time all flow has stopped. Then the flow reverses and continues its outward movement until $t = 3L/2a_0$, at which time the pressure at the center of the pipe has reached its minimum value. Also note that the exact nature of the $s - u/a_0$ coupling with time depends on the location in the pipe.

where $k$ is a constant. With this assumption Eq. (5-71) becomes

$$\frac{\partial^2 s}{\partial t^2} - a_0^2 \frac{\partial^2 s}{\partial x^2} = k \frac{\partial s}{\partial t} \tag{5-75}$$

Equation (5-75) has a solution of the form

$$s = \text{Re}\left\{ \sum_{n=1}^{\infty} C_n \exp\left[n(\Lambda t + i\omega x)\right] \right\} \tag{5-76}$$

where $C_n$ and $\Lambda$ may be imaginary and $\omega$ is real. We choose this form because we note that Eq. (5-75) is separable and has a strictly harmonic solution in $x$. Substituting Eq. (5-76) into Eq. (5-75) for a particular value of $n$ and solving for $\Lambda$ yields

$$\Lambda = \frac{k}{2n} \pm \sqrt{\frac{k^2}{4n^2} - a_0^2 \omega^2}$$

If we assume that $k/2n \ll a_0 \omega$ (this is a good assumption for weak coupling) we obtain

$$\Lambda = \frac{k}{2n} \pm i\omega a_0$$

Therefore the solution is of the form

$$s = \exp\left(\frac{kt}{2}\right) \cdot \text{Re}\left[ \sum_{n=1}^{\infty} \{C_{n_+} \exp\left[in\omega\,(x + a_0 t)\right] + C_{n_-} \exp\left[in\omega\,(x - a_0 t)\right]\} \right] \tag{5-77}$$

Without resorting to any further simplifications, we note that if $k$ is a positive quantity this solution states that $s$ will grow exponentially with time, and under this condition the system will be dynamically unstable to any small-amplitude perturbation. Thus a positive value of $k$ for this assumed functional dependence of the heat-release rate represents a positive feedback mechanism for the system. Equation (5-77) may now be solved for any set of boundary and initial conditions. We note in particular that if we have a tube with both ends open, the real part of Eq. (5-77) becomes

$$s = \exp\left(\frac{kt}{2}\right) \sum_{n=1}^{\infty} \sin \frac{n\pi x}{L} \cdot \left( A_n \sin \frac{n\pi a_0 t}{L} + B_n \cos \frac{n\pi a_0 t}{L} \right) \tag{5-78}$$

where the coefficients $A_n$ and $B_n$ are functions of the initial conditions. This solution is harmonic in $x$ and quasi-harmonic in $t$, and it predicts the exponential growth (at the same rate) of all the standing acoustic modes of the tube. However, Eq. (5-78) may be applied only to the beginning phases of the growth process because only in these phases are the acoustic assumptions valid.

The application of these linearization principles to the analysis of combustion-driven instabilities will be discussed further in Chap. 14, and other linearization applications will be discussed and/or described in other portions of the text.

## 5-10 TURBULENCE IN A HOMOGENEOUS FLUID

Turbulent flow occurs when the boundary and initial conditions that are characteristic of the flow lead to the spontaneous growth of hydrodynamic instabilities which eventually decay to yield a random statistically fluctuating fluid motion. The production of turbulent flow is associated primarily with the occurrence of strong shear regions, wakes, boundary layers, jet edges, separated flow regions, or buoyancy-driven flows. The major properties of the turbulence that is produced in this manner have been summarized in a very succinct way by Tennekes and Lumley.[12] Their introductory statements, when paraphrased and extended, yield seven major observations about the nature of simple turbulent flows. These are:

1. Turbulence is a feature of fluid flow. It is not a property of the fluid per se.
2. The major characteristics of turbulence are not controlled by molecular properties of the fluid and even the smallest turbulent scales are orders of magnitude larger than molecular scales.
3. Turbulence always contains three-dimensional random motions, characterized by high levels of vorticity fluctuations. The term "chaotic" is often used to describe fully turbulent flow.
4. Turbulence is characterized by high levels of momentum, heat, and mass transport due to turbulent diffusivity. The last two are only important if there is an initial thermal or concentration gradient in the flow or if the flow is reactive. The implications of this will be discussed in the next sections and in Chaps. 11 and 14.
5. If energy is not continuously supplied to turbulent flow the turbulence will decay and the flow will again become laminar.
6. Turbulence appears spontaneously in regions of high shear. Empirically it has been observed that the local Reynolds number $Re = VL/v$, based on some characteristic length and a reference velocity, determines the regions in which laminar, intermittent, or fully turbulent flow occur. Specifically, for low values of Re the flow will be fully laminar and for high values fully turbulent, while in some intermediate range the flow is observed to be sporadically unstable. Furthermore, these critical transition values of Re depend on the geometry of the device or flow that is generating the shear.

12. H. Tennekes and J. L. Lumley, *A First Course in Turbulence*, MIT Press, Cambridge, Mass. (1972).

7. In the transition regions and at slightly higher Reynolds numbers turbulence is primarily initiated by the formation of two-dimensional vortices which then break down to a fully three-dimensional turbulent flow. It is in fact true that in many "turbulent" flows, regions of highly organized vortical structures can be identified and are the source of the "turbulent" energy in the flow.

   With this background, consider a homogeneous fluid (i.e., one containing no concentration or thermal gradients) which contains well-developed turbulence and is moving at some uniform mean velocity $U$ relative to an observer. An example of such a flow would be the flow downstream of a coarse grid of wires. It is well known that in this case the ordered wake structure produced when the Reynolds number, based on wire diameter, is greater than 60 breaks down to form a fully turbulent flow at distances greater than about 20 wire diameters downstream of the grid. In this "turbulent" region the local flow velocity will have an instantaneous value given by $(U + u, v, w)$ where $U$ is constant and $u$, $v$, and $w$ are fluctuating with time but have the property that their mean values $\bar{u} = \bar{v} = \bar{w} = 0$, where the bar over a quantity represents time average, i.e.,

$$\bar{u} = \frac{1}{t_0} \int_0^{t_0} u \, dt$$

However, the root mean square of the velocity fluctuations are *not* zero and allow one to define the dimensionless intensity of turbulence for each cartesian component of velocity

$$I_x = \frac{\sqrt{\overline{u^2}}}{U}$$

$$I_y = \frac{\sqrt{\overline{v^2}}}{U} \tag{5-79}$$

$$I_z = \frac{\sqrt{\overline{w^2}}}{U}$$

In general these intensities are not equal. If they are the turbulence is said to be isotropic. It has been found that the turbulence that occurs at least 20 mesh diameters downstream of a grid (at Re > 60) and at the centerline of a well-developed turbulent pipe flow (i.e., at Re > 40,000 based on pipe diameter) is isotropic.

   If the turbulence is isotropic one can also define two unique scales of turbulence. These are determined from the correlation between the velocity fluctuations at two neighboring points in the flow, relative to (0, 0, 0)

$$g(y) = \frac{\overline{u(0) \cdot u(y)}}{I_x(0) \cdot I_x(y) \cdot U^2} \qquad (x = z = 0)$$

or

$$f(x) = \frac{\overline{u(0) \cdot u(x)}}{I_x(0) \cdot I_x(x) \cdot U^2} \qquad (y = z = 0)$$

where the bars again refer to time-average values. Here both $g(y)$ and $f(x)$ approach unity as either $y$ or $x$ approach zero. For a random field $g(y)$ and $f(x)$ could be expected to approach zero as either $y$ or $x$ approach infinity. However, we note that these definitions allow for the occurrence of a negative correlation at some separation if either of the products $\overline{u(0) \cdot u(y)}$ or $\overline{u(0) \cdot u(x)}$ is negative at any $x$ or $y$ value.

The theory of isotropic turbulence (as summarized by Hinze,[13] for example) may be used to show that $g(y) = g(z)$, and that for small values of either $y$ or $x$ both $g(y)$ and $f(x)$ decrease with the coordinate separation in a manner given by the quadratic functions

$$g(y) \cong 1 - \frac{y^2}{\lambda_g^2}$$

and

$$f(x) \cong 1 - \frac{x^2}{\lambda_f^2}$$

where $\lambda_g$ and $\lambda_f$ are called the transverse and longitudinal microscales or Taylor scales of isotropic turbulence. In addition, it may be shown that for the case of isotropic turbulence

$$\lambda_f = \sqrt{2}\,\lambda_g$$

Another scale, the macro or integral scale of turbulence, is defined with the equations

$$\Lambda_g = \int_0^\infty g(y)\,dy \qquad \Lambda_f = \int_0^\infty f(x)\,dx$$

where $\Lambda$ represents the base length of a rectangle of unit height which has the same area as that under the correlation curve. A schematic representation of a correlation coefficient $g(y)$ vs. $y$ and the defined correlation lengths $\lambda_g$, $\lambda_f$, and $\Lambda_g$ are shown in Fig. 5-11.

The above description of the intensity and scale of isotropic turbulence allows one to characterize the turbulence with only two numbers. A more complete understanding of the nature of isotropic turbulence can be obtained by characterizing its energy spectrum $E(\omega, t)$ where $\omega$ is the wave-number vector

13. J. O. Hinze, *Turbulence*, McGraw-Hill, New York, 1962.

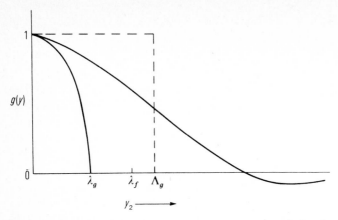

$y_2 \longrightarrow$

**Figure 5-11** The definition of various scales in an isotropic turbulent flow.

equal to $2\pi f/U$ and the spectrum is normalized to the mean square of the velocity fluctuations

$$\int_0^\infty E(\omega, t)\, d\omega = \tfrac{3}{2}\overline{u^2} \tag{5-80}$$

Thus, the quantity $2E(\omega, t)\, d\omega/3\overline{u^2}$ is the probability density function for the turbulent energy in isotropic turbulent flow. It should be noted that this approach is analogous to the kinetic theory analysis of the velocity distribution of random molecular motions, which yielded Maxwell's molecular velocity distribution function. However it must also be noted that this approach can be expected to be much more limited when applied to turbulence, because turbulence is always generated by a nonrandom (or not completely random) input due to the flow configuration. No such limitation exists in kinetic theory and the approach there can be purely stochastic. A typical 3-D spectrum for simple turbulence is shown schematically in Fig. 5-12.

While a detailed discussion of the implications of Fig. 5-12 relative to the behavior of even homogeneous isotropic turbulence is outside the scope of this text, some general observations can be made. First, eddies are formed as large wavelength (low frequency) disturbances in the flow, and then decay and interact to produce smaller and effectively higher and higher frequency shear regions. The majority of the kinetic energy of the turbulent fluctuations occurs in the low-frequency range, usually over an eddy size range that is somewhat dependent on geometric characteristics of the flow. These energy-containing eddies are those that define the "scales" of the turbulence that were introduced earlier. As the turbulence decays, smaller and smaller eddies are formed, and any well-developed turbulent flow has been found to have a detailed structure and decay behavior in the higher-frequency range which is independent of the initial conditions of formation. This "equilibrium" or "universal" region contains two subranges, the inertial and viscous subranges. At the lower frequencies of the

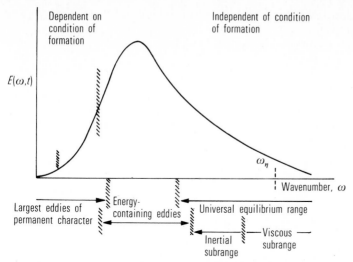

**Figure 5-12** A schematic representation of the energy spectrum for fully developed homogeneous isotropic turbulence. *(Adapted from Hinze,[13] with permission from McGraw-Hill, New York.)*

equilibrium subrange the local shear in the eddies is insufficient to cause significant viscous dissipation and the rate of energy loss is primarily due to the
breakup of these eddies to form smaller ones. When the eddy size (or the
reciprocal of the wave number of the spectrum, Fig. 5-12) reaches a critical
small dimension, the effect of viscosity starts to play a significant role in the
dissipation. The point at which this happens is a function of the total energy
stored in the turbulence associated with the flow and thus the total rate of
energy supply to the equilibrium range from the turbulence source. Following
Hinze's description (Ref. 13), if we set the rate of energy supply to the equilibrium range as

$$\varepsilon(t) = + \frac{3}{2} \frac{d(u)^2}{dt}$$

we find that we can define a length and velocity scale which are dependent only
on $\varepsilon(t)$ and $v$, the kinematic viscosity. These are

$$\eta = \left(\frac{v^3}{\varepsilon}\right)^{1/4}$$

and

$$v = (v\varepsilon)^{1/4}$$

where $\eta$ is called the Kolmogorov length scale of the turbulence. Usually the
dividing line between the inertial and viscous subranges is taken to be the wave
number at which

$$\frac{v\eta}{v} = \text{Re}_T \cong 1$$

In other words a "turbulent" Reynolds number, defined in terms of the rate of transport of energy to the equilibrium range from the eddy-forming range yields the significant length scale for viscous dissipation. This scale is generally very much smaller than the other scales defined earlier. However, it is still considerably larger than the molecular mean free path of the gas.

In the case of turbulence in a homogeneous fluid the only major gross consequence of the turbulent motion is the fact that the presence of turbulence markedly alters the rate of momentum transport through the fluid and to the boundaries of the flow. The classical way of handling this problem is to rewrite the Navier–Stokes equation (5-3) in terms of time-averaged values and averages of the fluctuating quantities. For example, the equation for the $x$ component of momentum for a turbulent flow in which the average vector velocity $\mathbf{V} = U$ may be written as

$$\frac{\partial U}{\partial t} + U\frac{\partial U}{\partial x} = -\frac{1}{\rho}\frac{\partial \bar{P}}{\partial x} + \nu\left[\frac{\partial^2 U}{\partial x^2} + \frac{\partial^2 U}{\partial y^2} + \frac{\partial^2 U}{\partial z^2}\right] - \left[\frac{\partial \overline{u^2}}{\partial x} + \frac{\partial \overline{uv}}{\partial y} + \frac{\partial \overline{uw}}{\partial z}\right]$$

where $\bar{P}$ is the time-average pressure. Similar equations may be written for the $y$ and $z$ components of momentum. In fact, in general, one can define the Reynolds stress tensor for turbulence as

$$
\begin{array}{ccc}
-\rho\overline{uu} & -\rho\overline{uv} & -\rho\overline{uw} \\[6pt]
-\rho\overline{vu} & -\rho\overline{vv} & -\rho\overline{vw} \\[6pt]
-\rho\overline{wu} & -\rho\overline{wv} & -\rho\overline{ww}
\end{array}
$$

Since this tensor is symmetric it contains six independent components. In this approach the six new terms in the equation are turbulence-related convective stress terms which arise from the averaging process. However, in order to solve the equations additional relations are needed between the coefficients of the convective stress terms and the description of the mean velocity field. This "closure" problem has been attacked for many decades and is still found to be essentially intractable in any general way even though useful results are obtained for specific cases. In the early decades of this century a simplification was made when the turbulent stresses were connected to the mean flow gradients by defining an "eddy" viscosity or characteristic mixing length whose value essentially replaces ordinary viscosity in the transport of momentum. The major impact of this work was the discovery of the "universal" nature of high Reynolds number isotropic turbulence in homogeneous fluids which was briefly outlined above. The interested reader is referred to the bibliography as a guide to the large volume of work that has been reported on experimental and theoretical studies of turbulence in a homogeneous fluid.

# 5-11  NONREACTIVE TURBULENT MIXING

Turbulent flow and turbulence descriptions become more complex when the fluid that is being excited into turbulent motion is initially inhomogeneous because it contains either significant thermal gradients or significant concentration gradients. Under these circumstances the turbulence can transport not only momentum but also the scalar quantities of temperature and/or species concentration.

The simpler of the two cases is when only thermal gradients exist in the incident flow. This is because, for this case, one is usually interested only in gross heat transport and not fluctuations. Thus within the spirit of the concept of an eddy diffusivity, Reynolds analogy may be used to discuss these heat fluxes by assuming that the turbulent Prandtl number $\Pr_T = v_T/\alpha_T$ is approximately equal to one. With this approximation energy transport and momentum transport in the turbulent flow can be assumed to be essentially equivalent.

When concentration gradients are initially present in the flow the situation becomes more complex. If we are simply interested in time-average species concentrations the assumption that the turbulent Schmidt number $\mathrm{Sc}_T = v_T/D_T$ is approximately equal to one is as adequate as any model based on eddy diffusivity. However, in most cases, when the mixture is initially unmixed we are primarily interested in learning when mixing has proceeded to the molecular level and this type of information cannot be obtained by the application of eddy diffusivity concepts. Because of this the concept of intensity and scale of segregation have evolved. Their definitions are quite analogous to the definitions of intensity and scale of turbulence based on fluctuating velocity components. Specifically when a binary mixture of species $A$ and $B$ are mixing the instantaneous value of $x_A$ and $x_B$ will fluctuate with time even though $x_A + x_B = 1$ at all points in the fluid. If we define $\bar{x}_A$ and $\bar{x}_B$ as the time-average mole fractions of $A$ and $B$, we can define an intensity of segregation $I_s$ as

$$I_s = \frac{\overline{a'^2}}{\bar{x}_A \cdot \bar{x}_B} \tag{5-81}$$

where

$$a' = x_A - \bar{x}_A \tag{5-82}$$

The quantity $I_s$ has the property that it ranges from zero to unity as the degree of molecular mixing ranges from completely mixed to completely unmixed. Also note that the intensity of segregation (or the degree of unmixedness to the molecular level) can only be determined by measuring time-resolved concentrations in the flow. Average concentrations, which can be determined using extensions of the eddy-viscosity approach that was discussed above, can never yield information on the intensity of segregation.

We may also define a scale of segregation by writing the correlation between the instantaneous concentrations at two points as

$$g_s(r) = \frac{\overline{a'(r_0) \cdot a'(r_0 + r)}}{\overline{a'^2}}$$

Based on this definition we can define either a linear scale of segregation

$$L_s = \int_0^\infty g_s(r) \, dr$$

or a volumetric scale of segregation

$$V_s = \int_0^\infty 2\pi r^2 g_s(r) \, dr$$

These are indicative of the average eddy size that contains different concentrations.

As in the case of turbulence in a homogeneous fluid we have now defined two numbers—the intensity and scale of segregation—which can be used to characterize the local state of completeness of mixing in a turbulent mixing situation. However, just as in the case of the intensity and scale of the velocity fluctuations for turbulence in a homogeneous fluid, these two terms are really not fully descriptive of the processes that are occurring. Note that in a homogeneous fluid this is not too serious a problem because we are only interested in gross transport. Nevertheless, in a mixing situation details of the mixing process can be important and cannot be adequately handled by the simpler concepts derived above.

Recently it has become possible to study the stochastic structure of turbulent mixing flows; in other words to simultaneously observe time-resolved values of the flow velocity components and species concentration at a point in the flow. Because of this, probability density functions (pdf's) are gaining wide use in the description of turbulent flows with mixing and there has been a surge of theoretical effort to interpret the observed behaviors using stochastic techniques. Figure 5-13 contains a schematic of the pdf's of the dimensionless mole fraction of the species for three different turbulent mixing situations showing how they change as the observation point is moved. The simplest case is represented in Fig. 5-13c where high-intensity mixing causes a relatively average concentration to exist throughout the region of interest. In this case the pdf's are essentially gaussian and decay with distance from the entrance plane. In the other two cases the distributions are definitely not gaussian and can contain a Dirac step function at either end because completely unmixed material (either 1 or 2) is present for some portion of the time. Figure 5-14 shows some actual measurements made inside a pipe in which air was flowing and helium was diffused inward through a porous wall section. This figure shows that the velocity component fluctuations are essentially gaussian throughout the flow but

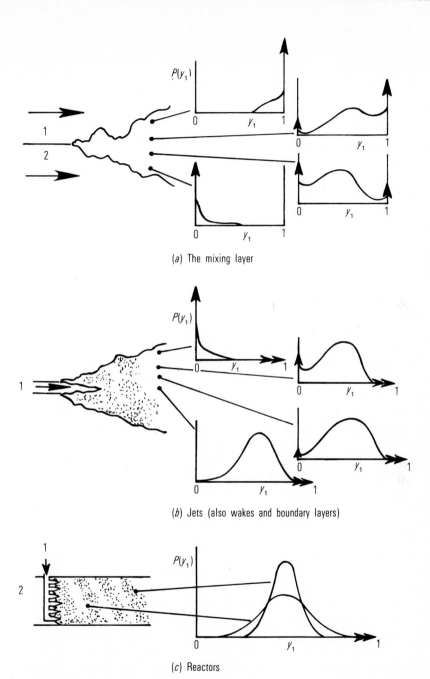

(*a*) The mixing layer

(*b*) Jets (also wakes and boundary layers)

(*c*) Reactors

**Figure 5-13** Some typical pdf's for the species mass fraction $y_1$ of species 1 in the mixture at various locations (schematic). (*Adapted with permission from R. W. Bilger, "Turbulent Flows With Non-Premixed Reactants" Topics in Applied Physics, Vol. 44, Turbulent Reacting Flows, Ed.: P. A. Libby and F. A. Williams, Springer Verlag, Berlin (1980).)*

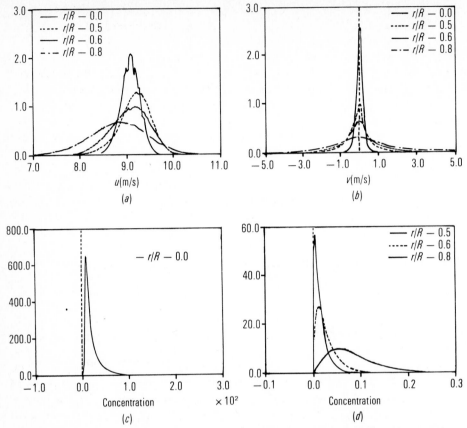

**Figure 5-14** The pdf's for velocity and concentration at the exit of a pipe in which air is flowing and helium is being diffused through the walls: (*a*) *u* component of velocity; (*b*) *v* component of velocity; (*c*) and (*d*) the helium concentration at various radial locations. *(Adapted with permission from R. A. Stanford and P. A. Libby, Phys. Fluids, vol. 17, pp. 1353–1361 (1974).)*

that the concentration fluctuations are definitely not gaussian during the mixing process.

It must be emphasized that the pdf's of Figs. 5-13 and 5-14 represent only the time-averaged probability of finding a specific concentration at a specific location. No information is given on the frequency of the fluctuation (which is related to scale) or the size of the sample volume (which will determine the sharpness of the resolution). Thus even with an experimental pdf available, it is almost impossible to determine how far mixing has progressed to the molecular level.

Another problem arises in the case when the two gases that are mixing have either widely disparate molecular weights or widely disparate temperatures. In this case, large-density fluctuations also occur in the flow. One technique that has been used to simplify the formalism of the derived equations for the fluctuation and shear terms is the use of Favre-averaging in which the coupled density velocity fluctuations $\rho'u'$, $\rho'v'$, and $\rho'w'$ are averaged, rather than the fluctuating velocities themselves. The concept of a probability density function for a Favre-averaged quantity has also been introduced.

There is another problem that is even more serious, relative to the description of turbulent mixing processes. This is due to the formation of large-scale vortical structures at the beginning of the mixing process and the fact that they can retain their identity during a good portion of the mixing process. An example of this behavior is shown in Fig. 5-15 for a planar mixing shear layer. In more complex general flows this structural behavior is the source of intermittency and the details of this intermittent behavior are not stochastic and are dependent on the structure of the shear region that is generating the turbulence. This means, of course, that it is impossible to develop a generally valid description of the mixing process based on a completely stochastic approach. The question of the generation and ultimate fate of the large-scale structures must be included if one can ever hope to understand nonreactive turbulent mixing in a general way.

The reader is referred to the bibliography for detailed descriptions of current research efforts on turbulent mixing flows.

## 5-12 TURBULENT MIXING WITH REACTION

When two streams that are mixing in a turbulent way are capable of reacting with each other or when a premixed turbulent flow is, by itself, capable of supporting a chemical reaction process, the description of the reaction process and how it is influenced by and/or interacts with the turbulent flow is a much more complex problem than those discussed in the previous two sections. We will break down the complex of possible behaviors into a number of distinctive limit types for discussion.

By far the simplest turbulent reactive flows are mixing flows in which the chemical reaction is either thermally neutral or almost thermally neutral. This is usually achieved by high dilution, and analysis of these types of reactive mixing situations is normally applicable only to pollutant-laden flows where the pollutant is reactive. We can gather some insight about the expected behavior of these flows by considering cases where the kinetic rate of the chemical reaction is either very rapid or very slow. The criteria which define these two limit behaviors are the magnitudes of the Damköhler numbers based on either eddy transport or diffusion (i.e., the integral scales or the Kolmogorov dissipative scale). Here scale and intensity allow one to define a characteristic eddy transport time which is ratioed to the chemical time. Since

$$(Da_1)_I \gg (Da_1)_K$$

where the subscripts $I$ and $K$ refer to the integral and Kolmogorov scale of the turbulence, only very rapid reactions will satisfy the requirement that

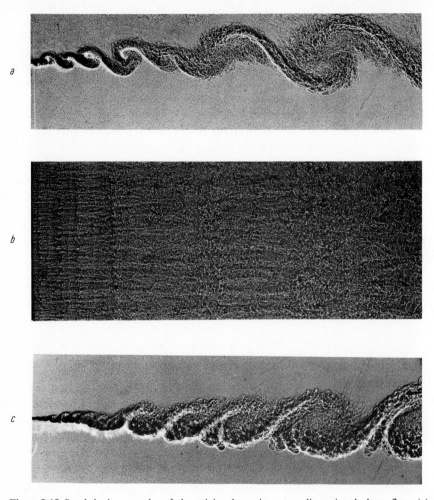

**Figure 5-15** Sparkshadow graphs of the mixing layer in a two-dimensional shear flow. (*a*) View through shear layer. (*b*) View perpendicular to shear layer. (*a*) and (*b*) Mixing layer in uniform density flow. ($\rho_2 = \rho_1$) $P = 0.8$ MPa, $U_1 = 3.8$ m/s, $U_2 = 10$ m/s, Re $= 3.3 \times 10^6$/m. (*c*) View through shear layer $P = 0.8$ MPa, $U_1 = 10.2$ m/s (helium), $U_2 = 3.8$ m/s (nitrogen), $Re_{N_2} = 2.3 \times 10^6$/m, $Re_{He} = 7.0 \times 10^5$/m. All photos are at the same scale. (*Courtesy of Professor Anatol Roshko, California Institute of Technology; (a) and (b) originally appeared in Konrad, J. H., "An Experimental Investigation of Mixing in Two-dimensional Turbulent Shear Flows with Application to Diffusion-Limited Chemical Reactions," Project SQUID Technical Report CIT-8-PU, December, 1976; Ph.D. Thesis, California Institute of Technology, 1977; (c) originally appeared in Rebollo, Manuel, "Analytical and Experimental Investigation of a Turbulent Mixing Layer of Different Gases in a Pressure Gradient," California Institute of Technology, Pasadena, California (1973).) (With permission of Cambridge University Press, originally in G. L. Brown and A. Roshko, J. Fluid Mech., vol. 64, Figs. 3 and 20, Plates 2 and 7 following p. 817 (1974).)*

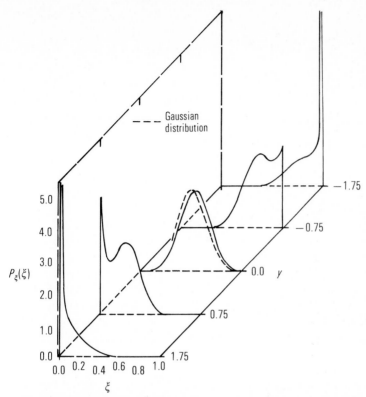

**Figure 5-16** Probability density functions for a conserved scalar in a two-dimensional mixing layer. Interpreted from the data of Alber and Batt.[14] *(Adapted from Bilger[15] with permission of Pergamon Press Ltd.)*

$(Da_l)_K \rightarrow \infty$. In this limit the chemistry is everywhere in equilibrium and the rate of reaction is strictly mixing-controlled. In the limit as $(Da_l)_l \rightarrow 0$ mixing is complete before the reaction occurs to any extent and the rate is kinetically controlled. A pdf analysis of essentially thermally neutral reactive mixing when the chemistry is relatively rapid was performed by Alber and Batt[14] using a cold nitrogen stream containing a small amount of slightly dissociated $N_2O_4$. When this stream mixed with warm room air in a shear layer $NO_2$ was formed by dissociation. A number of pdf's of a conserved scalar ($NO_2$ and temperature) interpreted from their work by Bilger[15] are shown in Fig. 5-16 for various locations on a cross section through the shear layer. The tops of the intermittency spikes are not shown in this figure.

Alber and Batt point out that for their experiment $(Da_l)_l$ was large (of the order of 500) but $(Da_l)_K$ was much smaller. They found that a local equilibrium

14. I. E. Alber and R. G. Batt, "Diffusion Limited Chemical Reactions in a Turbulent Shear Layer," *AIAAJ*, **14**: 70–76 (1976).

15. R. W. Bilger, "Turbulent Jet Diffusion Flames," *Prog. Energy Combust. Sci.*, **1**: 87–109 (1976).

model based on the local average temperature, $\bar{T}$, overestimated the peak average $NO_2$ concentration when compared with their pdf theory. They also found that the pdf theory yielded quite good agreement with measured $NO_2$ concentrations.

The next increase in complexity of the reactive-turbulent interaction problem occurs when the reaction becomes highly exothermic. In contrast to the previous case we can now distinguish three unique limit behaviors. These are (1) fast (hypergolic) chemistry, (2) slow chemistry throughout, and (3) slow chemistry when cold and fast chemistry when hot. Only the second of these two possibilities is currently tractable in an acceptable manner. The plug-flow reactor that will be discussed in the next chapter is based on the principle that when $(Da_1)_I \to 0$ the rates of the chemical reactions are kinetically controlled. Of the other two cases, the first, where all reactions are rapid, is the simplest to treat. This is because the turbulent mixing rate completely controls the rate of conversion. However, this case is much more complicated than the thermally neutral case that was discussed above because the heat that is released by the reaction markedly alters the turbulent flow itself and introduces large temperature and density fluctuations. The third case represents turbulent diffusion flames in which the chemistry has a significant temperature dependence. This case is extremely complicated because mixing can occur in cool regions without immediate reaction or in hot regions with immediate reaction. Thus the heat release from the reaction would be expected to have a *different* effect on the turbulence, density fluctuations, and temperature fluctuations than it has in case 1. A discussion of these two cases will be deferred to Chaps. 11 and 14 because they represent combustion processes.

There is one other type of turbulent reactive flow. In this case the reactants are initially premixed and the temperature is sufficiently low such that reaction cannot occur. Then a heat source or a holding region is placed in the flow and a highly exothermic chemical reaction occurs. The two limit cases that have been recognized for this type of flow are dependent on the intensity of the turbulence in the flow. For low-intensity turbulence a wrinkled, statistically fluctuating flame sheet can be identified in the flow. In the other case, where the intensity of the turbulence is extremely high, no flame sheet per se can be identified and the flow is said to be "well stirred." The theory for the well-stirred reactor will be discussed in the next chapter and other turbulent and reactive premixed flows will be discussed in Chaps. 11 and 14.

The interested reader is specifically referred to two recent reviews, one by Libby and Williams[16] and the other by Boris and Oran[17] which summarize the current status of the art of modeling all types of turbulent reactive flows. The bibliography contains other important current references.

16. P. A. Libby and F. A. Williams, "Turbulent Flows Involving Chemical Reaction," *Ann. Rev. Fluid Mech.*, **8**: 351–376 (1976).

17. J. P. Boris and E. S. Oran, "Modeling Turbulence: Physics or Curve Fitting," *Prog. Astronaut and Aeronaut.*, *AIAA*, New York, **76**: 187–210 (1982).

## PROBLEMS

**5-1** Derive the Froude number using the technique of Sec. 5-2.

**5-2** Use Eq. (5-24) to evaluate the relative importance of density and enthalpy change for the stoichiometric reaction $2H_2 + O_2 \rightarrow 2H_2O$. Assume that the oxygen comes from air that contains 79% $N_2$ and that there is no dissociation of the product $H_2O$. If the gamma of the gases is 1.2 what would the value of $q$ be for use in Eq. (5-27)?

**5-3** Show that at constant-mass flow the heat-addition model for steady one-dimensional flow exhibits a local minimum in the dimensionless entropy production $(S_2 - S_1)/R$ at the upper CJ point and a local maximum at the lower CJ point. Show that $(S_2 - S_1)/R$ is always positive for positive $q$ at both the lower and upper CJ points.

**5-4** The Chapman–Jouguet detonation velocity and pressure for stoichiometric hydrogen–air and methane–air mixtures initially at 101.35 kPa and 298.15 K are $V_{CJ} = 1968$ m/s and $P_{CJ} = 1.584$ MPa, and $V_{CJ} = 1802$ m/s and $P_{CJ} = 1.742$ MPa respectively. Calculate these properties using the heat-addition model and the values in Table 5-2 to compare with the actual values.

**5-5** Show that Eq. (5-58) reduces to an equation containing Reimann invariants in the unreactive case. Show that the same Reimann functions become variables in the heat-addition model. Why is it impossible to make such a simplification in the general case?

**5-6** Use Fig. 5-4 to determine and plot the locus of pressure and temperature when the velocity of sound decrement for an ideal dissociating gas is 6 percent. Use Fig. 4-7 to determine the degree of dissociation along this locus. Note that there is a maximum in the temperature. Using Table 4-1 determine this temperature and the attendant pressure for $Cl_2$, $H_2$, and $N_2$.

**5-7** For air, as the temperature is raised, we may assume that oxygen dissociation is the first important reactive process to occur. Using the dimensionless plots (4-9), (4-10), and (5-4), estimate the highest temperature (as a function of pressure) where nitrogen dissociation may be neglected (to 1 percent in composition and to 1 percent in its effect on temperature and equilibrium velocity of sound).

**5-8** The Hugoniot equation (5-33) also represents an integrated equation for a thermodynamically reversible (i.e., static) transformation from station 1 to station 2. Discover the path in the $(P, V)$ plane for this thermodynamically reversible transformation.

**5-9** Rewrite Eq. (5-33) (the Hugoniot relationship) in terms of the internal energy $e$ rather than the enthalpy $h$.

**5-10** Consider an infinitely long pipe containing an inviscid gas with $\gamma = 1.197$ throughout. One meter of this gas, $0 \le x \le 1$, is combustible with a normal (and invariant) burning velocity of 100 m/s and $q/c_v T = 7.819$. This mixture is ignited instantly at $x = 0$ at $t = 0$ and the flame propagates to the right at constant velocity until all the gas is consumed. Construct the $(x, t)$ wave diagram and the $(P, u)$ phase plane diagram for the process, including the postflame behavior.

*Hint:* The wave diagram should contain $R$ and $L$ facing shock waves, a deflagration wave moving to the right, and a contact surface moving to the left. When the flame reaches the end of the slug of combustible mixture it will extinguish and will generate $R$ and $L$ facing rarefaction waves and a new contact surface between hot combustion products and the gas to the right of the products.

**5-11** Work Prob. 5-10 for the case where the pipe is closed at $x = 0$ and extends to $+\infty$.

**5-12** Work Prob. 5-10 for the case where the flame is replaced by a Chapman–Jouguet detonation wave.

**5-13** Work Prob. 5-11 for the case where the flame is replaced by a Chapman–Jouguet detonation wave.

**5-14** Show that in the limit of weak shock strength, i.e., where $M = 1 + \varepsilon$ with $\varepsilon \ll 1$ and $q \equiv 0$, Eqs. (5-39) and (5-60) for simple one-dimensional traveling shock waves can be reduced to the acoustic relations for simple acoustic waves, Eq. (5-72).

**5-15** Derive an expression for the resonant acoustic frequencies of a tube that has one end open and one end closed.

# BIBLIOGRAPHY

### General equations and inviscid flow

Batchelor, G. K., *An Introduction to Fluid Dynamics*. University Press, Cambridge (1967).
Liepman, H. W., and A. Roshko, *Elements of Gas Dynamics*. Wiley, New York (1957).
Rudinger, G., *Wave Diagrams for Nonsteady Flow in Ducts*. Van Nostrand, Princeton, N.J. (1955).
Vincenti, W. G., and C. H. Kruger, Jr., *Introduction to Physical Gas Dynamics*. Wiley, New York (1965).
Williams, F. A., *Combustion Theory*. Addison-Wesley, Reading, Mass. (1965).
Wood, W. W., and J. G. Kirkwood, "Hydrodynamics of a Reacting and Relaxing Fluid," *J. Appl. Phys.*, **28**, 395–398 (1957).

### Turbulence in homogeneous fluids

Bradshaw, P., *An Introduction to Turbulence and its Measurement*. Pergamon Press, London (1971).
Bradshaw, P., (ed.), "Turbulence," *Topics in Applied Physics*, vol. 12. Springer Verlag, Berlin (1978).
Bradshaw, P., T. Cebeci, and J. H. Whitelaw, *Engineering Calculation Methods for Turbulent Flows*. Academic Press, New York (1981).
Lanford, O. E., III, "The Strange Attractor Theory of Turbulence," *Ann. Rev. Fluid Mech.*, **14**, 347–364 (1982).
Launder, B. E., and D. B. Spalding, *Mathematical Models of Turbulence*. Academic Press, New York (1972).
List, E. J., "Turbulent Jets and Plumes," *Ann. Rev. Fluid Mech.*, **14**, 189–212 (1982).

### Nonreactive turbulent mixing

Brodkey, R. S., *The Phenomena of Fluid Motions*, chap. 14. Addison-Wesley, Reading, Mass. (1967).
Cantwell, B. J., "Organized Motion in Turbulent Flow," *Ann. Rev. Fluid Mech.*, **13**, 457–515 (1981).
Fiedler, H., (ed.), "Structure and Mechanisms of Turbulence," I & II, vols. 75 and 76, *Lecture Notes in Physics*. Springer Verlag, Berlin (1978).
Schetz, J. A., "Injection and Mixing in Turbulent Flow," *Prog. Astronaut. Aeronaut.*, **68**, *AIAA*, New York (1980).

### Turbulent mixing with reaction

Bray, K. N. C., "The Interaction Between Turbulence and Combustion," *Seventeenth Symposium (International) on Combustion*, The Combustion Institute, Pittsburgh, Pa., pp. 223–233 (1979).
Brodkey, R. S., (ed.), *Turbulence in Mixing Operations*. Academic Press, New York (1975).
Bowen, J. R., N. Manson, A. K. Oppenheim, and R. I. Soloukhin (eds.), "*Combustion in Reactive Systems, III Turbulence*," *Prog. Astronaut. Aeronaut.*, *AIAA*, **76**, 187–380 (1981).
*Combust. Sci. and Tech.*, **13** (1975), Special issue on turbulent reactive flows.
Kennedy, L. A., (ed.), "Turbulent Combustion," *Prog. Astronaut. Aeronaut.*, *AIAA*, **58** (1978).
Libby, P. A., and K. N. C. Bray, "Variable Density Effects in Premixed Turbulent Flames," *AIAAJ*, **15**, 1186–1193 (1977).
——— and F. A. Williams, "Some Implications of Recent Theoretical Studies in Turbulent Combustion," *AIAAJ*, **19**, 261–274 (1981).
——— and F. A. Williams, (eds.), "Turbulent Reacting Flows," *Topics in Applied Physics*, vol. 44. Springer Verlag, Berlin (1980).
Murthy, S. N. B. (ed.), *Turbulent Mixing in Nonreactive and Reaction Flows*. Plenum Press, New York (1975).

# SIX

## COMBUSTION CHEMISTRY

### 6-1 INTRODUCTION

Combustion processes in general involve the interaction of chemistry and fluid mechanics (primarily gas dynamics). We will find in the rest of this text that this interaction is of paramount importance and that its diversity leads to the diversity of the combustion processes that have been observed in nature or produced by man. In all combustion processes, chemistry plays the central role even though in many cases, in an overall sense, fluid mechanics may dominate and drive the phenomena. This is because without the occurrence of chemical reactions there would be no large change in the temperatures of the system and therefore no combustion process per se.

Because of the innate complexity of most combustion reactions, simple models were used in the past to represent the apparent heat release that accompanies these reactions. Easy access to large digital computers plus the recent increase in our understanding of the basic kinetic steps involved in certain classes of combustion reactions have caused combustion kinetics modeling to become quite sophisticated during the past decade. Furthermore, in many cases sophisticated chemical modeling has been very successful in a predictive way—far surpassing the ability of the earlier modeling techniques. This leads to the observation that sophisticated predictive modeling will be more widely applied to combustion problems in the future. This is why the subject of combustion chemistry (and its modeling) is being treated as a separate and distinct chapter in this text.

In Chap. 4 we introduced the concept of an explosion temperature for either constant-volume or constant-pressure systems. In this chapter we will discuss the details of the "homogeneous" explosion process for various limit cases, first using the classical (and relatively simple) techniques developed decades ago. We use these explosions to discuss combustion chemistry because fluid dynamics plays only a minor role in this particular class of combustion processes. After this introduction to explosion processes, a number of experimental techniques that are used to develop kinetic rate data will be presented to indicate the experimental source of the data that is now being used in detailed kinetic schemes. Then the integration of a detailed combustion kinetic scheme will be presented using a simple example and the results of such detailed calculations for a number of simple combustion systems will be given. Finally, a hierarchy of techniques for modeling combustion chemistry will be presented which includes those already used in this and previous chapters.

## 6-2 ADIABATIC THERMAL EXPLOSIONS

Consider the reaction of a gaseous mixture of fuel and oxidizer or a gaseous substance which is exothermic and unstable (such as ozone or nitric oxide). Assume that this mixture or substance is placed in an adiabatic vessel of constant volume at some initial temperature $T_0$ and pressure $P_0$. Since there is no heat loss the exothermic reaction will cause the temperature of the contents of the vessel to rise and this will cause the reactions to accelerate. This self-acceleration process will proceed slowly at first until, after some finite delay time, the system will appear to "explode," or very quickly form a mixture of equilibrium products at a temperature and pressure that are very much higher than the initial temperature and pressure.

The first reasonable analyses of vessel "thermal" explosions were presented by Semenov in 1928 and Todes in 1933. The adiabatic vessel explosion process which is being treated in this section is a limit case of the vessel thermal explosion process to be discussed in the next section. In order to reduce the problem to a tractable form both treatments are based on a number of assumptions that are generalizations of those made by Semenov and Todes. They are:

1. The vessel contains an ideal gas mixture of fuel $F$, oxidizer $O$, and inert $I$.
2. The exothermic combustion or decomposition reaction is modeled using an Arrhenius expression for the overall rate of fuel disappearance of the form

$$\frac{d[F]}{dt} = -A[F]^n[O]^m[I]^l \exp\left(\frac{-E^*}{RT}\right) \tag{6-1}$$

where $A$, $n$, $m$, $l$, and $E^*$ are empirically determined constants.
3. The rate of energy release by the reaction is equal to the rate of disappearance of the fuel multiplied by a "heat of combustion" of the fuel.
4. The heat capacity of the gas in the adiabatic vessel is constant.

The last two of these assumptions lead to an energy-balance equation for the adiabatic system

$$\rho c_v \frac{dT}{dt} = -Q_{fu} \frac{d[F]}{dt}$$ (6-2)

where the units are: $\rho$ (kg m$^{-3}$), $c_v$ (J kg$^{-1}$ K$^{-1}$), $Q_{fu}$ (J/mole of fuel), and $[F]$ (moles m$^{-3}$). Since the gas mixture is ideal the initial concentrations of the fuel, oxidizer, and inert can be written in terms of the initial pressure and temperature of the system and their respective mole fractions

$$[F] = \rho n_{fu} = \frac{(P_{fu})_0}{RT_0} = x_{fu} \frac{P_0}{RT_0}$$

Substitution yields an expression for the rate of temperature rise

$$\frac{dT}{dt} = \lambda \exp\left(\frac{-E^*}{RT}\right)$$ (6-3)

where $\lambda$ is given by the expression

$$\lambda = \frac{AQ_{fu}}{c_v}\left(\frac{P_0}{RT_0}\right)^{n+m+l} x_{fu}^n(1 - x_{fu} - x_{in})^m x_{in}^l$$ (6-4)

The convention for $Q$ in Eq. (6-4) is the same as we introduced in Sec. 2-6, namely it is taken to be a positive constant. Thus for the initial conditions, $\lambda$ is also positive. Furthermore, for ordinary explosive mixtures we assume, as Semenov and Todes did, that the heat release on explosion is so large relative to the heat capacity of the system that the temperature rise will "run away" before the initial concentration of the reactants is changed appreciably. Thus $\lambda$ can be treated as a constant in Eq. (6-3) and it may be rearranged and integrated from $t = 0$.

$$\lambda t = \int_{T_0}^{T} \exp\left(\frac{E^*}{RT}\right) dT$$

We define $\theta = T/T_0$ and $\varepsilon = E^*/RT_0$ and change the integration variable to obtain the expression

$$-\frac{\lambda t}{T_0 \varepsilon} = \int_{\varepsilon}^{\varepsilon/\theta} \frac{e^{\varepsilon/\theta}}{(\varepsilon/\theta)^2} d\left(\frac{\varepsilon}{\theta}\right)$$

We note that $\varepsilon$ and $\varepsilon/\theta$ are usually large (i.e., $> 15$) and we integrate by parts to obtain a series in $\varepsilon/\theta$

$$-\frac{\lambda t}{T_0 \varepsilon} = \left[\frac{e^{(\varepsilon/\theta)}}{(\varepsilon/\theta)^2}\left(1 + \frac{2!}{(\varepsilon/\theta)} + \frac{3!}{(\varepsilon/\theta)^2} + \cdots + \frac{(n+1)!}{(\varepsilon/\theta)^n} + \cdots\right)\right]\Bigg|_{\varepsilon}^{\varepsilon/\theta}$$

which when evaluated yields

$$\frac{\lambda t}{T_0 \varepsilon} = \frac{e^{\varepsilon}}{\varepsilon^2}\left[1 + \frac{2}{\varepsilon} + \frac{6}{\varepsilon^2} + \cdots\right] - \frac{e^{(\varepsilon/\theta)}}{(\varepsilon/\theta)^2}\left[1 + \frac{2}{(\varepsilon/\theta)} + \frac{6}{(\varepsilon/\theta)^2} + \cdots\right]$$

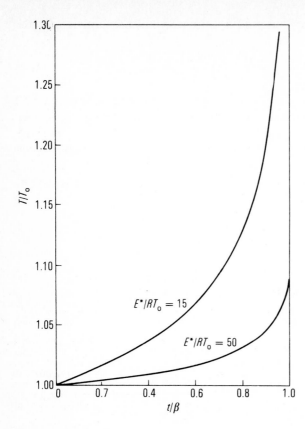

**Figure 6-1** Explosion behavior of an isolated system undergoing a pure "thermal explosion."

This may be written in a different form by dividing both sides by the constant term that appears on the right-hand side.

$$\frac{\lambda t \varepsilon}{T_0[1 + (2/\varepsilon) + (6/\varepsilon^2) + \cdots]e^{\varepsilon}} = 1 - \theta^2 e^{(\varepsilon/\theta - \varepsilon)} \cdot \left[1 + \frac{2}{\varepsilon}(\theta - 1) + \cdots\right]$$

Substituting the original variables we obtain an expression for the temperature-time history of an adiabatic explosion

$$\frac{t}{\beta} = \frac{t}{(RT_0^2/E^*\lambda)[1 + (2RT_0/E^*) + (6R^2T_0^2/E^{*2})]e^{E^*/RT_0}}$$

$$= 1 - \left(\frac{T}{T_0}\right)^2\left[1 + \frac{2RT}{E^*} \cdot \left(\frac{T}{T_0} - 1\right) + \cdots\right] \cdot \exp\left[\frac{E^*}{RT_0}\left(\frac{T_0}{T} - 1\right)\right]$$

where $\beta$ is a constant whose value is dependent upon the values of $T_0$, $\lambda$, and $E^*$. The dimensionless time $t/\beta$ is plotted versus the dimensionless temperature $T/T_0$ in Fig. 6-1 for two values of $E^*/RT_0$. Notice from this plot that $T/T_0$ increases only slowly at first but increases extremely rapidly as $t/\beta$ approaches 1. At $t/\beta = 1$ the system is said to explode and the time to explosion is found to be unique. Since for most systems $E^*/RT_0 > 15$, the ignition delay is given by the expression

$$t_{ig} = \beta \cong \frac{RT_0^2}{E^*\lambda} \exp\left(\frac{E^*}{RT_0}\right) \tag{6-5}$$

We have, therefore, shown that any isolated system containing a finite-rate exothermic reaction with a positive temperature sensitivity will always "explode" in a homogeneous manner. We further note that the initial assumptions are probably quite valid for this type of explosion. We base this on the observation that complete reaction in an exothermic system usually increases the temperature by a factor of 2 or more. Thus a 10 percent temperature rise represents at most only a 10 percent depletion of the reactants in the mixture.

## 6-3 VESSEL EXPLOSIONS, THERMAL

Equation (6-5) may be used to investigate the overall order and activation energy of thermal explosions only in systems which do not lose heat to the surroundings. If the system is not isolated, but is allowed to transfer heat to its surroundings, Eq. (6-2) must be modified to allow for this heat flow. The results now change drastically because the availability of a mechanism for heat loss leads to the possibility that under certain conditions no "thermal explosion" as such will occur in the system, even though it meets all the other requirements which lead to Eq. (6-5).

As a first approximation, let us assume that the system's temperature is uniform throughout, but that it loses heat to its surroundings (i.e., to the vessel walls at a temperature $T_0$) by simple conduction or by convection with a constant coefficient. Eq. (6-2) thus becomes

$$\rho c_v \frac{dT}{dt} = -Q_{fu} \frac{d[F]}{dt} - \frac{s}{V} h(T - T_0) \tag{6-6}$$

for a vessel of volume $V$ where $s$ is the surface area of the vessel and $h$ is an effective heat-transfer coefficient which is assumed to be independent of the pressure (units: joules per meter$^2$-second-kelvin). We immediately note that Eq. (6-6) predicts the attainment of a steady system temperature under conditions where the coefficient of the second term is large. However, since the first term on the right-hand side is always finite and increases with increasing system temperature, there is also the possibility that $dT/dt$ will never approach zero or go negative. In this case, an explosion will occur. Equation (6-6) thus predicts that, in a vessel, one should see "thermal explosion limits" which are a function of the pressure, temperature, and the fuel and oxidizer concentrations as well as the vessel size and geometry and the effective heat transfer coefficient. Let us examine one facet of the theoretical dependence of these limits. Figure 6-2 is a plot of the two terms on the right in Eq. (6-6) and illustrates that for a particular $T_0$ value, the explosion limit will be pressure-sensitive because the chemical

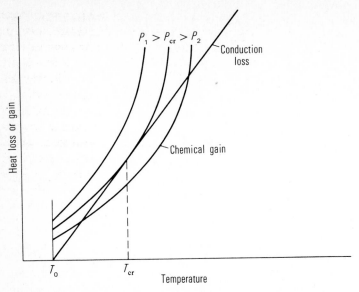

**Figure 6-2** Heat losses and gains in a finite vessel with exothermic reaction. In this case $P_1 > P_{cr} > P_2$. Thermal explosion only.

rate equations have a pressure dependence. We define the pressure and temperature for tangency in Fig. 6-2 as $T_{cr}$, $P_{cr}$ and write the heat-flow balance equation at tangency as

$$-Q_{fu} \frac{d[F]}{dt} = \frac{S}{V} h(T_{cr} - T_0)$$

and the condition for tangency as

$$\frac{d}{dT}\left(-Q_{fu} \frac{d[F]}{dt}\right) = \frac{d}{dT}\left[\frac{S}{V} h(T_{cr} - T_0)\right]$$

Substituting the kinetic rate equation and solving these equations simultaneously yields

$$\frac{RT_{cr}^2}{E^*} = T_{cr} - T_0$$

or

$$T_{cr} = \frac{1 \pm [1 - 4(RT_0/E^*)]^{1/2}}{2R/E^*}$$

We choose the minus sign to obtain the lower value of $T_{cr}$ since our analysis of a pure thermal explosion showed that the reaction rate for an explosion in an isolated system became very rapid after only a slight temperature rise. Once again we assume that $E^*/RT_{cr} \gg 1$ and therefore that the expression for $T_{cr}$ may be expanded to yield

$$T_{cr} = T_0\left(1 + \frac{RT_0}{E^*}\right)$$

Substituting into Eq. (6-6) for $T = T_{cr}$ and $dT/dt = 0$ therefore yields the explosion limit equation for a thermal explosion:

$$\ln\left\{\frac{P}{T[1 + 2/(n + m + l)]}\right\} = \frac{E^*}{(n + m + l)RT} + \text{const} \qquad (6\text{-}7)$$

Equation (6-7) is represented schematically in Fig. 6-3. We see from Fig. 6-3 that this simple theory predicts that the slope of the explosion limit line in the proper $\ln P$ vs. $1/T$ plot will directly yield an effective activation energy of the preexplosion chemical reaction if the overall order of the reaction is known.

Equation (6-6) can be numerically integrated to yield the temperature change with time in a "thermal" vessel explosion. The effect of initial pressure on the temporal behavior of the vessel contents when all other initial conditions remain the same is shown schematically in Fig. 6-4. We notice that for $P < P_{cr}$ the temperature of the vessel contents asymptotically approaches some value of temperature in the range $T_{cr} > T > T_0$. Under these conditions reaction is occurring but heat losses are sufficient to maintain a constant vessel temperature. This temperature is also shown in Fig. 6-2 as the lower intersection of the $P_2$ chemical gain time with the conduction loss term.

Figure 6-4 shows that when $P > P_{cr}$ an "explosion" occurs in that the temperature increases very rapidly after some apparent delay time. Figure 6-4 also lowest dissociation energy of the three molecules involved in this system, and therefore its collisional dissociation is more likely than dissociation of either a hydrogen molecule or a halogen acid molecule. Process 9a is more unlikely than

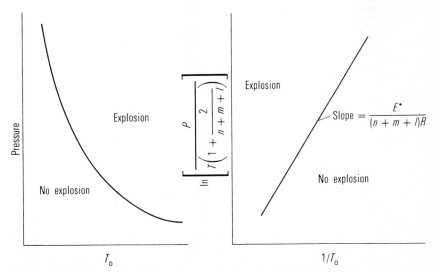

**Figure 6-3** Explosion-limit behavior in a simple thermal explosion with heat loss.

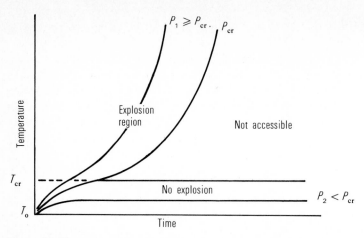

**Figure 6-4** The temperature–time history of a vessel explosion showing how the initial vessel pressure causes the system's behavior to change from slow reaction to explosion. *(With permission from W. E. Baker, P. A. Cox, P. E. Westine, J. Kulesz, and R. A. Strehlow, "Explosion Hazards and Evaluation," Elsevier, Amsterdam (1983).)*

shows that the delay to explosion increases as $P \to P_{cr}$. Furthermore, the explosion delay time at $P = P_{cr}$ has a unique value which is dependent on the system properties. This vessel ignition delay time is somewhat similar to the ignition delay time in an adiabatic system, Eq. (6-5). However, it is more complex because it contains vessel and heat-loss parameters. The experimental significance of these behaviors will be discussed more fully later in this chapter.

## 6-4 COMBUSTION CHEMISTRY—CHAIN REACTIONS

After Van't Hoff introduced his postulates concerning order and activation energy, an attempt was made to force or at least to catalog overall reactions in accordance with these elementary mechanisms.

However, the application of "thermal" theories to combustion reactions led to "overall" mechanisms in which $n$, $m$, and $l$ were found to be fractional powers. Even worse, $P$-$T$ plots of explosion limits were universally more complex than the simple curve predicted for thermal explosions (Fig. 6-3).

During the second quarter of the twentieth century, chemists slowly began to realize that every overall, macroscopically observable combustion reaction is the net result of a (sometimes large) number of independent elementary steps occurring at the molecular level, each with its own order, rate, and temperature-dependence. The aggregate of these steps is called a *kinetic mechanism*, and the "mechanism" leads to the observed macroscopic behavior. In combustion virtually all the kinetic mechanisms that have been verified at the present time involve very reactive unstable intermediate radicals or atoms. Typically these are generated in super-equilibrium concentrations during the reaction and are the main driving force for the conversion of reactants to products. Kinetic mechanisms that involve this type of radical or atom behavior are commonly called *chain reactions*.

Since chain mechanisms in general tend to be quite complex, we will analyze one simple system in great detail to show not only how the totality of possible reactive collision processes can be reduced to a reasonable mechanism, but also how the endothermicity or exothermicity of the detailed steps affect their importance to the overall reaction. We choose to study the reaction of a halogen with hydrogen to form the appropriate halogen acid. The overall process stated in terms of stable species is

$$H_2 + X_2 \rightarrow 2HX$$

where $X$ represents either F, Cl, Br, or I. We assume that the unstable atoms H and $X$ exist in the reacting mixture. In this case we have a system which contains five species. Let us discuss the prognosis of each possible binary-collision process in this system to determine how chain processes originate. Since there are five species, there are fifteen possible binary-collision processes which we must consider. Three of these are atom–atom collisions and therefore must be third-order processes if they are to yield products.

1. $H + H + M \rightarrow H_2 + M$
2. $H + X + M \rightarrow HX + M$
3. $X + X + M \rightarrow X_2 + M$

These three-body recombinations occur only slowly and tend to remove reactive species from the mixture. Six of the other possible binary-collision processes occur between an atom and a diatomic molecule. These may be listed with their possible results

4. $H + H_2 \begin{array}{l} \rightarrow H + H + H \\ \searrow H_2 + H \end{array}$  (energetically difficult)  
  (no change)

5. $H + X_2 \begin{array}{l} \rightarrow HX + X \\ \searrow H + X + X \end{array}$  (possible process)  
  (formation of halogen atoms)

6. $H + HX \begin{array}{l} \rightarrow H_2 + X \\ \searrow H + H + X \end{array}$  (possible process)  
  (energetically difficult)

7. $X + H_2 \begin{array}{l} \rightarrow HX + H \\ \searrow X + H + H \end{array}$  (possible process)  
  (energetically difficult)

8. $X + X_2 \begin{array}{l} \rightarrow X + X + X \\ \searrow X_2 + X \end{array}$  (formation of halogen atoms)  
  (no change)

9. $X + HX \begin{array}{l} \rightarrow H + X_2 \\ \searrow X + X + H \end{array}$  (energetically more difficult than 5a, 6a, and 7a).  
  (energetically difficult)

These six distinct collision processes therefore yield three possible reaction paths and two dissociation paths on the basis of energetics. Notice that the only dissociation process that is being considered is the collisional dissociation of the halogen molecule. Table 6-1 shows that the halogen molecule has by far the lowest dissociation energy of the three molecules involved in this system, and therefore its collisional dissociation is more likely than dissociation of either a hydrogen molecule or a halogen acid molecule. Process 9a is more unlikely than either process 5a, 6a, or 7a because, in this case, we are decomposing a molecule with a relatively strong bond and forming a molecule with a relatively weak bond. Therefore, since this reaction must be highly endothermic, it must have a high activation energy and a low probability of occurrence. It is required for detailed balancing late in the reaction process because it is the reverse of 9a. However it does not significantly decrease the rate of HX formation until equilibrium is approached.

The last six possible collision pairs are molecule–molecule collisions and therefore offer more possible paths.

10. $H_2 + H_2 \longrightarrow H + H + H_2$       (energetically difficult)
         $\searrow H + H + H + H$       (energetically extremely difficult)
11. $H_2 + X_2 \longrightarrow HX + HX$       (direct conversion)
         $\searrow H_2 + X + X$       (formation of halogen atoms)
         $\searrow H + H + X_2$       (energetically difficult)
         $\searrow HX + H + X$       (energetically difficult)
         $H + H + X + X$       (energetically extremely difficult)
12. $H_2 + HX \longrightarrow HX + H_2$       (no change)
         $\searrow H + H + HX$       (energetically difficult)
         $\searrow H + X + H_2$       (energetically difficult)
         $H + H + H + X$       (energetically extremely difficult)
13. $X_2 + X_2 \longrightarrow X + X + X_2$       (formation of halogen atoms)
         $\searrow X + X + X + X$       (energetically difficult)
14. $X_2 + HX \longrightarrow X + X + HX$       (formation of halogen atoms)
         $\searrow X_2 + H + X$       (energetically difficult)
         $\searrow X + X + H + X$       (energetically extremely difficult)
         $HX + X_2$       (no effect)
15. $HX + HX \longrightarrow H_2 + X_2$       (direct dissociation)
         $\searrow H_2 + X + X$       (formation of halogen atoms)
         $\searrow HX + H + X$       (energetically difficult)
         $H + X + H + X$       (energetically extremely difficult)

Notice that, of this large number of possible processes arising from molecule–molecule collisions, we are left with only halogen atom formation (processes 11b, 13a, 14a, 15b), direct conversion (process 11a), or direct dissociation (process 15a) as possible reactive processes. Thus the fifteen conceivable collisions result in the following set of possible kinetic processes for the reaction

$$X_2 + M \xrightarrow{k_1} 2X + M \qquad \text{(chain initiation)}$$

$$X + H_2 \xrightarrow{k_2} HX + H \qquad \text{(reaction 7a)}$$

$$H + X_2 \xrightarrow{k_3} HX + X \qquad \text{(reaction 5a)} \quad \text{(chain reaction)} \qquad (6\text{-}8)$$

$$H + HX \xrightarrow{k_4} X + H_2 \qquad \text{(reaction 6a)}$$

$$M + 2X \xrightarrow{k_5} X_2 + M \qquad \text{(three-body termination or deactivation)}$$

or the alternate route represented by the equation

$$H_2 + X_2 \xrightarrow{k_6} 2HX$$

The chain process [Eq. (6-8)] may be analyzed using the steady-state assumption by setting the overall rates $d[X]/dt$ and $d[H]/dt = 0$ relative to their rate of production and disappearance in the individual reaction steps. This produces a kinetic rate equation for the formation of halogen acid which has the form

$$\frac{d[HX]}{dt} = \frac{2k_2(k_1/k_5)^{1/2}[H_2][X_2]^{1/2}}{1 + (k_4/k_3)[HX]/[X_2]} \tag{6-9}$$

The simple direct combination reaction yields a rate expression of the form

$$\frac{d[HX]}{dt} = k_6[H_2] \cdot [X_2] \tag{6-10}$$

All the halogen–hydrogen reactions have been studied experimentally. It will be instructive to examine the heats of reaction for the steps in the mechanism [Eqs. (6-8)] individually for each of the different halogens and compare the expected behavior with the observed behavior. Table 6-1 contains a listing of the heats of formation of all the species as well as the heats of reaction for the steps in the mechanism at 300 K. Recall that if a heat of reaction is negative the reaction is exothermic and will have a very small activation energy. On the other hand reactions with positive heats of reaction are endothermic and their activation energies will be slightly larger than the magnitude of their heat of reaction. Also note from Table 6-1 that all the halogen–hydrogen reactions are overall exothermic and that their exothermicity decreases markedly as one goes from fluorine to iodine in the series.

We now examine each of the halogen reactions in detail. In the HF system both the primary chain steps (reactions 7a and 5a) are highly exothermic and we would expect this reaction to go rapidly. Experimentally it is observed that

**Table 6-1 Heats of formation and reaction for halogen–hydrogen systems at 300 K (kJ/mole)**

| Type | Reaction equation | Halogen | | | |
|------|-------------------|---------|------|------|------|
| | | F | Cl | Br | I |
| Formation | $\frac{1}{2}X_2 \rightarrow X$ | 79.1 | 121.0 | 111.8 | 76.2 |
| Formation | $\frac{1}{2}H_2 + \frac{1}{2}X_2 \rightarrow HX$ | −271.3 | −92.1 | −36.4 | −5.9 |
| Formation | $\frac{1}{2}H_2 \rightarrow H$ | 218.1 | 218.1 | 218.1 | 218.1 |
| Reaction 7a | $X + H_2 \rightarrow HX + H$ | −132.3 | 5.0 | 69.9 | 136.0 |
| Reaction 5a | $H + X_2 \rightarrow HX + X$ | −410.3 | −189.2 | −142.7 | −147.8 |
| Reaction 6a | $H + HX \rightarrow X + H_2$ | 132.3 | −5.0 | −69.9 | −136.0 |
| Overall reaction | $H_2 + X_2 \rightarrow 2HX$ | −542.6 | −184.2 | −72.8 | −11.8 |

the $H_2 + F_2$ system reacts rapidly on contact even at room temperature. In the case of chlorine, reaction 7a is only slightly endothermic and an $H_2 + Cl_2$ mixture is stable at room temperature in the dark. However, light causes $Cl_2$ to photo dissociate and once Cl atoms are formed the reaction is rapid at room temperature. For bromine the reaction is very slow at room temperature but proceeds at a reasonable rate above room temperature. Extensive studies of the $H_2 + Br_2$ reaction before chain reaction concepts were fully developed yielded an empirical rate expression for this reaction which is identical to Eq. (6-9). Note that for bromine, reaction 6a, which destroys HBr, is exothermic and therefore rapid. Also note that the inclusion of this reaction in the scheme yields the prediction in Eq. (6-9) that an excess of the product HBr will inhibit the forward reaction. This behavior can be taken to be direct evidence that a chain mechanism is operating. In the case of iodine, for many years the formation of hydrogen iodide was thought to be by the direct process, Eq. (6-10), and this reaction was in fact used as an example of a simple bimolecular reaction. It is true that the endothermicity of reaction 7a is so high that it has a sufficiently high activation energy to stop the chain mechanism. Thus the bimolecular mechanism was favored until 1967 when it was found, using photochemically produced iodine atoms, that the reaction occurs by either of the following two mechanisms[1]

$$H_2 + 2I \rightarrow 2HI$$

or

$$M + I + H_2 \rightarrow IH_2 + M$$
$$IH_2 + I \rightarrow 2HI$$

These steps are chemically indistinguishable from the direct process, Eq. (6-10), in a thermal system since thermal dissociation of iodine is rapid. Note that the species $IH_2$ was not considered in our discussion.

## 6-5 VESSEL EXPLOSIONS, BRANCHING CHAIN

There is another type of chain reaction which is extremely important in combustion processes. This is the *branching-chain reaction*. There are certain systems capable of reacting with a chain mechanism but which have the added feature that each cycle of the chain produces a net increase in the number of reactive chain carriers (either radicals or atoms). These branching-chain reactions can be completely uncontrollable because radical concentrations tend to rise exponen-

1. J. H. Sullivan, *J. Chem. Phys.*, **46**:73 (1967).

tially with time. A typical example is the hydrogen–oxygen branching-chain explosion

$$H_2 + O_2 \rightarrow 2OH \qquad \text{(initiating)}$$

$$H + O_2 \rightarrow OH + O \qquad \text{(branching)}$$

$$O + H_2 \rightarrow OH + H \qquad \text{(branching)} \qquad \text{(branched chain)}$$

$$OH + H_2 \rightarrow H_2O + H \qquad \text{(propagating)}$$

$$H + H + M \rightarrow H_2 + M \qquad \text{(terminating)}$$

One cycle of this reaction produces a net increase of two hydrogen atoms in the reactive system, and if terminating processes are not sufficiently rapid to use up two hydrogen atoms in the same length of time, this reaction will increase its rate exponentially with time and behave as a homogeneous explosion.

The general conditions for a branching-chain explosion may be discussed by considering a general reaction of the branching-chain type. Let $C$ represent a chain carrier; then the reaction set

$$nA \xrightarrow{k_1} C \qquad \text{(initiation)}$$

$$A + C \xrightarrow{k_2} \text{products} + \alpha C \qquad \text{(chain cycle)}$$

$$C \xrightarrow{k_w} \text{products} + \text{stable molecules} \qquad \text{(wall termination)}$$

$$C + M + M \xrightarrow{k_g} \text{products} + \text{stable molecules} \qquad \text{(gas termination)}$$

represents a typical branching-chain reaction. In the steady state

$$\frac{d[C]}{dt} = 0 = k_1[A]^n - k_2[C][A] + \alpha k_2[C][A] - k_g[C][M]^2 - k_w[C] \tag{6-11}$$

Therefore at steady state

$$[C] = \frac{k_1[A]^n}{k_2[A](1 - \alpha) + k_g[M]^2 + k_w} \tag{6-12}$$

In order for $[C]$ to be finite and positive it is obvious that

$$\alpha < 1 + \frac{k_g[M]^2 + k_w}{k_2[A]}$$

Therefore if

$$\alpha \geq 1 + \frac{k_g[M]^2 + k_w}{k_2[A]} = \delta \tag{6-13}$$

the reaction will become a branching-chain explosion.

We first note that Eq. (6-12) does not contain the initiation reaction. This is because the denominator of Eq. (6-12) represents the competition between the

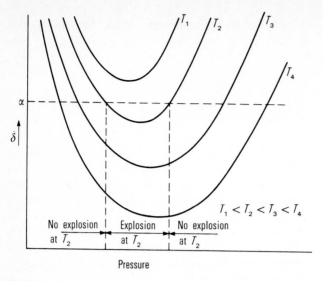

**Figure 6-5** The value of $\delta$ in Eq. (6-13) plotted against pressure for various temperatures; points outside the range of each curve represent no explosion for a particular value of $\alpha$.

production and destruction of radicals once they are present at high concentration in the system. We next note that in Eq. (6-13) the quantity $\delta$ is a function of pressure and temperature and that at fixed temperature the value of $\delta$ approaches infinity as $P$ approaches either zero or infinity. Furthermore, $\delta$ exhibits a minimum which is related to the relative rates of the three competing reactions. We also note that both the wall and gas-phase recombination reactions have a low-temperature dependence, while the rate-controlling reaction in

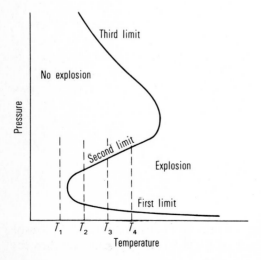

**Figure 6-6** The explosion peninsula behavior of a chain-branching explosion as constructed from Eq. (6-13) and Fig. 6.5.

the chain cycle must be an endothermic reaction with a relatively large activation energy, and therefore this reaction will be quite temperature-dependent. Accordingly, as the temperature is increased the quantity $k_w/k_g$ will be relatively constant in Eq. (6-13), but the quantity $k_g/k_2$ will decrease rapidly. This leads to the temperature dependence of $\delta$ shown schematically in Fig. 6-5. We now recall that $\alpha \geq \delta$ is the criterion for the occurrence of a spontaneous chain-branching explosion in the vessel, and that $\alpha < \delta$ is the condition for a slow reaction with a relatively constant concentration of the chain carrier $C$. Figure 6-5 may therefore be used to construct a typical explosion-limit diagram for a chain-branching reaction. This is shown schematically in Fig. 6-6. Note the occurrence of an explosion "peninsula" and the existence of three distinct pressure limits over a short temperature range. These limits are usually called the *first limit* (due to wall recombination), the *second limit* (due to gas-phase recombination), and the *third* (or high-pressure) *limit*. Chain-branching theory does not predict the occurrence of this third limit; it is simply the ordinary thermal explosion limit governed by Eq. (6-7). However the third-limit behavior of the $H_2$–$O_2$ reaction in this high-pressure–low-temperature region can also be explained by the production of the hydroperoxyl radical, $HO_2$, by the reaction

$$H + O_2 + M \rightarrow HO_2 + M$$

This will be discussed in more detail in Sec. 6-10-2. The occurrence of an explosion peninsula has been observed in many vessel experiments for a large variety of chemical systems. Its occurrence is usually considered to be unambiguous evidence that chain branching is occurring during the preexplosion period.

If an explosion occurs in a system capable of chain branching, the delay may be measured as a function of temperature, pressure, and the reactant concentrations, and it is proper to ask how these measured delays are related to the fundamental processes leading to the explosion. We observe, however, that the construction of an appropriate model is necessary before ignition delays may be related to the reaction parameters, and that for the general chain-branching reaction the analysis might be expected to be even more complex than for a simple thermal explosion. Because of this complexity, we will develop an expression for the explosion delay of a chain-branching explosion for only one limiting case, the low-pressure, high-temperature case where the terminating reactions become very slow relative to the chain-branching step. If this is the case, we find that after the initiation reactions cease to be important the increase in concentration of the chain carrier is governed by the chain-branching steps

$$\frac{d[C]}{dt} = (\alpha - 1)k_2[A][C] \qquad (6\text{-}14)$$

We note in passing that the assumption that radicals destruction is slow, which led to Eq. (6-14), is equivalent to the "adiabatic" assumption for the thermal explosion discussed in Sec. 6-2. If we assume that $[A]$ is constant, Eq. (6-14) may be integrated to yield the equation

$$\frac{[C]}{[C_0]} = \exp \left\{ (\alpha - 1)k_2[A]t \right\}$$

The assumption that $[A]$ is constant is usually found to be quite good. We now assume that for an explosion to occur, the ratio $[C]/[C_0]$ must become extremely large. This is equivalent to assuming that the reaction goes through many cycles before $[C]$ increases appreciably. This yields the result that

$$\ln ([A]t_{\text{ig}}) = \frac{E^*}{RT_0} + \text{const} \tag{6-15}$$

where $E^*$ is the activation energy of the rate-controlling step in the branching-chain cycle. Equation (6-15) has been verified experimentally for many high-temperature, low-pressure homogeneous explosions involving a chain-branching mechanism.

Lastly, we note that the assumption that $[C]/[C_0]$ becomes large does not mean that the temperature increases without bound, because it is quite possible for the chain cycle to be nearly thermally neutral. If the cycle is indeed thermally neutral, we obtain the simplest type of chain-branching explosion in which the chain-carrier concentration becomes very large before recombination reactions liberate heat energy to the system. In this case, the state variables of the system shown an extremely slow variation with time during the induction period. After this time, the exothermic recombination reactions cause a relatively rapid temperature rise and the formation of products.

## 6-6 UNIFIED THEORY OF VESSEL EXPLOSIONS

In 1969 B. F. Gray and C. H. Yang[2] presented a unified mathematical model for vessel explosions which can be applied to many complex chain-carrier processes. This theory also has the potential of being generalized to multiple chain carriers and to adiabatic vessel explosions. In its simplest form the model considers only a generalized chain branching and generalized chain-terminating reaction including their respective heats of reaction. Using the notation of the previous sections one may write Eq. (6-11) as

$$\frac{d[C]}{dt} = [(\alpha - 1)k_b - k_t][C] \tag{6-16}$$

where for any initial conditions the background concentrations of $A$ and $M$ are assumed to be invariant with time and are included in the values of $k_b$ and $k_t$ as appropriate. Also in this notation $k_b$ and $k_t$ represent composite rates and may

2. B. F. Gray and C. H. Yang, *J. Phys. Chem.*, **73**:3395 (1969); C. H. Yang, *ibid.*, 3407. Also see A. L. Berlad, *Combust. Flame*, **21**:275–288 (1973).

each contain more than one kinetic path. The energy-balance equation written in the spirit of thermal theory [Eq. (6-6)] becomes

$$\frac{dT}{dt} = -\frac{1}{\rho c_v}[(\alpha - 1)k_b \, \Delta E_b + k_t \, \Delta E_t][C] - \frac{s}{V\rho c_v}h(T - T_0) \qquad (6\text{-}17)$$

where $T_0$ is the vessel wall temperature and $[\Delta E_b]$ and $[\Delta E_t]$ are composite internal energy changes during reaction. As they stand, Eqs. (6-16) and (6-17) may be numerically integrated for any assumed initial value of $T$ and $[C]$ to determine the temperature and concentration time histories in the system and, if the system explodes, the delay to explosion.

Unfortunately, Eqs. (6-16) and (6-17) are highly nonlinear and are, therefore, essentially intractable mathematically. However, if they are rewritten as a ratio to eliminate time, the resulting equation in *phase space*,

$$\frac{dT}{d[C]} = \frac{1/\rho c_v[(\alpha - 1)k_b \, \Delta E_b + k_t \, \Delta E_t][C] - (sh/V\rho c_v)(T - T_0)}{[(\alpha - 1)k_b - k_t][C]} \qquad (6\text{-}18)$$

can be analyzed by examining it for the location and nature of its singularities by linearizing the equation in the neighborhood of each singularity that is discovered. The complete analysis of the behavior of (6-18) for all complex types of vessel explosion behavior is outside the scope of this text. It is sufficient to point out that in the simplest case that Gray and Yang analyzed Eq. (6-18) had two singularities in the ($[C]$, $T$) plane. One of these is a stable nodal point and the other a saddle point. This behavior is illustrated in Fig. 6-7, taken from Yang[2].

Here the separatrices of the saddle point are shown as dashed lines and the arrows represent the timewise behavior of the system obtained by integrating Eqs. (6-16) and (6-17). The stable nodal point is labeled $N$ and it represents the final steady-state slow-reaction condition under the specified initial vessel conditions. Note that the reactive fluid temperature is higher than the vessel wall

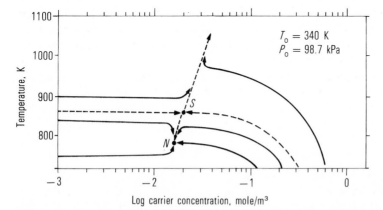

**Figure 6-7** The behavior of a vessel explosion on the ($[C]$, $T$) phase plane. Note the existence of a stable nodal point $N$ and the saddle point $S$. Here the ordinate $T$ is the gas-phase temperature in the vessel and $T_0$ is the wall temperature. (*Adapted from Gray and Yang*[2] *with permission of the American Chemical Society.*)

temperature just as predicted by thermal theory and that at the stable nodal point there is also a finite and constant "steady-state" radical concentration. Also note that above and to the right of the separatrix that starts at $T \cong 860$ K and continues through the saddle point to $[C] \cong 0.3$ moles/m$^3$ all initial conditions lead to explosion, while below and to the left all initial conditions lead to the stable slow oxidation solution. Thus, the unified theory predicts that even for the simplest case, explosion behavior can be triggered even at low initial temperature if sufficient reactive radicals are generated initially. This behavior is simply a generalization of the behavior of the $H_2$–$Cl_2$ system discussed in Sec. 6-4.

The unified theory can be made as sophisticated as one desires but even in its simplest form it contains the major important features of the real explosion process. In short, it illustrates that a real system explodes not simply because thermal or chain-branching mechanisms are operative, but because *both* are contributing during the induction period. It also illustrates that for supercritical conditions the path to explosion in the ($[C]$, $T$) plane quickly approaches the "explosion" separatrix before the actual "explosion" occurs. We will see in the next section that it also predicts the more complex explosion behaviors that are observed in certain combustion systems.

## 6-7 EXPERIMENTAL TECHNIQUES—VESSEL EXPLOSIONS

A large variety of experimental techniques have been used since Le Châtelier's time to try to deduce the kinetics of combustion reactions. In the early history of this field, the emphasis was on the determination of overall rates using Arrhenius kinetic modeling. More recently the emphasis has been on the determination of the rates of the elementary reaction steps that make up a "mechanism" and the subsequent forward numerical integration of this "mechanism" to determine the course of the combustion process. Even though theoretical approaches can be used to estimate preexponential factors and the temperature dependence of elementary reactions,[3] the actual determination of either the overall rate of a combustion process or the rate of one of the detailed steps in that process has remained primarily an experimental science. The major experimental techniques that have been or are being used are:

1. Vessel explosion studies
2. Plug-flow reactor studies
3. Chemical shock tube studies
4. Well-stirred reactor studies
5. Flame-structure studies
6. Radical excitation techniques

3. S. W. Benson, *Thermochemical Kinetics*, 2d ed., Wiley, N.Y. (1976).

Of the above six techniques, flame-structure studies will be discussed in Chap. 8 on flames; radical excitation techniques are a specialty of the kineticist and are concerned with the production of conditions such that one specific reaction can be studied without too much interference from other reactions. They are considered to be outside the scope of this text. The other four techniques will be summarized briefly in this and the next two sections to establish their principles and limitations.

Vessel explosion studies in gas-phase systems date back to Le Châtelier's time, and were undoubtedly conceived to determine an "ignition" temperature for his thermal theory of flame propagation. These experiments truly opened a Pandora's box of experimental results, the consequences of which have not been fully resolved as yet.

Two major types of vessel experiments have been used for the study of homogeneous or quasi-homogeneous explosions and the reaction kinetics leading to these explosions. These are (1) introduction into a preheated vessel, and (2) adiabatic compression by a piston device. These two experimental techniques have been used to study the explosion behavior of virtually all available exothermic systems with varying degrees of experimental sophistication. Four types of observations are usually made in studies of this type:

1. Does an explosion occur for a certain set of experimental conditions?
2. If an explosion occurs, how much time has elapsed between the onset of heating and the time of its occurrence; i.e., what is the explosion delay?
3. Unusual events, such as the appearance of cool flames during the preexplosion period, may be recorded.
4. The rate of change of concentration of the reactants and the pressure and temperature history of the mixture during the preexplosion and explosion period may be studied.

The introduction method is the oldest technique for studying vessel explosions and has yielded by far the most complex results. The earlier work using this technique led to the discovery of low-pressure and/or high-pressure peninsulas in many explosive mixtures. This is illustrated in Fig. 6-8.

The qualitative behavior of the low-pressure peninsula of Fig. 6-8 is adequately explained by the competition between branching and terminating reactions discussed in the previous two sections. The high-pressure peninsula and cool-flame behaviors illustrated in Fig. 6-8 do not occur when the fuel is hydrogen, carbon monoxide, or methane and are caused by degenerate branching which leads to the formation of high concentrations of very reactive organic peroxides. Experimentally, it is observed that when the concentrations of these species become sufficiently high, a visible "cool flame" flashes through the mixture and reduces their concentrations without markedly altering either the pressure or temperature. After the propagation of this cool flame the primary chain cycle continues even though its rate is possibly modified. The "cool flame" always appears at low temperatures and at pressures which are above

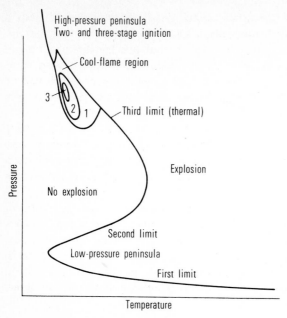

*similar in 6-26.*

**Figure 6-8** A schematic diagram showing the relative location of explosion, no explosion, and cool-flame regions for a fuel–oxygen–nitrogen mixture. Numbers in the cool-flame regions indicate the number of cool flames observed.

the chain-branching peninsula, along the thermal-explosion boundary in a $(P, T)$ plot. Multiple cool flames may occur before the slow reaction goes to completion or an explosion occurs. Also multiple-step explosions are observed in the upper peninsula.

This behavior has been extensively investigated and with the proper assumptions about the branching and terminating reactions is predicted by the unified theory of Gray and Yang[2]. Specifically for proper mechanism assumptions they observe that as many as four singularities can appear in phase space. This is shown in Fig. 6-9. The two new singularities that occur in Fig. 6-9 are a new saddle point and a limit cycle, which essentially indicates that cool-flame behavior is due to a thermokinetic oscillation in the system. Note that the limit cycle predicts that the system should exhibit a coupled $([C], T)$ oscillation for the proper initial conditions.

The introduction method has been used extensively to study explosion limits. However, in studies of this type both the first and third limits are intimately related to the vessel geometry and the first limit is also very sensitive to the effectiveness of the vessel surface in capturing radicals. Kineticists have found, for example, that it is necessary to "condition" a new explosion vessel by running many explosions to obtain reproducibility before taking quantitative data. Also, in many cases, the walls must be coated with a substance which has a rather low radical-capture efficiency to sufficiently retard the wall reactions so that the nature of the homogeneous second limit may be studied. Because of these effects, second-limit studies are by far the most fruitful and they have been used to determine the mechanism of homogeneous radical capture in some of the simpler systems. In general, the study of explosion limits in an introduction vessel must be accompanied by the results of other experimental techniques, before they may be used to help determine rate constants for the elementary steps of the chain cycle.

The introduction technique is generally not suited to the measurement of ignition-delay times. This is because, when the delay times are long, wall effects play an important role, while for short delay times, the time for heating the gas (mainly by conduction) becomes an appreciable fraction of the measured delay. In this connection we note that the simple theories discussed in the previous section completely ignore the initial transients in the system, and therefore represent only a first approximation to the real experimental behavior. This can lead to particularly serious errors when the introduction method is used for explosion studies.

There are three types of piston-compression devices that have been used to study homogeneous explosions. It is quite common to study explosion behavior in a spark ignition or diesel engine in which one cylinder has been modified to allow the introduction and compression of an experimental sample. There is also the specialized inertial-piston device in which a piston is driven into a closed tube by compressed air or helium and is allowed to bounce. The third

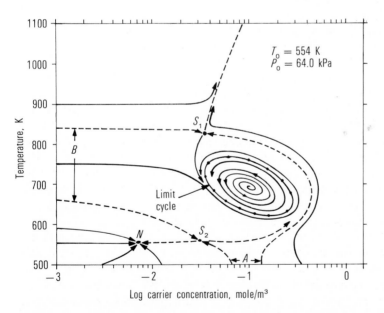

**Figure 6-9** The behavior of a more complex vessel explosion illustrating two saddle points, a nodal point, and a limit cycle. The limit cycle is a thermokinetic oscillation which can be physically interpreted as the multiple cool-flame behaviors observed in vessel explosion studies. Systems that have initial conditions in region $B$ (low radical concentration and intermediate temperatures) and region $A$ (low temperature and intermediate radical concentrations) end up in the limit cycle. Above and to the right of these ranges the system explodes while below and to the left it undergoes slow reaction at the nodal point $N$. (*Adapted from Gray and Yang*[2] *with permission of the American Chemical Society.*)

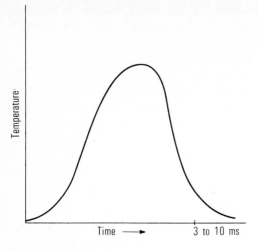

**Figure 6-10** Heating cycle in an adiabatic compression device without a locking device (schematic).

type is a device that mechanically accelerates a piston toward the back wall of a tube and then locks the piston in the neighborhood of its maximum displacement.

The use of the first two types of devices to study either explosion limits or delay times produces the heating cycle shown schematically in Fig. 6-10 only if the piston motion is slow enough to preclude the formation of strong shock waves during the compression portion of the cycle. The occurrence of an explosion during this cycle is usually recorded either by following the luminosity of the mixture through a window or by recording the presssure rise associated with the explosion by using fast-response pressure gauges. Schlieren observations have also been made.

Experiments of this type are fundamentally different from ordinary introduction experiments. In the first place, the time scale is markedly different. In ordinary vessel explosions the delay to explosion may be as much as 10 s while in adiabatic compression devices the total heating time is hardly ever greater than 10 ms. Thus the no-explosion area of the usual $(P, T)$ explosion limit plot will be increased because of the extremely short times available for explosion. However, one observes all the features ordinarily seen when using introduction techniques, such as the occurrence of cool flames and explosion peninsulas normally associated with degenerate and simple chain-branching reactions. One also observes a new phenomenon, the occurrence of detonation. However, multiple cool flames are not usually observed in adiabatic compression devices.

In these devices the short adiabatic heating time allows a determination of delay times because the heating is due entirely to a relatively homogeneous compression of the gas sample. Therefore the measured delay time need not be corrected for a conductive heat lag. However, if one wants to understand the significance of delay times, the second major difficulty becomes important. In this type of experiment the heating cycle always represents a significant portion of the total possible delay time, and the measured delays must therefore be related to explosion kinetics with a model that includes the finite heating rate. Because of this difficulty these types of piston devices have been used mostly to obtain engineering type data on fuels in connection with the actual design and operation of spark ignition or diesel engines. There have been few fundamental rate studies in explosion systems using these devices.

In contrast to the single-cycle devices, the piston-catching devices have the added capability of measuring a delay time which is significantly longer than the heating-cycle time. Thus they have the capability of studying overall rates in the intermediate temperature range where delays to explosion are somewhat longer than milliseconds but shorter than seconds.

## 6-8 EXPERIMENTAL TECHNIQUES— PLUG-FLOW REACTOR AND SHOCK TUBE

The second technique, the plug-flow or one-dimensional reactor, replaces the temporal history of an explosion by a spatial distribution. The apparatus consists of a heated tube in which heated oxidizer and fuel are mixed at an inlet and allowed to flow toward an outlet. The mixing is designed to be rapid and the flow down the tube is designed to be turbulent so that the flow velocity is relatively constant across the tube diameter. Also the temperature of the reactor is low enough so that $(Da_1)_I \to 0\ddagger$ and reaction kinetics controls the rate of conversion to products.

A shielded and coated thermocouple is used to sense the local gas temperature and the wall-temperature profile is adjusted to match the gas-temperature profile. Probes are used to extract samples of stable species and in some instances optical techniques are used to monitor transient species at stations along the tube. Since the flow velocity is low and pressure in the tube is almost constant, and the temperature and major species concentration profiles are known along the tube, a simple one-dimensional mass balance can be used to calculate the local rate of production of the species from their mole-fraction profiles

$$\frac{dx_i}{dt} = u \frac{dx_i}{dx} \qquad (6\text{-}19)$$

where $u = \rho_0 u_0 / \rho = \rho_0 u_0 M T_0 / M_0 T$. Here, the subscript 0 refers to initial conditions and $M$ is the mixture molecular weight based on analysis of the stable products. Most frequently the studies are carried out at relatively high dilution in a carrier gas such as nitrogen or argon and the molecular weight will be almost invariant down the length of the reactor.

The plug-flow reactor has been found to be useful for studying oxidation and pyrolysis reactions in the temperature range 1000 to 1900 K. Under these

‡ See Sec. 5-12 for further discussion.

conditions the reaction zone can be spread out to about one meter in length using relatively low subsonic flow velocities, so that compressibility effects are not important. This means that reactions with characteristic reaction times as short as three milliseconds can be studied in such a device.

The third device that is used extensively to study high-temperature reaction kinetics is the chemical shock tube. A conventional constant-area shock tube is a long pipe in which incident and reflected shock waves are produced in a gas or gas mixture by bursting a thin diaphragm separating a high- from a low-pressure region. Figure 6-11 contains an $(x, t)$ diagram illustrating the operation of an ideal one-dimensional shock tube, and temperature-time curves representing the behavior of gas samples at two initial positions in the expansion

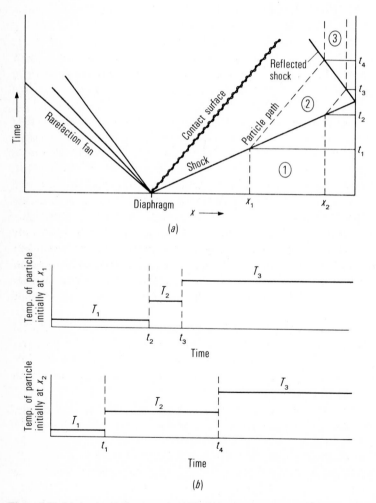

**Figure 6-11** (a) An $(x, t)$ diagram of an ideal shock-tube flow. (b) The time–temperature history of gas samples initially at positions $x_1$ and $x_2$ for the shock-tube flow shown in Fig. 6-11a.

chamber after the diaphragm burst. The state properties of regions 2 and 3 in Fig. 6-11 may be calculated using the flow-patching techniques discussed in Sec. 5-8.

The usefulness of the shock tube in kinetic studies of either endothermic or exothermic reactions is based primarily on the temperature-time history illustrated in Fig. 6-11. It has been shown that in a step shock wave the translational temperature of an element of the gas is raised rapidly (in less than $10^{-7}$ s) to a high value and then may be held at this temperature for relatively long times (up to 4 ms). Thus with the use of rapid-response instrumentation a relatively simple shock tube can become a high-temperature vessel for studying kinetic rates. Furthermore, because of the nature of the apparatus, it is relatively easy to obtain temperatures as high as 5000 K for these studies.

During the past 25 years there has been a tremendous amount of research on chemical systems using shock-tube techniques which cannot be adequately discussed in this text. The reader is referred to the review books listed in the bibliography for further information on specific kinetic studies using shock tubes. It is appropriate however, to make a few general observations concerning the technique and its relation to the study of explosion phenomena.

1. The gas is heated by a traveling shock wave and this is not a strictly homogeneous process. Furthermore, since the flow behind the shock wave is always subsonic relative to the shock, the occurrence of an exothermic or endothermic reaction behind the front will affect the behavior of the front. In highly exothermic systems rapid reactions lead to a rapid acceleration of the front and the shock wave eventually becomes a detonation wave. Thus the shock tube is not easily adaptable to the study of explosion delays in highly exothermic mixtures. There is one exception to this rule; at the back wall the initial conditions for the heating cycle are well defined, and homogeneous explosions have been observed in this region under certain circumstances. However, away from the back wall this explosion quickly becomes a detonation. Detonation behavior is discussed in Chap. 9 and detonation initiation is discussed in Chap. 13.
2. The shock wave leaving the back wall is not nearly as ideal as the incident shock wave. The primary cause of this effect is the reflected shocks interaction with the boundary layer in the flow behind the incident shock. This type of shock nonideality is least restrictive in gas mixtures with high specific-heat ratios.
3. The "vessel," even for back wall explosions, does not have a constant volume in a shock-tube experiment, because gas dynamics allows expansion when a local pressure rise occurs.
4. Because of the extremely short heating times and relatively short observation times available in the tube, the bulk of the gas sample does not contact the walls of the tube during the experiment. Thus rate constants obtained from experiments in the shock tube are fundamentally more useful in the interpretation of overall explosion kinetics than those obtained from ordinary vessel

experiments. This usefulness is augmented by the technique of working at high dilution. Quite frequently a reaction is studied in the presence of 95 to 99% argon as an inert carrier gas. The large excess of inert gas acts effectively as a thermostat bath for the reaction and allows one to study kinetics in even highly exothermic systems under conditions which are essentially isothermal, thus simplifying interpretation of results. Also, detonation cannot occur at high dilution.

## 6-9 EXPERIMENTAL TECHNIQUES—
## THE WELL-STIRRED REACTOR

The fourth experimental technique is based on the well-stirred or perfectly stirred reactor concept. In this concept one assumes that a chamber contains inlet and outlet ducts and is thermally insulated from its surroundings. It is further assumed that the fuel and oxidizer are introduced in such a manner that high-intensity turbulent mixing causes the contents of the reactor to be a spatially uniform mixture of reactants and products. This means that the rate of conversion of reactants to products must be controlled by the chemical rates of conversion and not by the mixing process. Under these conditions, as the flow rate through the reactor is increased, a point will be reached at which the chemical reactions are no longer fast enough to go to completion during the residence or stay time in the reactor. Thus it is observed that each fuel-oxidizer combination exhibits a unique "blowout" mass flow above which the combustion process will not stabilize in the reactor.

This technique has important practical implications relative to the performance of continuous combustors and bluff body flame holding in high Reynolds number flows. These applications will be discussed in Chap. 11. However, because of the complex nature of the flow in the reactor (rapid reaction is obtained by high-intensity turbulent mixing), the apparatus does not lend itself to the study of the rates of fundamental kinetic steps. For these two reasons, studies in well-stirred reactors have been mainly limited to hydrocarbon–air mixtures, and the modeling of the reactor behavior has been used to determine kinetic parameters for overall rate equations similar to Eq. (6-1).

Consider a well-stirred adiabatic chamber of volume $V$ with the global kinetics of Eq. (6-1) occurring everywhere at the same rate throughout its internal volume under conditions which are identical to the output conditions. Since it is assumed that pressure is constant and kinetic energy of the flow is small, we need consider only the species conservation equation for the fuel entering and leaving the reactor and the truncated energy equation $h = h_0$ where the subscript 0 refers to entrance conditions. The reaction itself will be assumed to be represented by the expression $F + vO \rightarrow (v + 1)P$, where $v$ is a stoichiometric coefficient not related to the exponents $n$ or $m$ of Eq. (6-1) because the reaction is an "overall" reaction. This is because $v$ is based on the overall stoichiometry of the reaction, but the overall order $n + m$ is based on experimental measurements of overall rates. Note that with this stoichiometry we can also assume that all species have the same molecular weight. For simplicity we assume that the entering mixture is fuel lean, i.e., that $(v + e)(x_{fu})_0 = (x_{ox})_0$ where $e$ is the excess oxidizer present in the incoming stream and we write the fuel species balance equation for the reactor as

$$\frac{J[(x_{fu})_0 - x_{fu}]}{V} = \frac{d[F]}{dt} \tag{6-20}$$

where $J$ is the total inlet flow rate with units of moles per second. Substituting Eq. (6-1) into (6-20) and defining $\eta = 1 - x_{fu}/(x_{fu})_0$ yields the expression

$$\frac{J[(x_{fu})_0 - x_{fu}]}{VP^{(n+m+l)}} = \frac{A(x_{fu})_0^n(1 - x_{fu} - x_{in})_0^m(x_{in})_0^l}{(RT)^{(n+m+l)}} \cdot (1 - \eta)^n [v(1 - \eta) + e]^m \exp\left(\frac{-E^*}{RT}\right) \tag{6-21}$$

Defining $\theta = T/T_0$ and $\varepsilon = E^*/RT_0$ yields the relationship

$$\eta = \mathrm{Da}_1\{(1 - \eta)^n \cdot [v(1 - \eta) + e]^m\} \cdot \frac{\exp(-\varepsilon/\theta)}{\theta^{(n+m+l)}} \tag{6-22}$$

where Damköhler's first ratio for the reactor as a whole is defined as

$$\mathrm{Da}_1 = \frac{A(x_{fu})_0^{n-1}(1 - x_{fu} - x_{in})_0^m(x_{in})_0^l}{(RT_0)^{(n+m+l)}} \Bigg/ \frac{J}{VP^{(n+m+l)}} \tag{6-23}$$

and represents a ratio of the characteristic stay time in the reactor to a characteristic time for the chemical reaction.

The energy balance equation can be written within the framework of the assumptions of Sec. 6-2 by assuming constant heat capacities and equating the temperature rise to the heat released by the reaction, using the $Q$ of Eq. (6-2). When this is done we obtain the relationship $\eta = (\theta - \theta_0)/(\theta_m - \theta_0)$ where

$$(\theta_m - \theta_0) = \frac{Q(x_{fu})_0}{C_p} \tag{6-24}$$

Equations (6-22) and (6-24) are relationships between the dimensionless extent of reaction $\eta$ and temperature $\theta$. For any specified initial mixture Eq. (6-24) is a straight line on an $(\eta, \theta)$ plot while Eq. (6-22) is a more complex curve whose locus depends on the value of $\mathrm{Da}_1$. Any intersections of the two curves represent possible solutions for the operation of a well-stirred reactor. Equations (6-22) and (6-24) are plotted in Fig. 6-12a. Here, as the flow through the reactor is increased, $\mathrm{Da}_1$ decreases and Eq. (6-22) can be progressively represented by the curves labeled $A$, $B$, $C$, and $D$ in Fig. 6-12a. All these curves have a physically unrealistic solution at $\eta \cong 0$. Curves $A$ and $B$ have a dynamically unstable solution at intermediate temperatures and a stable high temperature solution. Curve $C$ has a tangency solution at high temperature and $D$ has no high temperature solution. As can be seen from Fig. 6-12a there is a range of Damköhler numbers which yield three distinctly different values for $\eta$. This

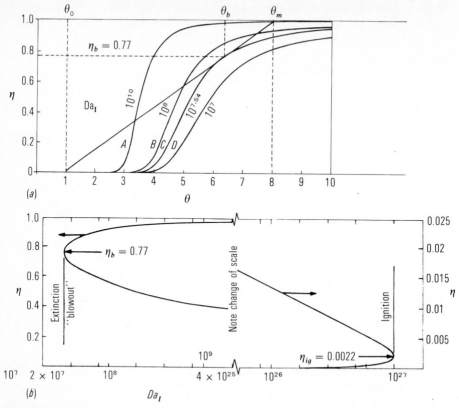

**Figure 6-12** The behavior of a well-stirred reactor for $E^* = 175$ kJ/mole, $e = l = 0$, $v = n = m = 1$, and $\theta_m/\theta = 8$. (a) Plots of Eqs. (6-24) and (6-22) for various values of $Da_I$. (b) The behavior of $Da_I$ vs. $\eta$ showing extinction.

behavior is shown in Fig. 6-12b which was calculated using Eqs. (6-22) and (6-24) for the same conditions that were used to construct Fig. 6-12a. This type of behavior has been observed in other highly exothermic combustion processes such as a counterflow diffusion flame. Usually the theorists say that the two extreme values of $Da_I$ represent extinction and ignition. However, the lower branch is not physically real and has never been observed experimentally. All the high-temperature solutions represent combustion solutions, and the tangency point on curve $C$ represents the maximum throughput velocity at which high-temperature combustion can occur in the reactor. Above this velocity theory predicts that the reactor will "blowout" under conditions where combustion is not complete (i.e., $\eta_b < 1$ and $\theta_b < \theta_m$, where the subscript $b$ represents blowout conditions). Thus combustion efficiency at blowout is less than one. This behavior is observed experimentally.

Blowout conditions are usually determined as a function of equivalence ratio and correlated to the overall activation energy $E^*$ and the overall order of the reaction $n + m + l$. Such a correlation for octane–air is shown in Fig. 6-13.

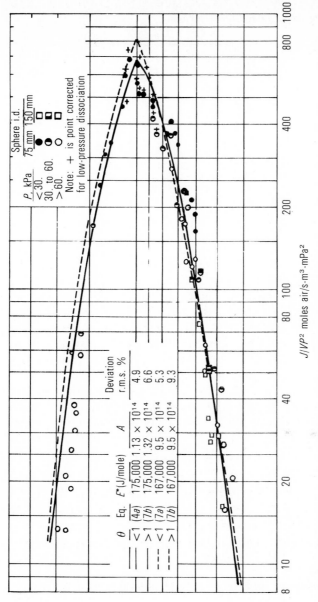

**Figure 6-13** Blowout data correlation for second-order reaction corrected to 400 K inlet temperature. *(Adapted from J. P. Longwell and M. A. Weiss, Ind. Eng. Chem., vol. 47, p. 1634 (1955), with permission of the American Chemical Society.)*

**Table 6-2‡ Combustion intensity of various combustion chambers**

| Apparatus | Heat release rate, $J\ m^{-3}\ s^{-1}\ kPa^{-2}$ |
|---|---|
| Domestic boiler | $0.01\text{--}0.03 \times 10^3$ |
| Navy boiler | $0.20\text{--}0.25 \times 10^3$ |
| Industrial gas turbine | $0.70\text{--}3.00 \times 10^3$ |
| Aircraft gas turbine | $4.0\text{--}11.0 \times 10^3$ |
| Well-stirred reactor | $\approx 170 \times 10^3$ |
| Premixed laminar flame zone | $\approx 300 \times 10^3$ |

‡ Data with permission from "Literature of Combustion of Petroleum," *Advances in Chemistry Series No. 20*, American Chemical Society (1958), p. 33.

Note that the overall reaction order is very close to 2. The conditions listed in Fig. 6-13 were approximated to construct Fig. 6-12. These are $E^* = 175$ kJ/mole, $e = 0$, $v = n = m = 1$, $l = 0$, and $\theta_m/\theta_0 = 8$.

Additionally, experiment has shown that the overall equation cannot model the process correctly for both lean and rich blowout. Specifically, for lean blowout very few hydrocarbons are present at incipient blowout and $CO \rightarrow CO_2$ oxidation is the rate-controlling step. For rich blowout, on the other hand, hydrocarbon fragments and free hydrogen are present at blowout, and earlier oxidation steps must be rate-determining.

It is interesting to note that the intensity of combustion (units $J\ m^{-3}\ s^{-1}\ kPa^{-2}$) in well-stirred reactors approaches that of the very narrow hot-reaction zone of a premixed gas flame. Typical combustion intensities are tabulated in Table 6-2.

The literature contains many analyses of the effect of imperfect mixing, of heat loss through the walls, and of complex chemistry on the behavior of well-stirred reactors. The reader is referred to the bibliography for further information on this subject.

## 6-10 SPECIFIC SYSTEMS—DETAILED KINETICS

### 6-10-1 The Technique and the Ozone Decomposition Reaction

In the last decade or so there has been a tremendous increase in both our understanding of the elementary steps of certain relatively simple combustion processes and our ability to numerically integrate complex kinetic schemes with speed and accuracy. In this section we first examine in considerable detail the full numerical integration of a very simple system, the ozone decomposition explosion, and then examine the results of such analyses for more complex systems. Ozone decomposition kinetics can be adequately modeled using just two irreversible and one reversible kinetic rate expressions. These are

$$O_3 + M \xrightarrow{k_1} O + O_2 + M \tag{6-25}$$

$$O_3 + O \xrightarrow{k_2} O_2 + O_2 \tag{6-26}$$

and

$$O + O + M \underset{k_{3r}}{\overset{k_3}{\rightleftharpoons}} O_2 + M \tag{6-27}$$

where the rate constants suggested by Baulch et al.[4] are

$$k_1 = 2.48 \times 10^8 \exp(-11{,}430/T) \text{ m}^3/\text{mole-s}$$

with the relative third-body efficiency of $O_2 = 1.0$ and $O_3 = 0.4$

$$k_2 = 5.2 \times 10^6 \exp(-2090/T) \text{ m}^3/\text{mole-s}$$

and

$$k_3 = 19.0 \exp(900/T) \text{ m}^6/\text{mole}^2\text{-s}$$

with the relative third-body efficiency of all species equal to one. In this scheme the third reaction is taken to be the only reversible reaction and its reverse rate is determined using Eq. (4-20) and the equilibrium constants of formation of O atoms from App. B

$$k_{3r} = \frac{101{,}325 K_{pf}^2 k_3}{8.31434\ T}$$

By combining Eqs. (2-16) and (4-5) and applying them to the three rate equations we obtain expressions for the rate of change of moles of each species per unit mass of the mixture

$$\frac{dn_{O_3}}{dt} = \frac{1}{\rho} \left[ -k_1[O_3]\{[O_2] + 0.4[O_3]\} - k_2[O_3][O] \right] \tag{6-28}$$

$$\frac{dn_{O_2}}{dt} = \frac{1}{\rho} \big[ k_1[O_3]\{[O_2] + 0.4[O_3]\} + 2k_2[O_3][O]$$

$$+ k_3[O]^2\{[O_3] + [O_2] + [O]\}$$

$$- k_{3r}[O_2]\{[O_3] + [O_2] + [O]\} \big] \tag{6-29}$$

and

$$\frac{dn_O}{dt} = \frac{1}{\rho} \big[ k_1[O_3]\{[O_2] + 0.4[O_3]\} - k_2[O_3][O]$$

$$- 2k_3[O]^2\{[O_3] + [O_2] + [O]\}$$

$$+ 2k_{3r}[O_2]\{[O_3] + [O_2] + [O]\} \big] \tag{6-30}$$

where $\rho$ is the density of the mixture at time $t$.

4. D. L. Baulch, D. D. Drysdale, J. Duxbury, and S. Grant, *Evaluated Data for High-Temperature Reactions*, vol. 3, University of Leeds, Leeds, England (1976).

In addition to these mass-balance equations we also need an energy-balance relationship if we are to follow the explosion process. We will consider only the adiabatic constant-pressure explosion case. Thus the energy-balance requirement becomes $h(t) = h_i$.

For any specific set of thermodynamic and species concentration conditions during the reaction the right-hand side of Eqs. (6-28), (6-29), and (6-30) are known. This means that the rate of change (moles per kilogram) can be calculated for each species at any time. The simplest way to determine the progress of the reaction is to use a forward numerical integration technique. For a small but finite step we can write Eqs. (6-28), (6-29), and (6-30) in the form

$$(n_{O_3})_{t_{n+1}} = (n_{O_3})_{t_n} + \left(\frac{dn_{O_3}}{dt}\right)_{t_n} \Delta t \qquad (6\text{-}31)$$

$$(n_{O_2})_{t_{n+1}} = (n_{O_2})_{t_n} + \left(\frac{dn_{O_2}}{dt}\right)_{t_n} \Delta t \qquad (6\text{-}32)$$

$$(n_{O})_{t_{n+1}} = (n_{O})_{t_n} + \left(\frac{dn_{O}}{dt}\right)_{t_n} \Delta t \qquad (6\text{-}33)$$

where

$$\Delta t = t_{n+1} - t_n$$

for the ($n$th) forward step. Once the new composition at time $t_{n+1}$ has been determined the new temperature of the system is determined from the JANAF tables by requiring that $h_{n+1} = h_i$ for the constant-pressure systems we are considering. For a constant-volume system the ideal gas law could be used to determine a new pressure and concentrations would be unchanged. For a constant-pressure system the ideal gas law is used to calculate a new density and therefore new species concentrations at time $t_{n+1}$. The new values of the state variables and concentrations at time $t_{n+1}$ can now be used to reevaluate the right-hand terms of Eqs. (6-31), (6-32), and (6-33) for the next forward step.

Unfortunately, even though the simple forward integration scheme described above illustrates how a set of coupled ordinary differential equations may be solved it is usually not possible to perform the integration in this manner. This is because the accuracy of each step of the numerical integration is determined by the time constant of the fastest reaction and in most combustion systems the fastest reaction is many orders of magnitude more rapid than the overall process that is being studied. Thus, to follow the overall process accurately using ordinary integration techniques requires a large amount of computer time. Systems of ordinary differential equations that contain an inordinately large range of time constants are commonly called *stiff* equations. Fortunately, Gear and others[5] have recently developed numerical

---

5. C. W. Gear, *Numerical Initial Value Problems in Differential Equations*, Prentice/Hall, N.Y. (1971); L. Lapidus and J. H. Seinfeld, *The Numerical Solution of Ordinary Differential Equations*, Academic Press (1971); also see bibliography.

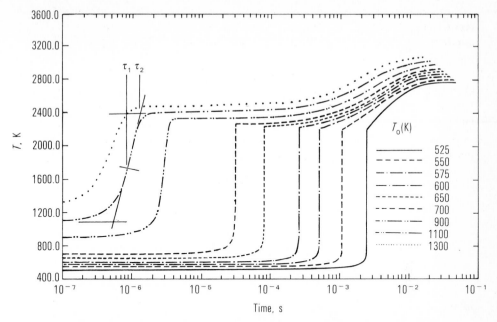

**Figure 6-14** Temperature–time adiabatic, constant pressure explosion histories in the ozone system for various initial temperatures $T_0$. No oxygen atoms in the system initially. The pressure is 101.325 kPa. The times $t_1$ and $t_2$ are explained in the caption of Fig. 6-15.

techniques for rapidly integrating stiff ordinary differential equations. A discussion of the details of these modern numerical integration techniques is outside the scope of this text and the reader is referred to the references for further information.

In the pure ozone system at one atmosphere initial pressure the equations are initially "stiff" if the initial oxygen atom concentration is zero and the initial temperature is below about 700 K. They then again become "stiff" when the ozone concentration becomes very small. Thus, even in this combustion system, some type of sophisticated technique for numerical integration is required over the entire range of initial and final conditions.

Figure 6-14 is a plot of temperature–time histories for the adiabatic constant-pressure explosion of ozone at 101.325 kPa and various initial temperatures for the case where the oxygen atom concentration is initially zero. This, as well as the other ozone explosion figures in this section, were obtained by numerically integrating Eqs. (6-28), (6-29), and (6-30), and the enthalpy-balance equation using a modified Gear technique.‡ Note in Fig. 6-14 that the temperature rise becomes increasingly sharper as the initial temperature is decreased. Also note that after the explosion has occurred the temperature rises

‡ Program prepared by Mr. Russell Skocypec, graduate student in Mechanical Engineering, U.I.U.C.

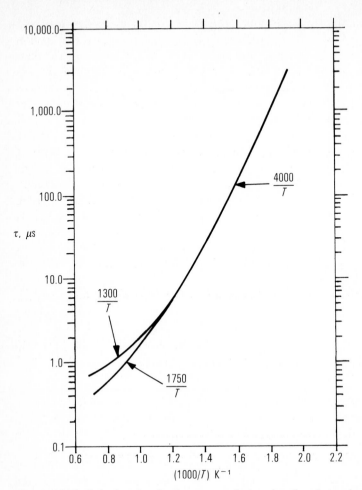

**Figure 6-15** The induction time for the ozone explosion plotted against reciprocal temperature. The lower line is $\tau_1$ in Fig. 6-14, the upper $\tau_2$. Note that for this explosion system the apparent activation energies for the overall reaction do not agree with those of the individual steps [Eqs. (6-25), (6-26), and (6-27)].

slowly to a final equilibrium value due to the relatively slow equilibration of the $O_2 \rightleftarrows 2O$ reaction. This means that the O atom concentration must have exceeded its final equilibrium value during the explosion process. This type of behavior is typical of both straight-chain and chain-branching explosions.

Figure 6-15 is a plot of the ignition delay time or induction time, as a function of reciprocal temperature. Effective activation energies for the explosion based on the measured delays are listed on the figure. Notice that at high temperatures there can be ambiguity about the definition of $\tau$ because of the slow final rise to explosion. This shows that the effective activation energy of the overall process can be affected by the way the delay time is measured in certain cases.

Figure 6-16 is a plot in the ([O], $T$) plane, which is the plane that was used to visualize the behavior of an exothermic radical branching-chain reaction when we discussed the unified theory of explosions in Sec. 6-6. Notice the similarities between Figs. 6-16 and 6-7. There is no separatrix in Fig. 6-16 because an adiabatic system will explode for *all* initial conditions. Figure 6-16 does show, however, that the ([O], $T$) trajectory follows almost the same path irrespective of the initial conditions, except when the initial [O] is so high that it overwhelms the normal explosion process. This behavior was observed in the unified theory and is typical of explosion processes, even when the chemistry is relatively complex. Figure 6-16 also shows an essentially discontinuous change in the slope of the ([O], $T$) curve at high [O], for the low initial temperature runs. This behavior is real for this simple system and represents the fact that the explosion process, which yields a nonequilibrium end state in which the [O] is higher than the final equilibrium value, is much more rapid than the final equilibration process. As Fig. 6-16 shows, the final end states lie on an "equilibrium composition line" for the cases studied. This is because the pressure is the same for all the runs.

Figure 6-17 is a representation of the rate of appearance or disappearance of the individual species due to each of the individual reaction processes occurring in the system. In this diagram, the width of the arrows is roughly proportional to the rate of conversion by that process. This type of representation is usually more useful when the reaction is more complex. It will be used again later in this section.

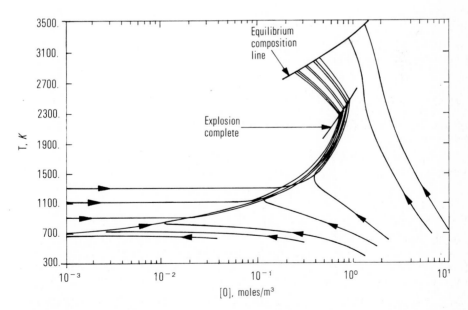

**Figure 6-16** The adiabatic constant pressure ozone explosion in the ([O], $T$) plane for various initial conditions of temperature and oxygen atom concentration.

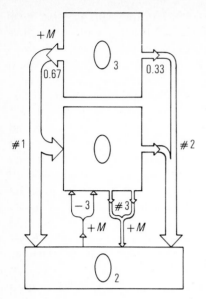

**Figure 6-17** Relative rates of the three reactions which are involved in the ozone decomposition explosion. Temp = 1500 K, $T_i$ = 900 K. Note that O atom recombination is consuming about 50 percent of the O atoms produced by reaction 1. Also note that at this temperature the conversion to the "explosive" state is about 50 percent and that the system has reached the explosion path of Fig. 6-16.

## 6-10-2 The Hydrogen–Oxygen Reaction

This reaction mechanism has been extensively studied during the past half-century and is now considered to be the classic example of a potentially explosive chain-branching reaction.[6] More recently it has become apparent that the hydrogen chain-branching mechanism is central to the mechanisms that are operative during the oxidation of carbon monoxide which contains traces of hydrogen-bearing compounds as well as the high-temperature oxidation of organic compounds of all types.

The primary branching chain with one example-initiating and one example-terminating reaction was introduced in Sec. 6-5. More complete mechanisms have been proposed which use up to 22 forward reaction kinetic steps and allow for the formation of the hydroperoxyl radical $HO_2$ and hydrogen peroxide $H_2O_2$.

Only one reaction in the chain-branching set is endothermic. This is the reaction

$$H + O_2 \rightarrow OH + O \tag{6-34}$$

It has been found that at high temperature and low pressure the delay to explosion follows Eq. (6-15), that is, $\ln([O_2] \cdot \tau) = E^*/RT + \text{const}$, where the slope of the explosion delay line yields $E^*/R$ where $E^*$ is approximately the activation

6. G. Dixon-Lewis and D. J. Williams, "The Oxidation of Hydrogen and Carbon Monoxide," *Comprehensive Chemical Kinetics*, vol. 17, eds. C. H. Bamford and C. H. F. Tipper, pp. 1–248 (with 523 references), Elsevier, Amsterdam (1977).

energy of reaction (6-34). It has also been found that as temperature is lowered and pressure is increased the reaction

$$H + O_2 + M \rightarrow HO_2 + M \tag{6-35}$$

starts to compete directly with reaction (6-34).

Figure 6-18 illustrates the rate of the conversion processes when recombination and the $HO_2$ formation reaction are insignificant. In this figure recombination is illustrated as the $H + OH + M \rightarrow H_2O + M$ and the $OH + O + M \rightarrow O_2 + H + M$ reactions and they are insignificant because only 10 percent of the hydrogen has formed radicals by the chain mechanism. The $H + O_2 + M \rightarrow HO_2 + M$ reaction is insignificant because the temperature is high and the pressure is low. Also note that almost all the OH radicals produced by the endothermic reaction $H + O_2 \rightarrow OH + O$ and the rapid exothermic reaction $O + H_2 \rightarrow OH + H$ form the product, water, by the reaction $OH + H_2 \rightarrow H_2O + H$, thus maintaining the chain-branching process. It should also be noted that even though about 10 percent of the molecular hydrogen has been consumed and a significant quantity of water vapor has been formed, the temperature has risen only about 5 K. This is because the main hydrogen–oxygen chain is almost thermally neutral, the exothermicity of $H_2O$ formation being offset by the formation of the endothermic radicals H, O, and OH in the chain. Thus the high-temperature–low-pressure hydrogen–oxygen reaction is also the primary example of a thermally neutral chain-branching reaction in which the radical concentrations first increase exponentially without causing a significant temperature change. The final temperature rise occurs primarily by recombination of the excessive amounts of radicals produced by the chain process.

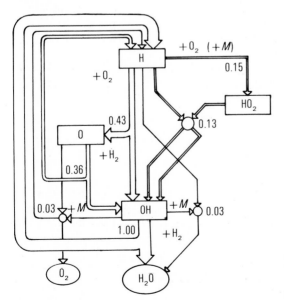

**Figure 6-18** Relative rates of the major reactions during the induction period of the hydrogen–oxygen explosion. Temp = 1105 K, $T_i$ = 1100 K, 10 percent hydrogen consumption. The reverse rate of the $H + O_2 \rightarrow O + OH$ reaction is about 10 percent of the forward rate. All others forward only. *(Calculation performed and drawing supplied by Dr. J. R. Creighton, Lawrence Livermore National Laboratory, Livermore, California.)*

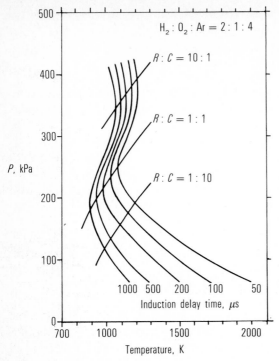

**Figure 6-19** Calculated induction delay times for the hydrogen–oxygen explosion reaction. The iso $R:C$ lines are for constant ratios of the rate of the $H + O_2 + M$ recombination reaction to the $H + O_2$ chain-branching reaction. *(Adapted, with permission, from Fig. 1 of "Weak and Strong Ignition, II, Sensitivity of the Hydrogen–Oxygen System," by E. S. Oran and J. P. Boris, Combust. Flame, vol. 48, 149–161, (1982). Also appeared as NRL M.R. 4671, Dec. 7, 1981, Naval Research Laboratory, Washington, DC.)*

At low temperatures and high pressures the nature of the hydrogen–oxygen reaction changes markedly. Under these conditions, reaction (6-35) becomes as fast or faster than the rate-controlling chain-branching step, reaction (6-34), and the net rate of the reaction is controlled primarily by the fate of the $HO_2$ radical. Figure 6-19 is a plot of iso-explosion-delay lines adapted from the calculations of Oran et al.[7]

Since reaction (6-35) is a three-body reaction, its rate is proportional to pressure squared and it has only a very weak temperature-dependence. On the other hand, (6-34) has a highly temperature-dependent rate which is proportional to only the first power of pressure. This means that the ratio of the rates of these reactions is a function of temperature and pressure. The lines on Fig. 6-19 labeled $R:C = 10:1$, $1:1$, and $1:10$ represent the conditions where the ratios of the rates of the recombination reaction (6-35) to the chain-controlling reaction (6-34) have those constant values. Note that the region below and to the right of $R:C = 1:10$ is where the typical "high" temperature reaction mechanism dominates while the region above and to the left of $R:C = 10:1$ is where $HO_2$ radical behavior dominates the mechanism. Also note that in the region where $R:C = 1:1$ the delay to explosion is *extremely* sensitive to the initial conditions. This behavior is typical of all real hydrocarbon–air explosion systems. In all cases one finds a high-temperature–low-pressure region in which

---

7. See caption of Fig. 6-19 for reference.

the explosion process proceeds by a rather straightforward chain mechanism and a low-temperature–high-pressure region where complex reaction processes tend to slow the explosion rate drastically and at the same time produce a complex set of intermediates that are not present at high temperatures. In general, these complicating reactions are radical destroying and when the temperature becomes sufficiently low the explosion process no longer occurs. We note that most fuel–oxygen mixtures can be stored indefinitely at room temperature even though they are thermodynamically in a metastable state. This is because the overall oxidation process has a zero rate at room temperature even though the rates of all the chain steps are still finite.

## 6-10-3  Carbon Monoxide Oxidation

The direct oxidation of carbon monoxide by oxygen has an extremely high activation energy and therefore is an extremely slow process even at high temperatures. It has been found that in systems that contain even small quantities of hydrogen-containing compounds, the primary processes that control the rate of oxidation of carbon monoxide are the production of OH radicals by the endothermic rate-controlling reaction, Eq. (6-34), and the subsequent formation of $CO_2$ by the rapid exothermic exchange reaction

$$OH + CO \rightarrow CO_2 + H \qquad (6-36)$$

Thus, the CO oxidation process can be modeled by adding reaction (6-36) to the hydrogen–oxygen chain mechanism. In this case reaction (6-36) competes with the other exothermic and rapid kinetic step $OH + H_2 \rightarrow H_2O + H$. However, even though the rate equations in the mechanism are the same, the overall reaction process is uniquely different because of the low concentrations of the hydrogen-containing species. This is shown in Fig. 6-20. During the induction period of this reaction, i.e. for times less than about one millisecond, all the radical concentrations as well as the $CO_2$ concentration are growing exponentially at the same rate and therefore maintaining their concentration ratios. This is typical of any high-temperature explosion induction period. At the end of the induction period the radical concentrations all change relatively slowly and the dominant process is the conversion of CO to $CO_2$ with the liberation of energy. Thus, in contrast to the hydrogen–oxygen reaction where recombination processes caused the temperature to rise, in the oxidation of carbon monoxide the final equilibration occurs because of the straight-chain reduction of $O_2$ to OH by the endothermic rate-controlling reaction (6-34) coupled to the rapid CO oxidation process, reaction (6-36). This is shown in Fig. 6-21, which is a conversion rate drawing for this system when about 50 percent of the CO has been consumed (or at about $t = 4$ ms in Fig. 6-20). In this diagram the oxygen atom is responsible for about half of the rate of production of $HO_2$, and most of the oxygen atoms produced by the primary OH formation reaction eventually react to again form oxygen molecules.

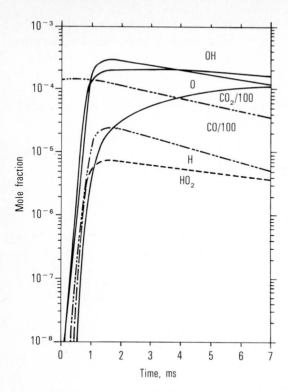

**Figure 6-20** The oxidation of moist CO in air at constant pressure, $T_0 = 1100$ K. Initial mole fractions: CO = 0.0135. $H_2O = 0.0248$. Note that initially all radical concentrations increase at the same exponential rate, but have different absolute magnitudes. This is typical of chain-branching induction zone behavior. *(Calculation performed and graph supplied by Dr. J. Creighton, Lawrence Livermore National Laboratory, Livermore, California.)*

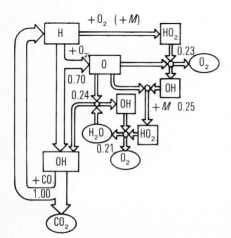

**Figure 6-21** Relative rates of the major reactions during moist CO oxidation, same conditions as Fig. 6-20. Drawn for $t = 4$ ms, that is, when about half of the CO has been consumed. Note that the $H + O_2$ reaction supplies the major amount of OH, which in turn supplies H atoms as a result of CO oxidation by the OH + CO reaction. *(Calculation performed and figure supplied by Dr. J. Creighton, Lawrence Livermore National Laboratory, Livermore, California.)*

The reaction [Eq. (6-36)] is dominant only at high temperatures and low pressures in homogeneous explosions. At low temperatures and high pressures the reaction

$$CO + HO_2 \rightarrow CO_2 + OH$$

becomes competitive and in the preheat zone of rich CO flames at high pressure the reactions

$$H + CO + M \rightarrow HCO + M$$

can become important. This is because of the high diffusivity of the H atom.

## 6-10-4 Methane Oxidation

Methane is a unique hydrocarbon. Not only is it the simplest hydrocarbon structurally, it is also the only hydrocarbon that has no C—C bond. This causes the mechanism of its oxidation to be uniquely different from any of the higher hydrocarbons. This uniqueness has been known for a long time. Methane is known to exhibit a delay to explosion, in shock tube experiments, which is almost 10 times the explosion delay for the higher aliphatics. It also has an unusually high autoignition temperature. The key to the methane oxidation reactions has been found to be the fact that the initial step must involve breaking a C—H bond to form a methyl radical, $CH_3$. The subsequent fate of the methyl radical then determines how the oxidation process proceeds. Specifically Warnatz[8] has summarized the high-temperature methane oxidation scheme as shown in Fig. 6-22. In this figure the first attack on methane is by O, H, or OH radicals to produce $CH_3$ which can then either be oxidized to formaldehyde, $CH_2O$, or condensed to form either ethane, $C_2H_6$, or the ethyl radical, $C_2H_5$. He reports that even in lean systems 30 percent of the $CH_3$ radicals undergo these condensation reactions and that in rich systems the amount rises to 80 percent. If the $CH_3$ is oxidized to formaldehyde this species is further oxidized to the formal radical CHO and eventually forms carbon monoxide, CO. The rate of these reactions at high temperatures is relatively rapid but the final oxidation of CO to $CO_2$ is relatively slow and depends primarily on the size of the OH radical pool in the system. Notice from Fig. 6-22 that once the condensation process occurs the oxidation becomes much more complex and the mechanism contains many more complex intermediate compounds. Mechanisms have been constructed and integrated for $CH_4$ oxidation that contain as many as 30 to 35 species and 80 to 100 distinct elementary reaction steps. Thus even this simplest hydrocarbon has an extremely complex oxidation mechanism, primarily because C—C bonds can form during the oxidation process.

8. J. Warnatz, *Eighteenth Symposium (International) on Combustion*, The Combustion Institute, Pittsburgh, Pa., pp. 369–384 (1981); ———, "Chemistry of Stationary and Non-Stationary Combustion," pp. 162–188, *Modelling of Chemical Reaction Systems*, eds. K. H. Ebert, P. Deuflhared, and W. Jäger, *Chemical Physics* 18, Springer-Verlag, Berlin (1981).

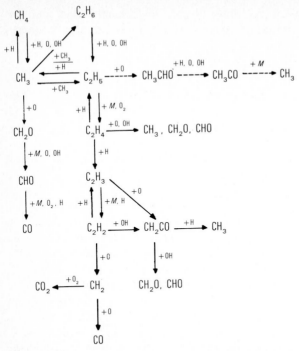

**Figure 6-22** Mechanism of combustion of $C_1/C_2$ hydrocarbon species at high temperature *(after Warnatz (Ref. 8), with permission of Verlag Chemie GMBH, Weinheim, Federal Republic of Germany)*.

Even though the reaction is overall quite complex it has been shown that the $CH_3$ radical has a temporal behavior similar to that of any reactive radical in a simpler mechanism. This is shown in Fig. 6-23 which is a phase-plane plot of the mole fraction of the $CH_3$ radical vs. temperature. This plot was produced by integrating a rather complex mechanism for $CH_4$ oxidation.

## 6-10-5 Acetylene and Ethylene Oxidation

These two hydrocarbons are also unique because they are both exothermic compounds (i.e., have a positive heat of formation), and do not contain an aliphatic bond. Their overall activation energy based on induction delay times is about 75 J/mole, which is very close to the activation energy of reaction (6-34), the primary rate-controlling reaction in the oxidation of hydrogen. Radical attack of the parent molecules occurs by reactions of the type (illustrated for acetylene)

$$O + C_2H_2 \rightarrow OH + C_2H$$

$$H + C_2H_2 \rightarrow H_2 + C_2H$$

$$OH + C_2H_2 \rightarrow H_2O + C_2H$$

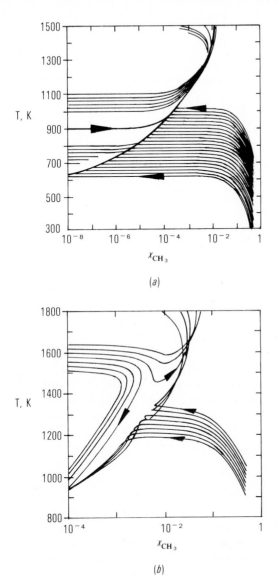

**Figure 6-23** The methane explosion in the phase plan. Equivalence ratio = 0.5. The arrows show the temporal behavior. (*a*) The adiabatic explosions case. (*b*) The explosion when heat loss is considered. Note that the separatrix is not drawn in this figure even though its behavior can be inferred from the lines that are present. (*Adapted from Guirguis, et al.*[9] *with permission.*) Note the similarities between these figures and Figs. 6-7 and 6-16.

9. R. H. Guirguis, A. K. Oppenheim, I. Karasalo, and J. Creighton, "Thermochemistry of Methane Ignition Combustion in Reactive Systems," ed. J. R. Bowen, *Prog. Astronaut. Aeronaut., AIAA*, New York, **76**: 134–153 (1981).

as well as

$$O + C_2H_2 \rightarrow CH_2 + CO$$

$$O_2 + C_2H \rightarrow HCO + CO$$

$$O_2 + CH_2 \rightarrow HCO + OH$$

In both these systems the induction zone reactions are exothermic, overall. As in all other cases the concentration of carbon monoxide goes through a maximum before the slower CO oxidation process produces $CO_2$. Details of the kinetics are still not well understood.

### 6-10-6 Oxidation of Higher Hydrocarbons

The oxidation of more complex hydrocarbons is still far from being understood in an unambiguous way. It is known that C—C single bonds are weaker than either double, triple, or C—H bonds, and that the first hydrocarbon fragment radicals are produced by the rupture of such a bond if one is available. As with the simpler molecules the high-temperature–low-pressure oxidation reactions are considerably simpler than those that occur at low temperature and high pressure. Also, within the constraints of the discussion in Sec. 6-10-3, final oxidation of CO to $CO_2$ still involves only reaction (6-36) and is relatively slow, because of the slow rate of production of OH by reaction (6-34) which is endothermic. It is also true that the number of intermediate compounds that appear during the oxidation process becomes truly large, and these compounds consist not only of hydrocarbon fragment radicals but also include a whole array of partially oxidized hydrocarbon fragments which may also be radicals. Pollard[10] presents an extensive discussion of the oxidation process in the neighborhood of the cool-flame region of Fig. 6-8. The reader is referred to the bibliography for more complete descriptions of reaction kinetics studies applicable to the oxidation of hydrocarbons.

### 6-11 COMBUSTION CHEMISTRY MODELING

Unless one can unambiguously identify all the reaction steps and their rates for any specific combustion problem one must always "model" the actual chemistry in some manner. We have thus far used a number of specific "models" for chemistry as appropriate to the sophistication of the description that was desired. It now seems appropriate to briefly categorize the major types of "chemistry" models that are used in combustion studies to identify their limitations and their relationship to each other. Those that are discussed here are

10. R. T. Pollard "Hydrocarbons," in *Comprehensive Chemical Kinetics*, **17**, eds. C. H. Bamford and C. H. F. Tipper, pp. 249–368, Elsevier, Amsterdam (1977).

really only basic examples of the large variety of combustion models that exist in the literature.

By far the simplest way to model chemistry is by legislated heat addition to a working fluid with specified (and usually rather simple) properties. In the heat-addition–working-fluid model discussed in Sec. 5.5, the heat is added infinitely rapidly as a constant-velocity heat-addition wave at a plane in one-dimensional steady flow (units of joules per square meter-second). In the general case the heat-addition rate for any fluid element [the $Dq/Dt$ of Eq. (5-25), with units of joules per kilogram-second] can be given any desired functional form and arbitrary wave velocity to hopefully model, in an effective way, some observed combustion process. The total heat that is added is usually dictated by observations of the state change associated with some specified completed combustion process (see, for example, Fig. 2-3 and Table 5-2). It must be emphasized that there is *no real physics* implied with this type of heat addition. The heat is simply legislated to appear with no associated composition change in the fluid. Thus it is possible to draw *too* many physical interpretations from results obtained using these types of heat-addition–working-fluid models.

The next hierarchy of complexity occurs when one models the combustion process by specifying an overall irreversible chemical reaction of the type

$$v_{fu} F + v_{ox} O \rightarrow (v_{fu} + v_{ox})P$$

with the rate of oxidation of fuel given by an Arrhenius temperature dependence

$$\frac{D[F]}{Dt} = -A[F]^m [O]^n [I]^l \exp\left(\frac{-E^*}{RT}\right)$$

where $I$ is an inert species and the exponents $m$ and $n$ are not necessarily related to the stoichiometric coefficients $v_{fu}$ and $v_{ox}$. We note that in many cases, actual values of $A$, $m$, $n$, $l$, and $E^*$ are determined from explosion-induction delay measurements on specific systems. For many fuel–air mixtures it is found that $n + m \cong 2$ and $l \cong 0$. Using this model the local heat-release rate is usually related to the fuel-consumption rate through a heat of combustion $Q_{fu}$ (units of joules per mole of fuel)

$$\frac{Dq}{Dt} = \frac{-Q_{fu}}{\rho} \frac{D[F]}{Dt}$$

where $Q_{fu}$ is a positive number and $D[F]/Dt$ is negative. Thus $q$ represents heat addition to the working fluid.

Using this approach one can model first-order kinetics by setting $n = v_{fu} = 1$, and $m = l = v_{ox} = 0$. In a similar manner for a general second-order reaction one can set $n = m = v_{ox} = 1$ and $l = 0$. It is also quite common when using this approach to set $M_{fu} = M_{ox} = M_{prod} = M_{in}$ and $c_{p_{ox}} = c_{p_{fu}} = c_{p_{prod}}$, with units of joules per kilogram. In this case mole fractions are equal to mass fractions at any point in the fluid and $y_{fu} + y_{ox} + y_{prod} + y_{in} = 1$. In many analyses the model is made even simpler by assuming that the temperature-dependence of the reaction rate is very large (large activation energy asymptotics, mentioned in Chap. 5 and to be discussed in more detail in Chap. 8). In some cases where diffusion controls, the rate is legislated to be infinite and the reaction zone occupies a "flame sheet" which separates a mixture of fuel, product, and (possibly) inerts from a mixture of oxidizer, products, and (possibly) a different inert concentration. Furthermore, on occasion these models have been simplified even further by assuming that the reaction itself liberates no heat, i.e., that $Q = 0$!

These models, which are based on overall irreversible kinetics, capture more of the physics of the combustion process than those that simply legislate heat addition. Their use is still almost inevitable in certain applications, such as the well-stirred reactor of Sec. 6-9. They have also been used effectively to capture the essence of combustion behavior in many other applications. However, our recently acquired and more sophisticated understanding of the detailed kinetics of certain combustion processes has led to an increased use of either complete kinetic schemes to follow, or sophisticated truncated kinetic schemes to model a large variety of combustion processes. This is because the recently introduced techniques of solving the "stiff" equations that were mentioned in the previous section have greatly increased our ability to handle complex systems.

A large variety of schemes have been devised for truncating the kinetic relationships in an attempt to retain the essence of the process without having to deal with the complexity of the full kinetics mechanism. Because of this, only four of these schemes will be discussed here. The reader is referred to the bibliography for further discussion of other truncation schemes.

One of the earliest schemes that was devised is the steady-state approximation, first invoked by Lindeman in his explanation of the apparent first-order decomposition of certain molecules. This was introduced in Chap. 4 and can be assumed to be valid if two conditions are met. These are (1) the concentration of the steady-state species must be small and (2) its rate of production and destruction by competing steps must be large relative to its overall rate of production or destruction. If this is true the approximation yields good results even though the concentration of the steady-state species changes slowly with time as the reaction proceeds.

The second approach uses the partial equilibrium concepts developed in Chap. 4. This approach works particularly well in the high-temperature–low-pressure heat-release phase of the chain-branching hydrogen–oxygen reaction because under these conditions the chain-branching reactions are fast, and are essentially in equilibrium after the induction period. Then, the slower recombination reactions that release the heat represent only a slight perturbation to this equilibrium condition. Thus, the concentration of the species responsible for recombination can be determined by calculating the partial equilibrium concentration of all the species under the assumption that the molecular weight of the mixture is constant. Then one can calculate the change in molecular weight due to recombination using an aggregate recombination relationship that combines all the individual recombination reactions.

The third technique involves the judicious choice of composite reactions that retain the essence of one portion of a complex kinetic scheme without jeopardizing other aspects of the scheme. The success of this type of effort is strongly dependent on the insight that the investigator brings to the problem. In many cases sensitivity analysis may be used to help gain such insight. In sensitivity analysis one studies the effect of changing the rate of individual reactions or groups of reactions on the overall results of the calculation to determine how sensitive the entire process is to the accuracy of that particular reaction rate or group of reaction rates.

The fourth and last truncated scheme which will be discussed here is the radical pool scheme introduced in the discussion of the unified theory in Sec. 6-6. In this approach the observation that radical concentrations tend to parallel each other during the induction period of an explosion (see Fig. 6-20) leads to the conclusion that a simpler scheme containing only a single reactive radical should yield the same gross effects. While this technique is the least sophisticated of the truncation schemes it does allow one to follow the details of explosion behavior in a relatively realistic way. Note that simple overall kinetic schemes in which one assumes that there is an irreversible reaction between the "fuel" and "oxidizer" does not satisfactorily model real behavior. Double radical pools have also been suggested so as to gain more flexibility from the model. Double radical pools are probably most useful in cases where radical diffusion is important. This is because of the importance of the hydrogen atom in any chain scheme and the fact that hydrogen atoms are such a low molecular weight species that they (as well as diatomic hydrogen) diffuse much more rapidly in any gradient situation. Therefore it would be incorrect to lump H radicals with the other radicals, if gradient-driven diffusion is important.

As was stated above, the diversity and relative degree of sophistication of different chemistry models in the literature is so large that it is almost impossible to characterize the entire range of models that have been used. Unfortunately, in many cases truncation of the chemistry by inadequate modeling has led to incorrect conclusions, just as when one truncates the equations of motion. Thus any conclusions about the combustion behavior of the system must always be viewed in light of the assumptions used in the model. Also, the reader must be cautioned *never* to extrapolate a truncated scheme outside the range of temperature, pressure, and concentration gradient situation for which it was developed without somehow independently verifying that the scheme is still valid in this new regime. This type of extrapolation can lead, and has led, to many grossly incorrect models of combustion processes.

# PROBLEMS

**6-1** Rework the equations for a thermal explosion in a vessel to illustrate an experimental technique for directly determining the orders of the overall rate equation, $m$ and $n$.

**6-2** Seery and Bowman, *Comb. Flame,* **14**:37 (1970) reported the following explosion delays behind a reflected shock in a shock tube. These can be assumed to be adiabatic explosion-delay times. The mixture composition was (in mole fractions) 0.091 $CH_4$, 0.182 $O_2$, and 0.727 Ar, and the reported data are as follows:

| $P$, atm | $T$, K | $\tau$, $\mu$s |
|---|---|---|
| 2.24 | 1724 | 85 |
| 1.73 | 1660 | 160 |
| 1.85 | 1559 | 465 |
| 1.85 | 1486 | 940 |
| 1.60 | 1428 | 1760 |

They empirically correlated these points and about 60 other data points over the temperature range of 1200 to 1900 K and equivalence ratio of $0.2 \leq \Phi \leq 5.0$ to obtain the equation

$$\tau[O_2]^{1.6}[CH_4]^{-0.4} = A \exp\left(\frac{E^*}{RT_0}\right) \qquad (6P\text{-}1)$$

where $A = 7.65 \times 10^{-18}$ (moles/cm$^3$)$^{1.2}$ s and $E^* = 51.4$ kcal/mole. Plot their points and Eq. (6P-1) as $\ln \tau[O_2]^{1.6}[CH_4]^{-0.4}$ vs. $10^3/T$, K. Then use Eqs. (6-5), (6-4), and (6-1) to determine an appropriate value of $A$ in Eq. (6-1). You should retain the exponents 1.6 and $-0.4$. However, note that $E^*$ should change because of the different temperature-dependence contained in Eq. (6-6). Do all your calculations in SI units and use $q$ and $a$ at the mean temperature of 1600 K.

**6-3** The hydrogen–oxygen chain-branching reaction is almost thermally neutral. Calculate an induction delay time for a $2 : 1 : 4 = H_2 : O_2 : Ar$ mixture at 1500 K and 100 kPa assuming that the concentration of H atoms starts at $10^{-8}$ MF and that the reaction is thermally neutral and proceeds at its initial rate (where $[O_2]$, $[H_2]$, and $[Ar]$ can be assumed to be constants) until the partial equilibrium value of H atom concentration is reached. Assume that only the three reactions

$$H + O_2 \rightarrow OH + O \qquad k_1 = 1.86 \times 10^8 \exp\left(\frac{70,250}{RT}\right)$$

$$O + H_2 \rightarrow OH + H \qquad k_2 = 1.82 \times 10^4 \, T \exp\left(\frac{37,240}{RT}\right)$$

and

$$OH + H_2 \rightarrow H_2O + H \qquad k_3 = 2.19 \times 10^7 \exp\left(\frac{21,550}{RT}\right)$$

occur initially and that these are the reactions that reach partial equilibrium with no recombination.

*Hint:* First determine the initial concentration of OH and O radicals using the relative rates of these reactions.

**6-4** The oxidation of moist CO occurs primarily by the step

$$OH + CO \rightarrow H + CO_2 \qquad (6P\text{-}2)$$

Use Fig. 6-20 to determine the rate constant for this reaction at $t = 2$ ms for the conditions of (6-20). Compare your answer to the literature value given by the expression $k_{6P\text{-}2} = 12.88 \, T^{1.3} \exp(-3220/RT)$.

**6-5** For the ozone decomposition explosion, note that irrespective of the initial conditions and the delay to explosion, when the explosion is nearly complete the phase plane path, i.e., the path in the $([O], T)$ plane, as shown in Fig. 6-19 is virtually independent of the initial conditions. Estimate the rate of production and destruction of oxygen atoms at 1500 and 2000 K during the explosion process using the information that is available in this text.

**6-6** For the well-stirred reactor, solve for the condition of tangency to determine the functional dependence of $\eta_b$, the dimensionless temperature at blowout, on the kinetic parameters.

**6-7** Derive Eq. (6-9) using the steady-state approximation for H and $X$ atoms.

# BIBLIOGRAPHY

**Experimental methods**

Belford, R. L., and R. A. Strehlow, "Shock Tube Technique in Chemical Kinetics," *Ann. Rev. Phys. Chem.*, **20**: 247–272 (1969).

Bradley, J. N., *Shock Waves in Chemistry and Physics*, Wiley, New York (1962).

Crosley, D. R., ed., *Laser Probes for Combustion Chemistry*, ACS Symposium Series, No. 134, American Chemical Society, Washington, D.C. (1980).

Essenhigh, R. H., "An Introduction to Stirred Reactor Theory Applied to Design of Combustion Chambers," *Combustion Technology*, eds. H. B. Palmer and J. M. Béer, pp. 373–415, Academic Press, New York (1974).

Gaydon, A. G., and I. R. Hurle, *The Shock Tube in High Temperature Chemical Physics*, Reinhold, New York (1963).

Greene, E. F., and J. P. Tonnies, *Chemical Reactions in Shock Waves*, Academic Press, New York (1965).

Lifshitz, A., ed., *Shock Waves in Chemistry*, Marcel Dekker, New York, N.Y. (1981).

**Kinetics**

Benson, S. W., "The Kinetics and Thermochemistry of Chemical Oxidation with Application to Combustion and Flames," *Prog. Energy Combust. Sci.*, **7**: 125–134 (1981).

Gardiner, W. C., Jr., and D. B. Olson, "Chemical Kinetics of High Temperature Combustion," *Ann. Rev. Phys. Chem.*, **31**: 377–400 (1980).

Indritz, D., M. Maday, R. Sheinson, and W. Shaub, *Construction of Large Reaction Mechanisms*, NRL Report 8498, Naval Research Laboratory, Washington D.C., 27 pp., Oct. 9, 1981.

Kaufman, F., "Chemical Kinetics and Combustion: Intricate Paths and Simple Steps," *Nineteenth Symposium* (International) *on Combustion*, pp. 1–10 (1983).

McKay, G., "The Gas-Phase Oxidations of Hydrocarbons," *Prog. Energy Combust. Sci.*, **3**: 105–126 (1977).

Palmer, H. B., "Equilibria and Chemical Kinetics in Flames," *Combustion Technology*, eds. H. B. Palmer and J. M. Béer, pp. 1–33 Academic Press, New York (1964).

"Symposium on Reaction Mechanisms, Models and Computers," *J. Phys. Chem.*, **81**(25): 2309–2586 (1977).

Westbrook, C. K., and F. L. Dryer, "Chemical Kinetics Modeling in Hydrocarbon Combustion," *Prog. Energy Combust. Sci.*, **10** (in press) (1984).

**Modeling**

Ebert, K. H., P. Deuflhard, and W. Jager, *Modelling of Chemical Reaction Systems*, eds. K. H. Ebert, P. Deuflhard, and W. Jager, *Chemical Physics* 18, Springer Verlag, Berlin (1981).

Edelson, D., "Computer Simulation in Chemical Kinetics," *Science*, **214**(4524): 981–986 (1981).

Pratt, D. T., "Mixing and Chemical Reaction in Continuous Combustion," *Prog. Energy Combust. Sci.*, **1**: 73–86 (1975).

Rabitz, H., M. Kramer, and D. Dacol, "Sensitivity Analysis in Chemical Kinetics," *Ann. Rev. Phys. Chem.*, **34** (in press) (1984).

CHAPTER
# SEVEN

## TRANSPORT-CONTROLLED COMBUSTION PROCESSES

### 7-1 INTRODUCTION

We have already observed that chemical composition changes must accompany combustion processes and that all such processes are also accompanied by a large increase in the temperature of the system. In the previous chapter we studied a combustion system in which the rate of conversion to products was controlled entirely by reaction kinetics, namely the isolated (adiabatic) explosion. We also found that in vessel explosions transport of heat and mass (free radicals) to the walls could markedly influence the course of the explosion. We further discussed two adiabatic systems (the plug-flow reactor and the shock tube) in which wave-heating triggers the chemistry,‡ which then proceeds as an adiabatic "explosion" traveling through the fluid.

In general we will find that there are three fundamental rate-controlling processes in combustion. These are reaction kinetics, transport of heat and mass (either laminar or turbulent), and wave-heating. These will always be combined in some unique way which is dependent on the initial and boundary conditions to produce all the unique and distinguishable combustion processes that have been observed.

---

‡ In the plug-flow reactor rapid mixing at the heated inlet is equivalent to rapid wave-heating of a cold reactive mixture to the inlet temperature.

In the previous chapter on combustion chemistry, the processes that were discussed were carefully selected because their behavior is primarily controlled by the reaction kinetics that always occurs during combustion. In this chapter we wish to examine a few relatively simple combustion processes that are primarily dominated by transport. Then in the next two chapters, we will describe in some detail the processes associated with the propagation of two simple *combustion waves*, namely, premixed gas flames and detonation waves. Following those chapters, a selection of more complex interactive and practical combustion processes will be introduced and discussed. It is hoped that this approach will allow any current (and possibly complex) combustion study to be placed in its proper context.

## 7-2 GAS-PHASE DIFFUSION FLAMES—PARALLEL FLOW

Diffusion flames have been known since antiquity. Natural fires are predominately a complex of diffusion-controlled flames, and the candle or oil-wick flames that early man used for artificial lighting are relatively simple examples of diffusion flames. The diffusion flame in its simplest form consists of an exothermic reaction zone separating relatively pure samples of oxidizer and fuel gases. As such it does not exhibit a characteristic propagation velocity. However, the characteristics of diffusion flames are markedly dependent on the aerodynamics of the particular flow situation, and the complete spectrum of diffusion-flame behavior cannot be discussed without invoking aerodynamic considerations.

In this and the following sections we will discuss six diffusion-flame configurations that are particularly simple aerodynamically. Three of these simple flames are enclosed flames driven by forced laminar convection such that buoyancy of the hot product gases can be neglected to a good first approximation. For the fourth case, droplet combustion, the assumption that gravity is unimportant is usually made in the analysis, even though it can be important. For the fifth case, the jet diffusion flame, buoyancy is included in the analysis. The sixth case is catalytic combustion—a process that is controlled by diffusion.

The first two "enclosed" diffusion flames to be discussed are both examples of parallel forced-flow laminar diffusion flames. Consider either a thin-walled tube containing fuel flowing coaxial to a longer and larger tube containing an oxidizer flow as shown in Fig. 7-1*a*, or a short thin splitter plate which separates fuel and oxidizer flow in a rectangular channel as shown in Fig. 7-1*b*. If the flow velocities and concentrations are in the correct range, the tube or splitter plates will "hold" a diffusion flame in the region above their trailing edges. This "holding" process will be discussed in Chap. 10. Depending upon the relative concentrations and flow velocities of the fuel and oxidizer, the flame that is observed will be either underventilated or overventilated. This is illustrated schematically in Fig. 7-1. In the underventilated flame the oxidizer is eventually completely consumed and excess fuel or partially oxidized products

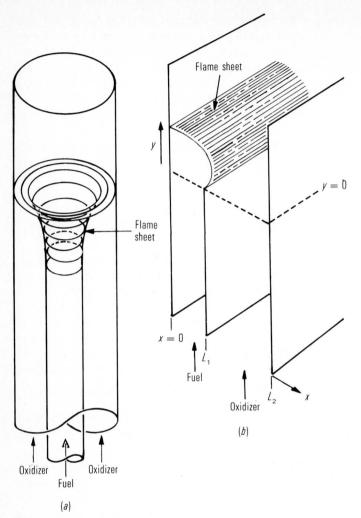

**Figure 7-1** The forced convection diffusion flame. (*a*) Cylindrical geometry (underventilated). (*b*) Planar geometry (overventilated).

leave the burner. In the overventilated flame, combustion is complete and only fully oxidized products and excess oxidizer leave the burner. In both cases it has been observed experimentally that the active "flame" sheet is relatively thin, and that diffusion of reactive species toward, and products away from, the location of the flame sheet determine the rate of the combustion process in this geometry. This fact was first used by Burke and Schumann[1] in 1928 in their

1. S. P. Burke and T. E. W. Schumann, "Diffusion Flames," *Proceedigns of the First Symposium on Combustion* (held at the Seventy-Sixth Meeting of the American Chemical Society, Swampscott, Mass., Sept. 10–14, 1928), *First and Second Symposium on Combustion*, The Combustion Institute, Pittsburgh, Pa., pp. 1–11 (1965).

now classical treatment of the enclosed parallel-flow diffusion flame for both geometries of Fig. 7-1. We will consider only the splitter-plate geometry of Fig. 7-1*b* because the solution takes the form of a trigonometric rather than Bessel function series. No generality is lost when this is done.

Consider the geometry of Fig. 7-1*b*. The following assumptions were made in Ref. 1:

1. The flow is steady.
2. Fuel is flowing up a channel of width $L_1$ and oxidizer up a channel of width $L_2 - L_1$.
3. A diffusion flame is being held at the point $(L_1, 0)$ by the infinitely thin splitter plate which terminates at $(L_1, 0)$. This plate initially separates fuel and oxidizer.
4. The velocity up the tube is *everywhere constant* and equals $v$. Also there is no horizontal velocity component anywhere (i.e., $u \equiv 0$). This assumption violates the ideal gas law because expansion due to heat release will cause an increase in temperature, and mass balance will cause $v$ to increase up the tube. Also $u$ can never be identically zero everywhere. In other words, this assumption is tantamount to assuming that the heat of reaction is zero and that the flame is isothermal.
5. The reaction rate between fuel and oxidizer at their interface is infinitely rapid. In other words, the second Damköhler number is infinite. This has the effect of removing the highly nonlinear rate equations from the problem and replacing them with a boundary condition at a "flame surface."
6. Transport of fuel and oxidizer is by diffusion in the $x$ direction only and the diffusion coefficient is a constant. Note that since the diffusion coefficient increases with temperature, this partially offsets the problems caused by assumption 4.
7. The reaction is irreversible and there is no mole change on reaction, that is, $F + vO \rightarrow (v + 1)P$. In other words, the molecular weights of the reactants and products are equal.

We will now proceed to simplify the general equations of motion on the basis of these assumptions. We first note that assumption 5 means that the flame sheet separates regions that either contain only oxidizer and products or fuel and products. Fuel and oxidizer never coexist in the flow. Next we notice that at the flame sheet $x_{ox} = x_{fu} = 0$, and Eq. (5-9) for the diffusion flux reduces to an equation that defines the boundary conditions at the flame sheet

$$\left. \frac{dx_{fu}}{dx} \right|_{F_-} = -\frac{1}{v} \left. \frac{dx_{ox}}{dx} \right|_{F_+} \tag{7-1}$$

Equation (7-1) simply states that the rate of transport of fuel to the flame sheet from the left (negative side) must be equal to the stoichiometric rate of transport of the oxidizer to the flame sheet from the right (positive side) if there is to be no net accumulation of fuel or oxidizer at the location of the flame sheet.

This means that this boundary condition allows one to define a new universal concentration variable

$$x_r = x_f = \frac{-x_{ox}}{v} \tag{7-2}$$

which is applicable to all the gradient regions associated with this flame system. The presence of the outer wall at $x = L_2$ and the inner wall at $x = 0$ creates the additional boundary conditions

$$\frac{\partial x_r}{\partial x} = 0 \qquad (x = 0, L_2) \tag{7-3}$$

The assumptions listed above mean that we need not consider Eqs. (5-1), (5-3), or (5-4), the overall mass conservation equation, the momentum equations, or the energy equation, and that the $\partial/\partial t$ terms drop out of Eqs. (5-2), the species conservation equations. Furthermore, since we have replaced the fuel and oxidizer mole fractions by the fictitious species $r$, using the flame boundary condition, Eqs. (5-2) reduce to the simple relationship

$$\frac{\partial x_r}{\partial y} = \frac{D}{v} \frac{\partial^2 x_r}{\partial x^2} \tag{7-4}$$

Notice that one can obtain the ordinary diffusion equation

$$\frac{\partial x_r}{\partial t} = D \frac{\partial^2 x_r}{\partial x^2} \tag{7-5}$$

from Eq. (7-4) by substituting $y = vt$ when $v$ is a constant. The solution to Eq. (7-4) for $y = (+)$ and the boundary conditions at $y = 0$, $x = 0$, and $x = L_2$ is

$$\frac{x_r}{(x_r)_0} = \frac{(x_{fu})_0}{(x_r)_0} \frac{L_1}{L_2} - \frac{(x_{ox})_0}{v(x_r)_0} \left( \frac{L_2 - L_1}{L_2} \right)$$

$$+ \frac{2}{\pi} \sum_{n=1}^{\infty} \frac{1}{n} \sin \frac{n\pi L_1}{L_2} \cos \frac{n\pi x}{L_2} \exp \left( \frac{-yn^2\pi^2 D}{v L_2^2} \right) \tag{7-6}$$

where

$$(x_r)_0 = (x_{fu})_0 + \frac{(x_{ox})_0}{v} \tag{7-7}$$

Equation 7-6 expresses a dimensionless mole fraction of reactant $x_r/(x_r)_0$ in terms of three dimensionless quantities, namely a burner configuration parameter $L_1/L_2$, and the dimensionless $x$ and $y$ position given by the variables $x/L_2$ and $\zeta = y\pi^2 D/v L_2^2$.

Since the locus of the flame is given by $x_r = 0$ and since the first two terms on the right-hand side of Eq. (7-6) are constants dictated by the conditions at $y = 0$, the infinite series must have a constant value at the flame sheet equal to

$$E = \frac{(x_{ox})_0}{v(x_r)_0} \frac{L_2 - L_1}{L_2} - \frac{(x_{fu})_0}{(x_r)_0} \frac{L_1}{L_2} \tag{7-8}$$

The functional form of the series is dependent on the burner configuration, that is, $L_1/L_2$, as well as $x/L_2$ and $\zeta$. A typical plot of $E$ as a function of $\zeta$ for various values of $x/L_2$ is shown in Fig. 7-2 for the case when $L_1/L_2 = 0.25$. Four typical flame shapes are shown in Fig. 7-3. When $\zeta = 0$, that is, at the burner rim, the series $E$ yields a square wave and one obtains

$$E = \frac{L_2 - L_1}{L_2} \quad \text{and} \quad x_r = (x_{fu})_0 \quad \text{for} \quad 0 \le x \le L_1 \quad \zeta = 0$$

and

$$E = -\frac{L_1}{L_2} \quad \text{and} \quad x_r = \frac{-(x_{ox})_0}{v} \quad \text{for} \quad L_1 < x \le L_2 \quad \zeta = 0$$

When the burner is stoichiometric $E = 0$ and $(x_{fu})_0 L_1 = [(x_{ox})_0/v](L_2 - L_1)$. When there is excess fuel $E$ is negative and the flame is underventilated. When

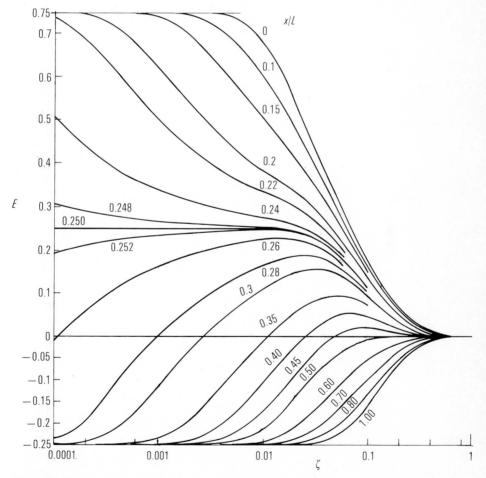

**Figure 7-2** The $(E, \zeta)$ behavior of a Burke–Shumann planar flame for $L_1/L_2 = 0.25$. Lines of constant $x/L_2$ are plotted.

there is excess oxidizer $E$ will be positive and the flame will be overventilated. Notice that in actuality the Burke–Schumann model predicts five types of flame behavior. Referring to Figs. 7-2 and 7-3, if $L_1/L_2 = 0.25$ a value of $E$ in the range $0.25 < E < 0.75$ will produce a simple overventilated flame with no bulge, i.e., with $x$ always less than $L_1$. At $E = 0.25$ the flame sheet is vertical at $\zeta = 0$ and $x$ is still single-valued in $\zeta$. In the range of $0.0 < E < 0.25$ the overventilated flame will bulge, i.e., will have a locus in which $x$ is sometimes greater than $L_1$ and is double-valued. At $E = 0.0$ the flame is stoichiometric and theoretically asymptotes $x/L_2 = 0.5$ as $\zeta \to \infty$. For the range $-0.25 < E < 0.0$ the

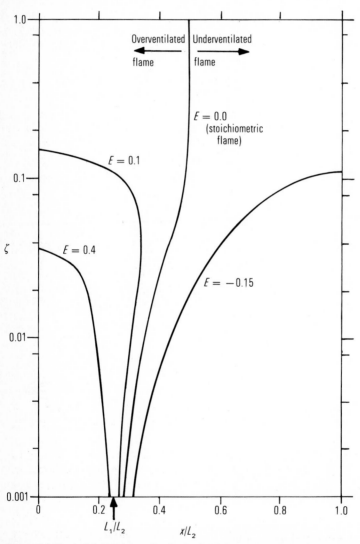

**Figure 7-3** Typical flame shapes for the Burke–Schumann model where $L_1/L_2 = 0.25$.

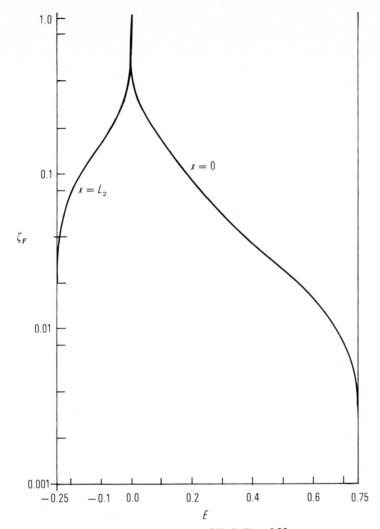

**Figure 7-4** Flame height $\zeta_F$ as a function of $E$; $L_1/L_2 = 0.25$.

flame will be underventilated and $x$ will again be single-valued in $\zeta$ with $x$ always greater than $L_1$. Also note that the maximum value of $x$ in any "bulged" overventilated flame is $L_2/2$.

Figure 7-4 is a plot of the flame height $\zeta_F$ for the case $L_1/L_2 = 0.25$. Notice that it goes to $+\infty$ at $E = 0$, that is, when the flame becomes stoichiometric. One could also determine concentration profiles for this flame by plotting the locus of $x_r = \text{const} \neq 0$.

It must be noted that the structure of Fig. 7-2 is independent of the overall equivalence ratio of the flame and is only determined by the $L_1/L_2$ ratio of the burner. Specifically, the left-hand side of this figure, for small values of dimensionless height $\zeta$ [or time in a stationary diffusion problem, Eq. (7-5)], shows symmetrical behavior about the value of $E = (E_{min} + E_{max})/2$ and represents diffusion in the neighborhood of the holding region *before* the influence of the walls at $x = 0$ and $L_2$ are felt. The symmetric behavior of the curves on the right-hand side of this figure around $E = 0.0$ represents the final diffusion stage when all terms but the first in the series become small and the concentration profile reduces to a simple cosine function of $x/L_2$ centered about $x/L_2 = 0.5$.

If one changes the $L_1/L_2$ ratio the nature of Fig. 7-2 changes. Specifically if $L_1/L_2 = 0.5$, $E$ will range from $-0.5$ to $+0.5$ and the curves for $x/L_2 = \text{const}$ will be completely symmetric about $E = 0.0$. In this case the model predicts that the flame will not bulge even though over- and underventilated flames are still predicted for positive and negative values of $E$ respectively. As $L_1/L_2$ approaches zero the range on $E$ approaches $-0.0$ to $1.0$, and the range of $E$ over which a bulged overventilated flame can occur is larger. Also underventilated flames occur only over a very small range.

A number of the above predictions of the Burke–Schumann model are limited by the assumption that the flow velocity $v$ is everywhere constant. This simply means that two different flow velocities cannot be used to change conditions. Another restriction is the assumption that the chemistry is infinitely rapid. The effect of this assumption has been investigated and it has been found to be quite adequate in the main body of the flame but inadequate in the "holding" region near the inner rim. Detailed structure of the flame is also strongly influenced by the fact that the chemistry is not a simple one-step affair. Additionally, it has been found to be very difficult to eliminate edge effects in the holding region and developed boundary layers upstream of the holding region. Therefore experimental studies of this type of parallel-flow diffusion flame are difficult.

The emphasis in this section has been to show the development of a classic and very simplified flame-sheet theory which has had some success in explaining the physics of a diffusion flame. In the next section complications caused by the inclusion of the heat of reaction will be presented using the classical flame-sheet analysis of single-droplet combustion. Then the actual structure of counterflow and parallel-flow diffusion flames will be discussed to illustrate the results of a more modern and detailed study of diffusion-flame structure.

## 7-3 SINGLE-DROPLET COMBUSTION

Another combustion process in which transport dominates is the spherical diffusion flame surrounding a single droplet of liquid fuel. This process has been studied extensively for many years. Experimentally, three major types of studies have been performed, none of which really maintain the ideal spherical geometry very well because the buoyancy of the hot products almost always causes a convective plume to rise from the burning drop. Drops have been suspended on a thin filament, ejected, and ignited while moving, or modeled by supplying fuel to the center of a porous sphere at a rate equal to the combustion rate. Except in the latter case, one usually observes three distinct phases during droplet combustion. In the ignition and droplet heating phase the cold droplet is being heated by the flame and two factors cause only a small change in droplet size. The first is thermal expansion of the liquid. The second is due to the fact that most of the heat transferred to the drop is heating the drop and thus not causing evaporation. During the second phase, to a good first approximation, the droplet can be assumed to have reached its evaporation temperature. In this state virtually all the heat that is transferred to the drop is used to supply the latent heat of evaporation. This phase of burning accounts for close to 90 percent of the total droplet lifetime. The final or burnout phase occurs after the droplet has completely evaporated. During this period the vapor that remains is almost completely consumed by the combustion process.

The second phase of droplet combustion has been studied more extensively than either the first or last phase. Experimentally it has been observed that during this phase of combustion the diameter of the droplet decreases at a rate given by the equation $d^2 = d_0^2 - kt$, where $k$ is a constant that has some theoretical significance. The complete theory for this phase of the combustion process is complicated by many factors, such as circulation in the droplet, finite-rate chemistry in the diffusion flame that surrounds the droplet, nonsteady accumulation of fuel between the surface and the flame, etc. The interested reader is referred to two recent reviews by C. K. Law,[2] in which the effects of these complications are discussed. Here we will only present the classical, quasi-steady flame-sheet theory of droplet combustion. This does predict the $d^2$ law and also captures many of the other features of the droplet combustion process. The simple theory also has the advantage that it delineates the dominant processes that are occurring during droplet combustion.

To perform the classical analysis we make the following simplifying assumptions.

1. The geometry is strictly spherical. This means that all spatial gradients are in the $r$ direction only.
2. The liquid density is much higher than the gas-phase density. This means that the rate of radius change of the droplet, $dr_L/dt$, will be small when compared with gas-phase diffusion velocities in the neighborhood of the drop. This further means that the entire flow outside the drop can be treated as a steady flow if the rate of evaporation is constant.
3. The drop is at its boiling temperature. This means that heat transferred to the drop causes evaporation by supplying only the constant latent heat of vaporization, $\ell$ (joules per kilogram). Thus, if the flow is steady the rate of evaporation will also be steady. These first three assumptions mean that all

2. C. K. Law, "Mechanisms of Droplet Combustion," *Proceedings of the Second International Colloquium on Drops and Bubbles*, ed. E. H. Le Croissette, JPL Pub. 82-7, pp. 39–53 (1982); "Recent Advances in Droplet Vaporization and Combustion," *Prog. Energy Combust. Sci.*, **8**: 171–202 (1982).

time-dependent terms can be eliminated from the equations of motion and thus that the equations become ordinary differential equations in the region $r_{L_+} < r < \infty$ where $r_{L_+}$ is the gas-phase side of the droplet surface.

4. Flow velocities are low. Therefore pressure is constant. This means that the momentum equation is not needed in the analysis.

5. The combustion reaction is a single-step irreversible reaction with the stoichiometry $F + vO \rightarrow (v + 1)P$ and it is infinitely rapid at the "flame surface." It can be shown that this means that the region outside the droplet can be divided into two regions $r_{L_+} < r < r_{F_-}$ and $r_{F_+} < r < \infty$ where $r_F$ is the flame radius. It also means that the mass fractions of fuel and oxidizer are $y_{fu} = 0$ in the range $r_{F_+}$ to $\infty$, $y_{ox} = 0$ in the range $r_{L_+}$ to $r_{F_-}$ and therefore they are both zero at the flame sheet. This assumption set is equivalent to that used in the Burke–Schumann treatment of a diffusion flame in the previous section. This assumption set also collapses the highly nonlinear reaction terms into a flame-sheet discontinuity which therefore also defines "flame" boundary conditions for the problem.

6. The combustion reaction is exothermic with a constant heat of reaction of $q_{fu}$ (in joules per kilogram of fuel). Note that heat is released only at the "flame sheet;" this means that temperature profiles will be included in this model. This is in contrast to the Burke–Schumann model in which the temperature was assumed constant throughout.

7. The droplet is pure fuel and contains no product. Also the oxidizer atmosphere has a specified temperature $T_\infty$ and oxidizer mass fraction $y_{ox,\,\infty}$ at $r = \infty$.

8. The molecular weights $M$, heat capacities $c_p$, thermal conductivities $\kappa$, and the quantity $\rho D$ are constants.

9. The Lewis number, Le, equals 1.

Using the assumptions 1 to 8, the conservation equations for the gas phase $(r_{L_+} < r < \infty)$ become:

*Overall mass conservation*

$$\frac{1}{r^2} \frac{d}{dr} (r^2 \rho u) = 0 \tag{7-9}$$

*Species conservation*

$$u \frac{dy_i}{dr} - \frac{\rho D}{\rho r^2} \frac{d}{dr} \left( r^2 \frac{dy_i}{dr} \right) = 0 \tag{7-10}$$

where $i = $ fu for fuel, $i = $ ox for oxidizer, and $i = $ prod for product.

*Momentum conservation*

$$P = \text{const}$$

*Energy conservation*

$$u \frac{dT}{dr} - \frac{1}{\rho r^2} \left( \frac{\kappa}{c_p} \right) \frac{d}{dr} \left( r^2 \frac{dT}{dr} \right) = 0 \tag{7-11}$$

Note that if one assumes that $\mathrm{Le} = 1$ (assumption 9) Eqs. (7-10) and (7-11) are in the Schvab–Zel'dovich form, even if the infinitely rapid reaction of assumption 5 was allowed to have a finite rate. This is because subtraction of any pair of equations to eliminate the rate term will yield, for example, the coupling variables

$$\beta_{\mathrm{fu}} - \beta_{\mathrm{ox}} = (y_{\mathrm{fu}} - y_{\mathrm{ox}}/v)$$

or

$$\beta_{\mathrm{fu}} - \beta_T = y_{\mathrm{fu}} - \frac{c_p(T - T_\infty)}{q_{\mathrm{fu}}}$$

and a homogeneous ordinary differential equation. Thus even with finite reaction rates and a relatively thick reaction zone the major features of the droplet-burning process can be derived analytically. We will, however, take a simpler approach by retaining the flame-sheet approximation in the following analysis.

We also have made the assumption that $\mathrm{Le} = 1$. Thus Eqs. (7-10) and (7-11) become identical

$$\rho u \frac{d\beta}{dr} = \frac{\rho D}{r^2} \frac{d}{dr} \left( r^2 \frac{d\beta}{dr} \right) \tag{7-12}$$

where $\beta$ equals either $y_{\mathrm{fu}}$, $y_{\mathrm{ox}}$, $T$ or one of the appropriate coupling functions mentioned above. Equation (7-9) states that the mass flow through any spherical shell is constant. Thus there is a constant mass-flow rate, $\dot{m} = 4\pi r^2 \rho u$, and we use this to define a mass-flow rate normalized to the diffusion coefficient

$$\dot{M} = \frac{\dot{m}}{4\pi \rho D} = \frac{r^2 \rho u}{\rho D} = \mathrm{const} \qquad \text{(in meters)} \tag{7-13}$$

If we further define $x = d\beta/dr$, Eq. (7-12) becomes

$$\frac{\dot{M}}{r^2} \, dr = d[\ln r^2 x]$$

which can be integrated to yield

$$\frac{d\beta}{dr} = \frac{C_1}{r^2} \exp\left( \frac{-\dot{M}}{r} \right) \tag{7-14}$$

This in turn can be integrated again to yield a universal relationship for the variable $\beta$

$$\beta = \frac{C_1}{\dot{M}} \exp\left( \frac{-\dot{M}}{r} \right) + C_2 \tag{7-15}$$

where $C_1$ and $C_2$ must be evaluated using the boundary conditions once the variable $\beta$ and the region of interest is specified. We can now write the equations for the structure of the diffusion flame based on the boundary conditions at the droplet surface, at the flame surface, and at infinity, even though we do not know the location of the flame surface (its location will be a result of the analysis). To do this we define the dimensionless quantities $\mu = \dot{M}/r_L$ and $\sigma = r/r_L$. This means that the flame's location is given by $\sigma_F = r_F/r_L$.

We first note that there are two regions of interest: $1 < \sigma < \sigma_F$ where only fuel and product are present, and $\sigma_F < \sigma < \infty$ where only oxidizer and product are present. If $\beta = y_{ox}$ we obtain the two equations for the boundaries of the region $\sigma_F < \sigma < \infty$

$$y_{ox, F_+} = 0 = \frac{C_1}{\dot{M}} \exp\left(\frac{-\dot{M}}{\sigma_F}\right) + C_2$$

and

$$y_{ox, \infty} = \frac{C_1}{\dot{M}} + C_2$$

Evaluating $C_1$ and $C_2$ and substituting into (7-15) yields an expression for $y_{ox}$ in the region $\sigma_F < \sigma < \infty$

$$y_{ox} = y_{ox, \infty} \left[\frac{\exp\left(-\dot{\mu}/\sigma\right) - \exp\left(-\dot{\mu}/\sigma_F\right)}{1 - \exp\left(-\dot{\mu}/\sigma_F\right)}\right] \tag{7-16}$$

In a similar manner the temperature distribution in this region can be determined to be

$$T = \frac{T_F[1 - \exp\left(-\dot{\mu}/\sigma\right)] + T_\infty[\exp\left(-\dot{\mu}/\sigma\right) - \exp\left(-\dot{\mu}/\sigma_F\right)]}{[1 - \exp\left(-\dot{\mu}/\sigma_F\right)]} \tag{7-17}$$

In the range of $1 \le \sigma \le \sigma_F$ the solution of Eq. (7-14) with the appropriate boundary conditions yields the expressions

$$y_{fu} = y_{fu, L_+} \left[\frac{\exp\left(-\dot{\mu}/\sigma_F\right) - \exp\left(-\dot{\mu}/\sigma\right)}{\exp\left(-\dot{\mu}/\sigma_F\right) - \exp\left(-\dot{\mu}\right)}\right] \tag{7-18}$$

and

$$T = \frac{T_F[\exp\left(-\dot{\mu}/\sigma\right) - \exp\left(-\dot{\mu}\right)] + T_L[\exp\left(-\dot{\mu}/\sigma_F\right) - \exp\left(-\dot{\mu}/\sigma\right)]}{[\exp\left(-\dot{\mu}/\sigma_F\right) - \exp\left(-\dot{\mu}\right)]} \tag{7-19}$$

We note at this point that we do not know the quantities $y_{fu, L_+}$, $T_F$, $\dot{M}$, or, as was mentioned earlier, $\sigma_F$ in Eqs. (7-13) or (7-16) to (7-19). To determine these we must examine the temperature and species gradient requirements at the droplet surface and flame-sheet boundaries.

In the region between the droplet surface and the flame sheet there is no net motion of product, only fuel motion. Thus the mass flux at the surface is given by the relationship

$$(\rho u)_{L+} = \left[ \rho y_{\text{fu}} \left\{ u - \frac{D}{y_{\text{fu}}} \left( \frac{dy_{\text{fu}}}{dr} \right) \right\} \right]_{L+}$$

which can be solved to yield an expression for the gradient of the fuel-mass fraction at the surface

$$\left( \frac{dy_{\text{fu}}}{dr} \right)_{L+} = \frac{1}{\rho D} \left[ \rho u (y_{\text{fu}} - 1) \right]_{L+} \tag{7-20}$$

Since all the energy flow to the droplet is used to evaporate the liquid, energy conservation at the surface of the drop may be written

$$\kappa \left( \frac{dT}{dr} \right)_{L+} = \ell (\rho u)_{L+} \tag{7-21}$$

At the flame sheet the boundary condition on the fuel and oxidizer mass-fraction gradients are given by the same requirement as was imposed in the Burke–Schumann treatment

$$\left( \frac{dy_{\text{ox}}}{dr} \right)_{F+} = -v \left( \frac{dy_{\text{fu}}}{dr} \right)_{F-} \tag{7-22}$$

Since the reaction rate is infinite at the flame sheet the energy balance at the flame sheet states that energy conducted to the flame sheet by diffusion of the reactants must be dissipated by heat conduction away from the flame sheet

$$q_{\text{fu}} \left( -\rho D \frac{dy_{\text{fu}}}{dr} \right)_{F-} = \frac{q_{\text{fu}}}{v} \left( \rho D \frac{dy_{\text{ox}}}{dr} \right)_{F+} = \left( -\kappa \frac{dT}{dr} \right)_{F+} - \left( -\kappa \frac{dT}{dr} \right)_{F-}$$

Using our assumptions this becomes

$$q_{\text{fu}} \left( \frac{dy_{\text{fu}}}{dr} \right)_{F-} = \frac{-q_{\text{fu}}}{v} \left( \frac{dy_{\text{ox}}}{dr} \right)_{F+} = c_p \left( \frac{dT}{dr} \right)_{F+} - c_p \left( \frac{dT}{dr} \right)_{F-} \tag{7-23}$$

We note in passing that our solution for the region $\sigma_F < \sigma < \infty$ already satisfies the requirement that all gradients vanish as $r \to \infty$.

Differentiation of Eqs. (7-16) to (7-19) and evaluation at $L_+$, $F_-$, and $F_+$ yields the expressions for the gradients in terms of the universal solution.

*At $r = r_{L+}$, that is, $\sigma = 1_+$*

$$\left( \frac{dy_{\text{fu}}}{d\sigma} \right)_{L+} = - \frac{y_{\text{fu}, L+} \dot{\mu} \exp(-\dot{\mu})}{\exp(-\dot{\mu}/\sigma_F) - \exp(-\dot{\mu})} \tag{7-24}$$

and

$$\left( \frac{dT}{d\sigma} \right)_{L+} = \frac{(T_F - T_L) \dot{\mu} \exp(-\dot{\mu})}{\exp(-\dot{\mu}/\sigma_F) - \exp(-\dot{\mu})} \tag{7-25}$$

At $r = r_{F_-}$, that is, $\sigma = \sigma_{F_-}$

$$\left(\frac{dy_{\text{fu}}}{d\sigma}\right)_{F_-} = \frac{-y_{\text{fu}, L_+} \dot{\mu} \exp(-\dot{\mu}/\sigma_F)}{[\exp(-\dot{\mu}/\sigma_F) - \exp(-\dot{\mu})]\sigma_F^2} \tag{7-26}$$

and

$$\left(\frac{dT}{d\sigma}\right)_{F_-} = \frac{(T_F - T_L)\dot{\mu} \exp(-\dot{\mu}/\sigma_F)}{[\exp(-\dot{\mu}/\sigma_F) - \exp(-\dot{\mu})]\sigma_F^2} \tag{7-27}$$

At $r = r_{F_+}$, that is, $\sigma = \sigma_{F_+}$

$$\left(\frac{dy_{\text{ox}}}{d\sigma}\right)_{F_+} = \frac{y_{\text{ox}, \infty} \dot{\mu} \exp(-\dot{\mu}/\sigma_F)}{[1 - \exp(-\dot{\mu}/\sigma_F)]\sigma_F^2} \tag{7-28}$$

and

$$\left(\frac{dT}{d\sigma}\right)_{F_+} = \frac{-(T_F - T_\infty)\dot{\mu} \exp(-\dot{\mu}/\sigma_F)}{[1 - \exp(-\dot{\mu}/\sigma_F)]\sigma_F^2} \tag{7-29}$$

The gradients of the fuel mass fraction and temperature at the droplet surface, Eqs. (7-20), (7-24), (7-21), and (7-25) may be ratioed to yield an expression for the fuel mass fraction at the surface in terms of the flame temperature

$$\frac{1 - y_{\text{fu}, L_+}}{y_{\text{fu}, L_+}} = \frac{\ell}{c_p(T_F - T_L)} = B_1 \tag{7-30}$$

where $B_1$ is the transfer coefficient for droplet evaporation. The concept of such an evaporative transfer coefficient was first developed early in the nineteenth century, when the theory of the wet bulb thermometer was first developed.[3] In that case $T_F$ becomes the ambient temperature $T_\infty$. In this regard it is interesting to note that droplet evaporation during combustion is controlled by $T_F$, that is, the droplet evaporates as though it is in an atmosphere held at the flame temperature (which is yet to be determined).

When Eqs. (7-27), (7-28), and (7-29) are substituted into the energy-balance requirement at the flame sheet, one obtains the relationship

$$\frac{q_{\text{fu}} y_{\text{ox}, \infty}}{c_p \nu} = (T_F - T_L)\frac{[1 - \exp(-\dot{\mu}/\sigma_F)]}{[\exp(-\dot{\mu}/\sigma_F) - \exp(-\dot{\mu})]} + (T_F - T_\infty) \tag{7-31}$$

which may be written

$$\frac{q_{\text{fu}} y_{\text{ox}, \infty}}{c_p} = (T_F - T_L)\frac{y_{\text{ox}, \infty}}{y_{F, L_+}} + \nu(T_F - T_\infty) \tag{7-32}$$

When $y_{\text{fu}, L_+}$ is eliminated from (7-32) using (7-30) one obtains an expression for the flame temperature

$$T_F = \frac{[(q_{\text{fu}} - \ell)/c_p](y_{\text{ox}, \infty}/\nu) + T_\infty + T_L(y_{\text{ox}, \infty}/\nu)}{1 + (y_{\text{ox}, \infty}/\nu)} \tag{7-33}$$

3. J. C. Maxwell, "Diffusion," *Encylopaedia Britannica*, Samuel Hall, London, **7**:218 (1878).

If $T_L = T_\infty$ Eq. (7-33) reduces to the form

$$T_F - T_\infty = \frac{(q_{\mathrm{fu}} - \ell)}{c_p \left( \dfrac{v}{y_{\mathrm{ox}, \infty}} + 1 \right)} \qquad (7\text{-}34)$$

Since $v/y_{\mathrm{ox}, \infty}$ mass of oxidizer is required to oxidize a unit mass of fuel, Eq. (7-34) states that the flame sheet is at the stoichiometric adiabatic flame temperature after the heat of evaporation is taken into account. The extra terms in Eq. (7-33) simply represent the properly weighted background temperature if $T_L \neq T_\infty$. Since we know $T_F$ in terms of known quantities we can also determine the proper values of $B_1$ and $y_{\mathrm{fu}, L+}$.

In order to determine $\mu$ and $\sigma_F$ we first note that Eqs. (7-21) and (7-25) may be equated to yield the relationship

$$\exp\left(\frac{-\mu}{\sigma_F}\right) = \left(\frac{1 + B_1}{B_1}\right) \exp\left(-\mu\right) \qquad (7\text{-}35)$$

With this information Eq. (7-31) may be solved for either $\dot{M}$ or $\sigma_F$ to yield

$$\mu = \frac{\dot{M}}{r_L} = \ln\left[ 1 + \frac{q_{\mathrm{fu}}\, y_{\mathrm{ox}, \infty}}{\ell v} + \frac{c_p(T_\infty - T_L)}{\ell} \right] \qquad (7\text{-}36)$$

$$= \ln\left[ 1 + B_2 \right] \qquad (7\text{-}37)$$

where $B_2$ is called the transfer coefficient for droplet combustion, first defined by Spalding.[4] Finally we obtain

$$\sigma_F = \frac{r_F}{r_L} = \frac{\mu}{\ln\left[ 1 + (y_{\mathrm{ox}, \infty}/v) \right]} \qquad (7\text{-}38)$$

This is the location of the flame sheet.

If $T_L = T_\infty$ these expressions become quite simple.

$$T_F = \frac{(q_{\mathrm{fu}} - \ell)}{c_p[(v/y_{\mathrm{ox}, \infty}) + 1]} + T_\infty \qquad (7\text{-}39)$$

$$y_{\mathrm{fu}, L+} = \frac{(q_{\mathrm{fu}} - \ell)}{q + y_{\mathrm{ox}, \infty}\, \ell/v} \qquad (7\text{-}40)$$

$$B_1 = \frac{[(v/y_{\mathrm{ox}, \infty}) + 1]\ell}{(q_{\mathrm{fu}} - \ell)} \qquad (7\text{-}41)$$

$$B_2 = \frac{q_{\mathrm{fu}}\, y_{\mathrm{ox}, \infty}}{\ell v} \qquad (7\text{-}42)$$

4. D. B. Spalding, "The Combustion of Liquid Fuels," *Fourth Symposium (International) on Combustion*, The Combustion Institute, Pittsburgh, Pa., pp. 847–864 (1953).

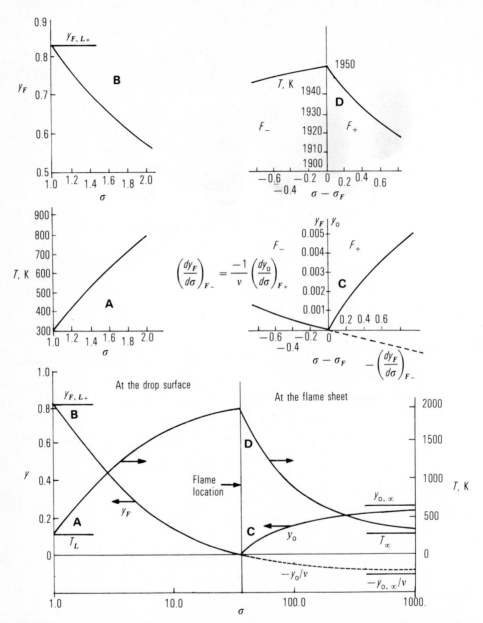

**Figure 7-5** Temperature and composition profiles predicted using the quasi-steady "flame sheet" theory of droplet combustion. Regions Ⓐ Ⓑ Ⓒ and Ⓓ of the main drawing are enlarged above to show details.

Figure 7-5 was constructed for the case where $q_{fu}/\ell = 100$, $q_{fu}/c_p = 35,000$, $T_L = T_\infty = 300$ K, $y_{ox, \infty} = 0.25$, and $v = 5$. For this case the flame temperature is 1950 K, $B_1 = 0.2121$, and $B_2 = 5.0$. Note that the dashed line in Fig. 7-5 is a smooth extension of the fuel mass-fraction curve and has a value $y_{fu} = -y_{ox}/v$ in the region where oxidizer exists. This is exactly equivalent to the behavior of the single "reactant" in the Burke–Schumann theory. The flame location therefore occurs on the universal function at the point where the slope of the oxidizer curve that just asymptotes $y_{ox, \infty}$ is exactly $-v$ times the slope of the fuel curve.

The temperature gradients at the flame surface are grossly different because in the steady-state case only sufficient energy is being transported to the fuel to just evaporate the liquid. All other heat liberated at the flame sheet is transported away from the drop to the ambient conditions at $r = \infty$.

Even though the quasi-steady theory developed above assumes a constant drop radius the drop must be getting smaller as it burns because there is an outward mass flow everywhere. If we write the equation for mass conservation for the drop we obtain the expression

$$-\frac{d}{dt}\left(\frac{\pi}{6}\rho_L d_L^3\right) = \dot{m} = 4\pi r^2 \rho u \tag{7-43}$$

However, recall that

$$\dot{M} = \frac{\dot{m}}{4\pi\rho D} = \dot{\mu}r_L = r_L \ln(1 + B_2)$$

Therefore, by integrating (7-43) and substituting, we obtain the relationship

$$-\frac{8r_L \rho D}{\rho_L} \ln(1 + B_2) = d_L^2 \frac{dd_L}{dt} \tag{7-44}$$

which can be rewritten as

$$-\frac{d(d_L^2)}{dt} = k = \frac{8\rho D}{\rho_L} \ln(1 + B_2)$$

This is the theoretical basis for the $d^2$ law mentioned earlier. Experimental values of $k$ for a number of different fuels are given in Table 7-1.

In actual droplet-combustion experiments it is found that the flame standoff distance is much less than that predicted by theory, and also that the value of $\sigma_F$ is not constant during the combustion process. Nevertheless the simple theory does predict reasonable values for $k$ and does capture the essence of the diffusion-controlled combustion of a drop. The interested reader is referred to the bibliography for more information on the processes that complicate the theory of droplet combustion.

**Table 7-1 The burning constant, $k$ for various liquid fuels (Oxidant: air at 20°C and 100 kPa)†**

| Fuel | $k_{calc} \times 10^{-7}$ m²/s | $k_{meas} \times 10^{-7}$ m²/s |
|---|---|---|
| Benzene | 11.2 | 9.9 |
| Toluene | 11.1 | 7.7 |
| o-Xylene | 10.4 | 7.9 |
| p-Xylene | 10.8 | 7.7 |
| Ethyl benzene | 10.8 | 8.6 |
| isopropylbenzene | 10.6 | 7.8 |
| n-Butylbenzene | — | 8.6 |
| tertiary-Butylbenzene | 10.4 | 7.7 |
| tertiary-Amylbenzene | — | 7.8 |
| Pseudocumene | 10.2 | 8.7 |
| Furfuryl alcohol | — | 7.2 |
| Ethyl alcohol | 9.3 | 8.5 |
| n-Heptane | 14.2 | 8.4 |
| iso-Octane | 14.4 | 11.4 |
| Tetralin | — | 7.6 |
| Decane | 11.6 | 10.1 |
| Amyl acetate | — | 8.0 |
| Petroleum ether (100–120°C) | — | 9.9 |
| Kerosene ($\rho = 0.805$) | 9.7 | 9.6 |
| Diesel oil ($\rho = 0.850$) | 8.5 | 7.9 |

† From A. M. Kanury, *Introduction to Combustion*, Gordon and Breach, New York, N.Y., p. 178 (1975), with permission.

## 7-4 COUNTERFLOW DIFFUSION FLAMES[5]

The parallel-flow diffusion flame described in Sec. 7-2 is of interest historically and does occur as a natural buoyancy-driven flame, such as a candle flame, for example. A buoyancy-driven flame will be discussed in the next section. There is another type of diffusion flame, namely the counterflow diffusion flame, which is important because it can easily be used to perform flame structure studies. Four different experimental geometries have been used to study such counterflow diffusion flames. They are illustrated in Fig. 7-6. Types 1 and 2 in this figure suffer from edge effects and type 3 has been used to model droplet combustion. Flames produced on a cylindrical burner of type 4 are very stable and have been used extensively to study flame structure and extinction. Extinction will be discussed in Chap. 12. In this section we will discuss the structure of a flame of type 4. The first important observation concerning this strictly two-dimensional flame is shown in Fig. 7-7. This photograph of a flame in which the incoming air flow was seeded with fine particles shows that the forward stagnation point

5. All the data discussed in this section was taken, with permission, from H. Tsuji, "Counter Flow Diffusion Flames," *Prog. Energy Combust. Sci.*, **8**: 93–119 (1982).

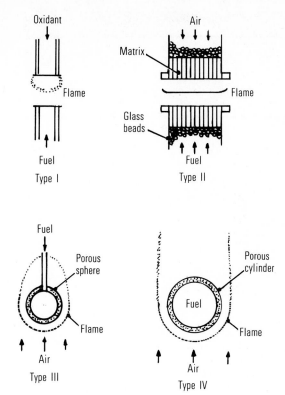

**Figure7-6** Classification of counter-flow diffusion flames into four types. *(With permission from Ref. 5.)*

of the flow is on the fuel-rich side of the thin blue luminous flame, and indeed this has been shown to be true for a number of fuels (hydrogen is the only exception). Under these conditions virtually all the chemistry occurs outside of the stagnation point region and is primarily diffusion-controlled when the flame is not near extinction. Figure 7-8 is a plot of the stream tube area and gas density at the center line of the flow, and Fig. 7-9 is a plot of the stable species concentrations for this flame. Figures 7-10 and 7-11 are plots of mass flux rates and net reaction rates for the major stable species. Figure 7-12 is a plot of the temperature profile in this flame. In Figs. 7-8 to 7-12 the distance variable $\eta$ is dimensionless and is defined as $(z/R)(2\,\text{Re})^{1/2}$ where $\text{Re} = V_z R/v$, $v$ is the mean kinematic viscosity in the flame and $R$ is the cylinder radius.

The major conclusions that one can draw from these observations are that:

1. The flame reactions are rather rapid, by comparison with diffusion. Note from Figs. 7-10 and 7-11 that the diffusive flux of methane from the fuel side is away from the cylinder $(-)$ while the flux of stable products inside the flame zone and oxygen outside the flame zone is toward the cylinder $(+)$.
2. The net flux of CO, $H_2$, and $C_2H_2$ are away from the cylinder just outside of the luminous zone.

**Figure 7-7** Particle-streak picture of a counterflow diffusion flame. Propane–air flame. Diameter of cylinder is 30 mm. The air flow is from below and is 0.8 m/s while the blowing rate of fuel through the cylinder walls is 86 mm/s. *(Courtesy H. Tsuji; originally appeared in Ref. 5. With permission.)*

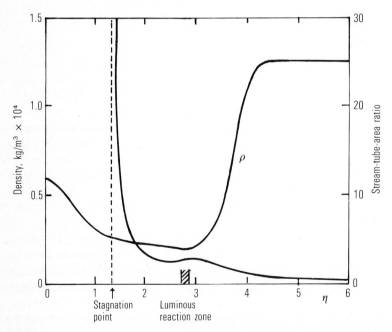

**Figure 7-8** Variations of gas density and stream-tube-area ratio across a methane flame in the stagnation region. *(From Ref. 5 with permission.)*

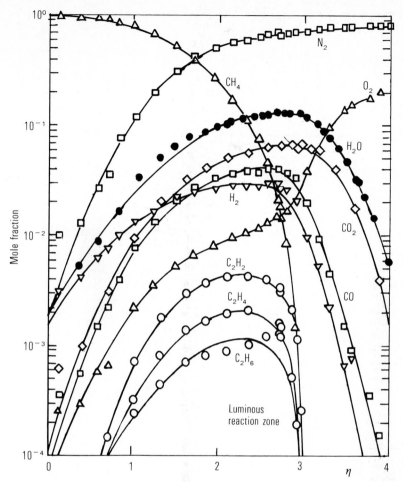

**Figure 7-9** Concentration profiles in the methane flame of Fig. 7-8. *(From Ref. 5 with permission.)*

3. Hydrogen and carbon monoxide are produced in the fuel-rich region and in the luminous zone but disappear rapidly on the oxygen side of the luminous zone in the region where the carbon dioxide production rate is most rapid. This separation of reaction zones is caused by the relatively slow oxidation of CO to $CO_2$, which proceeds almost entirely by the $CO + OH \rightarrow CO_2 + H$ reaction. This is because the OH radicals needed for this reaction are produced only slowly on the oxygen-rich side of the flame by the hydrogen branching-chain reactions discussed in Sec. 6-10. These are the same set of reactions that are operative during the "wet" oxidation of carbon monoxide.

4. The higher hydrocarbons that are produced in this flame are formed on the fuel-rich side and are destroyed in the luminous zone. It is obvious that the $CH_3$ radical condensation processes discussed in Sec. 6-10 are also occurring

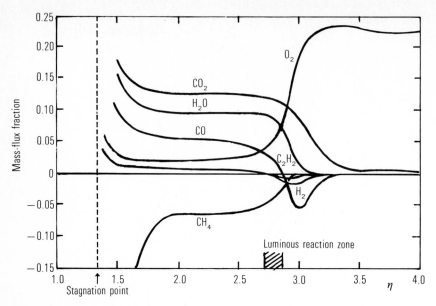

**Figure 7-10** Profiles of mass-flux fraction across the methane flame of Fig. 7-8 *(From Ref. 5 with permission.)*

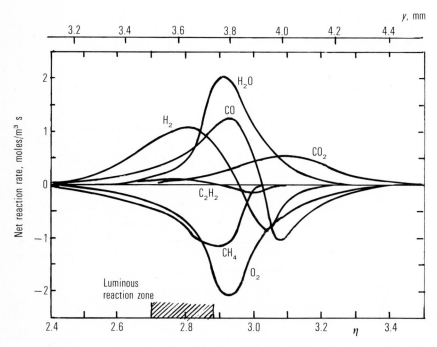

**Figure 7-11** Net reaction-rate profiles of various species in the methane flame of Fig. 7-8. *(From Ref. 5 with permission.)*

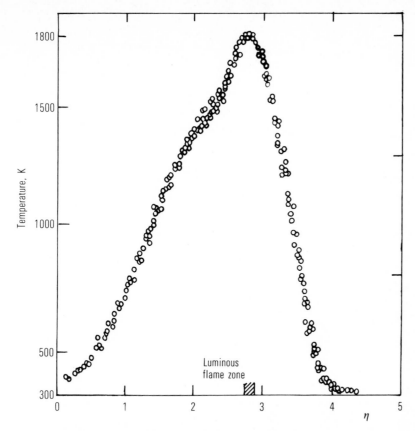

**Figure 7-12** Temperature profile in the methane flame of Fig. 7-8. *(From Ref. 5, with permission.)*

in this diffusion flame. Notice that even though these species are not thermo-dynamically stable at these high temperatures they are being produced at superequilibrium concentrations by the kinetic mechanisms discussed in Sec. 6-10.

Thus we see that the structure of a real diffusion flame of a hydrocarbon in air is much more complex than that predicted by the simple one-reaction-step theories. Nevertheless it is also apparent that the basic structural features of a diffusion flame as exemplified by the temperature profile, Fig. 7-12, are modeled in a reasonable manner by the simple theory.

## 7-5 THE JET DIFFUSION FLAME

The structure of a laminar diffusion flame held on a small tube has been calcu-lated using a full kinetic scheme, for a vertical hydrogen jet burning in still air

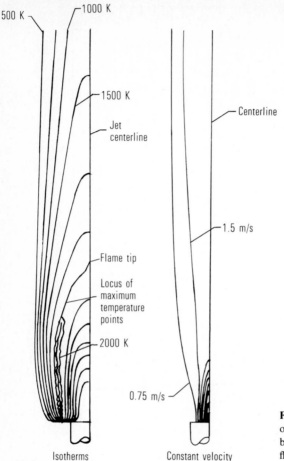

500 K

1000 K

1500 K

Jet centerline

Centerline

1.5 m/s

Flame tip

Locus of maximum temperature points

2000 K

0.75 m/s

Isotherms

Constant velocity contours

**Figure 7-13** Isotherms and the locus of maximum temperature for a buoyant laminar hydrogen diffusion flame. See text for inlet conditions. *(From Ref. 6, with permission.)*

in the earth's gravity field.[6] The authors used 16 forward reactions and their reverse reactions. The appropriate reverse rates were determined from the forward rates by using the equilibrium constants for each reaction. Their scheme included the species $HO_2$ and NO but not $H_2O_2$. The flame was held to the circular rim by starting with a stoichiometric mixture at the rim and placing a gaussian temperature profile, centered at the stoichiometric mixture, just above the rim. The height and width of this distribution were changed to determine its effect on the flow and chemistry downstream. The authors found little effect on the resulting flow field or species distribution downstream. Figure 7-13 is a plot of the isotherms, and the locus of maximum temperature for a 6 m/s hydrogen jet issuing from a 6.4 mm diameter tube and Fig. 7-14 is a plot

6. J. A. Miller and Robert J. Kee, *J. Chem. Phys.*, **81**:2534 (1977).

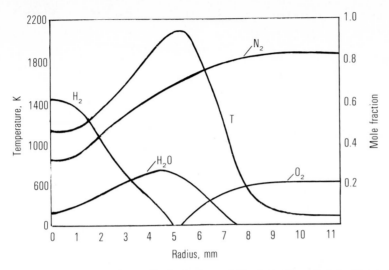

**Figure 7-14** Major species profiles in the flame of Fig. 7-13. *(From Ref. 6, with permission.)*

of the radial temperature and major species concentrations 10.7 mm above the rim. We note that the structure is reminiscent of simple one-step diffusion flames in which the chemistry is relatively rapid by comparison with the diffusion process. The two primary reactants, hydrogen and oxygen, exist together at only very small concentration levels and the temperature shows a broad peak at that location.

The most striking feature of this flame, however, is the calculated concentrations of the three primary-chain-carrying free radicals O, H, and OH. These are plotted as a function of radius at 10.7 mm above the rim in Fig. 7-15. We note from this figure that in the radius range from 3 to 6 mm, where the temperature is above about 1600 K, these three species are in partial equilibrium. Thus, their concentrations are almost an order of magnitude above what their concentrations would be if the system were fully equilibrated. As the authors point out, this means that the $HO_2$ radical plays an important part in the recombination and heat release that occur in this high-temperature region. This is because all the major $HO_2$ formation and destruction reactions are highly exothermic

$$H + O_2 + M \rightarrow HO_2 + M \qquad (\Delta H = -209.9 \text{ kJ at 1800 K})$$

$$OH + HO_2 \rightarrow H_2O + O_2 \qquad (\Delta H = -305.2 \text{ kJ at 1800 K})$$

and

$$H + HO_2 \rightarrow 2OH \qquad (\Delta H = -167.7 \text{ kJ at 1800 K})$$

This observation relative to partial equilibrium of the H atom chain-branching reactions simply reinforces the observations made in Sec. 6-10 relative to the ubiquity of this chain in hydrogen, carbon monoxide, and

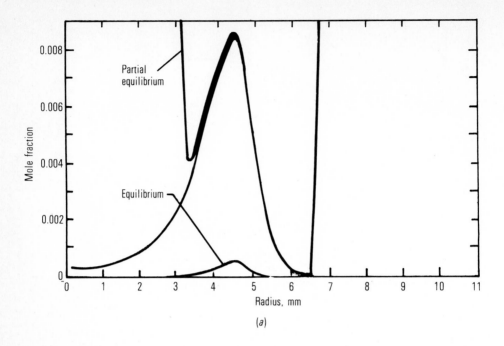

(a)

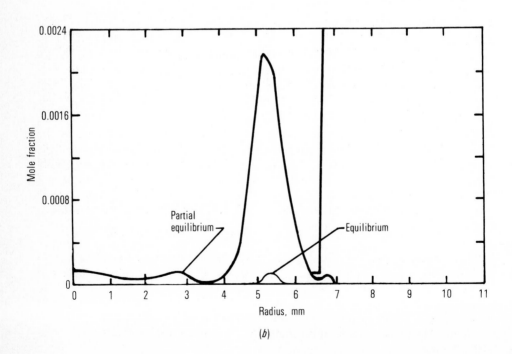

(b)

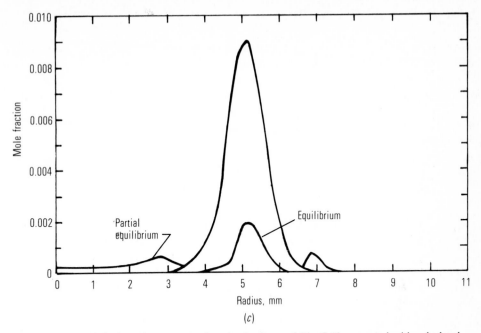

(c)

**Figure 7-15** The radical species concentrations in the flame of Fig. 7-13 compared with calculated partial equilibrium concentrations and full equilibrium concentrations. (a) H atom, (b) O atom, (c) OH radical. *(From Ref. 6, with permission.)*

hydrocarbon combustion processes. The fact that the major hot regions of the flame are in partial equilibrium and are only slowly reaching equilibrium due to recombination processes means that the maximum temperature in the flame is well below the predicted adiabatic temperatures ($T_{max} < 2200$ K, $T_{ad} = 2379$ K). This is because in contrast to a single-step reaction flame this is really a two-step flame. The first step is the formation of the partial equilibrium state, the second is the slower recombination process that occurs as more diluents diffuse into the flame zone.

Mitchell et al.[7] have made species measurements in a laminar axisymmetric forced-convection methane diffusion flame and their results also agree with the occurrence of partial equilibrium in the hotter regions ($T > 1800$ K) of their flames. In these flames they observed that once the hydrocarbon fragments were oxidized to CO the final oxidation of CO to $CO_2$ occurs by the reaction

$$OH + CO \rightarrow CO_2 + H \qquad (\Delta H = -90.27 \text{ kJ at } 1800 \text{ K})$$

which occurs in a region where the H atom chain is in partial equilibrium. This agrees with the net reaction-rate profiles measured in the counterflow diffusion flame and plotted in Fig. 7-11.

7. R. E. Mitchell, A. F. Sarofim, and L. A. Clomburg, *Combust. Flame*, **37**:227 (1980).

## 7-6 CATALYTIC COMBUSTION

Another type of combustion process in which diffusion plays an important role is catalytic combustion in a monolithic catalytic combustor. A combustor of this type consists of a bundle of ceramic tubes that are fused together or a honeycomb of holes which have a relatively large open area. The surface of the ceramic is treated with an appropriate catalyst and the combustible mixture is allowed to flow through the matrix. It is found experimentally that if the flow velocity is low enough and the catalytic bed is hot enough the combustor will "light off," and sustained combustion will occur in the bed. In a typical application when the approach flow is laminar the structure of the reaction zone is observed to behave as shown schematically in Fig. 7-16. In region 1 surface reactions predominate and there is a concentration profile in the boundary layer. In region 2 the flow and species distribution becomes relatively uniform across the tube and reactions in the gas phase are starting to dominate. Finally in region 3 the hot flame reactions are going to completion. This region, as well as region 2, behaves somewhat like a plug-flow reactor.

The advantages of catalytic combustion are many. The primary advantage in turbine applications is due to the fact that you can "light off" a properly designed catalytic combustor even with very lean mixtures. In fact the mixtures can be made so lean that the adiabatic flame temperature equals the design turbine inlet temperature. This means, of course, that the approach flow to the turbine will have a constant or almost constant temperature profile with no hot spots, as can occur when one turbulently mixes secondary air with the very hot primary products in a conventional combustor. It also means that pollutant production is very low because combustion is complete or almost complete in ultralean mixtures and the flame temperature never gets high enough to produce large quantities of the oxides of nitrogen. See Chap. 16 for a further discussion of combustion-generated pollution.

It has been observed that for any particular configuration and inlet conditions of fuel concentration, pressure and temperature, there is a maximum approach-flow velocity above which the reactor will not operate. It has also been observed that blowout occurs as a relatively slow process in which extinc-

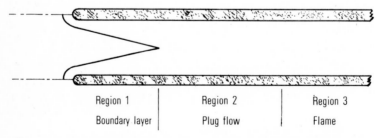

| Region 1 | Region 2 | Region 3 |
| Boundary layer | Plug flow | Flame |

**Figure 7-16** Typical combustion regions in a catalytic combustor tube. (schematic).

tion of the catalytic reactions first occurs at the leading edge and then washes back through the tube until extinction is complete.

There have been a number of sophisticated boundary-layer theories for the operation and stability of monolithic catalytic combustors. However, a relatively simple stagnation point flow theory of the catalytic combustor will be presented here to illustrate the processes that control blowout. We will assume, for simplicity, that the Lewis number, Schmidt number and Prandtl number of the gas are unity, that the molecular weights and heat capacities of all the reactants and products are equal and constant, and that the reaction kinetics at the catalytic surface are given by a one-step Arrhenius equation, similar to Eq. (6-1)

$$\mathcal{M}_c = A_s [F]^n [O]^m [I]^l \exp \left( \frac{-E^*}{R T_w} \right) \tag{7-45}$$

where the units on $A_s$ yield units of kilograms per square meter-second for $\mathcal{M}_c$ and the subscript $w$ on $T$ means wall temperature. We further assume that the catalytic wall is adiabatic and that its surface temperature is equal to the gas temperature at the surface. We now note that the assumption that $Le = 1$ means that the mole fraction at any point in the stagnation region can be expressed as a linear function of the temperature

$$x_{fu} = x_{fu_0} \left( \frac{T_b - T}{T_b - T_0} \right) \tag{7-46}$$

where $x_{fu_0}$ and $T_0$ are the inlet mole fraction and temperature of the fuel (lean combustion) and $T_b$ is the flame temperature.

We now wish to discuss mass transport to the surface in the stagnation point flow using a very simple model. The mass-transport coefficient $h_D$ is given by the expression

$$h_D = \frac{\rho D \, \mathrm{Sh}}{d} \qquad \mathrm{kg/m^2 \ s} \tag{7-47}$$

where $D$ is the diffusion coefficient, $\rho$ is the density, $d$ is the diameter of the leading edge, and $\mathrm{Sh}$ is the Sherwood number for mass transfer. Using this notation at the stagnation point

$$\mathcal{M}_D = h_D (x_{fu_0} - x_{fu_w}) \qquad \mathrm{kg/m^2 \ s} \tag{7-48}$$

As Kreith[8] points out the Nusselt number for the stagnation point on a cylinder perpendicular to an approach flow is proportional to the square root of the Reynolds number based on the cylinder's diameter. Also within the framework of our simplifying assumptions the Nusselt number for heat transfer is exactly equal to the Sherwood number for mass transport at the stagnation

8. F. Kreith, *Principles of Heat Transfer*, 2d ed., International Textbook Co., Scranton, Pa., pp. 408–410, p. 559, (1965).

point if reaction at the surface generates a concentration gradient in a forward stagnation region of the body. Thus, for every fixed geometry and inlet conditions of concentration, pressure, and temperature we find that

$$\mathscr{M}_D \propto (x_{fuo} - x_{fuo})\sqrt{V} = \eta x_{fuo}\sqrt{V} \tag{7-49}$$

where $\eta = (1 - x_{fuw}/x_{fuo})$. In other words the amount of mass transport to the surface is proportional to the square root of the approach-flow velocity and is directly proportional to the overall concentration difference between the inlet flow and the wall. Obviously for proper steady-state operation $\mathscr{M}_D = \mathscr{M}_c$. We now note that substitutions of the definition of $\eta$ into Eq. (7-46) where $T = T_w$ yields the relationship $\eta = (T_w - T_0)/(T_b - T_0)$. Thus the equations that govern the steady-state stagnation point temperature for a catalytic combustor with an adiabatic wall have a form that is identical to Eqs. (6-21) to (6-24) for the well-stirred reactor. This means, of course, that blowout of a catalytic combustor occurs at a critical minimum Damköhler number. Note that this is true, even though the physical processes that lead to sustained combustion are markedly different in these two devices. It appears that, in general, if flow conditions supply the fuel and oxidizer to a region of intense reaction too rapidly the reactions will be extinguished, and that this extinguishment behavior can be examined by forming a local Damköhler number based on a flow time and a characteristic heterogeneous kinetic time evaluated at the maximum local temperature in the system.

## PROBLEMS

**7-1** Consider a Burke–Schumann flame in which the fuel is a methane–nitrogen mixture and the oxidizer is 21% $O_2$ air. If the burner is two-dimensional as illustrated in Fig. 7-1, and $L_1/L_2 = 0.25$, calculate the percent $CH_4$ in the fuel stream to produce
  (a) an underventilated flame of height $\zeta = 0.1$
  (b) a "stoichiometric" flame
  (c) an overventilated flame of height $\zeta = 0.1$

**7-2** If pure methane and pure oxygen were burned in an $L_1/L_2 = 0.25$ flat Burke–Schumann diffusion flame what would the flame height $\zeta$ be? Would this flame be over- or underventilated, and if it is overventilated will it be bulged?

**7-3** Methane is rapidly released into a 10 m × 10 m square mine shaft that is horizontal and sealed at both ends. The methane rapidly layers to fill the top 25 percent of the shaft. Assume that at time $t = 0$ the concentration of methane is a square wave with all the methane on top of the air. Assume that only diffusion is occurring. Determine and plot the concentration profiles after 10 minutes and 2 hours. At what time would the concentration difference from top to bottom be 5 percent (i.e., ca. 22.5 percent at the bottom and 27.5 percent at the top)?

**7-4** A drop of benzene is 2 mm in diameter at time $t = 0$ and is burning at equilibrium in air $[T_\infty = 298K, (x_2)_\infty = 0.21]$. Calculate the location of the flame sheet, the flame temperature, and the constant $k$. Compare to the value given in Table 7-1.

**7-5** Determine if the water–gas equilibrium relation $CO + H_2O \rightarrow CO_2 + H_2$ is satisfied at the locations $\eta = 2$ and $\eta = 3$ in the methane-air counter-diffusion flame. Use Figs. 7-8 and 7-11 to make this determination.

**7-6** If the reaction $CO + OH \rightarrow CO_2 + H$ were the only reaction causing oxidation of CO to $CO_2$ and its rate expression were given by the equation $k_f = 10^{12.88}T^{1.3} \exp(-3222/RT)$ determine the concentration of OH radicals at $\eta = 3$ and $\eta = 3.1$ in this flame. Use Figs. 7-10 and 7-11 to make your determination.

# BIBLIOGRAPHY

**Droplet combustion**

Krier, H., and C. L. Foo, "A Review and Detailed Derivation of Basic Relations Describing the Burning of Droplets," *Oxid. Combust. Rev.*, **6**: 111–144 (1973).

Williams, A., "Fundamentals of Oil Combustion," *Prog. Energy Combust. Sci.*, **2**: 167–179 (1976).

**Catalytic combustion**

Kesselring, J. P., "Catalytic Combustion," *Advanced Combustion Methods*, ed. F. J. Weinberg, Academic Press, New York (1983) (in press).

# EIGHT

## FLAMES

## 8-1 INTRODUCTION

Even though flames in one form or another have been known and used since antiquity, a particularly simple form of combustion, the premixed gas flame, has been recognized and studied for only about 150 years. Its discovery was a side result of the researches of Berthollet, Dalton, Volta, and others (1776–1810) concerning the composition of combustible gases from different sources, particularly marsh gas (methane) and olefiant gas (ethylene). In their analyses, a combustible mixture of these gases with air was sparked in a bulb and the products of the premixed flame (explosion) were analyzed to determine the composition of the original combustible gas. During a later period (1805–1819) Davy's observation that a premixed flame would not propagate through a fine-mesh screen led to the invention of the safety lamp for mines. The "discovery" of premixed flames was essentially completed in 1855 by Bunsen's invention of a burner which allowed the stabilization of a premixed flame in a flow system.

We now know that a large collection of exothermic mixtures and pure substances which are capable of sustaining a homogeneous reaction will also support a subsonic "reaction wave" (or flame) which is self-propagating and has definite properties that are dependent only on the initial conditions in the gas. Unfortunately, these reaction waves are always rather thick and thus can be markedly influenced by curvature effects. Furthermore, they can be inherently unstable as a one-dimensional wave phenomena, and because of their low subsonic velocity they usually interact very strongly with the flow aerodynamics. In this chapter we wish to divorce our discussion from these complicating aerodynamic aspects as much as possible, reserving our discussion of those facets of flame behavior for Chaps. 10, 11, and 14. Thus in the following sections we will consider only the structure and inherent stability of laminar flames.

## 8-2 FLAME TEMPERATURE AND BURNING VELOCITY

Assume that Fig. 5-2 represents a strictly one-dimensional steady flame with a relatively low subsonic velocity. If this flame is to exist as a steady phenomena it must satisfy the equations derived in Chap. 5 at the two stations (1) and (2), where (2) is far enough downstream so that thermal and chemical equilibrium is reached in the hot gases. Therefore it must have two properties that are definable irrespective of the details of the processes occurring in the reaction zone— the burning velocity $S_u = u_1$ and the flame temperature $T_b = T_2$. In addition, we note from Chap. 5 that the momentum equation requires that $P_2 \neq P_1$ for a flow process with steady heat addition, and we first determine an upper limit for the pressure change associated with flames by solving Eq. (5-36) to yield

$$P_2 - P_1 = P_1 \gamma (1 - \varepsilon) M_1^2 \tag{8-1}$$

where $\varepsilon = v_2/v_1$ and $M_1 = S_u/a_1$. Since for all observable flames $M_1 < 0.02$, $v_2/v_1 < 15$, and $\gamma \leq 1.667$, Eq. (8-1) predicts that a maximum pressure change of $P_1 - P_2 \leq +0.01\ P_1$ may exist across any flame. We thus note that flame propagation is always accompanied by a slight pressure drop, and that ordinary one-dimensional flames with $M_1 \cong 0.001$ may be considered as essentially isobaric processes. This observation justifies the usual approximation of equating the flame temperature to an isenthalpic explosion temperature for the mixture. Also, for any specific system the isobaric assumption means that the flame temperature $T_b$ is relatively independent of the observed burning velocity and may be determined with reasonable accuracy without knowledge of the burning velocity. Measurements of flame temperatures have shown that the translational gas temperature rises monotonically through the flame and eventually reaches a value comparable to the isenthalpic values calculated using thermodynamics. Ordinarily the accepted value of a flame temperature is the calculated isenthalpic value because difficulties with the experimental measurement of flame temperature and the relatively high accuracy of present-day thermodynamic calculations cause one to place more confidence in the calculated value.

Figure 4.10 is a plot of constant-pressure explosion temperatures as a function of percent fuel in air and thus represents the flame temperatures of these fuels. Figure 8-1 illustrates the effect of initial pressure and temperature on the calculated flame temperature. Note from Fig. 8-1 that the flame temperature of hotter flames is relatively insensitive to the initial temperature of the mixture because dissociation reactions cause the flame gases to have a large effective heat capacity. Low-temperature flames are relatively more sensitive to changes in initial temperature because dissociation reactions are not as important at these temperatures. However, the flame temperature of a high-temperature flame

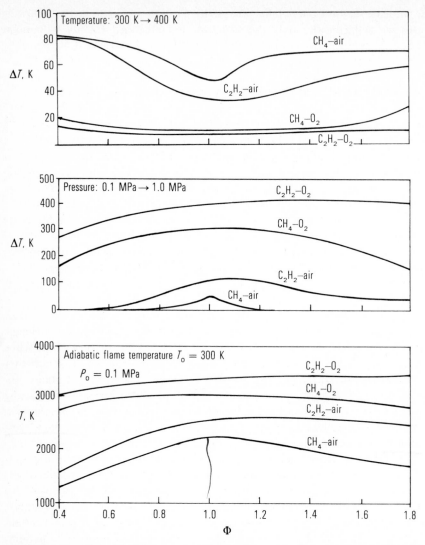

**Figure 8-1** The effect of initial pressure and temperature on the flame temperature of methane and acetylene flames.

is more sensitive to changes in the ambient pressure, because pressure shifts the dissociation equilibria at high temperatures but can cause only relatively small composition changes in low-temperature flames.

In addition to flame temperatures, an isenthalpic calculation always yields the equilibrium composition of the mixture some distance downstream from the flame. A typical equilibrium flame composition as a function of equivalence ratio is shown in Fig. 4-11. We note that this equilibrium composition may not be simply related to the composition in or near the reaction zone of the flame. Section 8-5 discusses internal structure in more detail.

We now turn our attention to the question of burning velocity. First, we note that our assumption that a flame is a steady-flow process in no way legislates the behavior of real flames. On the contrary, the justification for this assumption arises solely from our observations of flames under well-controlled conditions in the laboratory. It has been observed that in a variety of laminar-flow situations the flame behaves in a way best explained by assuming that it is a steady three-dimensional phenomena. The extrapolation of this three-dimensional real behavior to the strictly one-dimensional behavior required for the definition of a burning velocity is primarily a problem of careful experimental measurement and of properly assessing the effects of curvature on flame structure. Because of the difficulty of correcting for flame curvature in flows where the flame is positioned at an oblique angle to the flow, the most accurate burning velocities are obtained in situations where either the flame is strictly flat, even though oriented obliquely to the flow, or where the flame, even though slightly curved, is oriented strictly normal to the flow-velocity vector.

There are four measurement techniques which satisfy these requirements and which may be considered as the primary experimental techniques for determining burning velocity: (1) the slot-burner technique, (2) the soap-bubble technique, (3) the flat-flame burner technique, and (4) the bomb technique.

In the slot-burner technique a flame is stabilized above a rectangular nozzle (ratio of side lengths > 3), which has either a flat velocity profile generated by a two-dimensional Mache–Hebra nozzle, or a fully developed laminar pipe-flow profile generated by using a long entrance pipe of the same dimensions as the port. In this situation the flame is attached to the rim of the burner and completely burns the combustible mixture issuing from the port. The flame takes the appearance of a tent above the port and is therefore held at an oblique angle to the primary flow direction. It has been observed that a relatively flat oblique laminar flame may be transformed to a strictly one-dimensional flame by imposing a constant velocity transformation as shown in Fig. 8-2. Therefore the advantage of the slot burner is simply that the flame has rather large flat areas (both sides of the tent) in which curvature effects are minimized. Furthermore, since the flame completely encloses the incoming flow the incoming streamlines are not deflected until they enter the heat conduction zone. Thus the angle that the flat-flame surfaces make to the incoming flow can be measured accurately and used to evaluate the normal burning velocity by applying the simple equation

$$S_u = U \sin \alpha \qquad (8\text{-}2)$$

Experiments have shown that for this geometry the measurement of the local flow velocity and local flame angle at a point in the flame yields a burning velocity which is relatively constant over a large portion of the burner width. A typical evaluation is shown in Fig. 8-3. We will see later that the apparent increase in velocity at the tip of this flame is caused by an aerodynamic interaction related to the positive flame curvature in this region, and the apparent decrease of burning velocity near the rim of the burner is caused by heat losses to the wall and the negative flame curvature in this region.

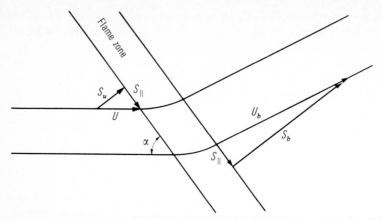

**Figure 8-2** The oblique one-dimensional flame. The flame appears as an oblique flame because of a simple velocity transformation at the velocity $S_{\parallel}$ along itself. The stream tube area parallel to the flame front does not change in this diagram.

The soap-bubble technique is another method which is known to yield accurate burning velocities without the application of an excessive number of corrections. In this technique a soap bubble is blown with a combustible mixture and ignited centrally by means of a capacitance spark. The flame burns as an outwardly propagating spherical flame, and since the bubble expands as combustion proceeds, the process occurs at constant pressure. In this case, after an initial propagation period the flame is only slightly curved and is oriented strictly normal to the flow direction. Its apparent propagation velocity is not $S_u$, however, but actually $S_b = u_2$ since the gas behind the flame is quiescent.

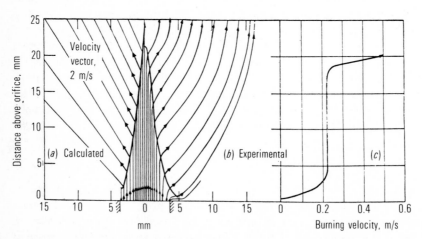

**Figure 8-3** Diagram of vertical center plane of natural gas–air flame on rectangular burner tube of $7.55 \times 21.9$ mm cross section. Mixture composition 7.5 percent natural gas; flow $204 \times 10^{-6}$ m³/s. (a) and (b) combustion zone and flow pattern, (c) burning velocity. (*With permission from B. Lewis and G. von Elbe, Combustion Flames and Explosions of Gases, 2d ed., Academic, New York, 1961.*)

This measured velocity may be converted to the normal burning velocity $S_u$ by noting that the continuity equation requires that $\rho_1 S_u = \rho_2 S_b$, and therefore that

$$S_u = \left(\frac{\rho_2}{\rho_1}\right) S_b$$

$$= \left(\frac{r_1}{r_2}\right)^3 S_b$$

where $r_1$ and $r_2$ are the initial bubble radius and final flame-ball radius respectively. Measurement of burning velocity using this technique thus requires the measurement of both the space velocity $S_b$ and the ratio of the initial to final radii of the ball of combustible gases. Since a relatively exact final radius is required, the best measurements are obtained when the initial bubble is surrounded by an inert atmosphere to prevent afterburning. Figure 8-4 is a typical example of an experimental measurement using a strip-film schlieren camera to observe the growth of the flame ball through a thin-slit aperture focused on the diameter of the flame ball. This technique is somewhat more difficult and less generally applicable than the slot-burner technique. It has the advantage, however, of requiring only small samples of the combustible mixture. Its disadvantages are:

1. Because the final radius is required rather accurately and because it is assumed that the final value of this radius represents gas which is completely in equilibrium, the technique is limited to soap bubbles of rather large initial diameter.
2. Soap bubbles are moist, and also are permeable to or react with certain gases. Thus mixtures whose burning velocity is markedly dependent on water-vapor content or which interact with the bubble material cannot be studied using this technique.
3. The accuracy of this technique is limited most seriously by the requirement that the final radius be determined accurately. However, measurements of flame-ball velocity vs. time for sufficiently large flame balls have shown that the apparent velocity approaches $S_b$ asymptotically as the diameter increases. Thus an alternative approach is to measure the asymptotic $S_b$ value for large flame balls and to divide it by the isobaric volume ratio calculated for a homogeneous constant-pressure process.
4. Mixtures with very high burning velocities are not amenable to measurement using this technique. As burning velocity increases, the expansion process association with flame growth sends out finite pressure waves which, on reflection from the walls of the chamber, cause a disturbance at the flame ball and complicate the measurement of final size. Also certain of these flames become hydrodynamically unstable and this precludes the measurement of a true normal burning velocity.

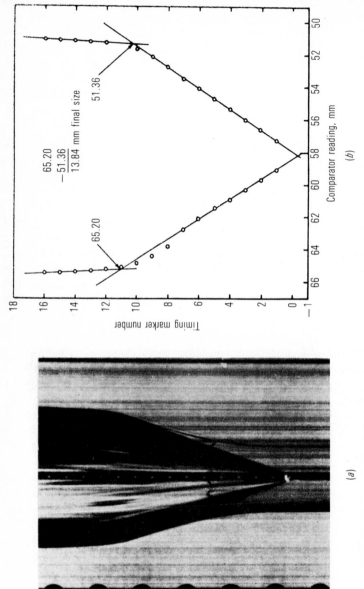

**Figure 8-4** A 6.53 percent ethylene–air mixture; argon atmosphere; timing light = 1000 flashes per second; standard distance, 177.88 mm at bubble equals 23.38 mm comparator reading; initial diameter = 6.91 mm. $(r_2/r_1)^3 = (13.84/6.91)^3 = 8.033$; space velocity $S_b$ (from graph) = $\Delta D/2 \, \Delta t = 12.87 \times 177.88/(23.38 \times 0.02 \times 1000) = 4.895$ m/s; $S_u = 4.895/8.033 = 0.609$ m/s. (*a*) Bubble explosion. (*b*) Graph of comparator reading. (*With permission from R. A. Strehlow and J. Stuart, Fourth Symposium (International) on Combustion, The Combustion Institute, Pittsburgh, Pa., p. 332 (1953).*)

258

The third precise technique, using the flat-flame burner, is strictly limited to very low-burning-velocity flames ($S_u < 200$ mm/s). In fact, flames observed using this technique usually cannot be observed as stable flames in other apparatus because of their low burning velocity. Figure 8-5 is a photograph of a flame stabilized on a flat-flame burner taken with illuminated particle tracks to show the velocity field ahead of the flame. In this type of apparatus, a rather large-diameter, low-velocity combustible stream is produced by passing the flow through an appropriate series of screens and honeycomb filters. The flame is stabilized as a flat flame by using a capping screen and a surrounding low-velocity, inert-gas flow. Experimentally, it is observed that a flame will be stabilized in this burner over only a very small flow range. Once the flame is stabilized at a reasonable distance from the port, the burning velocity may be determined by simply dividing the input volumetric flow rate by the total flame area. This apparatus has also been used to stabilize flames for structure studies.

The spherical-bomb technique is the fourth method that can be used to accurately measure burning velocity. In this method a flame is ignited by a spark at the center of a spherical bomb. The flame diameter is recorded as a function of time, usually using the visible flame front, and the pressure in the bomb is recorded with a fast-response pressure gauge. The flame velocity is then determined from these two pieces of information. This technique will be discussed in more detail in Chap. 14.

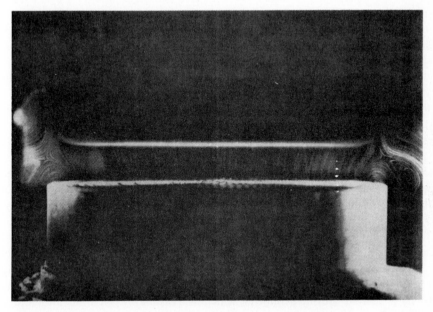

**Figure 8-5** Particle tracks in a flat ethylene–air flame. *(Courtesy of F. J. Weinberg, Imperial College, London. Originally from A. Levy and F. J. Weinberg, Seventh Symposium (International) on Combustion, The Combustion Institute, Pittsburgh, Pa., p. 296, 1959, with permission.)*

In addition to the above relatively precise methods for determining burning velocity, there are many others which have been used extensively. Four of these will be discussed briefly.

1. The burner method, using a flame-height measurement. It has been found that mixtures with very high burning velocity can be stabilized as laminar flames on relatively small burners. One technique for estimating the burning velocity of these small flames is to measure the total height of the inner cone and calculate the flame area, assuming that the flame has a simple conical shape. With this assumption the burning velocity of a flame on a circular port of radius $r$ is given by the equation

$$S_u = \frac{2V}{\pi r \sqrt{h^2 + r^2}}$$

where $h$ is the flame height and $V$ the volumetric flow rate of the combustible gas mixture. It is known that this technique does not yield an exact result, and in fact errors of 15 to 20 percent may be expected using this method.

2. The burner method, using the actual flame shape to calculate a flame area. This is a refinement of the above technique. In this case the flame, which is stabilized on a circular port, is photographed and the area is calculated from the photographic trace by numerical integration assuming cylindrical symmetry. Even though it would seem that this technique should yield a more precise burning velocity than the cone-height method, it is subject to many errors which do not adequately cancel each other. Aside from end effects, associated with curvature at the tip and heat losses at the rim, the technique suffers from the fact that the apparent flame position is different when measured using schlieren, shadowgraph, or visible-light photography, and the measured area therefore depends on the choice of optical technique.

3. Pressure-drop technique. Equation (8-1) states that a slight static pressure drop should be associated with flow through a steady flame. Therefore, a measured value of this pressure drop allows a calculation of the burning velocity, using (Eq. 8-1). This technique has been attempted in flames with a low enough final temperature so as not to melt a static-pressure probe. The results indicate that the burning velocity and pressure drop correlate well if the flame is sufficiently flat. However, the technique has never been used for the routine measurement of burning velocity.

4. The tube method. It has been observed that with the correct set of end openings certain flames will propagate down a horizontal tube at a relatively constant velocity. With proper calibration, a measurement of this steady velocity coupled with photographs of the shape of the propagating flame yields a relative burning velocity which is reasonably accurate and reproducible. Because of its simplicity and the fact that it requires relatively small quantities of combustible mixture, this technique has been used extensively to evaluate the relative burning velocity of various hydrocarbons in air.

However, the detailed behavior of a flame in a duct depends rather strongly on various aerodynamic considerations and is also greatly dependent on the presence of the walls. This technique has never been proposed as an absolute method of burning-velocity measurement.

The effect of mixture composition on the burning velocity for three fuel mixtures is shown in Fig. 8-6. It is interesting to note that the normal burning velocities for mixtures of $H_2 + CO$ and $H_2 + CH_4$ are approximately linearly related to fuel composition while $CO-CH_4$ mixtures show a distinct maximum in the burning velocity at 5% $CH_4$–95% $CO_2$. The reason for this will be discussed in detail in Sec. 8-5 on flame properties.

In this section we have seen that the flame temperature of a gaseous premixed flame may be determined without knowing the burning velocity or mechanism of propagation, and that, in general, the burning velocity is quite difficult to measure accurately. In this discussion we have ignored a possible complication—namely, that certain flat flames are thermodiffusional and/or hydrodynamically unstable and prefer to travel in other than the steady and strictly one-dimensional manner that we assumed. The behavior of these dynamically unstable flames will be discussed in Sec. 8-6 and Chap. 14.

## 8-3 LAMINAR FLAME THEORY—STEADY ONE-DIMENSIONAL FLOW

Soon after laminar premixed flames were stabilized in the laboratory, it was recognized that they represent a unique case of wave propagation; the steady self-sustaining propagation of an "exothermic reaction wave" at a low subsonic velocity in a direction normal to itself. The question of the propagation mechanism of this unique wave has been investigated extensively during the past 100 years. Early in this period, heat conduction was postulated as a primary mechanism of propagation. This "thermal" concept for flame propagation was first introduced by Mallard and Le Châtelier and culminated in the Zel'dovich treatment of the 1940s. In its simplest form the theory assumes that the flame:

1. Has a unique propagation velocity $S_u$.
2. Does not lose heat to the sides or downstream but does transmit heat to the upstream gas by thermal conduction.
3. Contains a relatively thin high-temperature reaction zone separated from a nonreactive and cooler conduction zone at a plane where "ignition" occurs. Thus this flame has a unique "ignition temperature" $T_{ig}$ which is closer to the flame temperature $T_b$ than to the unburned gas temperature $T_u$.

For simplicity one must also assume that the heat of reaction $q$ (J/kg), the heat capacity $c_p$ (J/kg K), the thermal conductivity $\kappa$ (J/m s K), and the rate of combustion $\omega$ (kg/m$^3$ s), are constants. The structure of such a flame is shown in Fig. 8-7.

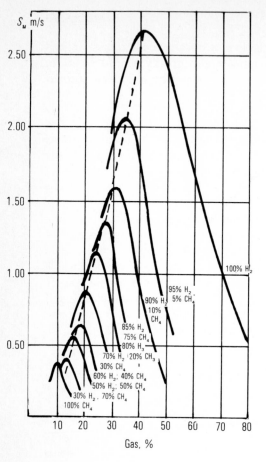

(a) $H_2 + CH_4$

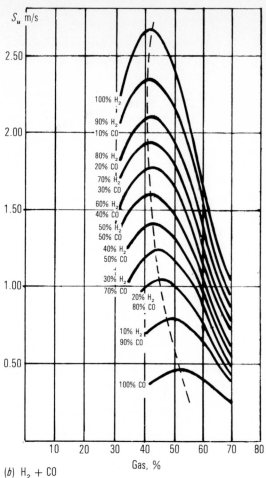

(b) $H_2 + CO$

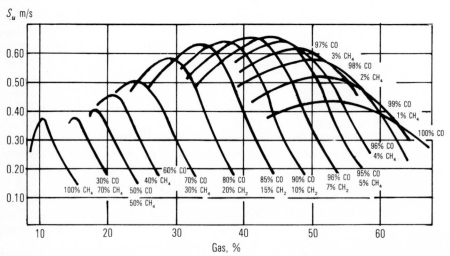

(c) $CH_4 + CO$

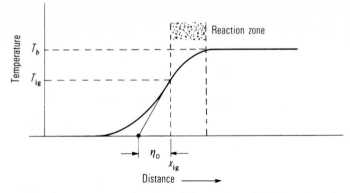

**Figure 8-7** The structure of a thermal theory flame. The distance $\eta_0$ is called the preheat zone thickness.

Within the framework of these assumptions we can immediately define the preheat zone thickness of this flame

$$\eta_0 = \frac{T_{ig} - T_u}{(dT/dx)_{ig}} \tag{8-3}$$

If we now apply energy conservation at the ignition plane we obtain the relationship

$$\kappa \left(\frac{dT}{dx}\right)_{ig} = \frac{\kappa(T_{ig} - T_u)}{\eta_0} = q\omega\eta_0$$

$$= c_p(T_b - T_u)\omega\eta_0 \tag{8-4}$$

which means (because $T_{ig} - T_u \cong T_b - T_u$) that the preheat zone thickness in such a flame is given approximately by the expression

$$\eta_0 \cong \sqrt{\frac{\kappa}{c_p\,\omega}} \tag{8-5}$$

Mass conservation states that $S_u \rho_1 = \omega\eta_0$ and therefore the thermal equation for the flame velocity becomes

$$S_u \cong \frac{1}{\rho_1} \sqrt{\frac{\kappa}{c_p}\,\omega} \tag{8-6}$$

Notice that we can eliminate the reaction rate from Eqs. (8-5) and (8-6) to obtain a relationship between the burning velocity and the preheat zone thickness of the flame

$$\eta_0 \cong \frac{\kappa}{\rho_1 c_p S_u} \cong \frac{\alpha}{S_u} \tag{8-7}$$

where $\alpha$ equals the thermal diffusivity of the gas.

**Figure 8-6** The normal burning velocity of different fuel mixtures with air. (*From L. N. Khitrin, Physics of Combustion and Explosion, National Science Foundation, Washington, D.C. (1962) with permission of the Keter Publishing House, Jerusalem, Israel.*)

Both the concept that a flame has a unique normal burning velocity and that it has a unique preheat zone thickness are important in industrial applications and in developing combustion safety criteria. It is important to note here that the normal burning velocity is essentially proportional to the square root of a thermal diffusivity multiplied by an effective rate of reaction and that the preheat zone thickness is uniquely related to the burning velocity by the thermal diffusivity of the gas.

The development of the thermal theory for flame propagation led to many vessel studies of reaction kinetics in exothermic systems. After these studies showed the prevalence of exothermic chain-branching reactions in combustion systems, forward diffusion of highly reactive chain carriers was postulated as an alternate mechanism of propagation. The subsequent proliferation of "simple" propagation theories greatly stimulated experimental flame research, and in particular led to many studies of burning velocity as a gross property which could possibly be correlated to the various theoretical predictions. However, it is only in the last decade or so that precise experimental probing techniques in conjunction with a correct and complete steady-flow theory have yielded good descriptive models of the highly coupled processes that cause stable flame propagation.

Accumulated experimental evidence indicates that an ordinary one-dimensional flame can propagate as a steady wave and that heat conduction, diffusion, and a coupled set of rapid exothermic chemical reactions all play important roles in the propagation mechanism. In the following development, we follow the approach of Hirschfelder and Curtiss[1] and include these three processes, but neglect the effect of radiative heat transfer, body forces, and second-order (i.e., coupled) transport effects, such as thermal diffusion. We also ignore nonstationary effects, such as flame acceleration or the possible occurrence of three-dimensional instabilities. That is to say, we will not consider the question of the inherent stability of a combustion wave at this time, but simply assume that the steady model represents a real flame.

Since the flame is essentially isobaric, we dispense with the momentum equation and discuss only mass and energy balances. To be complete we must discuss the mass balance of each species that appears in the flame. The velocity of each species at any plane in the flame is simply $u + v_i$ where $v_i$ is the average velocity due to diffusion of the $i$th species at that location and $u$ is the mass-flow velocity at that point. Since the total mass flow is constant at each plane, and since each species by definition has the same convective velocity, $u$, we find that the global mass-balance equation may be written as

$$\sum_{i=1}^{s} \rho_i(u + v_i) = \rho u = \mathcal{M}_p \tag{8-8}$$

1. J. O. Hirschfelder and C. F. Curtiss., *J. Chem. Phys.*, **17**: 1076 (1949).

which yields an equation relating the diffusion velocities of the individual species

$$\sum_{i=1}^{s} [I]M_i v_i = 0 \qquad (8\text{-}9)$$

where the symbol $[I]$ means concentration of the $i$th species. We may also define the diffusion velocity in terms of a diffusion coefficient

$$[I]v_i = -\mathcal{N} D_i \frac{dx_i}{dx} \qquad (8\text{-}10)$$

where $\mathcal{N}$ is moles of mixture per cubic meter, $x_i$ is the mole fraction of the $i$th species, and $D_i$ is the diffusion coefficient of the $i$th species in the local mixture subject to the constraint of Eq. (8-9). In general, these coefficients are not simply related to the binary-diffusion coefficients. The negative sign appears in Eq. (8-10) because a positive concentration gradient yields a negative diffusion velocity.

If we now consider a balance between two neighboring planes $x$ and $x + dx$, we see that the net rate of production of the $i$th species in the distance $dx$ per square meter of flame area is simply

$$[I](u + v_i)\Big|_{x+dx} - [I](u + v_i)\Big|_{x} \qquad (8\text{-}11)$$

However

$$[I](u + v_i)\Big|_{x+dx} = [I](u + v_i)\Big|_{x} + \frac{d}{dx}\{[I](u + v_i)\}\, dx \qquad (8\text{-}12)$$

and we may write

$$\left(\frac{d[I]}{dt}\right)_{\text{chem}} = \rho\left(\frac{dn_i}{dt}\right)_{\text{chem}} = \frac{d}{dx}\{[I](u + v_i)\} \qquad (8\text{-}13)$$

where $(dn_i/dt)_{\text{chem}}$ is defined as the net rate of production of the $i$th species at a station in the flame due to all the chemical reactions which are occurring at that location. Note that if $(dn_i/dt)_{\text{chem}}$ is not identically zero throughout the flame, concentration gradients will exist and cause diffusion of individual species to be superimposed on the motion in the flame. Thus the $i$ mass-balance equations become

$$\rho \sum_{j=1}^{p} v_{ij} \frac{d\lambda_j}{dt} = \frac{d}{dx}\{[I]u\} - \frac{d}{dx}\left(\mathcal{N} D_i \frac{dx_i}{dx}\right) \qquad (i = 1, 2, \dots, s) \qquad (8\text{-}14)$$

where there are $p$ chemical reactions occurring, each with their reaction coordinate $\lambda_j$ and rate $d\lambda_j/dt$. The quantity $\lambda_j$ has units of moles of reaction per kilogram of mixture.

We now construct an energy balance in the flame. Since the pressure is essentially constant and the flame is a low-speed subsonic wave, we write an enthalpy balance and therefore implicitly include the external work performed at constant pressure and exclude the flow energy. The enthalpy of the $i$th species is

$$H_i = \Delta H_{f_i}^{\circ} + \int_{T_0}^{T} C_{p_i} \, dT \tag{8-15}$$

where $\Delta H_{f_i}^{\circ}$ is the heat of formation at $T_0$ and $C_{p_i}$ the heat capacity per mole. The total enthalpy flux through a plane located in the flame is therefore

$$\sum_{i=1}^{s} [I](u + v_i)H_i \tag{8-16}$$

while the flux through a plane located in the equilibrium region downstream of the flame zone is given by the equation

$$\sum_{i=1}^{s} [I]_{eq} S_b H_{i_{eq}} \tag{8-17}$$

since no diffusion is occurring in this region. Here $S_b = u_2$ as defined earlier. The enthalpy gain of this volume element due to the mass flow is therefore

$$\sum_{i=1}^{s} [I](u + v_i)H_i - \sum_{i=1}^{s} [I]_{eq} S_b H_{i_{eq}} \tag{8-18}$$

Since the flow is steady, heat must be transported out at the rate of accumulation indicated in the above equation, and since heat is transported only in the flame zone we obtain the equation

$$\kappa \frac{dT}{dx} = \sum_{i=1}^{s} [I](u + v_i)H_i - \sum_{i=1}^{s} [I]_{eq} S_b H_{i_{eq}} \tag{8-19}$$

where $\kappa$ is the thermal conductivity of the mixture at $x$ and the first term on the right is evaluated at $x$. This equation may also be written as

$$\sum_{i=1}^{s} [I]_{eq} S_b H_{i_{eq}} = \sum_{i=1}^{s} [I]uH_i - \mathcal{N} \sum_{i=1}^{s} H_i D_i \frac{dx_i}{dx} - \kappa \frac{dT}{dx} \tag{8-20}$$

The three terms on the right in Eq. (8-20) are energy fluxes due respectively to convection, diffusion, and conduction in the flame zone. Equation (8-20) may also be written in the differential form

$$\frac{d}{dx} \sum_{i=1}^{s} ([I]uH_i) - \frac{d}{dx}\left[ \mathcal{N} \sum_{i=1}^{s} H_i D_i \frac{dx_i}{dx} \right] - \frac{d}{dx}\left( \kappa \frac{dT}{dx} \right) = 0 \tag{8-21}$$

Equation (8-21), the mass-flow equations (8-14), and relationships defining reaction rates, diffusion coefficients, thermal conductivities, and enthalpies may be used to determine the burning velocity and detailed structure of any premixed gas flame. Integrating this equation set to obtain concentrations, state properties, and fluxes in the flame zone requires a knowledge of the burning velocity

$S_u$ commensurate with specific assumptions for the reaction rates and transport properties and the upstream and downstream boundary conditions. Since this information is not available, we are left with the solution of a problem in which the burning velocity is determined as a characteristic or eigenvalue that satisfies the boundary conditions for the assumed properties.

We now wish to solve these flame equations for a particularly simple model to illustrate the coupled nature of the processes in an ordinary flame. We follow the approach first suggested by Friedman and Burke.[2] Consider the flame reaction to be an irreversible first-order decomposition of pure $A$ yielding only $B$ as product

$$A \rightarrow B$$

with the molecular weights $M_A = M_B = M$ and the enthalpies $H_A = C_p(T - T_u)$, $H_B = C_p(T - T_u) - Q$ where $C_p$ is taken to be a constant and $Q = C_p(T_b - T_u)$; therefore $H_B = C_p(T - T_b)$. Thus the convected enthalpy is zero at both the unburned ($u$) and fully burned ($b$) stations in this flame. We write the first-order rate equation as

$$\frac{d(\mathcal{N} x_B)}{dt} = {}^1 k \mathcal{N} x_A \exp\left(\frac{-E^*}{RT_b \tau}\right) \tag{8-22}$$

where

$$\tau = \frac{T - T_u}{T_b - T_u} \tag{8-23}$$

and $x_A$ and $x_B$ are mole fractions.

The introduction of $\tau$ in the rate equation forces the chemical rate to be zero at $T_u$ without significantly altering it in the hotter flame regions. This is reasonable, since most real flames propagate at a steady velocity into mixtures which have a vanishingly small decomposition rate at their initial temperature.

We now define a simplified binary diffusion coefficient $D = cT^2$ and thermal conductivity $\kappa = fT$ and thus find that the Lewis number, $\text{Le} = \kappa/\rho c_p D$, is a constant throughout the flame. With this set of assumptions, we are ready to transform the general flame equations to dimensionless form. We first note that in this binary gas mixture Eq. (8-9) yields

$$v_A = -v_B \left(\frac{x_B}{1 - x_B}\right) \tag{8-24}$$

and Eq. (8-20) therefore becomes

$$\frac{d\tau}{dx} = \frac{\mathcal{N} C_p u}{\kappa}(\tau - z) \tag{8-25}$$

2. R. Friedman and E. Burke, *J. Chem. Phys.*, **21**:710 (1953).

where $z$ is the dimensionless mass flow plus diffusive flux

$$z = \frac{(u + v_B)x_B \rho}{S_u \rho_u} \tag{8-26}$$

Equation (8-13) for the species $B$ may be written as

$$\frac{dz}{dx} = \frac{^1k\rho}{\mathcal{M}_p}(1 - x_B)\exp(-\varepsilon/\tau) \tag{8-27}$$

where $\varepsilon = E^*/RT_b$ is a dimensionless activation energy, and $\mathcal{M}_p$ is the mass-flow for this planar flame. Furthermore, the definition of a diffusion coefficient yields the equation

$$\frac{dx_B}{dx} = -\frac{x_B v_B}{D} \tag{8-28}$$

Since the flow is steady, these equations are parametric in $x$ and may be written in terms of $\tau$, $z$, and $x_B$ by dividing Eq. (8-25) into Eqs. (8-27) and (8-28) to eliminate $x$ as a variable. This yields the equations

$$\frac{dz}{d\tau} = \Lambda_1 \left(\frac{1 - x_B}{\tau - z}\right)\exp(-\varepsilon/\tau) \tag{8-29}$$

and

$$\frac{dx_B}{d\tau} = \mathrm{Le}\left(\frac{x_B - z}{\tau - z}\right) \tag{8-30}$$

with the boundary conditions $x_B = z = \tau = 0$ at station 1 (the cold boundary) and $x_B = z = \tau = 1$ at station 2 (the hot boundary). The constant $\Lambda_1$ is the dimensionless eigenvalue of the physically realizable solution to this equation set, and is given by the relation

$$\Lambda_1 = \frac{^1k\kappa\rho M}{S_u^2 \rho_u^2 C_p} \tag{8-31}$$

The value of $\Lambda_1$ must be determined by solving Eqs. (8-29) and (8-30) simultaneously. Notice that $\Lambda_1$ defines the burning velocity in terms of the reaction kinetic and transport properties of the combustible mixture. Also note that for our assumptions, the temperature-dependence of $\rho$ and $\kappa$ cancel so that $\rho\kappa = \rho_u\kappa_u$. Therefore, we find that $\Lambda_1$ is truly a constant throughout this flame. Thus we obtain the equation

$$S_u = \sqrt{\frac{^1k\kappa_u M}{\rho_u \Lambda_1 C_p}} = \sqrt{\frac{^1k\alpha_u M}{\Lambda_1}} \tag{8-32}$$

which defines the burning velocity.

Again, following Friedman and Burke's treatment, we note that Eq. (8-30) is directly integrable for two cases, $\mathrm{Le} = 1$ and $\mathrm{Le} = \infty$, yielding respectively $x_B = \tau$ or $x_B = z$. In the case where we assume $\mathrm{Le} = \infty$ we do so by setting the diffusion coefficient to zero, thereby producing a flame with $v_A = v_B = 0$. With these assumptions, we may rewrite Eq. (8-29) to yield

$$\frac{dz}{d\tau} = \Lambda_1 \frac{1-\tau}{\tau - z} \exp\left(-\varepsilon/\tau\right) \tag{8-33}$$

for Le = 1, or

$$\frac{dz}{d\tau} = \Lambda_1 \frac{1-z}{\tau - z} \exp\left(-\varepsilon/\tau\right) \tag{8-34}$$

for Le = ∞.

Figures 8-8 and 8-9 illustrate the structure of simple first-order decomposition flames for Le = 1 and Le = ∞. These figures were obtained by integrating Eqs. (8-33) and (8-34) using the following assumptions: $M = 0.032$, $P = 101,325$ Pa, $T_u = 300$ K (which means that $\rho_u = 1.348$ kg/m³), $C_p = 14.65$ J/mole K, $\kappa_u = 19.88 \times 10^{-3}$ J/m s K, $^1k = 10^{13}$, $E^*/R = 30,000$, and $T_b = 2000$ K. The calculated burning velocities using these assumptions are: $S_u = 0.851$ m/s for Le = 1 and $S_u = 2.546$ m/s for Le = ∞. Friedman and Burke found that, for flames having Le = 1 or Le = ∞, and these assumed properties, $\Lambda_1$ may be quite accurately represented by the equation

$$\Lambda_1 = \left[\frac{0.514}{\text{Le}} \varepsilon^2 + \left(1.0 + \frac{0.35}{\text{Le}}\right)\varepsilon + 0.20 + \frac{0.05}{\text{Le}}\right] \exp\left(\varepsilon\right) \tag{8-35}$$

Note that $\Lambda_1$, as given by Eq. (8-35), has a dependence on $\varepsilon = E^*/RT_b$ which is primarily exponential. The significance of this will be discussed after a flame with a more general kinetic expression is introduced.

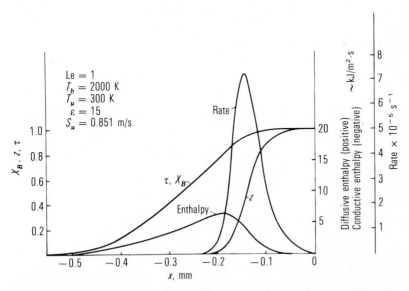

**Figure 8-8** Flame structure for a flame with Lewis number equal to 1, supported by a first-order reaction. Rate is the rate of disappearance of the species $A$, and equals $^1k \, (1 - x_B) \exp\left(-\varepsilon/\tau\right)$.

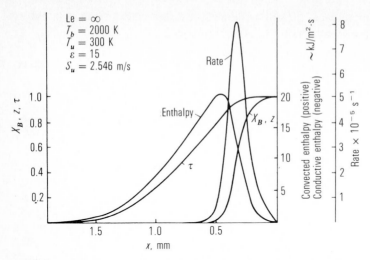

**Figure 8-9** Flame structure for a flame with a Lewis number which is equal to infinity because the diffusion coefficient is assumed to be zero. Reaction is assumed to be first-order in the species $A$. Rate is the local rate of disappearance of the specie $A$ and equals $^1k\,(1 - x_B)\exp(-\varepsilon/\tau)$.

The flame structure of Figs. 8-8 and 8-9 must be compared with that predicted by the simple thermal theory. Note that in the $\mathrm{Le} = 1$ flame the temperature and mole fraction curves for the product gas are coincident while in the $\mathrm{Le} = \infty$ flame (with no diffusion) the net flux and mole fraction curves for the product gases are coincident. This simply means that when $\mathrm{Le} = 1$, which is close to physical reality, the thermal conduction gradient is supported by the diffusion of the enthalpy of reaction carried with the reactive species and there is very little convected enthalpy change throughout this flame. Hence, thermal theory notwithstanding, the diffusion of species in the preheat zone of a flame is important. This is particularly true because reactive radicals and atoms diffusing from the hot reaction zone can trigger the chemistry and therefore augment the propagation rate. Also note that reaction does not start to be significant until the temperature has risen a significant amount. Thus the assumption of a high ignition temperature in the thermal theory appears to be justified.

We now wish to extend this simple theoretical model to investigate the effect of the order of the primary chemical reaction on the flame behavior. In this treatment we restrict ourselves to a flame with $\mathrm{Le} = 1$. The rate equation therefore becomes

$$\frac{d(\mathcal{N}x_B)}{dt} = {}^nk(\mathcal{N}x_A)^n \exp\left(\frac{-E^*}{RT_b\tau}\right) \tag{8-36}$$

where $n$ is the reaction order and $^nk$ has appropriate units. This yields a new conservation equation [replacing Eq. (8-27)]

$$\frac{dz}{dx} = \frac{{}^nk\rho^n}{\mathcal{M}_p M^{(n-1)}}(1 - x_B)^n \exp(-\varepsilon/\tau) \tag{8-37}$$

Since Le = 1 we again find from Eq. (8-30) that $x_B = \tau$, and therefore arrive at the final flame equation

$$\frac{dz}{d\tau} = \Lambda_n \left(\frac{1}{1 + \tau\phi}\right)^{n-1} \frac{(1 - \tau)^n}{\tau - z} \exp\left(-\varepsilon/\tau\right) \qquad (8\text{-}38)$$

where

$$\Lambda_n = \frac{{}^n k \kappa_u \, \rho_u^n}{\mathscr{M}_p^2 \, C_p \, M^{(n-2)}} \qquad (8\text{-}39)$$

and

$$\phi = \frac{T_b - T_u}{T_u} \qquad (8\text{-}40)$$

Therefore

$$S_u = \rho^{[(n/2)-1]} \sqrt{\frac{{}^n k \kappa_u}{\Lambda_n \, C_p \, M^{n-2}}} \qquad (8\text{-}41)$$

In this flame of order $n$, $\Lambda_n$ also shows a primary dependence on the activation energy and may be assumed to be roughly proportional to $\exp(\varepsilon)$ [see Eq. (8-35)]. Therefore the theoretical model predicts that to a first approximation

$$S_u \propto P^{[(n/2)-1]} \exp\left(\frac{E^*}{2RT_b}\right) \qquad (8\text{-}42)$$

Equation (8-42) could have been obtained from Eq. (8-6) of the thermal theory if one postulated that the reaction in the flame had a rate given by $\omega \propto P^n \exp(E^*/RT_b)$.

To complete our discussion of elementary flame theory we will now rederive Eq. (8-42) using large activation energy asymptotics.[3] The primary purpose of presenting this derivation is to illustrate this technique using a very simple example, namely a steady laminar flame. As was mentioned in Chap. 5 the technique can be applied to any three-dimensional nonsteady reactive flow in which the chemical reaction is observed to be confined to a narrow zone. Note from Fig. 8-8 that in the Friedman and Burke flame the region of rapid reaction is confined to that part of the flame where the dimensionless temperature $T$ is greater than 0.8 and that the lower temperature or preheat zone of this flame is essentially an unreactive flow. Furthermore, as the effective activation energy of the chemical reaction is increased the length of the reaction zone becomes smaller relative to the length of the preheat zone and the temperature change in the reaction zone also becomes smaller. Finally as $E^* \to \infty$ the length of the reaction zone approaches zero and the temperature range in the reaction zone

---

3. This presentation is based primarily on a private communication from C. K. Law, Northwestern University (1982).

becomes very small. Based on these observations, the principal aim of this technique is to develop simplified differential equations for the different zones of the flame, find their solutions, and then match properties and slopes at the boundaries by expanding the solutions in a geometric series of an appropriate small parameter to obtain a complete solution. Thus this technique, even though quite sophisticated, is based on a model which is quite similar to the very simple Mallard and Le Châtelier model introduced at the beginning of this section.

To illustrate this technique we will use a modified Friedman and Burke model for the flame. We will assume that the reaction is first order, that the Lewis number equals one, that the thermal conductivity is constant, and therefore that $\rho D = $ constant. Note that Le $= 1$ also means that $x_B = \tau$ throughout the flame. Also since the preheat zone will be legislated to be completely unreactive the exponential of Eq. (8-22) can be replaced by the simpler and more physically realistic expression $\exp(-E^*/RT)$ without encountering a "cold boundary" problem. Furthermore within the spirit of the large-activation-energy-asymptotics approach, one always identifies a small parameter that has physical importance to mathematically describe the behavior of the principal variables of interest. For this example we choose to use $\delta = 1/\varepsilon = RT_b/E^*$ and assume that $\delta \ll 1$.

Using the above assumptions we can differentiate Eq. (8-25) once and substitute Eq. (8-27) to obtain a single differential equation for the dimensionless temperature $\tau$, in the flame,

$$\frac{d^2\tau}{dx^2} = \beta \frac{d\tau}{dx} - \frac{\mathcal{N} C_p^{-1} k}{\kappa} (1 - \tau) \exp\left(-\frac{T_b}{\delta T}\right) \qquad (8\text{-}43)$$

where $\beta = \mathcal{N} C_p u/\kappa$. Note that since Le $= 1$, this equation fully describes the structure of and the mass flow into this flame.

We will now determine the mass flow analytically by applying the technique of large activation energy asymptotics to Eq. (8-43). The reactive and relatively thin region will be called the *inner* region while the nonreactive region that is upstream will be called the *outer* region. The hot flow downstream of the inner region must also be matched to the inner region; however, this region is relatively uninteresting mathematically because the flow is uniform, that is, $\tau = 1$ and $d\tau/dx = 0$.

We will first examine the inner region. Here the chemistry is fast and $\tau \cong 1$ because we are close to the hot boundary. We define the downstream boundary of this region as $x = 0$ and expand the dimensionless temperature $\tau$ in the small parameter $\delta$ by writing

$$\tau_{\text{inn}} = 1 - \delta\Theta(\chi) + O(\delta^2) \qquad (8\text{-}44)$$

Here the variable $\Theta(\chi)$ is assumed to be of order unity relative to $\delta$ and the new distance variable $\chi$ is a stretched coordinate defined by the equation $\chi = x/\delta$.

Note that the functional form of the variable $\Theta(\chi)$ has not been defined as yet. It must be determined by solving Eq. (8-43) in the inner region and matching that solution and slopes to the outer regions. To do this we must transform Eq. (8-43) using at least the leading order of Eq. (8-44). First we transform the derivatives. This yields

$$\frac{d\tau}{dx} = -\frac{d\Theta}{d\chi}$$

$$\frac{d^2\tau}{dx^2} = -\frac{1}{\delta}\frac{d^2\Theta}{d\chi^2}$$

also

$$1 - \tau = \delta\Theta$$

The exponential of Eq. (8-43) may be written as

$$\exp\left(-\frac{T_b}{\delta T}\right) = \exp\left(-\frac{\varepsilon T_b}{T}\right)$$

$$= \exp\left(-\varepsilon\right) \cdot \exp\left(-\frac{\phi\Theta}{\phi + 1}\right)$$

where $\phi$ is defined in Eq. (8-40). Using these relations we obtain an equation for $\Theta$ (a perturbed $\tau$ for $\tau \cong 1$) in terms of $\chi$ (a stretched $x$, for $x \cong 0$). For the inner region

$$-\frac{d^2\Theta}{d\chi^2} + \beta\delta\frac{d\Theta}{d\chi} = \frac{-\mathcal{N}C_p{}^1 k}{\kappa}\delta^2\Theta \exp\left(-\varepsilon\right) \cdot \exp\left(-\frac{\phi\Theta}{\phi + 1}\right) \qquad (8\text{-}45)$$

We now note that a simplification of this equation can be made. The left-hand side of Eq. (8-45) contains two terms. However this is a *thin* region in which gradients are changing rapidly. Therefore, since the gradient term is multiplied by $\delta$ (the small parameter) and the spatial rate of change of the gradient term has a coefficient of one, the spatial rate-of-change term must dominate. This approximation is realistic and it greatly simplifies the differential equation that must be solved to determine the structure of the inner region. We have in fact reduced the structure of the inner region to a form given by the solution of the equation

$$\frac{d^2\hat{\Theta}}{d\chi^2} = \left(\frac{\Gamma}{2}\right)\hat{\Theta} \exp\left(-\hat{\Theta}\right) \qquad (8\text{-}46)$$

where $\hat{\Theta} = \phi\Theta/(\phi + 1)$ and

$$\Gamma = \frac{2\mathcal{N}C_p{}^1 k}{\kappa}\delta^2 \exp\left(-\varepsilon\right)$$

The hot boundary of the inner solution has particularly simple boundary conditions. These are, as $\chi \to 0$,

$$\hat{\Theta}(0) = 0 \qquad (8\text{-}47)$$

and

$$\left(\frac{d\hat{\Theta}}{d\chi}\right)_0 = 0$$

We must now examine the functional form of the outer region. Since there are no chemical reactions in this region Eq. (8-43) becomes

$$\frac{d^2 \tau_{\text{out}}}{dx^2} - \beta \frac{d\tau_{\text{out}}}{dx} = 0 \qquad (8\text{-}48)$$

This equation has a solution of the form

$$\tau_{\text{out}} = C_1 + C_2 e^{\beta x}$$

The cold boundary conditions require that as $x \to -\infty$, $T = T_u$ and $\tau = 0$. Thus we may write

$$\tau_{\text{out}} = C_2 e^{\beta x} \qquad (8\text{-}49)$$

We now note that as $\delta \to 0$, the flame's reaction zone becomes *very very* thin such that the downstream boundary of the outer region is at $x \to 0$. This in turn means that $\beta_\chi \ll 1$ at this boundary. Thus we can expand this solution in $x\beta$ to yield

$$\tau_{\text{out}} = C_2(1 + x\beta) = C_2(1 + \chi\delta\beta)$$

We next take limits as $\chi \to -\infty$ or $\delta \to 0$ at fixed $x \ll 1$ to match the inner and outer solutions

$$\lim_{\chi \to -\infty} \tau_{\text{inn}}(\chi) = \lim_{\chi \to -\infty} \tau_{\text{out}}(\chi) \qquad (8\text{-}50)$$

and their slopes

$$\lim_{\chi \to -\infty} \frac{d\tau_{\text{inn}}}{d\chi}(\chi) = \lim_{\chi \to -\infty} \frac{d\tau_{\text{out}}}{d\chi}(\chi) \qquad (8\text{-}51)$$

When expressed in terms of the inner solution variable Eq. (8-50) becomes

$$\lim_{\chi \to -\infty} (1 - \delta\hat{\Theta}(\chi)) = \lim_{\chi \to -\infty} (C_2 + C_2 \delta\beta\chi)$$

Then by equating terms of equal order we obtain

$$\delta^0: \quad C_2 = 1$$

$$\delta^1: \quad \hat{\Theta}(\chi) = -\lim_{\chi \to -\infty} \left(\frac{\beta\chi\phi}{\phi + 1}\right)$$

Also Eq. (8-51) yields

$$\left(\frac{d\hat{\Theta}}{d\chi}\right)_{-\infty} = -\frac{\beta\phi}{\phi+1}$$

as the condition that balances the slope at the boundary.

We now solve the equation for the inner region by noting that

$$\frac{d^2\hat{\Theta}}{d\chi^2} = \frac{d}{d\chi}\left(\frac{d\hat{\Theta}}{d\chi}\right) = \left(\frac{d\hat{\Theta}}{d\chi}\right)\frac{d}{d\hat{\Theta}}\left(\frac{d\hat{\Theta}}{d\chi}\right) = \frac{1}{2}\frac{d}{d\hat{\Theta}}\left(\frac{d\hat{\Theta}}{d\chi}\right)^2$$

Thus Eq. (8-46) becomes

$$\frac{d}{d\hat{\Theta}}\left(\frac{d\hat{\Theta}}{d\chi}\right)^2 = \Gamma\hat{\Theta}e^{-\hat{\Theta}}$$

Integrating once yields

$$\left(\frac{d\hat{\Theta}}{d\chi}\right)^2 = \Gamma\int\hat{\Theta}e^{-\hat{\Theta}}\,d\hat{\Theta} + C_3$$

$$= -\Gamma(1+\hat{\Theta})e^{-\hat{\Theta}} + C_3$$

Using Eq. (8-47), we find that $C_3 = \Gamma$. Next we note that

$$\lim_{\chi \to -\infty}(1+\hat{\Theta})e^{-\hat{\Theta}} = 0$$

because of the dominance of the exponential term. Therefore we find that

$$\left(\frac{d\hat{\Theta}}{d\chi}\right)^2 = \Gamma$$

at the boundary between the inner region and the upstream outer region. This means that

$$\left(\frac{\beta\phi}{\phi+1}\right)^2 = \Gamma$$

Substituting for $\beta$ and $\Gamma$ yields the expression

$$\left(\frac{\phi}{\phi+1}\right)^2\left(\frac{\mathcal{N}C_p u}{\kappa}\right)^2 = \frac{2\mathcal{N}C_p{}^1 k}{\kappa\varepsilon^2}\exp(-\varepsilon)$$

Thus the square of the mass flux in this flame is

$$\rho_u^2 S_u^2 = \rho^2 u^2 = \left(\frac{\phi+1}{\phi}\right)^2\frac{2\kappa^1 k\rho^2}{\mathcal{N}C_p\varepsilon^2}\exp(-\varepsilon)$$

Using the ideal gas law this expression reduces to

$$S_u \propto P^{-1/2}\exp\left(-\frac{\varepsilon}{2}\right)$$

which is identical to Eq. (8-42) for a flame in which the reaction is first order.

The above example illustrates the use of large activation energy asymptotics. It uses the same principles as ordinary asymptotics except that it is applied to a combustion system in which the high exothermicity and effective activation energy of the reaction cause the reaction to occur in a thin zone. It is *not* a flame-sheet theory nor does it reduce to a flame-sheet theory as $E \to \infty$. The inner reactive region, though narrow, must still satisfy property- and slope-matching at both of its edges, and changes in the width of this region due to flow gradients or in nonsteady steady-flow situations can contribute to the predicted flame response. Results of asymptotic analyses will be presented in later chapters as appropriate, but the detailed analyses that led to those results will not be presented. The interested reader is referred to the bibliography for further information.

## 8-4 STEADY NONPLANAR AND SUPERADIABATIC FLAMES

It is instructive to look at the effect of flame curvature on a steady flame using an extension of the Friedman and Burke model. For this purpose we consider a steady source or sink flame in cylindrical or spherical geometry. For completeness, we also include the planar geometry which has already been considered. Figure 8-10 shows the geometry for such a steady source or sink flame. The vectors on this figure show the motion of the gas. We further assume that the flow velocity associated with the flame is so low that there are no pressure effects. We therefore remove from consideration any problems associated with the location of a sonic line or the singularity at the source or sink. Under these circumstances, the conservation of mass equation can be written in the following forms:

*Planar coordinates*

$$\rho u = \mathcal{M}_p$$

*Cylindrical coordinates*

$$2r\rho u = \frac{\mathcal{M}_s}{\pi}$$

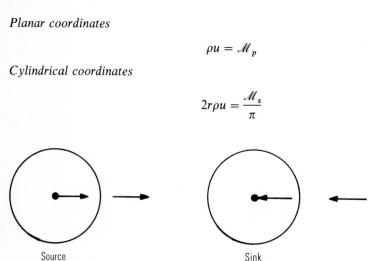

Source                                  Sink

**Figure 8-10** Geometry of a source or sink flame.

*Spherical coordinates*

$$4r^2 \rho u = \frac{\mathscr{M}_s}{\pi}$$

A general form can be obtained by defining $\mathscr{M}_p = \mathscr{M}_s/\pi$ to obtain the equation

$$(2r)^j \rho u^j = \mathscr{M}_p$$

In this equation, $j = 0$, 1, or 2 for planar, cylindrical, or spherical flows, respectively, and $\mathscr{M}_p$ is a positive number for a source flame and a negative number for a sink flame. The general equations become:

*For the conservation of species*

$$\frac{d}{dr}\left[(2r)^j [I](u + v)\right] = (2r)^j \left(\frac{d[I]}{dt}\right)_{\text{chem}}$$

*For the diffusion relation*

$$v_B = -\frac{D}{x_B}\frac{dx_B}{dr}$$

*For the conservation of energy*

$$\mathscr{M}_p \kappa \frac{dT}{dr} = \mathscr{M}_p \sum_{i=1}^{s} [I](u + v_i)H_i - (\mathscr{M}_p)_b \sum_{i=1}^{s} [I]_b S_b (H_i)_b$$

If we use the same assumptions that we used in developing the Friedman–Burke model, with the additional new definition of the variable $z$

$$z = \frac{x_B(u + v_B)}{u}$$

we obtain a new conservation-of-energy equation

$$\frac{d\tau}{dr} = \frac{(C_p/M)\mathscr{M}_p}{(2r)^j \kappa}(\tau - z) \tag{8-52}$$

a new diffusion equation

$$\frac{dx_B}{dr} = \frac{\mathscr{M}_p}{(2r)^j \rho D}(x_B - z) \tag{8-53}$$

and a new conservation of species equation

$$\frac{dz}{dr} = \frac{(2r)^j \rho^{-1} k(1 - x_B)}{\mathscr{M}_p} \exp(-\varepsilon/\tau) \tag{8-54}$$

We will solve these equations only for the case of Le $= 1$, which means as before that $x_B = \tau$. For this case, Eqs. (8-52) and (8-53) (the energy and diffusion equations) are identical and can be written in terms of the reduced temperature

$\tau$. We therefore have a single species equation and an energy equation to solve simultaneously. We now use the defined constant $\phi = (T_b - T_u)/T_u$ to obtain the relation $T/T_u = \phi\tau + 1$. However, since pressure and molecular weight are constant, we obtain $\rho T = \rho_u T_u$ and, therefore,

$$\rho = \rho_u \frac{1}{1 + \phi\tau}$$

Therefore, the species and energy equations may be written in the form

$$\frac{dz}{dr} = \frac{(2r)^j \rho_u{}^1 k}{\mathcal{M}_p} \frac{(1 - \tau)}{(1 + \phi\tau)} \exp\left(-\varepsilon/\tau\right) \tag{8-55}$$

$$\frac{d\tau}{dr} = \frac{\mathcal{M}_p(C_p/M)}{(2r)^j \kappa_u} \frac{(\tau - z)}{(1 + \phi\tau)} \tag{8-56}$$

Note that these equations cannot be divided into each other to eliminate $r$ when $j$ is equal to either 1 or 2 because the independent variable $r$ appears in the numerator and denominator, respectively. However, the pair of equations can be numerically integrated from the cold boundary where $\tau = z = 0$ by specifying that this point shall have some initial radius $r_i$. When this is done, the integration is carried out in real space with $r$ either increasing for a source flow or decreasing for a sink flow. The eigenvalue $\mathcal{M}_p$ must now be adjusted until $\tau = z = 1$ at the hot boundary. Notice that the actual structure of the flame is now dependent upon the choice of $r_i$ as well as the physical constants which are characteristic of the flame system.

The flow velocities associated with such source and sink flames are shown in Fig. 8-11. The most interesting feature of these flames is that they no longer exhibit the unique "burning velocity" of the one-dimensional laminar flame. Instead, for a sink flame the flow velocity simply continues to increase as one travels from some distance from the flame in the cold flow through the flame into the hot flow behind the flame. For the source flame the flow velocity decreases as one approaches the flame and then starts to increase somewhere in the preheat zone. Thus it exhibits a local minimum at some nonspecific location in the preheat zone. It then goes through a maximum velocity and finally continues to decrease in the hot gas region.

About the only characteristic location that one can actually define in these flames relative to their burning velocity is based on the burning velocity of an equivalent laminar flame. If we define such a burning velocity as $S_u$, we can then define a zero flame-thickness radius based on the value of the eigenvalue $\mathcal{M}_p$ for that particular flame

$$r_0 = \frac{1}{2}\left(\frac{\mathcal{M}_p}{\rho_u S_u}\right)^{1/j} \tag{8-57}$$

Once this value of $r_0$ is defined, we can relate other characteristic radii (such as the radius at which the chemical reaction rate is a maximum) to this eigenvalue radius for that particular flame.

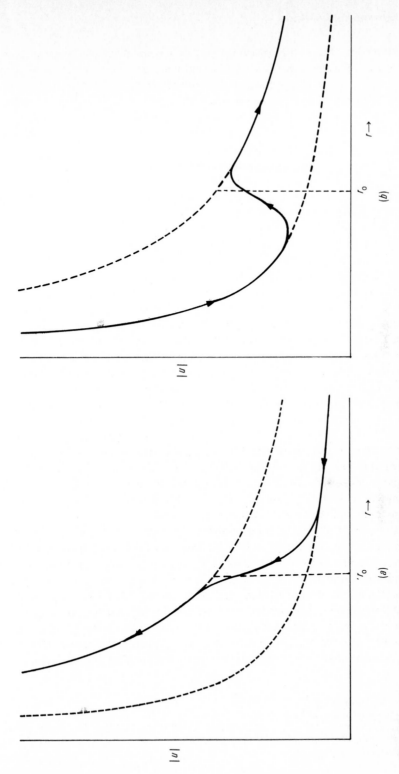

**Figure 8-11** Sink and source flames in a cylindrical or spherical geometry (schematic). (*a*) Sink flame. (*b*) Source flame.

As the eigenvalue radius of cylindrical or spherical source or sink flames becomes small relative to the one-dimensional laminar preheat zone thickness of that particular flame, the flame structure starts to alter significantly. Specifically, the major change is that for a source flame the preheat zone thickness becomes smaller, while for a sink flame the preheat zone thickness becomes considerably larger than that for an equivalent one-dimensional flame. These steady curved flames are strictly one-dimensional and adiabatic, in the sense that the direction of heat transfer is in the direction of the streamlines. Thus, there is no lateral loss of heat from each stream and the flame temperature remains invariant with curvature. It has been observed that in this circumstance the flow velocity through the plane where the chemical reaction rate is a maximum is almost invariant with radius of curvature, even for highly curved flames. This interesting result was first postulated by Fristrom[4] as a result of his study of the curvature of the tip of a Bunsen burner flame.

Another interesting departure from ordinary one-dimensional flame behavior occurs in the *excess enthalpy* or *superadiabatic* flame first studied experimentally by Hardesty and Weinberg.[5] This is an interesting stabilized flame that has important implications for ultralean mixture combustion. By judicious design of a burner, heat from the hot combustion process is transferred back to a region either deep in or well ahead of the location of the normal preheat zone from product gases which are well upstream of the zone of rapid combustion. When this is done correctly the combustion reactions in the flame occur at a temperature well above the normal adiabatic flame temperature of the mixture because the temperature temporarily overshoots the adiabatic flame temperature.

Takeno et al.[6] have theoretically investigated the structure of these flames. They used the Friedman and Burke[2] approach for a flame with an $A \to B$ first-order reaction with a slightly different exponential temperature-dependence in a system with the Lewis number, Le, equal to unity. We will summarize their approximations and the results of their analysis here.

They assumed that the flame holder is a noncatalytic porous block of length $L$ which has a sufficiently high thermal conductivity such that its temperature can be assumed to be a constant. They further assumed that once the block and flame come to "equilibrium" there is no heat loss to the surroundings. In other words all the heat transferred to the block from the hot regions of the flame is transferred back to the flowing gases in the cold regions. Thus some distance downstream of the block the product gases always reach the adiabatic flame temperature. They further assumed, for simplicity, that the pores in the block are much smaller than the preheat zone thickness of the flame and that

4. R. M. Fristrom, *Phys. Fluids*, **8**:273 (1965).

5. D. R. Hardesty and F. J. Weinberg, *Combust. Sci. Tech.*, **8**:201 (1974).

6. T. Takeno, K. Sato, and K. Hase, "A Theoretical Study on an Excess Enthalpy Flame," *Eighteenth Symposium (International) on Combustion*, The Combustion Institute, Pittsburgh, Pa., pp. 465–472 (1981).

the presence of the block does not change the cross-sectional area of the flow, i.e., the flow remains one-dimensional steady. With these assumptions the only change in Eqs. (8-22) to (8-28) is the addition of a heat-transfer term in the energy equation, which now becomes

$$\frac{d\tau}{dx} = \frac{\mathcal{N} c_p u}{\kappa} [d(\tau - z) - \mathcal{F}]$$  (8-58)

The new term due to heat transfer to the solid is defined as

$$\mathcal{F} = \frac{v \int_0^x \mathcal{H}(T_s - T)\, dx}{\mathcal{M}_p c_p (T_b - T_u)}$$  (8-59)

where $h$ is a heat transfer coefficient and $T_s$ is the solid temperature. Since the superadiabatic flame has a higher burning velocity than an ordinary flame burning in that mixture a quantity $d$ is defined as the mass-flow eigenvalue ratio $\mathcal{M}/\mathcal{M}_p$. This ratio is always greater than unity. Within the framework of these assumptions the energy-balance equation for the solid becomes

$$0 = \int_0^L \mathcal{H}(T_s - T)\, dx$$

and one can define a dimensionless solid temperature $\tau_s = (T_s - T_u)/(T_b - T_u)$, a dimensionless heat-transfer coefficient $\mathcal{K} = \kappa \mathcal{H}/\mathcal{M}_p^2 c_p^2$, and a dimensionless block length

$$\mathcal{L} = \int_0^L \frac{\mathcal{M}_p c_p}{\kappa}\, dx$$

The authors fixed the block temperature and numerically integrated the flame equations in phase space from the hot to the cold boundary. They found two limit behaviors in which the flame acted as though the block were simply a flame holder. In one limit $\tau_s = 1$, the block was sitting in the hot product gases and $\mathcal{M}/\mathcal{M}_p \to 1$. In the other limit case $\tau_s \to 0$ and the block was sitting in the preheat zone ahead of the major reaction zone of the flame. These two behaviors are illustrated in Fig. 8-12. Note that in both cases when $\mathcal{M}/\mathcal{M}_p \to 1.001$ there is no significant thermal overshoot. As $\tau_s$ is either decreased from one (for the upper branch) or increased from zero (for the lower branch) both $\mathcal{M}/\mathcal{M}_p$ and $\tau_m$, the maximum temperature in the flame, increase until at some critical value of $\tau_s$ both reach a maximum value. This behavior is shown in Fig. 8-13. The structure of various upper and lower branch flames are shown in Fig. 8-14.

It is interesting to note that these excess enthalpy flames have their maximum thermal overshoot when the flow velocity is maximum and that this thermal overshoot can be significantly higher than the adiabatic flame temperature. It is also interesting to note that increasing the approach-flow velocity above the critical value will cause these flames to blow out in a manner similar to the well-stirred reactor of Sec. 6-9 or the catalytic combustor of Sec. 7-6. Experimental observations of the behavior of these flames will be presented in Chap. 12.

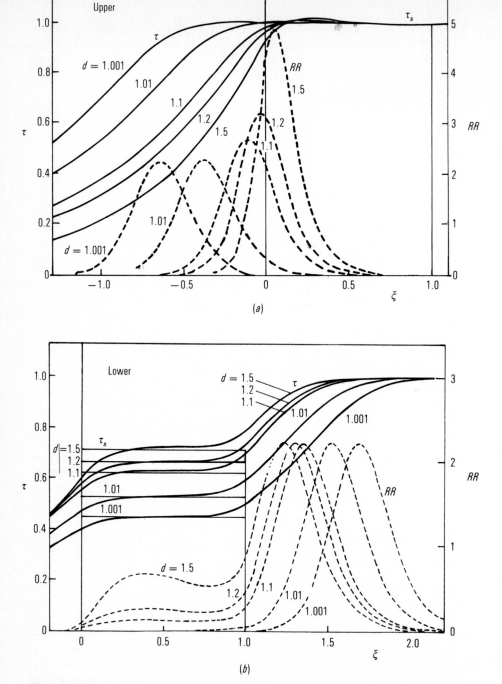

**Figure 8-12** The two-limit flame behaviors. (*a*) Flame is positioned upstream of the block. (*b*) Flame is positioned downstream of the block. In both figures $RR$ is an arbitrary local reaction rate and $\xi = x/L$ where $L$ is the block length. Notice that in these two limit cases there is hardly any overshoot of the temperature. (*With permission, from Ref. 6.*)

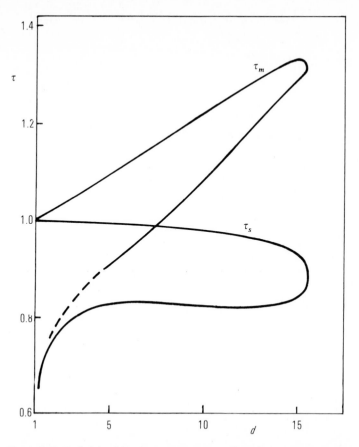

**Figure 8-13** Variation of block temperature $\tau_s$ and maximum gas temperature in the block $\tau_m$ with the mass flow rate $d = \mathcal{M}/\mathcal{M}_p$ for the analysis conditions of Ref. 6. The upper branch corresponds to the condition when the flame's reaction zone lies upstream of its position at the maximum. On the lower branch the flame lies downstream of that position, see Figs. 8-12 and 8-14. *(With permission, from Ref. 6.)*

## 8-5 FLAME PROPERTIES AND STRUCTURE

Figure 8-15 is an experimental evaluation of the pressure exponent of the burning velocity $(n/2 - 1)$ for a number of hydrocarbon flames. Note that over the burning velocity range of about 0.4 to 1.0 m/s the pressure exponent is 0. Thus Eq. (8-42) indicates that a second-order reaction is controlling the rate of propagation of these flames. Also note that above a burning velocity of about 1.0 m/s the exponent is positive and increases with burning velocity, and that for burning velocities less than about 0.4 m/s the exponent becomes increasingly negative as the burning velocity decreases. Both of these trends are explainable in terms of the complex kinetic processes that occur in these flames.

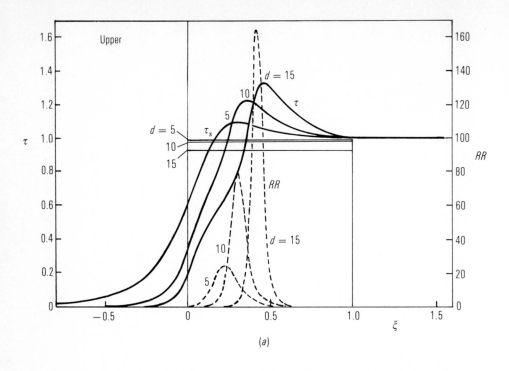

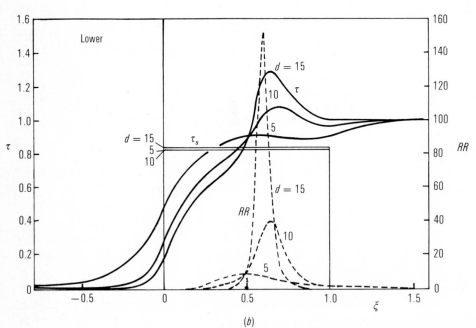

**Figure 8-14** Structure of some upper and lower branch excess enthalpy flames. (*a*) Upper branch. (*b*) Lower branch. $\xi = x/L$ and *RR* is a dimensionless reaction rate. (*With permission, from Ref. 6.*)

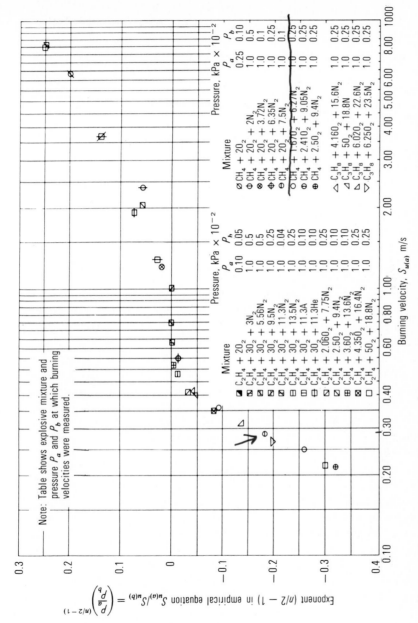

**Figure 8-15** The pressure exponent $(n/2 - 1)$ for a variety of hydrocarbon flames. *(Courtesy Dr. Bernard Lewis, Combustion and Explosions Research, Inc., Pittsburg, Pa.; originally appeared in Agard. Selected Combustion Problems, Butterworth, London, p. 177, 1954, with permission.)*

The higher value of the exponent for high-velocity flames can be explained by noting that they all have very high flame temperatures. Under these conditions, an increase in the ambient pressure causes a marked decrease in the degree of dissociation of the hot flame gases and, therefore, causes a marked increase in the flame temperature (see Fig. 8-1) and the rate of the chemical reactions in the flame. This causes the pressure exponent to be positive in the high-burning-velocity regime.

The situation is more complex in the low-burning-velocity regime. It has been shown recently that in most hydrocarbon air flames, the diffusion of hydrogen atoms into the preheat zone is the primary triggering mechanism for the chain-branching reactions which support flame propagation. It is well known that the hydrogen atom undergoes two primary competing reactions during any hydrocarbon oxidation process. These are the chain-branching reaction

$$H + O_2 \rightarrow OH + O$$

and the recombination reaction

$$H + O_2 + M \rightarrow HO_2 + M$$

Furthermore, the chain-branching reaction has a first-order pressure sensitivity (because it is a second-order reaction) and has a large activation energy, which means that its rate increases very rapidly with increased temperature. On the other hand, the recombination reaction has a second-order pressure sensitivity (because it is a third-order reaction) and has essentially no temperature sensitivity at all. In the preheat zone of the flame, hydrogen atoms diffuse toward the cold gas and at the same time react. In the higher temperature regions of the flame, the chain-branching reaction dominates and recombination is slow. However, as one travels toward the incoming gas, the temperature drops and at some point the two reactions become competitive. In the colder regions, the recombination reaction becomes dominant and destroys hydrogen atoms. If we now consider any flame system and we lower the flame temperature by dilution with an inert, we will lower the burning velocity and also move the point in the flame at which the rate of the recombination reaction and chain-branching reaction are equal toward the hot boundary. If we now take a specific mixture which has a low burning velocity and a low flame temperature, and increase the pressure level at which the flame is burning, we will increase the rate of the recombination reactions relative to the chain-branching reactions and, therefore, tend to lower the burning velocity. In other words, we should expect that for low-burning-velocity flames in which the competition between recombination and chain-branching reactions in the preheat zone is important, we should see a negative pressure exponent, and this is exactly what is observed in Fig. 8-15.

Increasing the initial temperature of the mixture should cause a marked increase in burning velocity for low-temperature flames and, because of the recombination/chain-branching competition in the preheat zone, this effect should be very strong for flames that have very low burning velocities. It should be moderately strong for flames that have burning velocities in the range of about 0.4 to 1.0 m/s and should have almost no influence on flames that have burning velocities above about 1 m/s. This is because the high effective heat capacities of the dissociated product gases at these high temperatures will cause the flame temperature and, therefore, the burning velocity, to change only very slightly as the initial temperature of the mixture is changed. Figure 8-1 shows the effect of initial temperature on flame temperature.

Experimental studies of real flames and theoretical modeling with relatively complete kinetic schemes have shown that in addition to the structural features of a flame illustrated by the simpler thermal and $A \rightarrow B$ reaction theories real flames contain a high-temperature region where the reactions slowly reach equilibrium. Simply stated, the final oxidation of carbon monoxide to carbon dioxide occurs primarily through the exchange reaction

$$CO + OH \rightarrow CO_2 + H$$

which is relatively slow. Therefore in most flame systems the carbon monoxide concentration goes through a maximum and finally decays to its equilibrium value somewhat downstream of the rapid reaction region of the flame. Since the oxidation of CO to produce $CO_2$ is exothermic this final equilibration causes the temperature to rise downstream of the fast reaction zone. These studies have also yielded excellent results for the structure and burning velocity of simpler systems such as the hydrogen–oxygen system, the carbon monoxide–oxygen system, and the methane–oxygen system. Specifically Dixon-Lewis[7] has shown that thermal diffusion of hydrogen molecules and atoms are important in an $H_2$–$O_2$–$N_2$ flame, and has also shown that partial equilibrium of the OH, H, and O radicals occurs in the higher-temperature regions of the flame, just as in the diffusion flames of Sec. 7-6. Warnatz[8] also comes to the same conclusions concerning partial equilibrium and describes how the $HO_2$ radical formation process in the colder regions of the preheat zone disturb that equilibrium. Warnatz has also calculated the pressure and temperature dependency of the burning velocity of the $H_2$–air flame, and his calculated results agree quite well with experimental data.

Warnatz[8] has shown that one can get good agreement between calculated burning velocities and experimental measurements when CO is the fuel, by simply adding the reactions

$$CO + OH \rightarrow CO_2 + H \tag{8-60}$$

and

$$CO + O + M \rightarrow CO_2 + M \tag{8-61}$$

7. G. Dixon-Lewis, *Philos. Trans. R. Soc. London*, **A292**:45–99 (1979).

8. J. Warnatz, *Combust. Sci. Technol.*, **26**:203–213 (1981); *Ber. Bunsenges. Phys. Chem.* **82**:643 (1978).

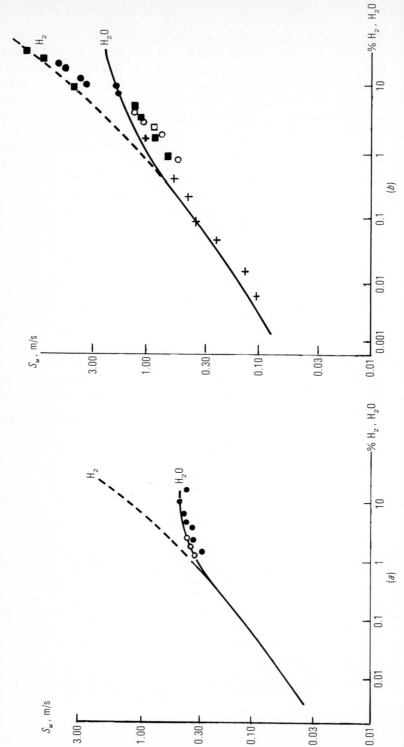

**Figure 8-16** The effect of $H_2$ or $H_2O$ addition on the normal burning velocity of stoichiometric $CO-O_2$ or $CO$–air flames. (*a*) $CO$–air flames, $H_2O$ added. (*b*) $CO-O_2$ flames, $H_2$ added. (*b*) $CO-O_2$ flames, $H_2$ added; open points and +'s, $H_2O$ added. (*Adapted from Ref. 8b, with permission.*)

to the hydrogen–oxygen reaction scheme that he used to study $H_2$–$O_2$ and $H_2$–air flames. The CO system is very interesting because the burning velocity depends very strongly on the concentration of hydrogen-containing compounds in the initial mixture. This is because when the hydrogen concentration becomes appreciable, reaction (8-60) coupled with the hydrogen chain dominates the oxidation of CO to $CO_2$ just as it does in the homogeneous explosion process discussed in Sec. 6-10-3. Thus, even though the flame temperature is essentially unaffected by the addition of small amounts of either $H_2$ or $H_2O$ to the initial mixture the normal burning velocity changes by an order of magnitude. Warnatz's[9] calculated burning velocities for stoichiometric CO–$O_2$ and CO–air mixtures with added $H_2$ or $H_2O$ and their comparison to burning velocities measured by five independent investigators are shown in Fig. 8-16.

The structure of methane–air flames has been studied by a number of investigators using "full" kinetic schemes. In general the results agree quite well with experimental measurements of the flame structure and burning velocity. A typical comparison is shown in Fig. 8-17. It is interesting to note that even though the homogeneous oxidation of methane has a considerably longer induction delay than the higher hydrocarbons, the burning velocity of a methane–air flame is almost the same as that of a higher hydrocarbon. The reason for this is that the hot reaction zone of this flame acts as a copious source of radicals and it is their diffusion toward the preheat zone and their reaction with stable species in that region that triggers the chemistry and causes the propagation rate to be similar to that of a flame involving a higher hydrocarbon.

As a matter of fact virtually all the hydrocarbons have similar burning velocities when mixed with air. The two major exceptions are acetylene and ethylene. Both of these compounds are exothermic and therefore have high flame temperatures (see, for example, Fig. 4-10) and this causes them to have higher burning velocities than the other hydrocarbons.

In general the burning velocity of a mixture of two fuels in air can be estimated by determining the effective equivalence ratio of that mixture with air and linearly averaging the burning velocity of the two fuels at that equivalence ratio. This is shown in Fig. 8-6, which also shows the one major exception to this rule. Notice that mixtures of methane and carbon monoxide have higher burning velocities than either pure methane or pure (wet) carbon monoxide. This synergistic effect should also be present in any hydrocarbon–carbon monoxide mixture. It is obviously due to an increase of the OH radical concentration caused by the hydrocarbon decomposition chain reaction. This enhances the rate of oxidation of carbon monoxide which in turn causes the observed increase in burning velocity.

9. J. Warnatz, *Ber. Bunsenges. Phys. Chem.*, **83**:950 (1979).

## 8-6 FLAME INSTABILITIES

The theory of flame propagation developed in Secs. 8-3 and 8-4 is based on the a priori assumption that flames are indeed one-dimensional steady waves propagating in a direction normal to their orientation. While it is true that such one-dimensional and apparently steady combustion waves can be generated and studied in the laboratory, it is also true that, in general, laminar flames can be inherently unstable.

This instability appears in two forms. There is a short-wavelength instability; flames that exhibit this instability are called cellular flames. There is also a long-wavelength hydrodynamic instability which is not observable in most laboratory experiments because it normally occurs only when the scale of the experiment is large. This latter instability will be discussed in Chap. 14.

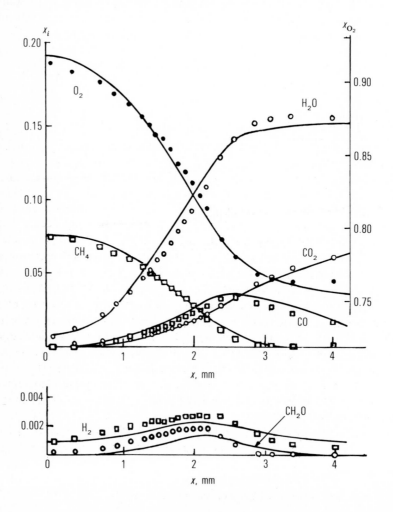

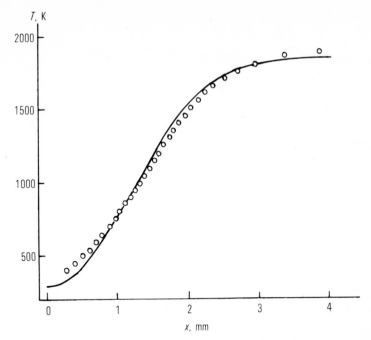

**Figure 8-17** Comparison of the calculated structure of a methane–oxygen flame with experimental measurements. Initial conductions are $P = 10$ kPa, $T = 350$ K, $S_u = 0.55$ m/s, $x_{CH_4} = 0.0785$, $x_{O_2} = 0.9215$. *(Points, data from R. M. Fristrom, C. Grunfelder and S. Favin, J. Phys. Chem., vol. 64, 1386 (1960). Lines calculated by J. Warnatz, Eighteenth Symposium (International) on Combustion, The Combustion Institute, Pittsburgh, Pa., p. 369 (1981).) (With permission.)*

Cellular flames have been observed for over 50 years. Markstein[10] summarized the experimental observations of their behavior in the 1960s. Visible light photographs of three-dimensional and two-dimensional cellular flames are shown in Figs. 8-18 and 8-19. The three-dimensional cellular flames were obtained by propagating a flame against a low-velocity approach flow in a transparent pipe. The two-dimensional cellular flame shown in Fig. 8-19 was obtained by stabilizing a flame on only one long edge of a slot burner, and three edge-on photographs of the sheet flame show the behavior of a two-dimensional cell. Note that there is a relatively large section of the flame of variable width that is convex toward the approach flow, separated by much more strongly concave sections whose radius of curvature is relatively invariant. The system is actually dynamic, because cells are constantly being formed and destroyed and also move along the burner rim. Figure 8-20 illustrates the dynamic nature of cellular instability and shows that cells can be formed or destroyed with time. The range of size shown in Fig. 8-19 illustrates the approximate range that is observed experimentally. If the spacing between troughs

10. G. H. Markstein, *Non-Steady Flame Propagation*, Macmillan, New York (1964).

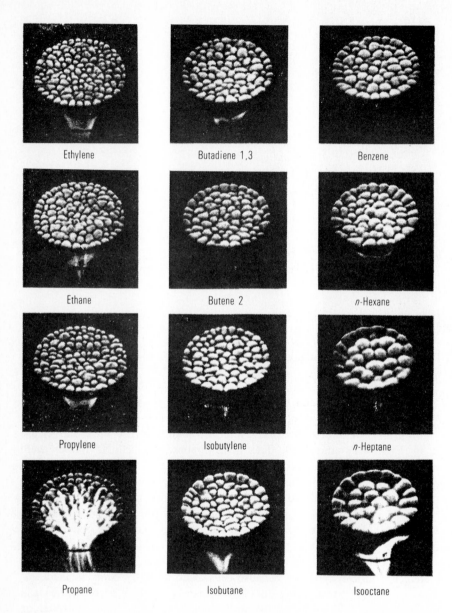

Figure 8-18 Cellular flames in fuel-rich hydrocarbon–air nitrogen mixtures at atmospheric pressure. Tube i.d., 100 mm. *(Courtesy G. H. Markstein, Factory Mutual Research, Norwood, Mass., originally appeared in J. Aero. Sci., vol. 18, p. 199, 1951.) (With permission.)*

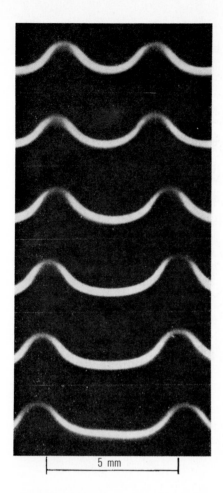

**Figure 8-19** Profiles of steady cellular slot-burner flames, *n*-butane–air mixture, equivalence ratio 1.47. Flow is upward. *(Courtesy H. G. Markstein, Factory Mutual Research, Norwood, Mass., orginally appeared in Combustion and Propulsion, Third AGARD Colloquium, Fig. 8 Facing page 178, 1958.) (With permission.)*

5 mm

becomes smaller than the spacing shown in the upper photograph in Fig. 8-19, the cell will disappear and the two neighboring troughs will merge. If the cell size gets any larger than that shown in the bottom photograph the flat region becomes dynamically unstable and a new cell is formed.

It has also been observed that spontaneous cellular instability primarily occurs in fuel-oxidizer mixtures when the Lewis number of the deficient species $(\alpha/D)$ is smaller than some critical value. In other words rich mixtures of heavy fuels or lean mixtures of light fuels are found to exhibit cellular instability. There are exceptions to this rule as, for example, the rich ethylene–air mixture of Fig. 8-18. Here, even though ethylene and oxygen have almost the same diffusion coefficient, the flame is cellularly unstable on the rich side.

Sivashinsky[11] has reviewed recent developments in the theory of flame

11. G. I. Sivashinsky, "Instabilities Pattern Formation and Turbulence in Flames," *Ann. Rev. Fluid Mech.*, **15**:179–199 (1983).

**Figure 8-20** Streak-camera record of nonsteady cellular slot-burner flame. Time increases from top to bottom. *n*-butane–air mixture, $\Phi = 1.56$ *(Courtesy G. H. Markstein, Factory Mutual Research, Norwood, Mass., originally in G. H. Markstein and D. Schwartz, Proceedings of the Gas Dynamics Symposium on Aerothermochemistry, Northwestern University, Evanston, Ill., 1956.) (With permission.)*

instability. He points out that cellular instability is not a hydrodynamic instability but instead is a thermodiffusive instability. The source of this instability had been called "preferential diffusion" ever since the first qualitative explanations of cellular structure appeared. Simply stated the thermal gradient in the preheat zone has a stabilizing effect because a convex section of flame would be expected to propagate more slowly than a concave section if thermal diffusivity only were operating in that region of the flame. However, when any sufficiently light species of the fuel-oxidant pair is sufficiently deficient, any concave perturbation of a flat flame shape will grow to become a deep trough, because preferential diffusion of the light and deficient species toward the reaction zone will deplete that species in the neighborhood of the perturbation and cause the local burning velocity to be reduced. This is shown schematically in Fig. 8-21. Note from this figure that a depression in the flame surface will always cause the flow at the centerline to become richer in the heavier species. In the case shown in Fig. 8-21 the fuel is assumed to be the species with the lower diffusion coefficient and, therefore, on the lean side the local burning velocity would be increased. This means that a heavy-fuel lean-flame will be stable to this type of perturbation while a heavy-fuel rich-flame would be expected to be unstable.

As Sivashinsky[11] points out, recent analyses of the stability of flames using large-activation-energy-asymptotics analysis techniques have reproduced virtually all the cellular behaviors that are observed in the laboratory. It has even been found that the fact that certain stoichiometric flames are cellularly unstable can be predicted theoretically using a suitably weighted average of the Lewis number of the fuel and oxygen. In all these analyses if the actual Lewis number is below some critical Lewis number the flame will be linearly unstable and break down to form a chaotic cellular structure. A typical cellular flame surface generated by computer is shown in Fig. 8-22.

Even though the new approach to nonsteady flame theory using large activation energy asymptotics has yielded extremely useful results, even predicting cell sizes that are close to those observed experimentally, it can never be expected to quantitatively predict cellular structure. This is because it necessarily uses a very simplified kinetic scheme while real flames contain very complex chemistry.

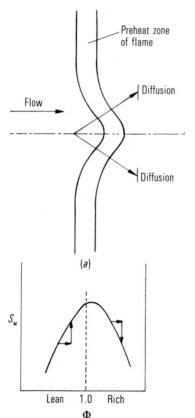

**Figure 8-21** Schematic of the qualitative "preferential diffusion" theory of cellular instability. Case for fuel heavier than oxygen. Rich mixtures are thermal-diffusive unstable in this case. (a) Diffusion behavior. (b) Effect on local burning velocity along centerline.

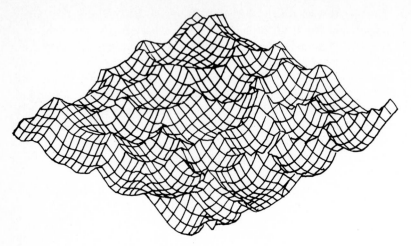

**Figure 8-22** Calculated cellular flame surface propagating downward. This structure is not steady but is in constant motion. *(Sivashinsky*[11] *with permission,* © *1983 by Annual Reviews Inc.)*

## 8-7 DECOMPOSITION AND UNUSUAL FLAMES

A few of the more important examples of unusual flame systems are given in Table 8-1 along with their maximum burning velocity. We note from Table 8-1 that there are many exothermic compounds which may be made to decompose in a spontaneously propagating flame front to yield less exothermic products. Certain of these decomposition flames have been studied quite extensively because they represent kinetically simple systems. This is particularly true of the ozone-decomposition flame whose burning velocity has both been studied as a function of oxygen dilution and calculated, using various approximations of the theory developed in Sec. 8-3. Another decomposition flame, that of nitric oxide, is unusual because it represents a flame that was first observed after its possible existence was predicted theoretically. Nitric oxide is a relatively stable exothermic molecule and it decomposes slowly even at 1500 K. Furthermore, its heat of reaction is sufficiently low so that its ordinary flame temperature is not far above 1500 K. In 1948 Henkel, Hummel, and Spaulding[12] predicted that a nitric oxide decomposition flame should be obtainable if the gas were preheated to at least 1000 K, and in 1952 Parker and Wolfhard[13] produced such a burner flame in preheated nitric oxide. They obtained the flame by first burning a nitric oxide–hydrogen flame and then slowly reducing the quantity of hydrogen until they obtained the pure decomposition flame. Of the other flame systems listed in Table 8-1 the hydrogen–bromine flame has been studied exten-

12. M. J. Henkel, H. Hummel, and D. B. Spaulding, *Third Symposium (International) on Combustion,* The Combustion Institute, Pittsburgh, Pa., p. 135 (1949).
13. W. G. Parker and H. G. Wolfhard, *Fourth Symposium (International) on Combustion,* The Combustion Institute, Pittsburgh, Pa., p. 420 (1952).

sively because the primary flame reactions represent an example of a simple chain reaction without branching.

Flames which contain either nitrogen dioxide or an excess of nitric oxide as the oxidizer hold a special interest because they exhibit two distinct luminous flame zones under the proper circumstances. This is caused by the relative inertness of the nitric oxide, which is either present in the initial mixture or appears as a consequence of the decomposition of nitrogen dioxide in the flame. These double flames are actually a fuel-oxidizer flame closely followed by a simple nitric oxide decomposition flame. This phenomenon has been observed in a number of other systems all of which have the property that two independent decomposition reactions with properly different rates are occurring in a tandem fashion in the flame. (The second reaction must occur at approximately the same velocity as the first at the flame temperature of the first.)

## 8-8 IONIZATION IN FLAMES

Most flames exhibit ion concentrations well above those theoretically available on the basis of equilibrium thermodynamic calculations. Since an investigation

**Table 8-1‡ Unusual flame systems**

| | Method of velocity determination | Burning velocity, $S_u$ m/s | Flame temperature, K |
|---|---|---|---|
| Decomposition flames | | | |
| Acetylene, $C_2H_2$ | Tube | 0.20 | — |
| Ethylene oxide, $C_2H_4O$ | Flat flame | 0.05 | 1200 |
| Methyl nitrate, $CH_3NO_3$ | Bomb | 2.50 | 1800 |
| Ethyl nitrate, $C_2H_5NO_3$ | | | |
| $T = 473$ K | Bomb | 0.10 | — |
| Methyl nitrite, $CH_3NO_2$ | Flat flame | 0.075 | 1300 |
| Hydrazine, $N_2H_4$   $P = 5$ kPa | Bomb | 1.10 | 1900 |
| Hydrogen azide, $HN_3$ | | | |
| $P = 6.6$ kPa | Bomb/burner | 9.50 | 3000 |
| Other flame systems | | | |
| Hydrogen–bromine, $H_2 + Br_2$ | Burner | 2.00 | 1650 |
| Diborane–hydrazine, | | | |
| $B_2H_6$–$N_2H_4$ | Burner | 50.00 | — |
| Hydrazine–nitrous oxide, | | | |
| $N_2H_4$–$N_2O$   $P = 5$ kPa | Bomb | 1.64 | 2650 |
| Hydrazine–nitric oxide, | | | |
| $N_2H_4$–NO   $P = 5$ kPa | Bomb | 2.45 | 2740 |
| Hydrogen–chlorine, $H_2$–$Cl_2$ | | | |
| (60% $H_2$) | Burner | 2.35 | 2100 |
| Perchloric acid–methane | | | |
| $HClO_4 + CH_4 + 1.5N_2$ | Burner | 1.22 | 2460 |

‡ Data from *Combustion Symposium* volumes—see Index in Volume X, The Combustion Institute (1965), for individual experimental values.

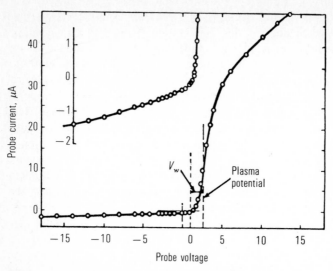

**Figure 8-23** Typical Langmuir probe curve in ethylene–oxygen flame; $P = 350$ Pa. *(With permission from H. F. Calcote, Ninth Symposium (International) on Combustion, The Combustion Institute, Pittsburgh, Pa., 1963.)*

of ions in flames must be performed in situ the exact determination of the composition of the ions is quite difficult. Three techniques have been used. These are (1) the Langmuir probe, (2) the mass spectrometer, and (3) microwave absorption. Of these, the third produces the least quantitative data since with it one observes the whole flame. Microwave absorption has been found to be most useful in observing the relaxation of ion concentrations in the postflame gases or studying the effect of external parameters on the ion concentration in the reaction zone of thick, low-pressure flames.

The Langmuir probe technique measures the current to a fine probe held at a particular voltage at a point in the flame zone. Figure 8-23 is a typical voltage-current plot for a Langmuir probe mounted in a low-pressure flame. The relatively high current that flows when the probe is charged positively and the net positive current at zero probe voltage may be explained by the very high mobility of one specific negative particle, the electron. Since the current at positive voltage is predominantly caused by electron capture, the voltage-current curve in this neighborhood may be used to determine the free-electron temperature in the flame. Values determined in a variety of low-pressure hydrocarbon–oxygen-diluent flames indicate that this temperature is always quite close to the translational gas temperature.

The Langmuir probe is not an appropriate tool for detailed studies of the positive ions in the flame because it cannot distinguish species. Also, because of the extremely high mobility of electrons the probe cannot be used to observe the presence of the negative ions in a flame. For the quantitative observation of ion concentrations a mass spectroscopic technique is necessary. With this technique both positive and negative ions have been observed in hydrocarbon flames and their concentrations throughout the flame have been mapped in certain cases. At present it is thought that the primary ion-producing reaction in hydrocarbon flames is the reaction

$$CH + O \rightarrow CHO^+ + e$$

The remaining ions are thought to be formed by a series of exchange reactions with either this ion or the free electrons. This mechanism of primary ion formation also explains the observation that hydrogen–oxygen flames have extremely low ion concentrations relative to the hydrocarbon flames of much lower temperature.

We note that even though ionization occurs in flames, flame structure calculations and measurements of neutral species concentrations agree reasonably well without the inclusions of ionization processes. However, ionization processes are important for soot formation in fuel-rich flames. This will be discussed in Chap. 16.

## 8-9 SPRAY DROPLET AND DUST FLAMES

A spray of any combustible liquid will burn in air. If the droplets are small enough and the liquid has a high enough volatility the flame that is produced when the flow is laminar looks exactly like a premixed gas flame. Such flames can be stabilized on a Bunsen burner, for example. The reason for this behavior is that highly volatile small drops will completely evaporate in the preheat zone of the flame and the hot reaction zone will essentially behave as though the approach flow were gaseous. As the volatility decreases and droplet size increases the drops will penetrate farther and farther into the flame before evaporating. For very large low-volatile droplets flame propagation proceeds more by the flame propagating from drop to drop than as a typical premixed gaseous flame. Sprays have been observed to ignite and burn outside the usual range of combustibility of that fuel. This can be dangerous in an accident situation. Flammability ranges are discussed in Chap. 12.

Any combustible dust can support a combustion wave. This includes all organic and most metal dusts. The mechanism by which large-scale dust flames propagate is markedly different from the mechanism by which liquid spray flames propagate. This is because, as the single droplet theory of Sec. 7-3 showed, the temperature of a liquid drop is limited by the boiling temperature. No such limit exists for dust flames and the dust particles can be heated to very high temperatures. Therefore dust flames are highly luminous and radiate a tremendous amount of their energy. It has been estimated that small-scale dust flames stabilized on a burner radiate as much as 20 percent of the combustion energy to the surroundings. Furthermore if one constructs a burner to maintain an adiabatic flame, i.e., a furnace whose walls are everywhere in equilibrium with the local radiation temperature of the dust particles, the burning velocity increases from about 0.1–0.2 m/s to about 0.9 m/s and the flame thickness increases from about 0.01 m to about 1.0 m. In this flame virtually all the energy is transported to the preheat zone by radiation from particle to particle and subsequent conductive transport to the gas.

A detailed discussion of the extensive recent work on spray droplet or dust flames is outside the scope of this text. The reader is referred to the bibliography for reviews and books on this subject.

## PROBLEMS

**8-1** Determine the effect of the reactions activation energy on burning velocity for the $A \rightarrow B$ flame if all other properties of the flame are held constant. Do the calculation for both an $Le = 0$ and an $Le = 1$ flame.

**8-2** Either Eq. (8-33) or (8-34) can be integrated from $\tau = z = 0$ to $\tau = z = 1$ or from $\tau = z = 1$ to $\tau = z = 0$ to determine the flames eigenvalue for any value of $\varepsilon$. However the variable coefficient term in front of the exponential is indeterminate at both $\tau = z = 0$ and $\tau = z = 1$. Resolve this problem and show how to start a numerical calculation from either the hot or cold boundary.

**8-3** A stoichiometric propane–air flame has a burning velocity of 0.43 m/s. Calculate its preheat zone thickness and plot the temperature profile in the preheat zone. Assume $\alpha = $ const at $(T_u + T_b)/2$.

**8-4** Calculate the low value of the heat of combustion for the 11 fuels of Fig. 4-10. Note that these explosion temperatures can be assumed to be flame temperatures. See if you can find any regular behavior between the maximum flame temperature, the percent fuel at the maximum flame temperature, the CH ratio (except for $NH_3$ and $C_2N_2$), and the heat of combustion. Comment on your results.

**8-5** It has been alleged that neglecting the kinetic energy change in the energy equation is a good assumption. Assume initial conditions of $P = 100$ kPa, $T = 300$ K, a burning velocity of 1.0 m/s, and a flame temperature of 2400 K, and calculate the effect of including the kinetic energy on the calculated flame temperature.

## BIBLIOGRAPHY

**Asymptotic analysis**

Buckmaster, J. D., and G. S. S. Ludford, *Theory of Laminar Flames*, Cambridge University Press (1982).
Clavin, P., "The Dynamic Behavior of Premixed Flame Fronts in Laminar and Turbulent Flows," *Prog. Energy Combust. Sci.*, **10**, in press (1984).

**Flame velocity and temperature**

Barnett, H. C., and R. R. Hibbard, *Basic Considerations on the Combustion of Hydrocarbon Fuels with Air*, NACA Report 1300, U.S. Government Printing Office, Washington, D.C. (1959).
Gaydon, A. G., and H. G. Wolfhard, *Flames, Their Structure Radiation and Temperature*, 3d ed., Chapman and Hall, London (1970).
Jost, W., *Explosion and Combustion Processes in Gases*, McGraw-Hill, New York (1946).
Khitrin, L. N., *Physics of Combustion and Explosion*, Israel Program for Scientific Translations, Jerusalem (1962).
Lewis, B., and G. von Elbe, *Combustion Flames and Explosions of Gases*, Academic, New York (1961).
Rallis, C. J., and A. M. Garforth, "The Determination of Laminar Burning Velocity," *Prog. Energy Combust. Sci.*, **6**: 303–329 (1980).

## Flame structure

Fenimore, C. P., *Chemistry in Premixed Flames*, Pergamon, New York (1964).
Fristrom, R. M., and A. A. Westenberg, *Flame Structure*, McGraw-Hill, New York (1965).
Markstein, G. H., *Non-steady Flame Propagation*, Macmillan, New York (1965).
Weinberg, F. J., *Optics of Flames*, Butterworth, London (1963).
Williams, F. A., *Combustion Theory*, Addison-Wesley, Reading, Ma. (1965).

## Mist and dust flames

Krazinski, J. L., R. O. Buckius, and H. Krier, "Coal Dust Flames: A Review and Development of a Model for Flame Propagation," *Prog. Energy Combust. Sci.*, **5**:31–71 (1979).
Palmer, K. N., *Dust Explosions and Fires*, Chapman and Hall, London (1973).
Smoot, L. D., and M. D. Horton, "Propagation of Laminar Pulverized Coal–Air Flames," *Prog. Energy Combust. Sci.*, **3**:235–258 (1977).

# NINE

# DETONATIONS

## 9-1 INTRODUCTION

The word *detonation* has been in common usage for at least 200 years. In earlier times it referred to the sudden decomposition of certain chemicals and mixtures with the production of considerable noise "like thunder," even though these materials were unconfined when they decomposed. For example, there is the report[1] that in 1793, a Dr. Wurzer was mixing $1\frac{1}{2}$ grains of three parts sodium chlorate and one part sulfur at Bonn in a mortar and pestle when "he obtained a detonation which rendered him deaf for several days." It was, in fact, well known at that time that detonation is the rapid decomposition of certain substances which occurs upon percussion or with frictional heating, and that these same substances can be decomposed quietly under the proper conditions.

At the present time the term *detonation* should be applied only to processes where a shock-induced combustion wave is propagating through a reactive mixture or pure exothermic compound. The phenomenon of detonation is known to be similar for solids or liquids and for gaseous substances and we will therefore discuss only gas-phase detonation in detail. The nature of detonation for a condensed phase and its similarities to and differences from gas-phase detonation are summarized in Chap. 15.

Since a shock wave is always observed to precede the reaction front in a detonation, the propagation velocity of such a combustion wave is always supersonic relative to the undisturbed media. The first experimental evidence for

1. *Nicholson's Journal*, **2**:473 (1798).

this supersonic nature of gaseous detonation waves was obtained in 1881, when Berthelot and Vieille and independently Mallard and Le Châtelier observed supersonic combustion waves while studying flame propagation in tubes filled with gaseous combustible mixtures. A detailed study of this phenomenon by these investigators in France and by Dixon in England produced the interesting observations that the propagation velocity of this newly discovered wave was very high, and that it was remarkably constant for any mixture composition and initial pressure and temperature. Specifically, it was observed that the tube's geometry had little effect on the propagation velocity of a detonation wave. This, of course, contrasts very strikingly with the propagation behavior of ordinary combustion waves, in which the tube geometry is extremely important. It was also observed that detonation could be initiated in a tube from a spark or other low-energy ignition source after a highly variable and quite geometry-dependent delay time during which an accelerating flame propagated down the tube.

The observation of a new phenomenon as unusual as detonation led to a number of models and attempted explanations. In 1899 Chapman[2] essentially stated the one-dimensional flow requirements summarized in Chap. 5 and compared the minimum theoretical velocity with the experimental results of Dixon. In 1905 Jouguet,[3] in a long theoretical monograph, independently restated Chapman's rule that the tangency point of the Rayleigh line and Hugoniot curve should represent the gross behavior of detonations and therefore should allow a calculation of the detonation velocity from *thermodynamic* and *hydrodynamic* considerations only. As we have seen, this point is now called the upper Chapman–Jouguet (CJ) point in the *P, V* plane. It is shown in Fig. 5-3. Close quantitative agreement between the theory and experimentally measured velocities was not obtained immediately, however, because of the paucity of thermodynamic and chemical data. For example, the existence of atoms and radicals such as H, O, N, and OH at equilibrium was not suspected and, in fact, the variation of heat capacity with temperature was only known qualitatively at the time of the first calculations. Chapman even suggested that detonation studies might be useful for the determination of high-temperature heat capacities.

We now know that the CJ criterion, when applied to detonations that are well removed from limit behavior, predicts velocities which agree with the experimentally measured propagation velocities within a few percent. A comparison of a few recently measured and calculated detonation velocities and some calculated CJ pressures and temperatures is given in Table 9-1.

This excellent agreement is a rather remarkable result because CJ theory is a strictly hydrodynamic theory in which state 2 is assumed to be in full chemical (thermodynamic) equilibrium. Thus the correct propagation velocity is predicted without invoking any chemical rate information. The proper justification for this behavior has never been found. The literature does contain many

2. D. L. Chapman, *Philos. Mag.*, **213**, Series 5, **47**:90 (1899).
3. E. Jouguet, *J. Pure Appl. Math.*, **70**, Series 6, **1**:347 (1905); **2**:1 (1906).

**Table 9-1 Detonation properties** $T_0 = 25°C$, $P_0 = 100$ kPa

| System | Measured | Calculated | | |
|---|---|---|---|---|
| | Velocity, m/s | Velocity, m/s | Pressure, MPa | Temperature, K |
| 80% $H_2$ + 20% $O_2$ | 3390 | 3408 | 1.80 | 3439 |
| 66.7% $H_2$ + 33.3% $O_2$ | 2825 | 2841 | 1.88 | 3679 |
| 25% $H_2$ + 75% $O_2$ | 1163 | 1737 | 1.42 | 2667 |
| $CH_4 + O_2$ | 2528 | 2639 | 3.16 | 3332 |
| $CH_4 + \frac{3}{2}O_2$ | 2470 | 2535 | 3.16 | 3725 |
| 41.2% $C_2N_2$ + 58.8% $O_2$ | 2540 | 2525 | 4.62 | 5120 |

attempts to justify the observed CJ behavior based on one-dimensional models of the wave. It has been well documented however that the detailed structure of any self-sustaining detonation propagating at CJ velocity is three-dimensional nonsteady. As it turns out the characteristic size of this three-dimensional structure has been found to be uniquely related to the rates of the chemical reactions occurring in the wave.

The modern era of detonation research was opened in the years 1940–1945 when Zel'dovich,[4] von Neumann,[5] and Döring[6] each independently formulated essentially the same one-dimensional model for the structure of a detonation wave. The true three-dimensional nature of all self-sustaining detonations was not suspected until the late 1950s, however, and a reasonable model for detonation initiation and failure did not appear until the early 1980s. Initiation and failure will be discussed in Chap. 13.

## 9-2 STRUCTURE OF A ONE-DIMENSIONAL STEADY DETONATION WAVE

To discuss the internal structure of a one-dimensional detonation wave we further manipulate the equations derived in Chap. 5 for the constant gamma-heat addition model. Equation (5-40) allows one to calculate the maximum amount of heat, $q_{CJ}$, which may be added to the flow for any specific incoming Mach number. It represents choked-flow heat addition for a one-dimensional steady flow. We now wish to examine flows for which $M_1$ is fixed but $q \le q_{CJ}$ and thereby determine the nature of possible intermediate flow states for various heat-addition models. We therefore define a heat-addition parameter

4. Ia. B. Zel'dovich, *J. Exptl. Theor. Phys.* (USSR), **10**: 542 (1940).

5. J. Von Neumann, O.S.R.D. Rep. No. 549 (1942), Ballistic Research Laboratory File No. X-122, Aberdeen Proving Ground, Md.

6. W. Döring, *Ann. Physik.*, **43**: 421 (1943).

$\phi = q/q_{CJ}$, where $0 < \phi < 1$ for any steady exothermic flow. With this definition, Eq. (5-38) reduces to the expression

$$\varepsilon = \frac{(1 + \gamma M_1^2) \pm (M_1^2 - 1)\sqrt{1 - \phi}}{(\gamma + 1)M_1^2} \tag{9-1}$$

We note from Eq. (9-1) that for $\phi = 0$ the plus-sign solution yields $\varepsilon = 1$ (i.e., the trivial case of no change between the stations) and the minus-sign solution yields the expression

$$\varepsilon = \frac{2 + (\gamma - 1)M_1^2}{(\gamma + 1)M_1^2}$$

which is the equation for the volumetric change across a nonreactive shock in a constant-gamma gas.

These equations, relating heat addition to the flow properties, were first applied to the discussion of one-dimensional detonation structure by Zel'dovich, von Neumann, and Döring in the period 1940–1943. In its simplest form, their model (the ZND model) assumes that a steady one-dimensional detonation consists of nonreactive shock discontinuity followed by a region of heat addition (due to chemical reaction) which continues until the flow becomes sonic at $\phi = 1$ for a CJ detonation or until $\phi$ reaches a value of

$$\phi_{max} < 1$$

for an overdriven detonation wave. The question of accessibility of various flow regimes in the ZND model for cases of unusual heat-addition processes has been discussed rather extensively, but we will content ourselves here with representing the detonation as a $\phi = 0$ (adiabatic) shock transition followed by a flow in which $\phi$ increases monotonically with time until a value of $\phi_{max} \leq 1$ is reached. Thus if we know a value of $\phi$ at some plane in this one-dimensional wave we will be able to calculate all the properties of the flow at that plane. The density may be obtained from Eq. (9-1) (using the negative sign on the radical) and the pressure from the Rayleigh equation. Using these, we may also calculate the local temperature ratio $T/T_1 = \gamma\varepsilon$, the velocity of sound ratio $a/a_1 = \sqrt{T/T_1}$ and the dimensionless flow velocity $u/a_1 = M_1\varepsilon$. The actual physical structure of this one-dimensional detonation is not given by simply listing these properties, however, but may be obtained by noting that at any $x$ position in this steady reactive flow

$$\frac{d\phi}{dx} = \frac{1}{uq_{CJ}} \cdot \frac{Dq}{Dt} \tag{9-2}$$

where the quantity $Dq/Dt$ is the local rate of heat addition at the plane of interest. The actual structure therefore depends on the reaction kinetic model assumed for the gas behind the shock front, and one may construct any appropriate detonation structure in the framework of the constant-gamma heat-addition model by using these equations and various simple or complex

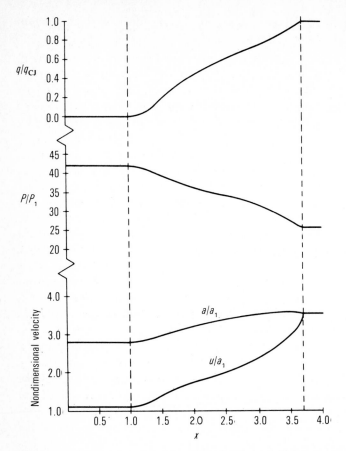

**Figure 9-1** Structure of a $M = 6$ one-dimensional detonation in a $\gamma = 1.4$ gas. Top curve, heat addition, $q/q_{CJ} = \phi$ vs. $x$. Middle curve is pressure profile, bottom two curves are dimensionless velocity of sound and particle velocity. This detonation has an induction zone between the shock (at $x = 0$) and the plane $x = 1$, and has its CJ plane at $x = 3.70$. The $\phi_{\parallel}$ plane is at $x = 2.04$ in this particular case. (See Fig. 9-15b).

reaction schemes. Ordinarily the structure is built by starting the calculation at the shock front, where $\phi = 0$ and integrating to the plane $\phi$ by determining the appropriate upper limit on the integral in the equation

$$\phi = \int_0^x \frac{1}{u q_{CJ}} \cdot \frac{Dq}{Dt} \, dx \qquad (9\text{-}3)$$

Numerical integration is usually required to determine $\phi(x)$ from this equation. The structure of such a typical detonation wave is shown in Fig. 9-1 for a particular heat-release model.

This one-dimensional steady approach is not limited to the case of heat addition in a constant-gamma, constant-molecular-weight fluid. In fact, it may be easily extended to the general case of a one-dimensional steady wave in a real gas with realistic caloric and thermal equations of state and complex kinetics. In this case one first calculates a CJ detonation velocity by simultaneously solving the Rayleigh and Hugoniot equations for tangency, assuming thermodynamic and chemical equilibrium at the CJ plane. This yields the lowest possible value for the steady detonation velocity. A knowledge of this velocity allows one to construct a shock-wave reaction-zone structure of any desired complexity by forward integration from the cold flow. In this type of calculation, one need not even assume a shock discontinuity but may integrate through the shock wave with the transport equations. One may thus produce a steady one-dimensional structure whose complexity is dependent on the complexity of the starting assumptions. One may also construct an overdriven wave using this technique by simply assuming that the cold gas flow velocity is above the calculated CJ detonation velocity of the mixture.

In conclusion, we note two important properties of the simple ZND model. First, as shown in Fig. 9-1, the model predicts a pressure maximum at the shock transition. This maximum is commonly called the *von Neumann spike*, and its prediction by the ZND model stimulated a considerable amount of experimental work on structure in the 1950s. Second, we observe that the ZND model is strictly a steady one-dimensional model and therefore cannot be used to discuss the question of stability. For this purpose the original equations must be modified to include time-dependent terms. However, we defer our discussion of detonation stability until we describe the experimental results concerning the actual three-dimensional nonsteady nature of self-sustaining detonation waves.

## 9-3 PHYSICAL PROPERTIES OF PREMIXED GASEOUS DETONATIONS

The first evidence that detonations might prefer to travel with a configuration which is locally three-dimensional and nonsteady was obtained in 1926 when Campbell and Woodhead[7] first observed spin in limit mixtures in circular tubes. At the time of its discovery (and for at least 30 years thereafter), the occurrence of multidimensional structure was thought to be a phenomenon peculiar to limit mixtures. A typical open-shutter photograph of a spinning detonation in an acetylene–oxygen mixture is shown in Fig. 9-2. It is now known that all self-sustaining detonation waves exhibit significant three-dimensional structure in a region which extends from the leading shock front through the reaction zone of the detonation. This structure is characterized by a nonplanar leading shock wave which at every instant consists of many curved shock sections which are convex toward the incoming flow. The lines of intersection of these curved shock segments are propagating in various directions at

7. C. Campbell and D. W. Woodhead, *J. Chem. Soc.*, p. 3010 (1926); p. 1572 (1927).

**Figure 9-2** Open-shutter photograph of a single-spin detonation in an acetylene–oxygen mixture diluted with argon. Photograph was taken through a 25 mm diameter glass tube. *(Courtesy G. L. Schott, Los Alamos Scientific Laboratory, Los Alamos, N.M.; originally appeared in Phys. Fluids, vol. 8, p. 850 (1965). (With permission.)*

high velocity across the front and actually consists of triple-shock interactions (i.e., Mach-stem configurations). The third shock of these interactions extends back into the reactive flow regime and is required for the flow to be balanced at the intersection of the two convex leading shock waves. In general, the flow in the neighborhood of the shock front is therefore quite complex. However, it has been observed that these propagating transverse waves exhibit a "preferred" spacing which is primarily related to the "thickness" of the reaction zone. Furthermore, even though in spherical detonations these waves appear in a completely random manner and are disoriented relative to each other, in those cases where a detonation tube is used they have been observed to couple with the geometry of the tube. It is true, of course, that a finite-amplitude transverse shock wave cannot exist behind a leading shock front without being attached to a transverse pressure wave propagating in the hot-gas column well behind the front. Therefore, coupling might be expected in those cases where the detonation's preferred spacing corresponds to the spacing required for a transverse standing or traveling mode in the hot-gas column downstream of the front. In fact, it has been found that the coupling phenomenon is so general that virtually every possible mode of coupled transverse-wave phenomena may be observed in self-sustaining detonations.

In describing the structure of real detonation waves we must therefore contend with three important non-one-dimensional properties of the detonation, each of which must be discussed in detail. These are (1) the nature of the coupling with the tube geometry, (2) the detailed structure of transverse-wave phenomena at the shock front and in the reaction zone, and (3) the relationship between the preferred spacing exhibited by the transverse phenomena and the initial conditions in the detonatable mixture.

It is interesting that the nature of the coupling was essentially explained before any details concerning the structure of the front were observed or surmised. Experimentally it was noted quite early that a particular limit detonation not only prefers to travel in a round tube as a spinning wave but that the pitch ratio (i.e., the ratio of the length of the tube required for the phenomenon to traverse one circumference to the diameter of the tube) is a constant for any initial conditions and tube geometry. In 1946 and 1952 Manson[8] and Fay[9] independently presented an acoustic theory of spin which produced excellent agreement with experimental $P/d$ values. In this theory an acoustic equation is solved to find the allowable transverse coupled modes in the hot-gas column downstream of the propagating detonation front. Manson and Fay both found that the first helical (rotational) acoustic mode of the tube yielded the correct pitch for the strong transverse wave at the front which causes the luminosity shown in Fig. 9-2. Fay also applied his theory to rectangular tubes, triangular tubes, and circular annular spaces, and showed that the application of acoustic theory predicted the correct pitch in all cases. However, he also showed that true spin occurs only as the helical transverse mode of a round tube, and that the acoustic modes which appear in rectangular or triangular geometries are simply standing transverse modes of the tube.

We now wish to discuss the detailed shock structure that occurs at the detonations shock front. ZND theory yields the conclusion that the region between the lead shock wave and the Chapman–Jouguet plane in a one-dimensional detonation is a region in which the flow is subsonic in a shock-oriented coordinate system. This means that it is possible for pressure waves to propagate across the front as the front itself propagates. It is in fact observed that *all* self-sustaining detonations contain propagating transverse pressure waves of finite amplitude which thus are really propagating shock waves. These shock waves interact with the lead shock to produce Mach stems which propagate across the detonation front. A schematic of the shock structure near the triple point of one such propagating Mach-stem wave system is shown in Fig. 9-3.

One interesting feature of these Mach stems is the ability of their triple point to write a line on a smoked surface. These thin lines were first observed by Antolik[10] in 1875 on soot-coated plates held near a spark discharge. Mach later interpreted them as being caused by the intersection of "sound waves," and in 1897 he observed the same type of writing when a bullet passed by obstacles placed near a coated surface. This work preceded by eight years his use of spark schlieren photography to observe the shock waves associated with a traveling bullet. In 1959, Denisov and Troshin[11] revived the technique and first applied it to the observation of transverse waves in detonations. Since that time it has been extensively applied to the study of detonation structure, mainly because of its simplicity. The current consensus is that wood smoke deposited in an almost opaque layer on the surface produces the best smoked-foil records. The foils may be "fixed" after firing by spraying with a clear lacquer.

8. N. Manson, *Compt. Rend.*, **222**:46 (1946).

9. J. A. Fay, *J. Chem. Phys.*, **20**:942 (1952).

10. K. Antolik, *Pogg. Ann.* **230**, Series 2, **154**:14 (1875).

11. Y. H. Denisov and Y. K. Troshin, *Dokl. Akad. Sci. SSSR*, **125**:110 (1959). (Translation *Phys. Chem. Sect*, **125**:217, 1960).

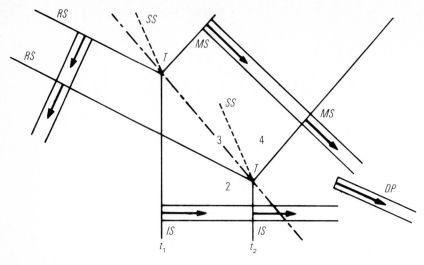

**Figure 9-3** A typical propagating Mach stem on a detonation front (schematic) shown at two times, $t_1$ and $t_2$. Triple point, $T$; incident shock, $IS$; reflected shock, $RS$; Mach-stem shock, $MS$; slip stream, $SS$. Region 2 is compressed by the incident shock. Regions 3 and 4 are compressed by the reflected shock and Mach-stem shock respectively and have the same pressure, which is higher than region 2. The arrows show that each shock propagates in a direction only normal to itself. The $DP$ arrow represents the approximate direction of propagation of the detonation front. Note that as time progresses the triple point is essentially generating Mach-stem and reflected shocks and over-riding or destroying the incident shock.

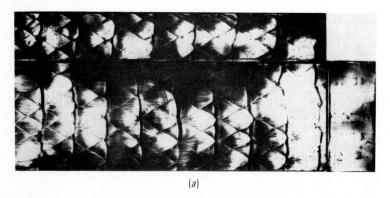

(a)

**Figure 9-4** Smoked-foil records of propagating detonations obtained in an $83 \times 38$ mm tube. (a) Rectangular mode, showing side, top, and end foil records; $2H_2 + O_2 + 3$ Ar, $P_0 = 14$ kPa, entire record shown. (b) Planar mode; $0.2\ H_2 + 0.1\ O_2 + 0.7$ Ar, $P_0 = 9.3$ kPa, entire record shown. (c) Very regular structure; $0.0625\ C_2H_4 + 0.1875\ O_2 + 0.75$ Ar, $P_0 = 13.33$ kPa. (d) Relatively regular structure; Acetylene–oxygen, $\Phi = 0.625$, $P_0 = 6.67$ kPa. (e) Irregular structure; $2H_2 + O_2$, $P_0 = 16$ kPa. (f) Irregular structure; $0.25\ C_2H_4 + 0.75\ O_2$, $P_0 = 5.33$ kPa. Note scale for (c), (d), (e), and (f). Propagation to the right in (a) and (b); to the top in (c), (d), (e) and (f).

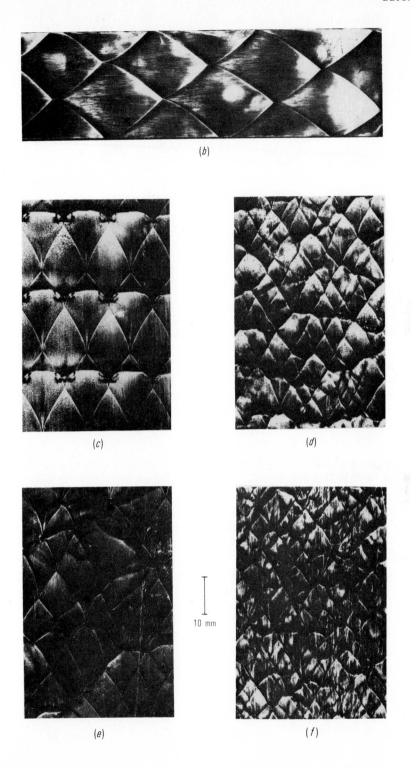

(b)

(c)

(d)

(e)

10 mm

(f)

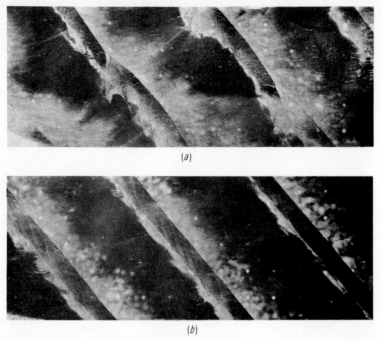

(a)

(b)

**Figure 9-5** Smoked-foil writing produced by single-spin detonation in a round 25 mm diameter tube, using a two-thirds acetylene–oxygen mixture diluted with argon. Detonation is propagating from left to right. (a) Spin with some bounce (i.e., with a nonconstant transverse velocity). (b) Constant-velocity spin. (*Courtesy G. L. Schott, Los Alamos Scientific Laboratory, Los Alamos, N.Mex.; originally appeared in Phys. Fluids, vol. 8, p. 850 (1965).) (With permission.*)

A few examples of smoked-foil records obtained with propagating detonations are shown in Figs. 9-4 and 9-5 and a record obtained from an adaptation of this technique which allows the orientation of the shock wave to be observed is shown in Fig. 9-6. This figure was obtained by sprinkling sand grains on the film after smoking and before firing the detonation. Each sand grain produces a small local and symmetric interaction with the leading shock wave of the detonation and thereby flags its orientation at that location.

In contrast to the single spin of Fig. 9-5, which contains only one rotating wave head, and can be rendered steady in a rotating coordinate system, all other detonations contain waves propagating in many directions. These waves are constantly intersecting each other and therefore in these cases the flow in the neighborhood of the detonation front is three-dimensional nonsteady. The detailed structure of this flow can be analyzed for only the simplest cases. Fortunately it was discovered that the hydrogen–oxygen and acetylene–oxygen systems, when diluted with more than 50 percent argon, yield a structure, in a rectangular channel, which is extremely regular and reproducible; see Figs. 9-6a, b and c. The regular structures of Figs. 9-6a and c are still three-dimensional nonsteady because the vertical lines on these smoked foil records are due to the reflection of "slapping" waves that are propagating toward the smoked surface and are thus orthogonal to the waves that are writing the diamond pattern. These two orthogonal wave sets are essentially uncoupled and therefore, to a good first approximation, the structural details of such a transverse wave system can be analyzed by considering only the two-dimensional nonsteady flow associated with one set of propagating waves. Such a structure is illustrated in Fig. 9-7 for the case when the 2-D system is symmetric.

The detailed analysis of even this structure is quite involved, but it has yielded considerable insight into the processes that are occurring during detonation propagation. First, a detailed examination of the structure of a single isolated Mach stem in a reactive system yielded the observation that the slipstream balance condition at the triple point is governed by an unreactive–unreactive $(P, \theta)$ balance requirement even though shock heating can lead to explosion delays that are less than 1 $\mu$s. This analysis is shown in Fig. 9-8. Note from this

Propagation
direction

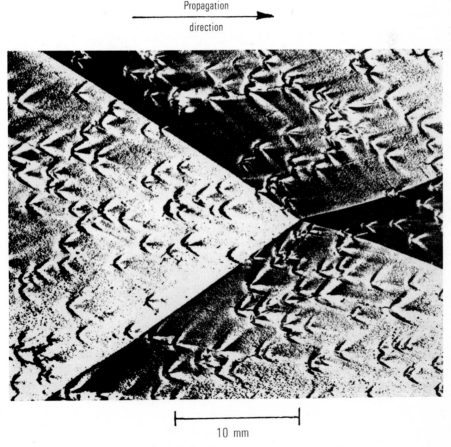

10 mm

**Figure 9-6** Smoked-foil record showing orientation of the incident shock waves by use of sand grains. Propagation from left to right: $0.3(2H_2 + O_2) + 0.7$ Ar; $P_0 = 13.3$ kPa.

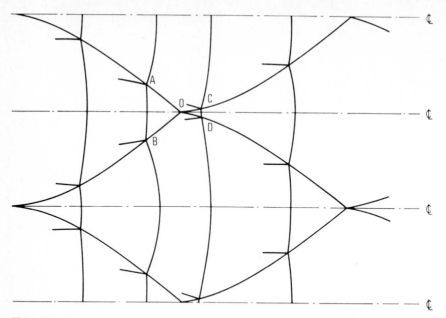

**Figure 9-7** Idealized two-dimensional detonation showing shock-wave structure at various times and associated smoked-foil pattern.

figure that the flow in region 2 is supersonic relative to the triple point $T$. Thus a reflected shock can appear as a steady shock in coordinates that are attached to $T$ and the intersection of this shock polar with the original shock polar produces a slipstream-balance condition that is unique for any initial value of $\beta$ and $w$, the approach-flow velocity relative to the triple point. The flow in region 4 is always subsonic and the pressure in this region is driving the triple shock structure.

The information that simple Mach-stem structure can be determined by using unreactive–unreactive $(P, \theta)$ balance requirements at the slipstream means that the observed shock intersection geometry obtained from smoked-foil records can be used to analyze the details of simple 2-D unsteady detonative flow. We will summarize the analysis that is based upon the dynamics of a symmetric intersection, because of its relative simplicity. Figure 9-9 is an enlargement of one of the intersections of Fig. 9-7 and shows schematically how the shocks of Fig. 9-6 are oriented immediately before and after an intersection. The two triple points labeled $A$ and $B$ share a common incident shock $A–B$, and because of symmetry the reflected shocks $A''–A$ and $B''–B$ are of equal strength, as are the Mach-stem shocks $A–A'$ and $B–B'$. The value of $\alpha$ that can be measured from a smoked-foil record must be equal to 90-$\beta$. The only piece of information that is needed for a complete calculation is the normal Mach number of the incident shock $A–B$ in the laboratory frame of reference.

Now let us examine the shock properties at time $t + \Delta t$, again referring to Fig. 9-9. Once again the shock $C–D$ is common to both triple point $C$ and $D$,

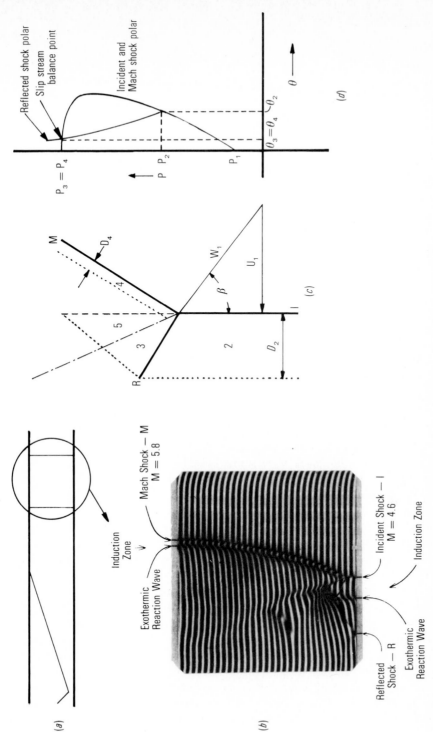

**Figure 9-8** Mach stem structure in a reactive gas. (*a*) Two-dimensional ramp geometry used to produce the mach stem illustrated in part *b*. (*b*) Spark interferogram of a reactive mach stem in a $2H_2 + O_2 + 2CO$ mixture, $P_1 = 2$ kPa. (*c*) Single Mach-stem structure showing angles of flow deflection in a coordinate system with point $T$ stationary and also showing location of reaction waves based on induction delay kinetics. $D_2$ and $D_4$ are the calculated delays to explosions. (*d*) Shock polars used to calculate Mach-stem structure shown in (*c*). Unreactive-unreactive slipstream balance assumed (*Photograph in* (*b*) *courtesy D. R. White, General Electric Research Laboratory, Schenectady, New York.*)

315

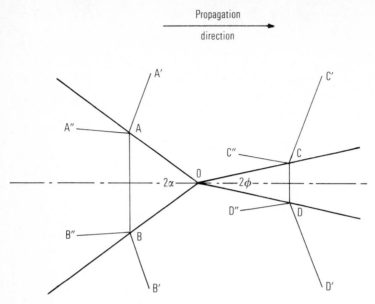

**Figure 9-9** Shock structure at an interaction point. Line $A'ABB'$ represents shock at time $t$ (before intersection). Line $C'CDD'$ represents shock at time $t + \Delta t$ (after intersection).

and the reflected shocks $C-C''$ and $D-D''$ must be of equal strength, as are the shocks $C'-C$ and $D'-D$. However, after the intersection of the triple points the new Mach stems $C$ and $D$ are structured such that the new normal shock $C-D$ is a Mach-stem shock while shocks $C-C'$ and $D-D'$ are incident shocks. However, shocks $A-A'$, $B-B'$, $C-C'$, and $D-D'$ are all of equal strength. This means that the relative Mach numbers or strengths of the shocks must be $M_{AB} < M_{AA'} = M_{BB'} = M_{CC'} = M_{DD'} < M_{CD}$.

Referring now to Fig. 9-7, the lead shock wave in any one detonation cell must start out as a strong Mach-stem shock produced by the intersection and must then continually decay as it propagates through the cell. At the instant that the next intersections occur halfway down the cell this lead shock becomes an incident shock for the two new triple points. Now recall that detonations, when propagating in tubes that are large relative to their structural size, propagate at or very close to the Chapman–Jouguet velocity. Thus the shock that disappears at the end of a cell must have a velocity which is less than $M_{CJ}$ and the new one that is formed by the intersection must have a velocity higher than $M_{CJ}$. An analysis of the symmetric intersection geometry of Fig. 9-9 using real gas enthalpies and a range of $M_{AA}$ from 0.75 to 0.95 $M_{CJ}$ yielded the information that the intersection angle $\alpha$ is a sensitive function of the strength of the reflected shock $(P_3 - P_2)/P_2$, and relatively insensitive to the Mach number $M_{AA}$. Furthermore an analysis of slightly unsymmetric interactions yielded the information that the refraction of one wave, i.e., the quantity $\alpha - \phi$ for the symmetric intersection of Fig. 9-9 is a sensitive function of the strength of the other opposing wave and not very dependent on the incoming shock Mach number. Thus, a smoked-foil record can be used to determine the strength of a single transverse wave as it propagates across the detonation while the detonation propagates down a tube. Figure 9-10 is a plot of the transverse wave strength of a single wave in a propagating detonation. Note that the wave strength fluctuates a considerable amount from intersection to intersection but nevertheless has an average value of about 0.2. This means that the reflected shock has a Mach number of about 1.4, that is, it is a relatively weak shock. This average strength for the reflected shock means that the lead shock wave of the detonation must decay from about 1.2 $M_{CJ}$ to about 0.85 $M_{CJ}$ as it propagates through each cell. This represents a sizable variation in shock strength and, therefore, in the entropy increase associated with different elements of the wave. This means of course that the shocked and reacted gases will contain large temperature and velocity of sound gradients after they reach pressure equilibrium somewhere downstream of the reaction zone. This is the primary reason why it is not useful to attempt to justify self-sustenance or CJ behavior using one-dimensional time-dependent arguments.

The last property of transverse structure which should be discussed is its preferred spacing. Experimental results have shown that for a specific mode of propagation, the spacing of transverse waves of the same family is intimately related to the mixture composition and the initial pressure. Table 9-2 contains transverse wave spacings or cell sizes reported in Refs. 12, 13, and 14.

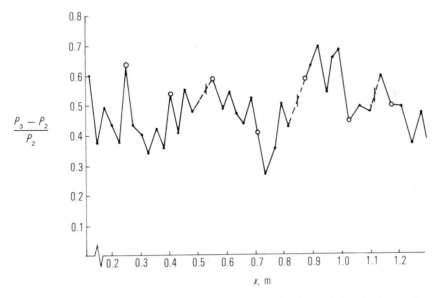

**Figure 9-10** The strength of a single transverse wave measured at each interior intersection and at wall intersections (∘). The dashed line means that the intersection geometry was not clear enough to be used to determine strength. *(Adapted from R. A. Strehlow, A. A. Adamczyk, and R. J. Stiles, Astronaut. Acta, vol. 17, pp. 509–527 (1972), with permission.)*

**Table 9-2 Detonation cell sizes or transverse wave spacings (mm)†**

For    Selected    Fuels:    Stoichiometric    Mixtures, $P = 101.325$ kPa, $T = 293$ K

| Fuel | Oxygen | Air | | |
|------|--------|--------|--------|--------|
|      |        | Ref. 12 | Ref. 13¶ | Ref. 14 |
| $H_2$ | 1.5‡ | 15 | 15 | 15 |
| $CH_4$ | 4.3‡ |  | 330 | $280 \pm 30$ |
| $C_2H_2$ | 0.2§ | 10 | 5.65 | 9.8 |
| $C_2H_4$ | 0.5§ | 25 | 26 | 28 |
| $C_2H_6$ | 1.14‡ |  | 53.5 | 54–62 |
| $C_3H_6$ | 0.56‡ |  |  |  |
| $C_3H_8$ | 1.1§ | 54 | 53.5 |  |
| $C_4H_{10}$ |  |  | 53.5 |  |

† Adapted, with permission, from Refs. 12, 13, and 14.
‡ Ref. 12—extrapolation to 101.325 kPa.
§ Ref. 12—extrapolation from air to oxygen.
¶ Minimum value—slightly rich mixtures except for $H_2$.

The implication of the data of Table 9-2 is that the transverse wave spacing is somehow related to the rate of the exothermic reactions that are occurring in the detonation. This contention is further supported by Fig. 9-11, which is a plot of the pressure-dependence of cell size, $Z$, normalized to the cell size of the stoichiometric fuel–oxygen mixture at one atmosphere, for all the fuels tabulated in Table 9-2. The dependence shown in Fig. 9-11 is almost first-order and that is what one would expect if the reactions whose rates controlled the cell size were second-order. This relationship will be quantified in the next section.

Figure 9-12 shows the effect of diluting a stoichiometric fuel–oxygen mixture with nitrogen until the oxidizer has the composition of air. Note that the three hydrocarbons that were tested all have the same sensitivity to dilution, but that hydrogen is less affected by dilution. Because of the identical behavior of all the tested hydrocarbons, their behavior was extrapolated to the other hydrocarbons of Table 9-2. It is felt that this extrapolation is reasonable because the hydrocarbons that were tested have markedly different minimum cell spacings.

12. J. H. S. Lee, R. Knystautas, and C. M. Guirao, "The Link Between Cell Size, Critical Tube Diameter, Initiation Energy, and Detonability Limits" in *Fuel–Air Explosions, S M Study Series No. 16*, eds. J. H. S. Lee and C. M. Guirao, University of Waterloo Press, Waterloo, Ontario, Canada, pp. 157–186 (1982).
13. R. Knystautas, C. Guirao, J. H. Lee, and A. Sulmistras, "Measurement of Cell Size in Hydrocarbon–Air Mixtures and Predictions of Critical Tube Diameter, Critical Initiation Energy, and Detonability Limits," Paper presented at the Ninth International Colloquium on Gas Dynamics of Explosions and Reactive Systems, Poitiers, France, July 3–8, 1983.
14. I. O. Moen, P. A. Thibault, J. W. Fink, S. A. Ward, and G. M. Rude, "Detonation Length Scales for Fuel–Air Explosives" Paper presented at the Ninth International Colloquium on Gas Dynamics of Explosions and Reactive Systems, Poitiers, France, July 3–8, 1983.

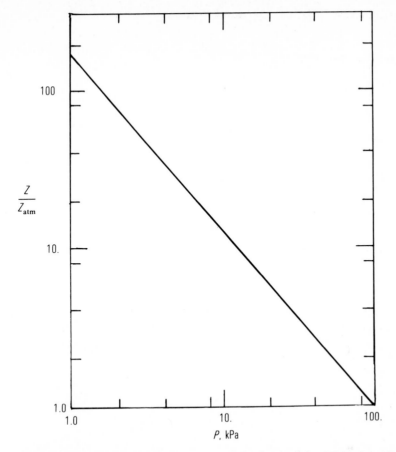

**Figure 9-11** The effect of initial pressure on cell size for the fuels of Table 9-2. All the data for all these fuels lie on the line to within about a factor of two. *(Data from Figs. 1 to 9 of Ref. 12, adapted with permission.)*

Figure 9-13 is a plot of the dependence of cell spacing on the equivalence ratio for the hydrogen–air system. It has the U shape that one expects, based on the effect of equivalence ratio on chemical reactivity.

## 9-4 THEORY OF STRUCTURE— STRUCTURAL SIZE AND KINETICS

Early investigations of the stability of Chapman–Jouguet flow were concerned primarily with a comparison of the properties of this flow with the properties of neighboring steady flows which also satisfy the Hugoniot and Rayleigh equations. However, transient stability cannot be fruitfully investigated using this technique, since this type of analysis does not allow for the occurrence of the

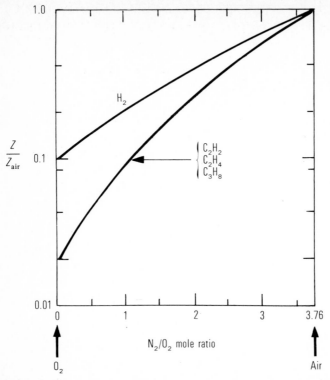

**Figure 9-12** The effect of nitrogen removal from air for four stoichiometric fuel–oxidizer mixtures. *(Data taken from Figs. 10 to 13 of Ref. 12, adapted with permission.)*

truly nonsteady flow which is a possible solution to the equations of motion. In other words, any analysis of stability to be realistic must specifically allow for nonsteady behavior.

Erpenbeck[15] has pioneered in the field of this nonsteady analysis and has shown, with a linearized treatment in Laplace transform space, that all detonations with realistic exothermic kinetics should be linearly unstable to transverse perturbation waves of the proper frequency for any arbitrarily overdriven detonation. In a sense, the results of his analysis are in agreement with the observation that transverse structure exists on all self-sustaining detonations. Unfortunately, because of the complexity of his analysis it is impossible to describe the nature of the flow that arises from an instability of his type. More recently Abouseif and Toong[16] have constructed a theory for one-dimensional instability which yields physical insight into the processes that are occurring.

In order to construct a physically meaningful theory of two-dimensional detonation stability we turn to the one-dimensional ZND model of detonation.

15. J. J. Erpenbeck, *Phys. Fluids*, **8**:1192 (1965); **7**:684 (1964).
16. G. E. Abouseif and T. Y. Toong, *Combust. Flame*, **45**:67–94 (1982).

We note that the entire reaction zone of this detonation is subsonic and there-fore that acoustic amplitude wave fronts may possibly propagate for long trans-verse distances in the reactive-flow region. We therefore consider in detail the propagation of high-frequency waves of acoustic amplitude and arbitrary initial orientation in this reactive flow. For our purposes, the frequency of the wave train is always assumed to be sufficiently high so that dispersion is not impor-tant. This frequency limit is determined by the requirement that the flow properties change only very slowly over a distance equal to a wavelength of the wave being considered. If this is true, the behavior of a coherent wave front may be approximated by writing the parametric equations for a sound ray,

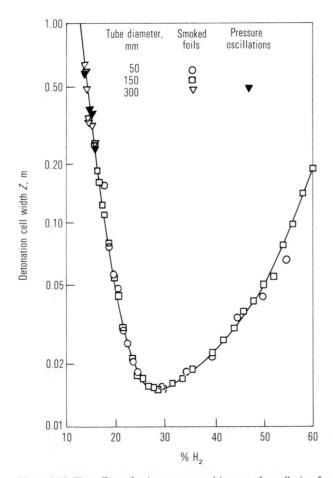

**Figure 9-13** The effect of mixture composition on the cell size for hydrogen–air detonations at ambient conditions. Note that the minimum in the curve occurs at stoichiometric. (*Taken from a paper entitled "Hydrogen–Air Detonations" by J. Lee, R. Knystautas, C. Guirao, W. B. Benedick, and J. E. Shepard presented at the Second International Workshop on the Impact of Hydrogen on Water Reactor Safety at Albuquerque, N.Mex. on Oct. 3–7, 1982, with permission.*)

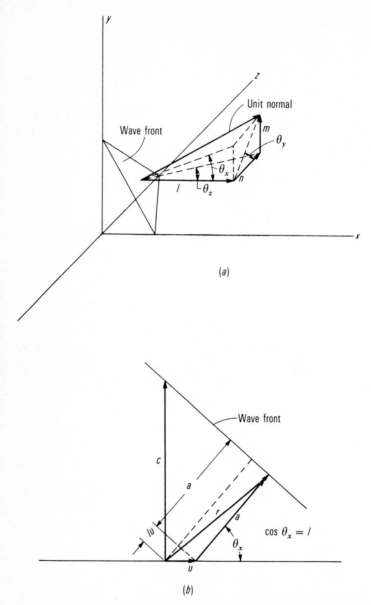

**Figure 9-14** (*a*) Direction cosine definitions for a three-dimensional orientation of the wave front. (*b*) Two-dimensional case. Unit normal is oriented in the direction of vector **a** but energy propagates in the direction of the vector **r**.

where a ray is defined as a curve whose direction at every point represents the direction of propagation of energy in a specific wave front passing through that point. Using the ZND detonation of Fig. 9-1 as a background flow for the acoustic treatment, we note that

$$a = a(x) \qquad u = u(x) \qquad v = w = 0 \qquad \frac{da}{dx} = \frac{da}{dx}(x) \qquad \frac{du}{dx} = \frac{du}{dx}(x)$$

while all other gradients are zero. Consider, therefore, a wave front whose orientation to the $x$ axis is given by the direction cosines $l$, $m$, and $n$ between its normal and the $x$, $y$, and $z$ directions as shown in Fig. 9-14a. Milne,[17] in 1921 first derived the ray equations for the general case. For the one-dimensional flow considered here these equations reduce to the set

$$\frac{dx}{dt} = la + u$$

$$\frac{dy}{dt} = ma \qquad\qquad (9\text{-}4)$$

$$\frac{dz}{dt} = na$$

and

$$\frac{1}{l}\left(\frac{dl}{dt} + \frac{da}{dx} + l\frac{du}{dx}\right) = \frac{1}{m}\frac{dm}{dt} = \frac{1}{n}\frac{dn}{dt} \qquad\qquad (9\text{-}5)$$

Equation (9-4) gives the direction of propagation at any instant and Eq. (9-5) gives the rate of change of direction of the normal to the front in terms of the gradients of the flow. We note that the second half of Eq. (9-5) reduces to

$$\frac{m}{n} = \text{const}$$

which means that the projection of any ray on a $y$, $z$ plane always yields a straight line. Thus, in complete generality, we may dispense with the direction cosine in the $z$ direction and look at ray propagation in the $x$, $y$ plane as shown in Fig. 9-14b. Since $n = 0$, we may write

$$m = (1 - l)^{1/2}$$

and the final parametric equations become

$$\frac{dx}{dt} = la + u \qquad\qquad (9\text{-}6)$$

$$\frac{dy}{dt} = (1 - l)^{1/2}a \qquad\qquad (9\text{-}7)$$

17. E. A. Milne, *Philos. Mag.*, 42, 96 (1921).

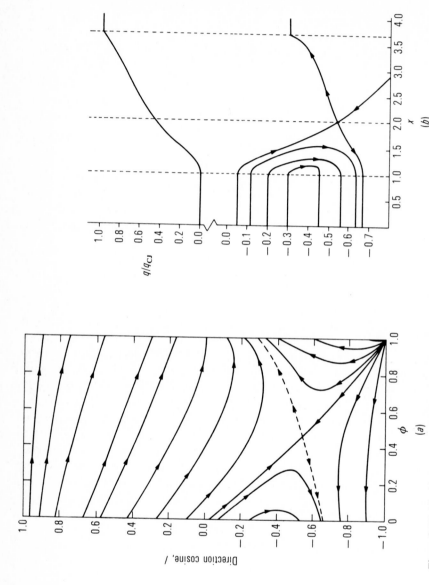

**Figure 9-15** $M = 6$, $\gamma = 1.4$ detonation of Fig. 9-1. (a) An $(l, \phi)$ plot of the ray behavior for a number of $c$ values. (b) An $(l, x)$ plot showing the mapping of the ray behavior into a physical plane for a specific heat-addition model. $x = 1$ at the end of the induction zone, the $\phi_{\parallel}$ plane occurs at $x = 2.04$ and the CJ plane occurs at $x = 3.70$. Arrows on the curves indicate direction of ray propagation.

and

$$\frac{dl}{dt} = (l^2 - 1)\left(\frac{da}{dx} + l\frac{du}{dx}\right)$$

(9-8)

However, since all the flow variables and gradients are functions of $x$ only, we may eliminate time by combining Eqs. (9-6) and (9-8) to obtain the equation

$$(la + u)\frac{dl}{dx} = (l^2 - 1)\left(\frac{da}{dx} + l\frac{du}{dx}\right)$$

(9-9)

Equation (9-9) has the solution

$$c = \frac{a + lu}{(1 - l^2)^{1/2}}$$

(9-10)

where $c$ is a constant along any ray as it propagates throughout the variable $u$, $a$ field. By inspection of the vector diagram, Fig. 9-14$b$, we note that geometrically $c$ is the apparent velocity of the wave front in the $y$ direction at fixed $x$ for any value of $u$, $a$ and $l$. We also see that Eq. (9-10) may be solved for $l$ to yield the equation

$$l_{\pm} = \frac{-au \pm c\sqrt{c^2 + u^2 - a^2}}{u^2 + c^2}$$

(9-11)

Thus at any position (and therefore fixed value of $u$ and $a$) there are either 0, 1, or 2 values of $l$, depending on the value of $c$. A plot of $l$ for a number of fixed values of $c$ as a function of $\phi$ or $x$ is shown in Fig. 9-15 for the detonation diagrammed in Fig. 9-1. The arrows on the lines of constant $c$ in this figure represent the behavior of the rays with time as given by Eqs. (9-6) and (9-8). The $(l, \phi)$ behavior is obtained by transforming the equations using Eq. (9-2). We note that a horizontal line in Fig. 9-15 represents a coherent wave front which initially has a fixed constant inclination in the flow field, while a vertical line in this figure represents a wave front which is initially a sound circle at a point in the flow. We also note that there are two singular points in this flow, one being the saddle point at the intersection of the $c$ equals constant curves, the other being at the point $(-1, 1)$ in the $(l, \phi)$ plot or $(-1, x_{CJ})$ in the $(l, x)$ plot.

We wish to investigate the nature of the saddle point in more detail. From Eq. (9-11) we note that for every value of $x$ there is a minimum value of $c = (a^2 - u^2)^{1/2}$ at which $l$ has but one value. At this point

$$l = -\frac{u}{a}$$

In other words, the ray of critical orientation has a direction cosine equal to the negative of the local Mach number of the flow. From Eq. (9-6) we see that for this orientation of the wave front and precisely at this point in the flow

$$\frac{dx}{dt} \equiv 0$$

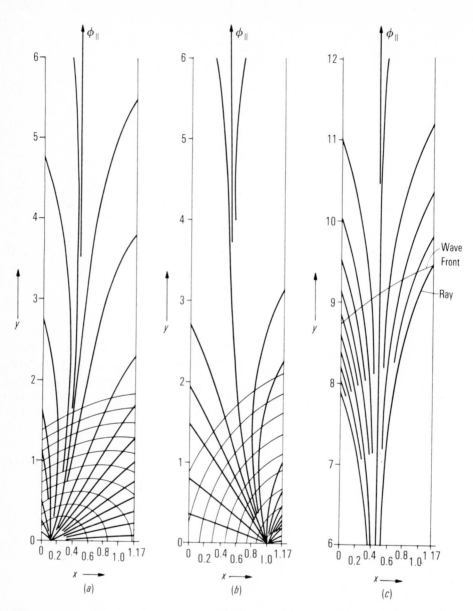

**Figure 9-16** The behavior of rays and wave fronts emanating from a sound source in the reactive region of a detonation with no induction zone. Reflection from the shock is not shown, ———— wave fronts, ———— rays. Figure (c) is the late behavior of the rays from either (a) or (b).

which means that the ray (or the energy trapped in the wave front) is propagating purely in the $y$ direction. Furthermore, substituting this unique value of $l$ into Eq. (9-8) yields the equation

$$\frac{dl}{dt} = \frac{(u^2 - a^2)}{2a^3} \frac{d}{dx} (a^2 - u^2)$$

which means that $dl/dt = 0$ at the point in the flow where

$$\frac{d}{dx} (a^2 - u^2) = 0 \qquad (9\text{-}12)$$

We note from Fig. 9-1 and from the nature of the equations that a one-dimensional detonation in which $Dq/Dt$ is everywhere positive has a plane at which the quantity $(a^2 - u^2)$ has a simple maximum and therefore that this detonation has a plane in the flow for which a wave front of orientation $l = -u/a$ will have $dl/dt = 0$ and $dx/dt = 0$. Therefore in this case a ray will propagate parallel to the shock for an infinitely long time without changing its orientation. The behavior of a number of the rays connected with a specific wave front emanating from a sound source in this type of flow field are shown in $x$, $y$ coordinates in Fig. 9-16.

Up until this point, even though we utilized the model of Fig. 9-1 to discuss the behavior of transverse waves, the resulting conclusions have been quite general and may be applied to any $u$, $a$ field constructed for a particular kinetic scheme in a gas with a complex thermal and caloric equation of state. However, in order to discuss the condition for the existence of a saddle point in the flow and the amplitude behavior of the transverse waves propagating in its neighborhood, we must now reduce mathematical complexity by simplifying the detonation model. We therefore return to the model of a constant-gamma gas with heat addition developed in Sec. 9-2 to obtain the ray equations

$$\frac{d\phi}{dt} = (l\mathscr{A} + \mathscr{U}) \frac{d\phi}{dx} a_1 \qquad (9\text{-}13)$$

and

$$\frac{dl}{dt} = (l^2 - 1) \left[ \frac{d\mathscr{A}}{d\phi} + l \frac{d\mathscr{U}}{d\phi} \right] \frac{d\phi}{dx} a_1 \qquad (9\text{-}14)$$

where $\mathscr{A} = a/a_1$ and $\mathscr{U} = u/a_1$. By solving for the conditions that $dl/dt = 0$ and $d\phi/dt = 0$, we find that the $\phi$ value for the plane of a ray propagating in the $y$ direction without curvature is

$$\phi_{\|} = 1 - \left[ \frac{1 + \gamma M^2}{2(M^2 - 1)} \right]^2$$

where $M$ is the approach flow Mach number, and that the orientation of such a ray is given by the relation

$$l_{\|} = -1/(2 + \gamma)^{1/2}$$

Thus we see that there is a lower Mach number for the occurrence of the plane $(d/dx)(a^2 - u^2) = 0$ in a constant-gamma detonation with strictly exothermic reactions and that this Mach number is always lower than $M = 3.0$.

Since a ray bundle which is properly oriented at the $\phi_{\parallel}$ plane remains at the $\phi_{\parallel}$ plane but is spreading with time its amplitude must be decreasing as time progresses because of this spreading. However, this acoustic signal is propagating in a reactive region which has other unique properties which also affect its amplitude behavior. If we assume that we can neglect the effects of crossflow, we find that we may discuss this behavior by following the behavior of an acoustic signal propagating in a truly homogeneous reactive flow with constant-background properties. When this is done the acoustic equation of Sec. 5-9 becomes

$$\frac{\partial^2 s}{\partial t^2} - a_{\parallel}^2 \frac{\partial^2 s}{\partial y^2} - k \frac{\partial s}{\partial t} = 0 \tag{9-15}$$

where $k$ is a constant that depends on the nature of the reactive flow and the degree of ray spreading at the $\phi_{\parallel}$ plane. The solution of Eq. (9-15) is of the form (see Sec. 5-9)

$$s = \exp\left(kt/2\right) \cdot C \exp\left[i\omega(y - a_{\parallel} t)\right] \tag{9-16}$$

Thus, if $k$ is a positive number, the wave bundle will increase its amplitude with time. It has been found that for all reasonable functional forms for the exothermic chemical reactions occurring at the $\phi_{\parallel}$ plane the net value of $k$ including its negative component due to ray-spreading is positive. Thus a simple ray-tracing theory, as well as Erpenbeck's more formal analysis shows that a one-dimensional ZND detonation wave is linearly unstable to any acoustic perturbation that occurs in the reaction zone. However we have still not justified the existence of the finite amplitude structure that is observed.

There are currently three theories of structural size which merit discussion. In the first place Barthel and Strehlow noted that the velocity of sound increases markedly as one travels into the reactive zone from the induction zone (see Fig. 9-17). This means that a wave which is initially a sound circle will distort and eventually form a strong cusp or caustic somewhere just beyond the end of the induction zone. This is illustrated for the ZND detonation of Fig. 9-1 in Fig. 9-17. In this figure the caustic was formed just prior to wave 400. Furthermore the occurrence of a caustic means that the acoustic energy is being focused and therefore that the local acoustic pressure is increased dramatically. Later Barthel[18] theorized that such a caustic would act as an acoustic source and send new acoustic signals in all directions. Thus a single acoustic perturbation will, after the detonation has traveled some distance, generate two new acoustic pulses at some fixed $\pm y$ location. Then, after another equal time interval, three new pulses will be generated, one of which will be at the original

18. H. O. Barthel, *Phys. Fluids*, **17**: 1547 (1974).

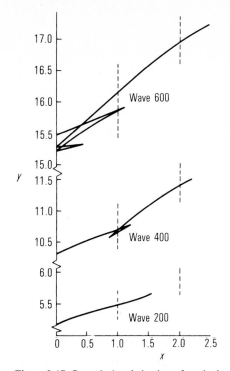

**Figure 9-17** Convolution behavior of a single wave front with time for the detonation structure shown in Fig. 9-1. Wave number refers to number of forward-step integrations performed for each wave drawn. At $t = 0$ the wave was assumed to be a sound circle located at the shock front at $y = 0$. *(Adapted from H. O. Barthel and R. A. Strehlow, Phys. Fluids, vol. 9, p. 1896 (1966), with permission.)*

pulse location. Since the actual transverse waves are relatively weak shock waves, he theorized that the calculated caustic spacing should be simply related to the observed transverse wave spacing. A plot of his comparison of spacings in the hydrogen–oxygen system, using the full kinetic scheme to calculate the ZND structure, is shown in Fig. 9-18.

The other two theories for structural size use the observed flow of a simple symmetric 2-D nonsteady Chapman–Jouguet detonation as the background flow. Strehlow[19] theorized that the extremely stable structure that is sometimes observed must be hydrodynamically stable to an acoustic perturbation. He analyzed the growth of an acoustic wave propagation across the middle of a cell and its subsequent decay when it meets the opposing reflected shock at the cell boundary. Using this theory he was able to determine the maximum cell size that would be stable to an imposed transverse acoustic perturbation. Barthel,[20] in another paper, was concerned with the mechanism that causes the transverse

19. R. A. Strehlow, *Astronaut. Acta*, **15**: 345 (1970).
20. H. O. Barthel, *Phys. Fluids*, **15**: 43 (1972).

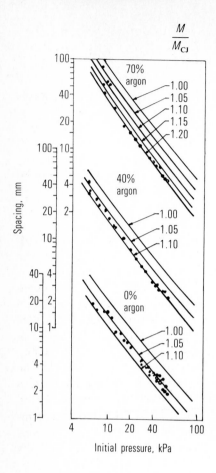

**Figure 9-18** Calculated transverse spacing using caustic theory[18] for the hydrogen–oxygen–argon system. The data points were reported by J. R. Biller, "An Investigation of Hydrogen–Oxygen–Argon Detonations" PhD. thesis UIUC, August 1973. Tech. Report AAE 73-5, UIUL-Eng-73-0505. *(With permission, from Ref. 18.)*

wave system to be sustained. He noted that the transverse waves must be continuously attenuating because of their curved nature in a real detonation and therefore that there must be some chemical mechanism for strengthening these waves if they are to exist for a long time. He noted that the intersection of two transverse waves produces a small triangular region (region *OCD* of Fig. 9-9) which is momentarily at a much higher pressure and temperature than any of the surrounding flow. He further postulated that this region "explodes" after some characteristic delay time and that this explosion sends a strong positive acoustic pulse toward the lead shock wave and the two reflected shocks that are traveling away from the apex of the cell. If this pulse reaches the triple point before the next intersection at the half-cell point it will reinforce the strength of the reflected shock. However, if this pulse does not reach the triple point before its next intersection the first transverse wave that it intersects will be traveling toward it. Under these circumstances that transverse wave will be positively decayed by the arrival of the pulse. Thus the second theory of Barthel predicts a minimum cell spacing that will remain stable.

This behavior has been verified experimentally. Two experiments were performed. In the first one a structurally stable $H_2$–$O_2$–Ar detonation was propagated through a diffusion interface into pure argon and the decay of the transverse wave structure was measured. The logarithmic decay of transverse wave strength was found to be 7 percent per cell length before the waves stopped writing on the smoked foil. In the second experiment the same structurally stable detonation was propagated through a diffusion interface into a mixture which would normally propagate a detonation with cells that were 50 percent larger. At first the spacings became somewhat irregular with no apparent change in strength. Eventually two waves of the same family became quite close together and when this happened the following wave completely disappeared within three cell lengths. Taki and Fujiwara[21] have also verified this behavior using hydrocode calculations.

Since the caustic theory, the acoustic maximum size theory, and the pressure pulse minimum size theory all predict different spacings, we can see why structural regularity is dependent on the hydrodynamic-reaction kinetics interaction, at least in a qualitative way. In stoichiometric hydrogen–oxygen detonations the structural regularity is excellent at high argon dilution and gets progressively poorer as the amount of argon in the mixture is reduced. Note from Fig. 9-18 that caustic theory predicts the correct cell size irrespective of regularity. However at high argon dilution the maximum cell size for instability is greater than the minimum cell size for destruction. Thus each wave in the system has an infinite lifetime and can couple strongly with the acoustic modes of the hot gas column downstream of the front. However at low argon dilution the reverse is true. Under these circumstances acoustic amplification is continuously generating new transverse waves, and the pressure pulses from previous intersections are continually destroying existing waves. Thus the lifetime of any specific transverse wave is finite, the waves can never excite the standing acoustic modes of the downstream column of hot gas, and the structure is irregular.

## 9-5 SUPERSONIC COMBUSTION AND STABILIZED DETONATIONS

The term *supersonic combustion* describes a process which is relatively new technologically. In a supersonic-combustion device the purpose is to utilize a combustion process to produce thrust without allowing the flow to become subsonic at any point in the engine. We choose to discuss the principles of this phenomenon in this chapter because of its similarity to the "reaction wave" combustion of a one-dimensional detonation.

21. S. Taki and T. Fujiwara, "Numerical Simulation of Triple Shock Behavior of Gaseous Detonation," *Eighteenth Symposium (International) on Combustion*, The Combustion Institute, Pittsburgh, Pa., 1671 (1981).

Let us again consider the Hugoniot and Rayleigh equations for one-dimensional flow with heat addition. As we showed earlier, these equations may be solved by eliminating the pressure ratio $y$ to yield Eq. (9-1) for the density ratio as a function of the extent of reaction and incident Mach number. These equations may also be solved for $y$ by eliminating $\varepsilon$ to yield an analogous equation for the pressure ratio between stations 1 and 2

$$y = \frac{1 + \gamma M^2 \pm \gamma(M^2 - 1)\sqrt{1 - \phi}}{(\gamma + 1)}$$

where $M$ is the Mach number of the approach flow. Once again we note that for $\phi = 0$ the solution obtained by using the negative sign is the trivial solution $y = 1$, while that obtained by using the positive sign is the pressure ratio for a simple shock transition from state 1 to state 2. We now wish to examine the flow properties of the solution which starts at the point $y = 1$ and continuously approaches the CJ point along the lower Rayleigh path in a $P, v$ plane as heat is being added. This corresponds to the classical "weak" detonation solution and represents a shockless transition in which the flow remains supersonic throughout.

We therefore see that strictly one-dimensional supersonic combustion would be obtained if we could trigger an exothermic reaction in a supersonic flow without the presence of a normal shock wave (which would simultaneously render the flow subsonic). One further requirement is that the initial Mach

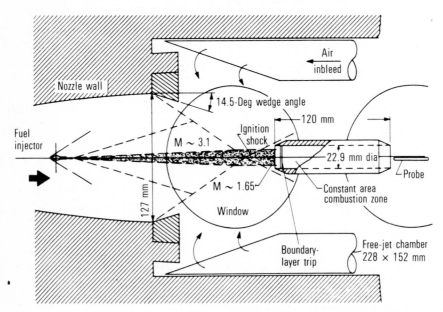

**Figure 9-19** Apparatus to produce supersonic combustion in a constant-area duct. Note the oblique ignition shock waves. *(USAF, Arnold Engineering Development Center, Tullahoma, Tenn.)*

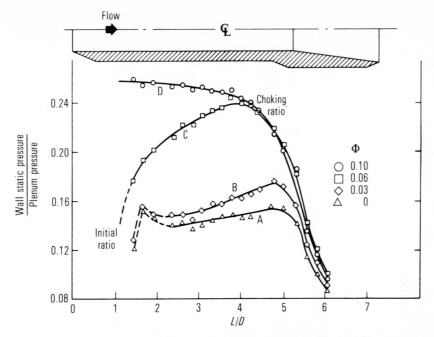

**Figure 9-20** Pressure records obtained in the apparatus shown in Fig. 9-16. Note the existence of a choking-equivalence ratio. Also note that for equivalence ratios below this critical value the pressure rise corresponds to the weak detonation solution and is a shockless transition to the final state, while above this critical value the pressure rise corresponds to the strong detonation solution and to an overdriven detonation. *(USAF, Arnold Engineering Development Center, Tullahoma, Tenn.)*

number be sufficiently high so that the full heat addition (by the chemical reaction) will not thermally choke the flow. In general, it has been observed that choking causes a shock to propagate upstream through the reaction zone and produce a detonation.

The real difficulty in producing supersonic combustion in a channel is that of providing the proper triggering mechanism. Two primary mechanisms of triggering have been found to be available, (1) an oblique shock, and (2) injection of fuel into a flowing stream whose static temperature is already well above the ignition temperature. The work of Rubins and coworkers[22] represents a good example of oblique-shock triggering in a one-dimensional steady flow. A drawing of their apparatus is shown in Fig. 9-19. In this experiment hydrogen is injected into the preheated air at a point where the static temperature is below the ignition temperature, and a pair of opposing oblique shocks are used to raise the temperature to the ignition temperature. To maintain one-dimensional flow the central portion of the flow (which contains a rather uniform hydrogen

22. P. M. Rubins and R. P. Rhodes, *AIAA J.*, **1**:2778 (1963); P. M. Rubins and J. C. Cunningham, "Shock-Induced Supersonic Combustion in a Constant-Area Duct," AEDC-TDR-64-266 (January 1965), AEDC Arnold Air Force Station, Tullahoma, Tenn.

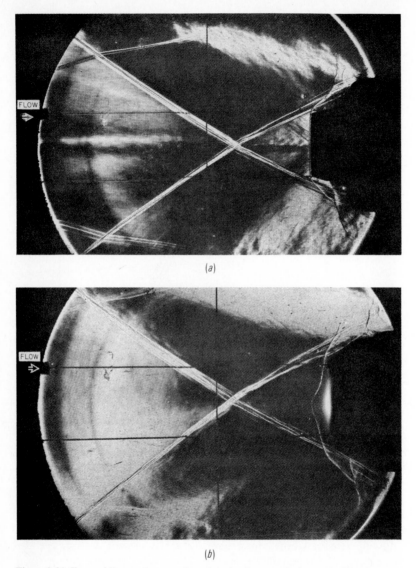

(a)

(b)

**Figure 9-21** Two schlieren photographs taken in the tunnel shown in Fig. 9-16. (a) Test section operating supersonic throughout. (b) Subsonic flow in test section. Notice the expelled normal shock wave. *(Courtesy P. Rubins, ARO Inc., Arnold Engineering Development Center, Tullahoma, Tenn.)*

concentration) is intercepted by a tube which prevents lateral expansion. Pressure records for a variety of hydrogen additions are shown in Fig. 9-20, and two photographs showing unchoked and choked flow in the tube are shown in Fig. 9-21.

The other process, that of local fuel injection in a high-temperature supersonic stream, is probably more desirable in an actual supersonic-combustion ramjet engine. In this case fuel is ejected at side ports and usually burns while mixing. An oblique shock is produced in this case and the flow is locally neither completely supersonic nor one-dimensional. However, as the flow remains primarily supersonic the pressure rise in the combustion chamber is kept to a minimum during heat addition.

In addition to the observation of true supersonic combustion, research in this area has led to the construction of "standing" detonation waves in supersonic streams. In this case the reaction is triggered by a normal shock wave in a flow containing fuel which has been injected upstream at a point where the static temperature is low. Three techniques for generating these normal shocks have been used. These are illustrated in Fig. 9-22. Both the normal shock producedin an axisymmetric shock bottle on an underexpanded jet (Fig. 9-22a) and the normal Mach-stem shock produced in a two-dimensional tunnel (Fig. 9-22b) have the disadvantage that lateral expansion of the flow is allowable downstream of the normal shock. Thus it is quite difficult to analyze these processes to determine or confirm kinetic data. However, Fig. 9-22c illustrates a technique

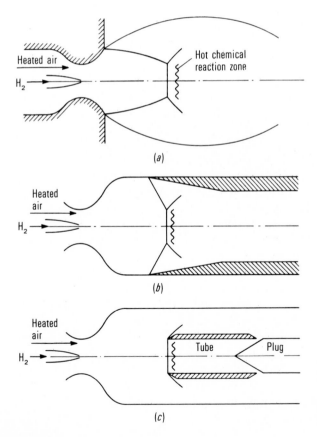

**Figure 9-22** Standing detonation experiments, (a) behind underexpanded shock bottle, (b) on Machstem of two oblique shocks, (c) behind normal shock held on lip of tube by a choking plug.

which allows one to reasonably maintain one-dimensional flow and simultaneously measure pressure at a series of stations in the reaction zone. A one-dimensional flow with no lateral expansion is obtained in this case by advancing a control plug sufficiently to just produce choked flow so that the shock "stands" on the lip of the test chamber. If properly designed this apparatus can maintain a steady detonation for long periods of time and therefore be used to conveniently determine pressure, temperature, and concentration profiles throughout the reaction zone.

In all the "standing detonation" studies illustrated in Fig. 9-22 there has never been any evidence of transverse phenomena occurring on the normal shock. All of these "detonations" are maintained by external means, however, and therefore represent overdriven waves. This fact, of course, simply negates the requirement for transverse waves but does not eliminate their possibility. It is quite probable, however, that their occurrence is suppressed because in all cases studied to date the active reaction zone has always had a considerably greater length than the width of the test apparatus.

## 9-6 SPRAY-DROPLET, WALL FILM, AND DUST DETONATIONS

It has been observed experimentally that sprays of even very low vapor pressure hydrocarbons in air can support a shock wave followed by combustion which has an overall steady velocity near the calculated Chapman–Jouguet velocity. Detailed observations of the interaction of a strong shock with a single droplet has shown that the high-velocity flow field that surrounds the droplet as it is being accelerated causes a fine mist to be stripped off the surface and to enter the wake behind the droplet. If the shock is strong enough the combustible mixture in the wake "explodes" and this explosion produces a shock wave which eventually catches the lead shock wave and reinforces it.

It has also been observed experimentally that a thin film of a very low vapor pressure hydrocarbon on the wall of a pipe can support a shock wave followed by combustion which has an overall steady velocity near the calculated Chapman–Jouguet velocity. The mechanism of dispersion in this case involves a Kelvin–Helmholtz instability of the surface of the film in the presence of a strong shear flow.

There is not nearly as much evidence that combustible dusts, when suspended in air, can support a detonation. However, the violence of certain grain elevator explosions indicates that some portions of the internal structure were loaded very rapidly during the explosion process and this can be taken as indirect evidence that combustible dusts will support a detonation wave if the conditions are correct. The effect of radiative transport of energy during a dust detonation is not known as yet, in contrast to our current understanding of its effects in dust flames.

# PROBLEMS

**9-1** Calculate the Hugoniot curve for the detonative decomposition of nitric oxide (NO) at 300 K. Assume that only nitrogen ($N_2$), oxygen ($O_2$), and nitric oxide (NO), are at equilibrium along the Hugoniot. Calculate the CJ detonation velocity for this process by determining the tangency condition for the Rayleigh line and Hugoniot curves. Note that this process is pressure-independent because the equilibrium composition is independent of pressure. (Use enthalpies and equilibrium constants from tables in App. B; see also Tables 4-2 and 4-3, and App. E2)

**9-2** As the Mach number of a CJ detonation decreases the $\phi_\parallel$ plane moves toward the shock wave. Calculate the Mach number for $\phi_\parallel = 0.0$ as a function of gamma for a constant-gamma heat-addition detonation.

**9-3** In the constant-gamma heat-addition model the weak detonation solution corresponds to a shockless flow transition during heat addition, while the strong solution corresponds to a relatively strong shock wave followed by heat addition. Show, for a slight overdrive ($\phi = 1 - \xi$ where $\xi \ll 1$), that the entropy rise is the same near the CJ point irrespective of the path (i.e., weak or strong) and that the first and second derivatives of the entropy with volume ratio $\varepsilon$ are continuous through the CJ point, even though the physical nature of the process changes in a discontinuous manner.

**9-4** Calculate the Mach-stem structure for a $\gamma = 1.4$ gas if $M_I = 5.0$ and $\beta = 50°$. Determine the pressure, temperature, entropy, shock Mach numbers, and the flow velocities and angles for all regions relative to the triple point $T$ if $P_1 = 100$ kPa and $T_1 = 300$ K.

**9-5** Assume that a detonation in a gas with $\gamma = 1.4$ has a Mach number of 7. Calculate the ZND structure of this detonation if the conversion rate is constant and has a value of $dq/dt = 10^{12}$ J/kg s. Determine $u$, $P$, and $T$ as a function of $x$ and plot your results.

**9-6** Acetylene is a highly exothermic compound. Calculate the CJ detonation velocity assuming that $H_2$ and C(s) graphite are the detonation products.

# BIBLIOGRAPHY

Fickett, W., and W. C. Davis, *Detonation*, University of California Press, Berkeley (1979).

Gruschka, H. K., and F. Wecken, *Gasdynamic Theory of Detonation*, Gordon and Breach, New York (1971).

Kirkwood, J. G., *Shock and Detonation Waves in Gases*, Gordon and Breach, New York (1967).

Soloukhin, R. I., *Shock Waves and Detonations in Gases*, Mono, Baltimore (1966).

Zel'dovich, Ia. B., and A. S. Kompaneets, *Theory of Detonation*, Academic Press, New York (1960).

# TEN

## COMBUSTION AERODYNAMICS— LAMINAR FLOW

## 10-1 INTRODUCTION

In most cases the shape of a flame and its propagation behavior are markedly influenced by external constraints such as the presence of walls, buoyancy forces, regions of shear, and nonsteady flow. In Chap. 8 we discussed the gross properties and internal structure of a very simple laminar premixed flame, under conditions where such aerodynamic forces do not play a dominant role. The primary reason for studying this simple one-dimensional flame first is to help the student distinguish between the coupled transport-chemical kinetic processes that are operative within a flame and the external forces that dictate the flame's propagation behavior.

We learned in Chap. 8 that a premixed flame propagates as a wave, and therefore propagates only in the direction normal to its orientation. We also learned that it propagates at a relatively low subsonic velocity, which is determined by a subtle interaction between transport and the rates of highly coupled chemical reactions that are overall exothermic. In this chapter we will investigate how certain aerodynamic constraints affect the propagation behavior of steady premixed laminar flames. In later chapters the more complex behavior of turbulent and nonsteady flames will also be discussed.

## 10-2 LAMINAR FLAME HOLDING

Premixed gas flames may be observed as steady waves in a variety of laminar-flow situations, and except for the case of the flat flame on a flat-flame burner, they will exist as a steady-wave phenomena only if the flow velocity of the main stream is well above the normal burning velocity of the mixture. This prevents the flame from propagating downstream into the apparatus. Thus all attached laminar flames are oriented obliquely to the flow, as indicated in Fig. 8-2 or Fig. 10-1. We see from these figures that even though a flame of this local geometry will appear steady to an observer, an element of this flame is, in reality, propagating along the flame at a velocity $S_\parallel$. Thus, as shown in Fig. 10-1, the flame can exist only as an apparently steady flame at some time $t_2$ at the point $a$, if at some earlier time $t_1$ it appeared steady at a point $b$ whose distance from point $a$ (along the flame in the upstream direction) is given by the expression $x_2 - x_1 = (t_2 - t_1)S_\parallel$. This implies that all steady flames must have an attachment region which continually reignites the oblique flame sheet, and that attachment must itself be steady if the flame is to appear steady.

From the vector relations in Fig. 10-1 we note that any holding region must necessarily always be at the farthest upstream point of the steady flame, and we further see that at this point the local flow-velocity vector and the local flame-velocity vector must be equal in magnitude, coincident in direction, and of opposite sign. Furthermore, as we travel away from the attachment point into the bulk of the combustible mixture, the local flow velocity must always exceed the local burning velocity. If for any particular geometry and flow velocity these conditions are not met, the flame will either "flash back," i.e., travel

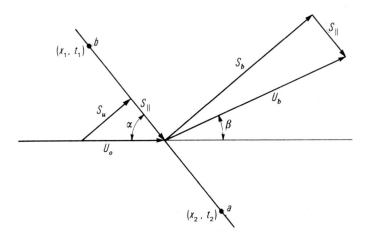

**Figure 10-1** Oblique flame orientation and flow-velocity relations if the flame is considered as an infinitely thin flat flame. (Compare with Fig. 8-2.)

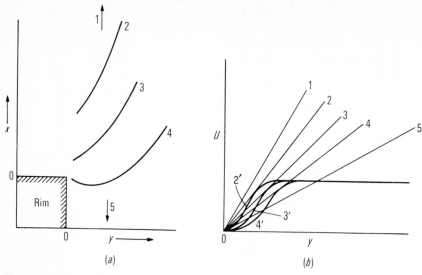

**Figure 10-2** Schematic diagram of holding behavior: (*a*) physical plane showing various stable flame positions; (*b*) $U$-$y$ plane showing the relationship between the local flow and flame velocity.

upstream, "blow off," i.e., be washed downstream by the flow, or exist as some type of nonsteady repetitive phenomena. All three of these nonsteady behaviors have been observed.

The mechanism of steady attachment may be discussed by considering the flow in the neighborhood of an idealized burner rim as illustrated in Fig. 10-2. Experimentally, it is observed that laminar flame attachment usually occurs within approximately 1 mm of the rim lip. The attachment point is therefore in the viscous boundary layer region for these low subsonic flows. For simplicity we assume that the velocity gradient of the flow near the attachment point is a constant that is a monotonically increasing function of the bulk-flow velocity. Typical velocity profiles just inside the rim are plotted as the straight lines 1 through 5 in Fig. 10-2*b*. The normal component of the burning velocity for three arbitrarily chosen flame positions and quenching behaviors are also plotted as the curves 2′, 3′, 4′. The differences in these curves are due to both the presence of the cold wall and the effect of diffusive mixing with the surrounding gas after the combustible jet issues from the tube. These are typical burning velocity curves for the flame positions shown in Fig. 10-2*a*. Curve 4′ represents the burning velocity for a laminar flame which is positioned entirely inside the rim of the burner. This curve is drawn tangent to the straight-line curve 4 which, therefore, represents a critical velocity gradient at the rim for flashback. Note that a lower velocity gradient, such as 5, allows the local burning velocity to exceed the local flow velocity and, therefore, will allow flashback to occur in the neighborhood of the rim. The velocity gradient represented by curve 3 will cause a flame in position 4 to blow off, but this flame will be stabilized in a new position, because, above the burner, the effect of rim quenching is diminished before the effect of diffusion causes excessive dilution of the peripheral combustible gases. Thus as the flow velocity is increased the flame will "seat" at progressively higher positions. However, its local burning velocity will eventually fall below the local flow velocity at all points in the flow and blowoff will occur. This blowoff limit behavior is illustrated with curve 2′ in Fig. 10-2.

This rather simple conceptual approach predicts that blowoff and flashback should correlate well with the velocity gradient in the neighborhood of the holder. This correlation has been verified experimentally and Fig. 10-3 is a typical summary of flashback and blowoff behavior (in this case for a natural gas flame) plotted against the velocity gradient $dU/dy$ with units of $s^{-1}$. Of these two critical behaviors, blowoff is the more complex. This is because at flashback the flame is always attached in the immediate neighborhood of the rim and its attachment is therefore controlled primarily by the nature of quenching at the rim or holding body. Blowoff, on the other hand, may occur when the flame is a considerable distance above the attachment body, and may therefore be controlled by factors other than simple quenching.

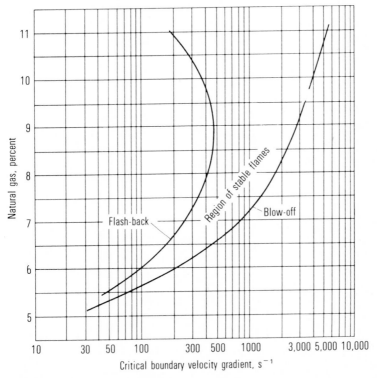

**Figure 10-3** Flashback and blowoff critical velocity gradient for a natural-gas–air flame. *(With permission from B. Lewis and G. von Elbe, Combustion Flame and Explosions in Gases, 2d ed., Academic, New York, 1961.)*

## 10-3 FLAME STRETCH

In 1953, Karlovitz and coworkers[1] introduced the concept of flame stretch. Flame stretch occurs when aerodynamic and boundary conditions are such that the area of an element of the flame front is caused to increase with time. Thus, the growing flame ball of the spherical flame of Sec. 8-2 is a stretched flame, while the steady curved flames of Sec. 8-4 are not. The magnitude of flame stretch is defined as the logarithmic rate of increase of flame area with time, with units of $s^{-1}$. Karlovitz nondimensionalized stretch by multiplying it by the characteristic time it takes the flame to burn through its preheat zone,

$$K = \frac{\partial(\ln \Delta A)}{\partial t} \frac{\eta_0}{S_u} \tag{10-1}$$

where $K$ is called the Karlovitz number. Laminar flame stretch occurs during many aerodynamic-flame interactions. One of the simplest of these interactions is the response of a flat oblique flame sheet to a one-dimensional gradient in the approach flow velocity, see Fig. 10-4.

In the geometry of Fig. 10-4 an element of the flame front is slipping along the flame front at the velocity $U_{\parallel} = S_u \tan \delta$. Since the geometry is two-dimensional the rate of area increase may be derived as follows. Consider two points on the flame front which are separated by a distance $\Delta s = s_2 - s_1$.

The area of a unit of depth $D$ of this flame section is

$$\Delta A = \Delta s D$$

and

$$\frac{\partial(\Delta A)}{\partial t} = D \frac{\partial \Delta s}{\partial t} = \left( (U_{\parallel})_0 + \frac{\partial U_{\parallel}}{\partial s} \Delta s - (U_{\parallel})_0 \right) D$$

Therefore in this case

$$\frac{\partial \ln (\Delta A)}{\partial t} = \left( \frac{\partial U_{\parallel}}{\partial s} \right)$$

or

$$K = \frac{\eta_0}{S_u} \left( \frac{\partial U_{\parallel}}{\partial s} \right) \tag{10-2}$$

Since $U_{\parallel} = U \sin \delta$ we obtain

$$K = \frac{\partial U}{\partial y} \frac{\eta_0}{U} \sin \delta + \frac{\partial U}{\partial y} \frac{\eta_0}{U} \cot \delta \cos \delta \tag{10-3}$$

or

$$= \frac{\partial U}{\partial y} \frac{\eta_0}{U} \frac{1}{\sin \delta} \tag{10-4}$$

1. B. Karlovitz, D. W. Denniston, D. H. Knappschaefer, and F. E. Wells, *Fourth Symposium (International) on Combustion*, The Combustion Institute, Pittsburgh, Pa., p. 613 (1953).

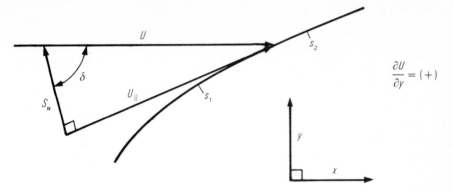

$$\frac{\partial U}{\partial y} = (+)$$

**Figure 10-4** Vector diagram for a stretched oblique flame in steady flow.

In another approach one can substitute $U_{\parallel} = S_u \tan \delta$ to obtain

$$K = \frac{\eta_0}{R} \sec^2 \delta$$

where $R$ is the local radius of curvature of the flame. The first term on the right-hand side of Eq. (10-3) is the result obtained by Karlovitz et al. However, we obtained Eq. (10-4) which, though markedly different for small $\delta$, approaches the same limit value for large $\delta$. The reason for this difference is that the original derivation did not recognize that the flame angle $\delta$ also changes as the oblique flame sheet slips along itself at the velocity $U_{\parallel}$.

We note that the Karlovitz number, as defined in Eq. (10-1), is actually a ratio of a chemical rate to a flow-induced rate of stretch. Thus, it is really the reciprocal of the first Damköhler number, and some critical maximum value of this number should represent the level of stretch which will just cause extinction of the flame as it propagates across the gradient region.

In the holding region formed by the wake of a thin flat plate, the flame is in a flow region that contains a large velocity gradient and is therefore stretched. At blowoff the flame is far enough away from the holder so that heat transfer to the holder is not important and the presence of two flame sheets means that the flow is symmetric and that there is no dilution of the mixture by entrained surrounding gases, as there is with a rim-held flame. Furthermore, as the bulk flow velocity is increased, flame stretch is increased. It has been found experimentally that under these conditions, the critical gradient for blowoff, when nondimensionalized with the flame's preheat zone thickness divided by the normal burning velocity, yields a dimensionless ratio whose value is approximately unity if the flat plate is thin.[2] While the fact that this ratio is constant does imply that stretch is operative at blowoff, the ratio is not a Karlovitz number. In other words, the quantity $U \sin \delta$ in Eq. (10-4) cannot be replaced

2. H. Edmondson and M. P. Heap, *Combust. Flame*, **15**:179 (1970).

by $S_u$. We note that the only place on an oblique flame sheet where this would be true is where $\delta$ approaches 45°, and that this implies that $U_{\parallel}$ equals $S_u$. It may be that for the circumstances of these particular experiments $U_{\parallel}$ was close to $S_u$, but this has not been proved. In other words, the mechanism of flame holding is not understood at the present time.

## 10-4 LAMINAR FLAME SHAPES

In this section we will consider the aerodynamics of attached or steady flames when the approach flow is laminar. Steady attached flames may be observed in a variety of laminar-flow situations. The multiple exposure of a series of Bunsen flames in Fig. 10-5 illustrates typical behavior for a flame stabilized on a cylindrical nozzle that produces a flow having a constant-velocity profile. This flame is burning at low pressure and is dominated by the tip and rim behavior. We have already discussed the attachment mechanism at the rim and now wish to discuss the aerodynamics of the tip. This is a region of negative stretch, in the Karlovitz sense, because the flame's radius of curvature is such that the streamtube area at the start of the preheat zone is less than at the start of the reaction zone. Fristrom[3] has shown that in well-behaved (i.e., noncellular) flames, the flow velocity profile through the hot-reaction zone of the flame is not affected by flame curvature for curvatures ordinarily encountered at the flame tip. However, as he indicates, this is not true for the preheat region, since it is always much thicker than the hot-reaction region (see Figs. 8-8 and 8-9). Thus in the typical enclosed flame front as produced on a Bunsen burner, divergence of the central stream tube as it passes through the preheat zone causes the flow velocity to lower until it matches the burning velocity locally at the start of the reaction zone.

If the molecular weight of the fuel and oxygen in the air are sufficiently different preferential diffusion can cause either an enhancement of the burning velocity or a localized extinction at the tip of the Bunsen cone. Law et al.[4] have observed this effect in rich methane–air and rich propane–air flames. Photographs of these flames are reproduced in Fig. 10-6. We note that in the tip regions preferential diffusion of the lighter methane toward the conical flame will shift the centerline equivalence ratio toward stoichiometric and thus enhance the burning velocity at the tip. It is evident that the tip luminosity is very high and the radius of curvature is relatively large in the flame of Fig. 10-6b and that this is due to this enhancement.

In the case of propane, just the opposite happens. Here the flame is extinguished at the tip when the concentration of propane in the approach flow is above 5.5 percent. Also note that extinction becomes so extensive at 6.4 percent

3. R. M. Fristrom, *Phys. Fluids*, **8**: 273 (1965).
4. C. K. Law, S. Ishizuka, and P. Cho, *Combust. Sci. Tech.*, **28**: 89–96 (1982).

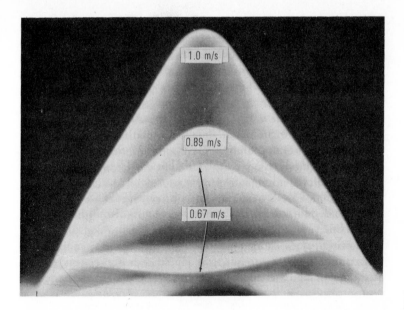

**Figure 10-5** Visible-light photograph of a low-pressure methane–air flame on a Mache–Hebra nozzle (constant velocity across duct) at various flow velocities, showing the effect of velocity on flame-front shape. *(Courtesy of R. M. Fristrom, Applied Physics Laboratory, Silver Spring, Md.; originally in Phys. Fluids, vol. 8, p. 273 (1965).) (With permission.)*

propane that a sooting diffusion flame replaces the normal blue premixed flame. This behavior is typical of all common hydrocarbons and in the gas appliance industry is called *yellow tipping.*

One can also support an attached flame on a thin plate placed in the flow so that it is oriented parallel to the flow. In this case, if the plate is thin enough, that is, equal to or less than $\eta_0$, the flame will be held by the boundary-layer flow over the plate and blowoff will be independent of the plate thickness. If the flame is steady it looks like a V. This flame markedly alters the flow field ahead of it because expansion of the gases downstream of the flame causes the streamlines ahead of the flame to diverge by a significant amount.

For low-approach velocities and near stoichiometric mixtures, with flat plate holders non-steady oscillatory flame holding is observed. This type of holding is characterized by flame oscillation between a forward and aft portion of the body. In a typical sequence, a flame seated near the aft end of the body will propagate toward the forward end through the developing boundary layer (from the previous cycle) and then increase its angle to the flow until it can no longer maintain its forward position. The entire flame is then washed downstream. After some time the flame again repropagates up the newly established boundary layer and the cycle is repeated. This nonsteady attachment can lead to rather violent oscillations of the entire oblique flame sheet if the circumstances are correct.

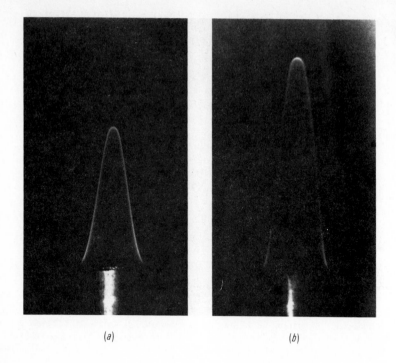

(a)          (b)

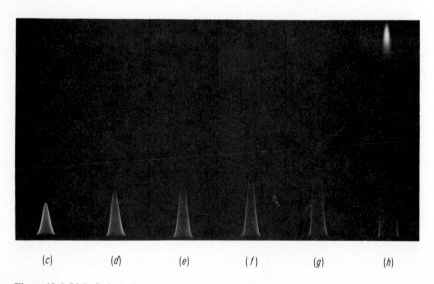

(c)        (d)        (e)        (f)        (g)        (h)

**Figure 10-6** Rich fuel–air burner flames illustrating enhancement and extinguishment at the tip. (a) 12.8% $CH_4$, (b) 14.9% $CH_4$, (c) 4.1% $C_3H_8$, (d) 5.3% $C_3H_8$, (e) 5.5% $C_3H_8$, (f) 5.7% $C_3H_8$, (g) 6.0% $C_3H_8$, and (h) 6.4% $C_3H_8$. (a) and (b) tube diameter = 10 mm, $v = 0.92$ m/s; (c) → (h) tube diameter = 12.7 mm; $U = 0.75$ m/s. *(Courtesy C. K. Law, Northwestern University, originals appear in Ref. 4.) (With permission.)*

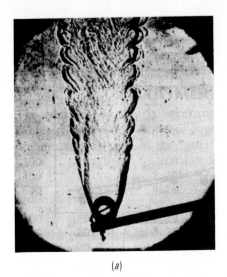 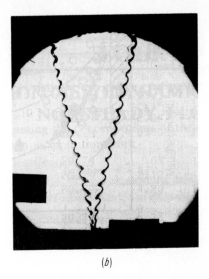

(*a*)　　　　　　　　　　　(*b*)

**Figure 10-7** Instabilities on a flame sheet. (*a*) Natural instabilities in a system showing cellular instability. Note the symmetrical nature of the fluctuations. (*b*) Instability in same mixture caused by holder vibrations. Note that the fluctuations on the opposing sheets are out of phase. (*Reproduced from H. E. Petersen and H. W. Emmons, Phys. Fluids, vol. 4, p. 456 (1961), with permission.*)

If an open attached flame in a strictly laminar approach flow appears "unstable" or nonsteady, the cause can be related to either one or both of two effects, cellular instability or nonsteady attachment. A quantitative assessment of the effect of these two sources of instability on the behavior of a particular unstable flame is usually quite difficult to make. However, a definitive study of this problem by Petersen and Emmons[5] has shed considerable light on the mechanism of instability growth, and the properties of these unstable flames. They studied V-flames in a laminar propane–air flow held on fine wires that could be *vibrated* normal to the flow direction with very small amplitudes at frequencies up to 1700 Hz. They were able to observe the effects of both natural instabilities as shown in Fig. 10-7a and "attachment" instabilities produced artificially by vibrating the wire, as shown in Fig. 10-7b. The instability patterns shown in Fig. 10-7 are traveling along the flame sheet away from the holder at approximately the velocity $S_{\parallel}$ which is defined in Fig. 10-1. Their work strikingly shows the importance of the concept that an element of the flame front always propagates normal to itself, and that flame-shape perturbations in the laminar stream are attached to the flame front and only secondarily affect or interact with the flow field. In addition, their results show that

1. Flame-front stability regions are intimately tied to the equivalence ratio of this propane–air flame, and driven instabilities can be made to grow outside

5. R. E. Petersen and H. W. Emmons, *Phys. Fluids*, **4**:456 (1961).

of the region of appearance of spontaneous instabilities. They also found equivalence ratio ranges where the flame refused to exhibit gross unstable behavior even though the holder was oscillated at relatively high amplitudes.

2. Both the spontaneous and driven instabilities exhibit a characteristic minimum wavelength or cell spacing which, when coupled with the flame's oblique angle, determine the upper limit on the wire frequency that will yield a growing flame-front perturbation. These driven instabilities were observed to have a slightly shorter characteristic wavelength than the spontaneously occurring instabilities.

3. The initial rate of growth of the instabilities was found to be exponential until an amplitude of about $0.3\,L$ was reached, where $L$ is the wavelength of the disturbance. This initial growth period is followed by a negative rate of growth due to cutoff of the deeply extended cusps of the flame by ordinary propagation. However, distortion of the flow field ahead of the flame, coupled with preferential diffusion effects, was found to prevent the return of the flame to simple planar geometry, and once they started to grow the cylindrical disturbances were always observed to equilibrate at an amplitude of about $0.2\,L$. Furthermore, these "steady" cusped flames always had their cusps oriented downstream, i.e., as in the spontaneous cellular flames of Sec. 10-6.

The authors also summarize data which show that the preferred wavelength of the disturbance in a flame varies with the equivalence ratio for any particular fuel–oxidizer mixture and is also dependent on the fuel type. They further point out that even in this simple geometry the two sides of the V-flame do not behave in a truly independent manner. Thus, as general as their results are, they are still quite definitely dependent on the gross flame geometry and cannot be easily extrapolated to other geometric situations.

We therefore see that an attached flame held in a laminar approach stream can exhibit steady behavior or a nonsteady cellular type of behavior. In addition, we find that gross aerodynamic interactions across the entire flame can sometimes determine the type of behavior that will occur. However, we also see quite clearly that all the observed flame-front instabilities are tied to the flame's previous surface shape by the requirement that each element of the flame front must propagate in a direction normal to its instantaneous orientation. We have thus incidentally introduced the concept of the "life" of an element of an attached flame as the time after leaving the attachment region for such an element to completely consume the gas available to itself through the normal propagation process. We note, in addition, that the overall effect of local flame-front curvature is quite complex, since it leads to the production of induced flow disturbances ahead of the flame, to preferential diffusion in the preheat zone of the flame and, possibly in certain circumstances, to "stretch" in the Karlovitz sense.

Another interesting steady flame is the flame that propagates from the

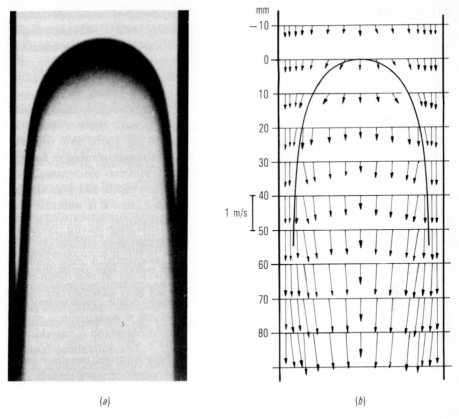

(a)                                                  (b)

**Figure 10-8** An upward propagating flame in a 51 mm diameter tube containing a 5.37 $CH_4$–air mixture. (a) A schlieren photograph of the flame. (b) The velocity vector field in flame-oriented coordinates.

bottom open end of a cylindrical tube toward the upper closed end. A schlieren photograph of such a flame in a 51 mm tube is shown in Fig. 10-8. In a fuel-lean mixture this flame is observed to propagate the entire length of the tube at constant velocity and with constant shape. Thus, in a coordinate system that is moving at the flame speed, the flame appears as a steady flame.

We note that the flow velocity at the flame's centerline is only about 50 mm/s and that this is the normal burning velocity of this flame. Nevertheless the entire flame structure is translating up the tube at about 230 mm/s.

This flame is being "held" at the centerline where it is oriented strictly normal to the flow velocity vector. At every other location in the tube the flame is oblique to the flow. This flame is a stretched flame even at the centerline because it has a finite radius of curvature there and the flow is necessarily divergent ahead of the flame.

# PROBLEMS

**10-1** Fully developed pipe flow has a parabolic velocity distribution. If the normal burning velocity of a fuel–air mixture is 0.45 m/s and the centerline flow velocity is 2.0 m/s in a 12 mm diameter smooth pipe, calculate the flame angle to the flow and the Karlovitz number at $r = 2$ and 4 mm. Assume that the flame is two-dimensional (i.e., neglect the radius of curvature of the flame due to its circular shape). How would this affect the Karlovitz number?

**10-2** A flame with a burning velocity of 0.45 m/s is held on the rim of a burner which has a flat velocity profile. Assume that the flame is infinitely thin and therefore has a conical shape. Determine the Karlovitz stretch factor due to the circular shape of the burner.

**10-3** What is the lifetime of an element of the flame of Prob. 10-2? Determine it as a function of burner diameter from 3 to 25 mm and flow velocity from 1.0 to 3.0 m/s. Cross-plot your results.

# BIBLIOGRAPHY

Beér, J. M., and N. A. Chigier, *Combustion Aerodynamics*, Halsted Press, Wiley, New York (1972).
Bartlma, F., *Gasdynamik der Verbrennung*, Springer-Verlag, Wein (1975).

# ELEVEN

## COMBUSTION AERODYNAMICS—
## TURBULENT FLOW

## 11-1 INTRODUCTION

In Chap. 5 we saw that the presence of highly temperature-dependent, highly exothermic reactions greatly increases the complexity of turbulent flows. In this chapter we will examine a few of the simpler turbulent flows that contain either a diffusion or a premixed flame. The treatment will be primarily descriptive, because the current theoretical treatment of these flows is not only outside the scope of this text, but is also quite controversial in many cases.

## 11-2 TURBULENT PREMIXED FLAMES—
## LOW-INTENSITY TURBULENCE

The most common type of turbulent premixed flame that has been studied in the laboratory is the rim-stabilized burner flame for which the approach flow is turbulent. Usually, this is accomplished by either placing a turbulence generating grid upstream of the rim or by generating fully developed turbulent pipe flow at Reynolds numbers above 2300. Figure 11-1 contains both long-duration visible light and very short duration schlieren photographs of rim-held premixed flames at three different approach-flow turbulence intensities. In these cases the turbulence was grid-generated isotropic turbulence. With flames of this type, the "turbulent burning velocity" is usually measured by determining the locus of

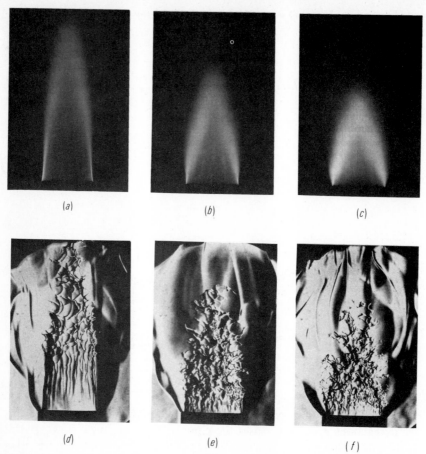

**Figure 11-1** (*a*) to (*c*) Direct long-exposure photographs and (*d*) to (*f*) short-duration schlieren photographs of turbulent premixed burner flames. Burner diameter is 54 mm. Propane–air, $\Phi = 1.1$. Flat velocity profile, $U = 4.5$ m/s. *(Courtesy H. Tsuji, University of Tokyo, Tokyo, Japan. Photographs originally appeared in Bulletin of the JSME, vol. 22, no. 167, p. 848 (1979).) (With permission.)*

maximum light intensity on a long exposure photograph. This is used to determine an "average" flame area, assuming cylindrical symmetry. Then this area is divided into the volumetric flow rate of the burner to yield a "turbulent burning velocity."

Unfortunately, turbulent burning velocities measured in this manner are apparatus-dependent. This effect, which was discovered by Damköhler in 1940, is shown in Fig. 11-2. We note from Fig. 11-2 that, irrespective of the fuel that is used, the flame stabilized on the larger burner has a larger turbulent burning velocity. This apparent increase of burning velocity with burner size is caused by the transient nature of the oblique flame sheets that make up the wave front. In this geometry each element of the flame front is held at the rim as a laminar oblique flame (see Fig. 11-1) and travels to the tip at the velocity $S_{\parallel}$. Thus, as can be seen in Fig. 11-1, the flame sheet is laminar near the holder and is responding to approach-flow turbulence as it propagates toward the tip. As Fig. 11-1 shows, when the intensity of the approach-flow turbulence is low the flame becomes a wrinkled laminar flame, but when the intensity is high small individual flamelets are formed in the neighborhood of the tip. Since each flame element has a longer lifetime on a larger burner, there is more time available for the flame to become wrinkled and, therefore, to exhibit a higher effective turbulent burning velocity.

The probability density function (pdf) of the temperature fluctuations in a premixed turbulent burner flame verify that the flame is simply a wrinkled

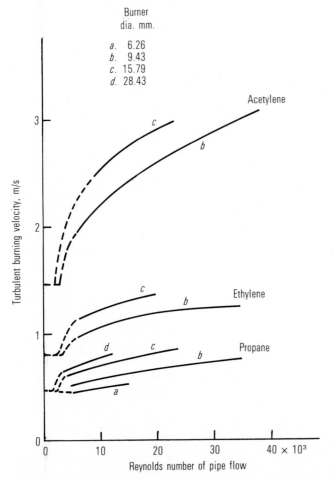

**Figure 11-2** Variation of the turbulent burning velocity with the pipe flow Reynolds number and the pipe diameter. *(Adapted from D. T. Williams and L. M. Bollinger, Third Symposium (International) on Combustion, Williams and Wilkins, Baltimore, 176 (1949). With permission from J. M. Beér and N. A. Chigier, Combustion Aerodynamics, Wiley, N.Y. (1972).)*

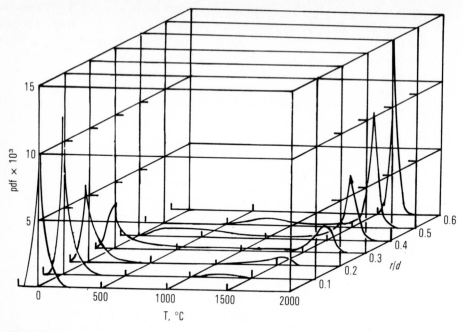

**Figure 11-3** Radial variations in the pdf of fluctuating temperature in a premixed turbulent flame. Burner diameter is 40 mm. Natural gas–air, $\Phi = 0.9$. Flat velocity profile, $U = 5.44$ m/s. Turbulence intensity is 6 percent and $x/d = 1.75$. *(With permission from A. Yoshida and R. Günther, Combust. Flame, vol. 38, p. 249 (1980).)*

laminar flame. Figure 11-3 shows that at a fixed elevation above the rim the pdf at the centerline is a delta function representing the cold-approach flow, and that as the radius is increased the pdf first becomes bimodal (i.e., contains two delta functions) and then becomes a single delta function representing the hot product gases. This is exactly what we would expect to see if a thin flame sheet that separated cold flow and hot products was fluctuating in position because of the approach-flow turbulence.

Photographs of turbulent burner flames show that the flame is responding to the turbulence by forming cellular type structures that are weakly convex toward the approach flow and have sharp cusps toward the product gases. Furthermore the characteristic size of these structures is obviously related to the thermodiffusive instability discussed in Sec. 8-6. Boyer, et al.[1] have used laser tomography to determine the spectral density of the $u$ component of the local flame velocity and laser doppler anemometry to measure the spectral density of the $u$ component of the turbulence just ahead of the flame. They used a special burner in which they could stabilize a nearly one-dimensional flame in a slowly

1. L. Boyer, P. Clavin, and F. Sabathier, "Dynamical Behavior of a Premixed Turbulent Flame Front," *Eighteenth Symposium (International) on Combustion*, The Combustion Institute, Pittsburgh, Pa., p. 1041 (1981).

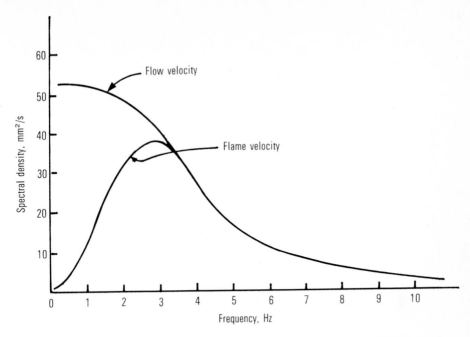

**Figure 11-4** Typical power spectra of the $U$ component of the flow velocity and flame velocity as determined in Ref. 1. *(With permission, from Ref. 2.)*

expanding duct. The smoothed spectra are shown in Fig. 11-4, taken from Clavin and Williams.[2] We note, as they did, that the flame responds to the turbulent fluctuations at high frequency but fails to follow the velocity fluctuations at frequencies that are below about 3 Hz. In other words the flame is acting as a high-pass filter. They point out that the thermodiffusive relaxation time of the cellular structure of a flame is of the order $D^2/\alpha$ where $D$ is the cellular wavelength and $\alpha$ is the thermal diffusivity of the gas. Since $D$ is of the order of 10 mm and $\alpha$ is of the order of 100 mm$^2$/s, the characteristic relaxation time is of the order of 1 s. Thus in the neighborhood of 1 Hz, the flame's stability response cannot be neglected and this reduces the response of the flame to the turbulent fluctuations. Furthermore their analysis, using large activation energy asymptotics, confirms this observation.

In other words, the high-frequency components of the turbulence do not really influence the flame structure. This observation is exactly equivalent to the observation of Petersen and Emmons (discussed in Sec. 10-4) that the flame would not wrinkle in response to the vibrating holder, when the frequency was high enough to produce wavelengths shorter than a characteristic length which was somewhat smaller than the normal spontaneous preferential diffusion cell length discussed in Sec. 8-6.

2. P. Clavin and F. A. Williams, *J. Fluid Mech.*, **116**:251 (1982).

## 11-3 TURBULENT JET-DIFFUSION FLAMES

Turbulent jet-diffusion flames exhibit a complex set of behaviors and their structure is markedly influenced by the geometry of the pipe delivering the flow to the jet, by the rim conditions, by the orientation of the pipe relative to the earth's gravity field, and by the nature of the surrounding boundary conditions. We will limit our discussion to vertical jets generated above the exit of long straight smooth pipes with or without a surrounding air flow and with or without a rim pilot flame.

We first look at the general holding behavior of flames attached to a burner rim when the gas fed to the jet is a fuel–air mixture. Figure 11-5 contains a schematic representation of the various holding behaviors that are observed. To the left we have the typical rim-held premixed flame discussed in Sec. 10-2. The dashed line shows the expected blowoff behavior if the surrounding atmosphere is inert. As we can see the presence of air around the jet markedly alters the holding behavior. In the first place one can hold a flame over the entire rich-fuel–air mixture range, and when the fuel concentration is above the inert-atmosphere rich-blowout limit, one can consider these flames to be diffusion flames. Secondly, we see that "lifted" flames are held at high jet velocity. In these flames the high-velocity turbulent jet mixes with the surrounding air and

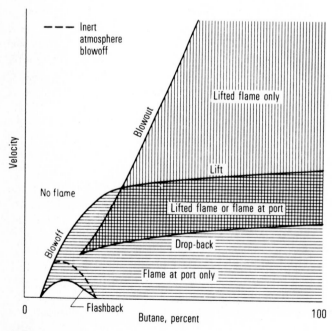

**Figure 11-5** Stability limits for attached and lifted flames for a port exhausting into air. *(With permission from K. Wohl, N. M. Kapp, and C. Gazley, Third Symposium (International) on Combustion, The Combustion Institute, Pittsburgh, Pa., 1949.)*

at some point above the burner rim the flame attaches itself to the flow and maintains itself in a broad region of turbulent mixed flow. The cross-hatched region of Fig. 11-5 is a region where hysteresis occurs. In other words, the type of flame that is actually stabilized depends on the previous history of the jet. As the flow velocity is increased from some low velocity, lift off will occur along the top edge of the cross-hatched region, then as the flow velocity to this lifted flame is decreased the flame will drop back to the rim at the lower edge of the cross-hatched region.

Lifted flames can remain attached at even very high approach-flow velocities. The most spectacular example of such a lifted flame is the disastrous flame which seats, after ignition, above a broken pipe leading to either a large high-pressure storage vessel or a high-pressure gas well. In these cases the nozzle flow is sonic and the region downstream may even contain supersonic flow, if the back pressure is sufficiently high. Nevertheless, a flame is still held and such a flame can be extinguished only by shutting off the flow of gas or by blowing out the flame by using a high-explosive charge to momentarily displace the holding region.

We will now concentrate on jets that contain no oxygen—just fuel and possibly an inert gas. As the jet velocity is increased from a very low value the behavior of the burning jet is controlled by a number of different dominant forces. For low-velocity jets the pipe flow will be laminar and the primary diffusion flame will also be laminar. For very low velocity jets the plume above the flame will be primarily driven by free convection. In other words, the momentum of the jet will be insufficient to influence the flow above the jet. A candle flame is a good example of such a flow. At higher velocities where the exit flow is still laminar the balance between inertial and gravity forces dictates the behavior of the plume. In other words the Froude number, which increases as $V^2$, is an important parameter in determining the jet length and structure. A typical example of how the length and structure of the diffusion flame depends on the approach flow is shown in Fig. 11-6. We note from Fig. 11-6 that for low flow velocity the main diffusion flame is laminar and the length increases markedly as the pipe-flow velocity is increased. At some point, however, the diffusion flame starts to become turbulent some distance from the port and, as the flow velocity is increased, the transition point approaches the burner rim while the total diffusion flame length decreases somewhat. Eventually in the fully turbulent region the gross diffusion flame behavior is almost independent of both the Froude number and the Reynolds number of the flow issuing from the port.

The nature of the transition process has been investigated in some detail for both hydrogen and acetylene jets issuing into a laminar coflowing air stream.[3] The location of the transition point was measured both for cold flow and with

3. T. Takeno and Y. Kotani, "Transition and Structure of Turbulent Diffusion Flames," *Turbulent Combustion*, ed. L. A. Kennedy, *Prog. Astronaut. Aeronaut.*, AIAA, New York, **58**:19 (1978).

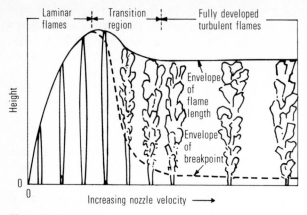

**Figure 11-6** Progressive change in flame type with increase in nozzle velocity *(With permission from H. C. Hottle and W. R. Hawthorne, Diffusion in Laminar Flame Jets, Third Symposium (International) on Combustion, The Combustion Institute, Pittsburgh, Pa., p. 254 (1949)).*

a diffusion flame present. Figure 11-7 shows how the measured transition distances depend on the Reynolds number of the fully developed approach flow in the central pipe. We note that cold jets make the transition to turbulent plume behavior well before the hot diffusion flames do. Also by the time the approach-flow Reynolds number reaches 2300 the transition length becomes independent of the type of fuel or the presence of the flame.

The authors of Ref. 3 point out that the most probable reason why transition is delayed in the hot jets is the fact that the kinematic viscosity at the flame temperature is larger by a factor of about 5 for $H_2$ and 50 for $C_2H_4$ when compared with the cold-gas kinematic viscosity. In other words the effective Reynolds number of the hot gas jet is much smaller than that of the cold gas jet, and this delays transition.

This has been verified by Takagi et al.[4] They studied a vertical $H_2$–$N_2$ jet issuing from a straight long pipe with fully developed turbulent flow at Reynolds numbers of 4200 and 11,000. Specifically they compared the turbulence levels and the stable species average concentrations for the cold jet to the turbulent diffusion flames. They found that at Re = 11,000 the maximum value of $\sqrt{u^2}$ decreased from 8.5 to 6.8 m/s, and the location of the maximum increased from 10 to 32 diameters. Radial profiles showed that turbulence levels at the average location of the flame sheet were higher when the flame was present than when the flame was absent, i.e., for a cold flow. However, the diffusion flame appeared to have about the same diameter as the cold jet even though the axial concentration profiles showed that the diffusion flame was longer, undoubtedly due to expansion caused by the combustion process.

Another interesting feature of the turbulent-jet diffusion flame is that they are in a sense similar to turbulent rim-held premixed flames, because they too

4. T. Takagi, H. Shin, and A. Ishio, *Combust. Flame*, **37**: 163 (1980); **40**: 121 (1981).

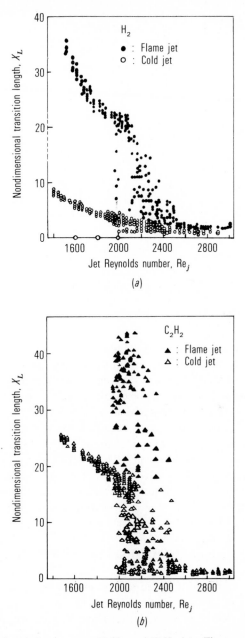

**Figure 11-7** Transition length for hot or cold $H_2$ and $C_2H_2$ jets. The transition length has been nondimensionalized using the inner pipe diameter. *(With permission from Ref. 3.)*

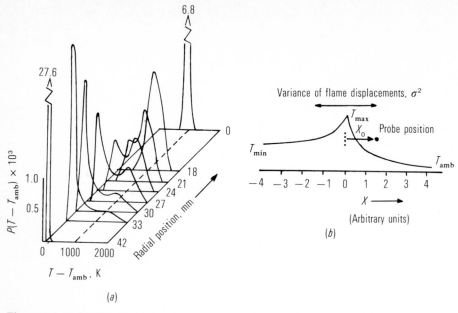

**Figure 11-8** Probability density functions at various radial positions at $x/d = 40$ for a pilot-held methane diffusion flame. $d = 6$ mm, Re = 9200. (*a*) The pdf's. (*b*) The temperature profile whose fluctuating position would cause the distributions observed in (*a*). (*With permission from Ref. 5.*)

can be described as having wrinkled flame sheet structures. This was pointed out by Roberts and Moss.[5] These authors used compensated thermocouples to measure the fluctuating temperatures in a turbulent diffusion flame. Figure 11-8 shows the radial pdf's that the authors measured. We note that at $r = 0$ the pdf is essentially a delta function centered about $T - T_{amb} = 1000$ K and that the distribution becomes bimodal as the radius increases and eventually at large radius becomes a single delta function at $T = T_{amb}$. We also note that the bimodal profiles are not as sharp as in Fig. 11-3 because a diffusion flame is always much thicker than a premixed flame. Nevertheless, there is no question but that this flame can be interpreted as being a wrinkled laminar diffusion flame. The inferred structure, at one instant of time, is shown in Fig. 11-8.

## 11-4 RECIRCULATION HOLDING AND SWIRL FLOW

In order to hold a premixed flame in a high-speed and, usually, highly turbulent flow, one must either use a small hot pilot flame or rely on recirculation holding. There are a number of ways that one can produce a recirculation zone

5. P. T. Roberts and J. B. Moss, "A Wrinkled Flame Interpretation of the Open Turbulent Diffusion Flame," *Eighteenth Symposium (International) on Combustion*, The Combustion Institute, Pittsburgh, Pa., p. 941 (1981).

in a duct flow to provide flame holding. Four such techniques are illustrated schematically in Figure 11-9. In the first two examples in that figure, flow separation is induced by the presence of a bluff body in the flow. Separation can also be induced by a sudden expansion of the duct area and by physically introducing a region of backflow using an opposed flow jet. A photograph of a flame held by the recirculation zone of a rod is shown in Fig. 11-10. Note that while the flame is visible the gases in the recirculation zone are not. Measurements of temperature in the recirculation zone show that the average temperature is about 0.9 of the theoretical flame temperature. The implication is that the recirculation zone is acting in a manner similar to that of a well-stirred reactor and that blowoff of the flame occurs for the same reasons that reactor blowout occurs. Since the well-stirred reactor has a loading at blowout of

$$\frac{J}{V P^n} \qquad \text{moles/s m}^3 \text{ Pa}^n$$

where $n$ is the overall kinetic order of the reaction, we may determine the critical loading of a cylindrical baffle by noting that the recirculation volume is proportional to $d^3$, where $d$ is the diameter of the baffle. Thus

$$\frac{J}{V P^n} \propto \frac{J}{d^3 P^n}$$

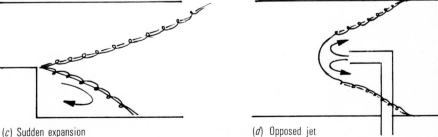

(*a*) Vee gutter

(*b*) Rod or sphere

(*c*) Sudden expansion

(*d*) Opposed jet

**Figure 11-9** Examples of different types of recirculation zone generators.

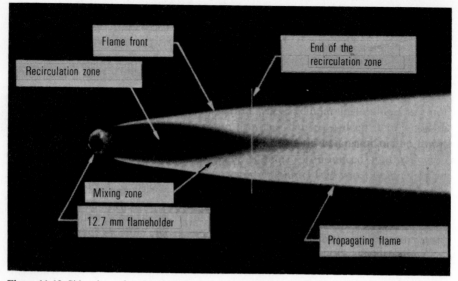

**Figure 11-10** Side view of rod-stabilized flame in gasoline–air mixture. *(Reproduced from E. E. Zukoski and F. E. Marble, Combustion Researches and Reviews, AGARD, Butterworth, London, 1955, with permission.)*

however $J/d^2 = \text{moles/m}^2 - \text{s} = \mathbf{V}[S]$ where $\mathbf{V}$ is the flow velocity at the baffle (including the effect of blockage in the duct) and $[S]$ is the total species concentration in moles per cubic meter. This can be written as

$$\frac{\mathbf{V}}{dP^{n-1}}\left(\frac{[S]}{P}\right) = \frac{\mathbf{V}}{dP^{n-1}}\left(\frac{1}{RT}\right)$$

Since the flame temperature for any fuel at any specific equivalence ratio is essentially a constant we obtain the final correlation for blowoff

$$\frac{\mathbf{V}}{dP^{n-1}} = \text{const} \tag{11-1}$$

It has been found that data for a specific holder shape is well correlated by Eq. (11-1) and that the effect of equivalence ratio is quite similar to that exhibited in Fig. 6-13 for a well-stirred reactor. It has also been observed that at incipient blowoff the overall combustion efficiency decreases. Thus it appears that recirculation holding in high-intensity turbulent flows can be modeled as a well-stirred reactor followed by a plug-flow reactor where the reactions go to completion.

The combustion of intensely turbulent fuel jets can also be modeled as though they contain a well-stirred region followed by a region which behaves like a plug-flow reactor. It has also been observed that the intensity of combustion in the well-stirred region can be enhanced by inducing swirl in the jet.

Three principal techniques have been used to generate swirl in a jet. These are:

1. Tangential entry of the fluid into the cylindrical duct that generates the jet
2. The use of guide vanes in an axial flow
3. Rotation of mechanical devices such as rotating vanes and grids or even the rotation of the supply tube itself

The swirling jet that is produced by such devices can be characterized by defining a swirl number. This is simply the ratio of the axial flux of angular momentum to the flux of axial momentum.[6] The two fluxes of interest are the angular momentum flux

$$G_\phi = \int_0^{R_0} (v_\phi r)\rho u 2\pi r \, dr = \text{const} \tag{11-2}$$

$$G_x = \int_0^{R_0} u\rho u 2\pi r \, dr + \int_0^{R_0} P 2\pi r \, dr = \text{const} \tag{11-3}$$

The swirl number is defined as

$$\mathscr{S} = \frac{G_\phi}{G_x R_0} \tag{11-4}$$

where $R_0$ is taken to be the exit radius of the burner nozzle. The swirl number as defined cannot be determined without an extensive survey of the velocity and pressure field in the jet. To a good first approximation, the swirl number can be calculated by knowing only the input velocity distribution in the swirl generator and by dropping the contribution of the axial pressure distribution term in Eq. (11-3)

$$\mathscr{S}' = \frac{G_\phi}{G_x' R_0}$$

$$G_x' = 2\pi \int_0^{R_0} \rho u^2 r \, dr$$

It has been observed experimentally that when the swirl number is less than 0.6 the axial pressure gradients induced by swirl are too small to cause internal recirculation along the axis of the jet, and in this case increased swirl simply increases the intensity of combustion in the well-stirred portion of the jet and reduces the residence time there. When the swirl number is increased to above 0.6 the swirling flow induces a recirculation zone along the centerline and the jet radius increases. This causes the residence time in the well-stirred portion to increase with increased swirl. This behavior is shown in Fig. 11-11. It has been observed that burners with swirl numbers of about 0.6 exhibit high-combustion

---

6. J. M. Beér and N. A. Chigier, *Combustion Aerodynamics*, Wiley, N.Y., p. 106 (1972).

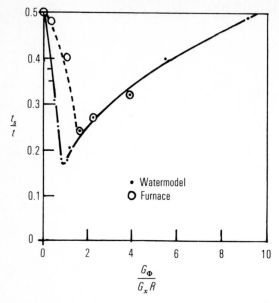

**Figure 11-11** Residence time in a stirred reactor as a fraction of total residence time as a function of swirl number. *(With permission from Ref. 6.)*

intensity and more efficient combustion. The reader is referred to the bibliography for further information on recirculation holding and swirl flames including spray droplet combustion in a highly turbulent environment.

## PROBLEM

**11-1** Use the data from Fig. 6-13 for stoichiometric octane–air to determine the blowout velocity for a flame held on a 20 mm diameter sphere at 100 kPa pressure.

## BIBLIOGRAPHY

**General**

Beér, J. M., "Combustion Aerodynamics," *Combustion Technology*, eds. H. B. Palmer and J. M. Beér, 61–89 Academic Press, New York (1974).
Gunther, R., "Turbulence Properties of Flames and Their Measurement" *Prog. Energy Combust. Sci.*, **9**: 105–154 (1983).
Libby, P. A., N. Chigier, and J. C. LaRue, "Conditional Sampling in Turbulent Combustion," *Prog. Energy Combust, Sci.*, **8**: 203–232 (1982).
Remenyi, K., *Combustion Stability*, Akademiai Kiado, Budapest (1980).
Sivashinsky, G. I., "Instabilities, Pattern Formation, and Turbulence in Flames," *Ann. Rev. Fluid Mech.*, **15**: 179–199 (1983).
Strahle, W. C., "Duality, Dilatation, Diffusion and Dissipation in Reacting Turbulent Flows," *Nineteenth Symposium (International) on Combustion*, 337–348 (1983).

Swithenbank, J., "Flame Stabilization in High Velocity Flow," *Combustion Technology*, eds. H. B. Palmer and J. M. Beér, 91–125 Academic Press, New York (1974).
"Turbulent Reactive Flows: Special Issue," *Combust. Sci. Tech.*, **13** (1976).

**Continuous combustors**

Edelman, R. B., and P. T. Harsha, "Laminar and Turbulent Gas Dynamics in Combustors—Current Status," *Prog. Energy Combust. Sci.*, **4**: 1–62 (1978).
Goulard, R., A. M. Mellor, and R. W. Bilger, "Air Breathing Propulsion—A Review," *Combust. Sci. Tech.*, **14**: 195 (1976).
Herbert, M. V., "Aerodynamic Influences on Flame Stability," *Progress in Combustion and Fuel Technology*, **1**, eds. J. Ducarme, M. Gerstein, and A. H. Lefebvre, Pergamon, New York, 145–182 (1960).
Khalil, E. E., *Modelling of Furnaces and Combustors*, Abacus Press, Tunbridge Wells, Kent, U.K. (1982).
Lefebvre, A. H., *Gas Turbine Combustor Design Problems*, Hemisphere, McGraw-Hill, New York (1980).
Mellor, A. M., "Semi-empirical Correlations for Gas Turbine Emissions, Ignition, and Flame Stabilization," *Prog. Energy Combust. Sci.*, **6**: 347–358 (1980).
Norster, E. R., ed., *Combustion and Heat Transfer in Gas Turbine Systems*, Pergamon, New York (1971).
Pratt, D. T., "Mixing and Chemical Reaction in Continuous Combustion," *Prog. Energy Combust. Sci.*, **1**: 73–86 (1975).

**Diffusion flames**

Bilger, R. W., "Turbulent Diffusion Flames," *Prog. Energy Combust Sci.*, **1**: 87–109 (1975).
Eickhoff, H., "Turbulent Hydrocarbon Jet Flames," *Prog. Energy Combust. Sci.*, **8**: 159–169 (1982).

**Premixed flames**

Chomiak, J. "Basic Considerations in the Turbulent Flame Propagation in Premixed Gases," *Prog. Energy Combust. Sci.*, **5**: 207–221 (1979).

**Spray combustion**

Chigier, N. A., "The Atomization and Burning of Liquid Fuel Sprays," *Prog. Energy Combust. Sci.*, **2**: 97–114 (1976).
Elkotb, M. M., "Fuel Atomization for Spray Modelling," *Prog. Energy Combust. Sci.*, **8**: 61–91 (1982).
Faeth, G. M., "Current Status of Droplet and Liquid Combustion," *Prog. Energy Combust. Sci.*, **3**: 191–224 (1977).
———, "Evaporation and Combustion of Sprays," *Prog. Energy Combust. Sci.*, **9**: 1–76 (1983).
Lefebvre, A. H., "Air Blast Atomization," *Prog. Energy Combust. Sci.*, **6**: 233–261 (1980).
Sirignano, W. A., "Fuel Vaporization and Spray Combustion Theory," *Prog. Energy Combust. Sci.*, **10**, in press (1984).

**Swirl flows**

Beér, J. M., and N. A. Chigier, *Combustion Aerodynamics*, Halsted Press, Wiley, New York (1972).
Lilley, D. G., "Swirl Flows in Combustion: A Review," *AIAA J.*, **15**: 1063–1079 (1977).

# TWELVE

## FLAME IGNITION AND EXTINCTION

### 12-1 INTRODUCTION

In Chaps. 7, 8, and 10 we discussed laminar diffusion flames, laminar premixed flames, and certain of their more important interactions with flow aerodynamics. However no mention was made of how such flames are ignited and thereby caused to propagate or how such flames can be extinguished. Also, even though they were alluded to in Chap. 8, the factors which influence the range of mixture composition within which a diffusion flame can be stabilized or a premixed flame is found to spontaneously propagate was not discussed. Both ignition and extinction at the limit are transient, i.e., time-dependent phenomena and as such deserve separate treatment. They will be discussed in this chapter.

### 12-2 EXTINCTION OF DIFFUSION FLAMES[1]

Early studies using the type I counterflow diffusion flame of Fig. 7-6 showed that as the flow rate is increased a point is reached where the flame "blows out," i.e., can no longer be stabilized. This mass-flow rate was called the *flame strength*. Furthermore, it was found that in most systems the flame strength can be directly correlated to the "strength" of a stoichiometric mixture of the same fuel in the same oxidizer mixture where that strength is defined as $\rho_u y_{fu} S_u$ kilograms per square meter-second. This correlation is shown in Fig. 12-1. This is

---

1. This section contains a summary of a portion of H. Tsuji, "Counter Flow Diffusion Flames," *Prog. Energy Combust. Sci.*, **8**: 93–119 (1982).

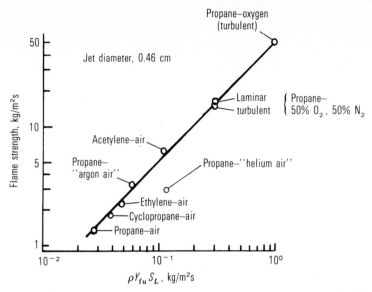

**Figure 12-1** Correlation between the "flame strength" of an opposed jet diffusion flame and the "strength" of a stoichiometric premixed flame. *(With permission from Ref. 1.)*

not surprising because a diffusion flame always has its most rapid reaction rate in the region where the mixture is near stoichiometric and the counterflow diffusion flame is stretched as the flow velocities are increased because this decreases the residence time of the gases in the flame zone.

The pressure dependency of the flame strength has also been determined and is plotted for one flame system in Fig. 12-2. The fact that this log plot has a slope of two supports the contention that flame strength is determined by a competition between residence time and a characteristic reaction time because the overall order of the flame reactions is known to be two.

More recently the phenomenon of counterflow diffusion flame extinction has been extensively studied using a burner of type IV in Fig. 7-6. The flame stability diagram for three typical fuels in air is shown in Fig. 12-3 where a dimensionless fuel injection rate $(V_{fu}/V_z) \cdot (Re/2)^{1/2}$ is plotted against the stagnation velocity gradient $2V_z/R_0$. Here $V_{fu}$ is the fuel injection velocity at the surface of the cylinder, $V_z$ is the approach flow velocity, $R_0$ is the radius of the cylinder, and $Re = V_z R_0/v$ where $v$ is the average kinematic viscosity in the flame. The lines plotted through the open points represent extinction conditions while those plotted through the solid points represent the condition of incipient soot formation and, as a consequence, the appearance of a luminous yellow zone on the fuel side of the flame. The extinction limits that are observed at low fuel injection rates are caused by excessive heat transfer back to the porous cylinder (which is the fuel source) and therefore are not very interesting. However the blowoff limit at high fuel blowing rate is observed to be only a

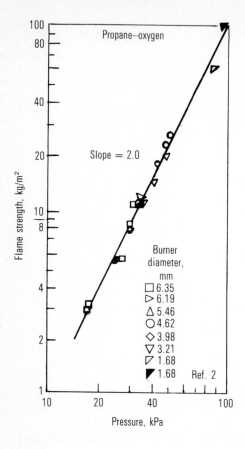

**Figure 12-2** Effect of pressure on the "flame strength" of an opposed jet diffusion flame. *(With permission from Ref. 1.)*

function of the stagnation flow velocity gradient $2V_z/R_0$. This critical gradient is equivalent to the critical flame strength described earlier in this section.

If one dilutes either the air or the fuel with an inert, one finds an absolute limit beyond which no flame can be stabilized irrespective of the fuel blowing rate or stagnation velocity gradient. In the case when air is diluted with nitrogen this critical oxygen concentration is defined as the ratio $[O_2]/([O_2] + [N_2])$ and is called the *limiting oxygen index* of the fuel. In cases where either the fuel is diluted or other inerts are used the limit is called the limiting fuel concentration limit $[F]/([F] + [I])$ or limiting oxygen concentration limit $[O_2]/([O_2] + [I])$.

The limiting oxygen index is important in fire safety. A few limiting fuel concentration and limiting oxygen concentration limits, and flame temperatures measured at the limits are presented in Table 12-1. Note that for all methane flames and for hydrogen in the artificial atmosphere in which helium is the inert, the measured flame temperatures agree at the limit for each limit fuel concentration and limit oxygen concentration pair. Also note that for all "stable" extinguishments the temperatures increase as the inert diffusivity

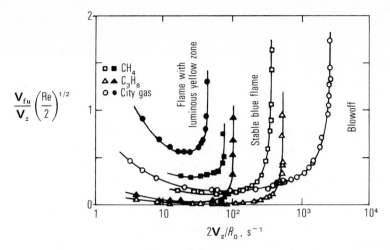

**Figure 12-3** Blowoff and incipient soot formation regimes for a type IV burner. *(With permission from Ref. 1.)*

increases. The constant temperature for any pair indicates that the absolute extinction limits are primarily controlled by kinetics. The variation of temperature with the inert species indicates that diffusion is also contributing to the limit behavior. The data also show that the very significant effects of preferential diffusion that are usually present when hydrogen is a fuel can be suppressed if a very light species such as helium is used as an inert.

Chung et al.[2] have analyzed the counterflow diffusion flame in the neighborhood of the stagnation point streamline using simple single-step Arrhenius kinetics. They were the first authors to define a local first Damköhler number

2. P. M. Chung, F. E. Fendell, and J. F. Holt, *AIAA J.*, **4**: 1020 (1966).

**Table 12-1 Limiting fuel and oxygen concentrations for methane and hydrogen using various inert diluents[†]**

| | Methane | | | | Hydrogen | | | |
| | Added to fuel | | Added to oxygen | | Added to fuel | | Added to oxygen | |
| Inert | $\frac{[CH_4]}{[CH_4]+[I]}$ | $T_f$, K | $\frac{[O_2]}{[O_2]+[I]}$ | $T_f$, K | $\frac{[H_2]}{[H_2]+[I]}$ | $T_f$, K | $\frac{[O_2]}{[O_2]+[I]}$ | $T_f$, K |
|---|---|---|---|---|---|---|---|---|
| Nitrogen | 0.165 | 1460 | 0.143 | 1480 | 0.114[‡] | 1120[‡] | 0.052 | 1010 |
| Argon | 0.076 | 1430 | 0.096 | 1440 | 0.081[‡] | 1110[‡] | 0.036 | 990 |
| Helium | 0.320 | 1610 | 0.149 | 1620 | 0.119 | 1080 | 0.044 | 1080 |

‡ These hydrogen flames were located on the fuel side of the stagnation point at extinction and at the limiting concentration became "striped" flames; i.e., they became cellularly unstable due to preferential diffusion effects.

† (With permission from Ref. 1.)

based on the flow velocity and kinetic rate at the hottest point in the flame. They found that a plot of $T$ vs. $Da_1$ had the same general shape as the curve that we derived earlier for $\eta$ vs. $Da_1$ for the well-stirred reactor (Fig. 6-12b). They also found that $Da_1$ could not be determined a priori. Over a large range of $Da_1$, three temperatures were predicted for the same value of $Da_1$ and the uppermost of these corresponded to the diffusion flame solution. Referring to Fig. 6-12b they also observed a broad minimum in $Da_1$ at relatively high temperature, which corresponded to extinction of the diffusion flame, and a sharp maximum at very low temperature. As was stated in Chap. 6, this lower branch is not physically real because the Arrhenius-rate expression is incorrect at these low temperatures. More recently Linan[3] has completed a comprehensive analysis of diffusion flame extinction.

The striking similarity of the behavior of the Damköhler number and the similar way that extinction is predicted for two markedly different exothermic flows leads us to a rather general conclusion. Namely, whenever a flow time can be adjusted independently of the chemical reaction rate in a highly exothermic, highly temperature-dependent reactive-flow situation, an examination of the behavior of the local Damköhler number should allow the prediction of extinction. We also note that another general conclusion is that even though extinction of inherently hot or high-temperature systems can be predicted using this approach, absolute compositional limits such as those tabulated in Table 12-1 cannot. These instead appear to be related to innate changes in the chemistry of the combustion process, i.e., a shift in the balance between chain-branching and recombination reactions. Finally we note that for flame-sheet theories $Da_1 = \infty$ under all circumstances and thus extinction behavior cannot be discussed in the framework of a flame-sheet theory.

## 12-3 FLAMMABILITY LIMITS AND EXTINCTION IN PREMIXED GASES

Virtually every fuel–oxidizer combination will support premixed flame propagation only when the fuel concentration is within a certain range, bounded by some upper and lower limit concentration. Outside of this range a flame will not propagate a long distance from an ignition source. From a safety standpoint, where the oxidizer is air, the most important of these limits is the lean or lower flammability limit, LFL, or the lean or lower explosion limit, LEL (these terms are used interchangeably). The upper flammability limit, UFL, or upper explosion limit, UEL, can be important under certain circumstances, but they are usually not important from a safety standpoint because dilution with more air can cause a rich nonflammable mixture to become flammable.

The U.S. Bureau of Mines in Pittsburgh, Pennsylvania, has identified one particular technique for determining flammability limits as being their

3. A. Linan, *Acta Astronautica*, **1**: 1007–1039 (1974).

"standard" technique.[4] In this technique a 51 mm internal diameter tube 1.5 m long is mounted vertically and closed at the upper end with the bottom end open to the atmosphere. The gaseous mixture to be tested for flammability is placed in the tube and ignited at the lower (open) end. If a flame propagates the entire length of the tube to the upper end, the mixture is said to be flammable. If the flame extinguishes somewhere in the tube during propagation, the mixture is said to be nonflammable. The choice of this technique as a standard is the result of a considerable amount of research. Specifically, it is known that upward propagation in a tube of this type exhibits wider limits of flammability than downward propagation. In fact, if one places a mixture whose composition is between that of the measured upward and downward propagation limits in a large vessel, and ignites it at the center, one finds that the flame propagates to the top of the vessel and burns only a portion of the material in the vessel before extinguishing, leaving a fair portion of the fuel–air mixture unburned. In other research, it has been found that as the tube becomes smaller the combustible range becomes narrower until one reaches the quenching diameter. At that point there is no mixture of that fuel with air which will propagate a flame through the tube. Fifty-one millimeters was chosen as the diameter for the standard tube because this is the diameter at which a further increase in tube diameter causes only a slight change in the limits. Some typical limits are tabulated in Table 12-2.

The initial pressure and temperature of the mixture affects the flammability limits somewhat. Increasing the temperature always widens the limits because it causes the flame temperature to increase. Increasing the pressure has little effect on the lower or lean limit but causes the upper limit to increase.

The addition of inert gases or inhibitors to the mixture causes the limits to narrow and eventually with sufficient added inert all fuel–air mixtures are nonflammable. This is shown for methane and a general higher hydrocarbon ($C_nH_{2n+2}$; $n > 5$) in Figure 12-4. In general the addition of an inert such as He, $N_2$, $H_2O$, or $CO_2$ to the mixture narrows the limits because such mixtures have a lower flame temperature than the original fuel–air mixture. However inhibitors such as $CCl_4$ and $CH_3Br$ tend to act as radical scavengers and thus alter the chemistry in the flame. It is interesting to note that methyl bromide can burn in air but it still acts as an inhibitor in rich methane flames. The dashed lines in Figure 12-4a represent the flammable range for $CH_4$–$CH_3Br$–air mixtures that lie outside the normal lean limit of $CH_4$–air mixtures. It is also interesting to note that the limit curves for methane–air–inert mixtures peak at approximately the stoichiometric line but gradually shift toward the carbon monoxide–hydrogen stoichiometric line as the number of carbons in the aliphatic chain increases. At $C_5$ and above the shift is essentially complete and the

4. H. F. Coward and G. W. Jones, *Limits of Flammability of Gases and Vapors*, Bureau of Mines Bulletin 503, 155 pp. (1952), also M. G. Zabetakis, *Flammability Characteristics of Combustible Gases and Vapors*, Bureau of Mines Bulletin 627, 121 pp. (1965).

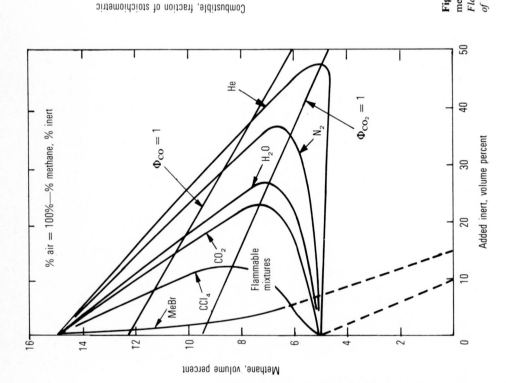

**Figure 12-4** The effect of inerts and inhibitors on the flammability limits of methane and alkanes. *(Adapted, with permission, from M. G. Zabetakis, Flammability Characteristics of Combustible Gases and Vapors, U.S. Bureau of Mines Bulletin 627 (1965).)*

**Table 12-2 Flammability limits of some common fuel–oxidizer mixtures (in mole %)‡**

| Fuel | Oxidizer | Lean limit | Rich limit |
|------|----------|------------|------------|
| Hydrogen | Air | 4.0 | 75.0 |
| Carbon monoxide (moist at 18°C) | Air | 12.5 | 74.0 |
| Ammonia | Air | 15.0 | 28.0 |
| Cyanogen | Air | 6.6 | — |
| Methane | Air | 5.0 | 15.0 |
| Ethane | Air | 3.0 | 12.4 |
| Propane | Air | 2.1 | 9.5 |
| Butane | Air | 1.8 | 8.4 |
| Ethylene | Air | 2.7 | 36.0 |
| Acetylene | Air | 2.5 | 100.0 |
| Benzene | Air | 1.3 | 7.9 |
| Methyl alcohol | Air | 6.7 | 36.0 |
| Ethyl alcohol | Air | 3.2 | 19.0 |
| Diethyl ether | Air | 1.9 | 36.0 |
| Carbon disulphide | Air | 1.3 | 50.0 |
| Hydrogen | Oxygen | 4.0 | 95.0 |

‡ Data taken from M. G. Zabetakis, *Flammability Characteristics of Combustible Gases and Vapors*, U.S. Department of Mines Bulletin 627 (1965) with permission.

limit behaviors for all higher hydrocarbons are similar enough to be plotted on one graph, Fig. 12-4b.

The excess enthalpy burner discussed theoretically in Sec. 8-4 allows one to stabilize a flame in ultralean mixtures which would ordinarily not support a propagating flame. Recently Kotani and Takeno[5] have stabilized such flames in a 50 mm circular burner that contained a close packed bundle of 1 mm o.d., 0.6 mm i.d., 30 mm long ceramic tubes. This burner block was well shielded from losses to make it as "adiabatic" as possible. The resulting stability diagram is shown in Fig. 12-5. Note that all types of burner behavior predicted by the theory were observed and that the lean limit was reduced to an equivalence ratio of 0.3 from its usual value of about 0.5.

In 1898 Le Châtelier and Boudouard[6] proposed a rule for determining the lean limit of a mixture of two combustible gases, from the known lean limits of the constituent species in the mixture. If we call $LFL_m$ the volume percent of the fuel mixture at the lean limit and $LFL_1$, $LFL_2$, ..., $LFL_n$ the volume percent lean flammability limits of the n constituent fuels and $x_1$, $x_2$, ..., $x_n$ the

5. Y. Kotani and T. Takeno, An Experimental Study on Stability and Combustion Characteristics of an Excess Enthalpy Flame, *Nineteenth Symposium (International) on Combustion*, The Combustion Institute, Pittsburgh, Pa., p. 1503, (1983).

6. H. Le Châtelier and O. Boudouard, *Compt Rend*, **126**: 1344–1347, 1898.

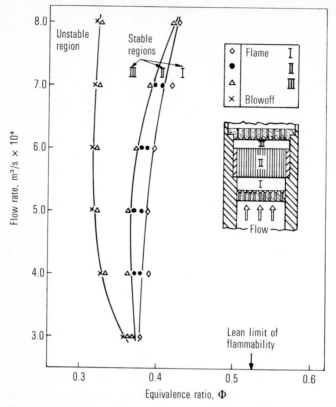

**Figure 12-5** Stability diagram for an excess enthalpy flame holder. Fuel is methane. *(Adapted from Ref. 5, with permission.)*

mole fractions of each constituent fuel in the fuel mixture, a generalized form of Le Châtelier's rule is given by the formula

$$LFL_m = \frac{1}{x_1/LFL_1 + x_2/LFL_2 + \cdots + x_n/LFL_n} \tag{12-1}$$

If the fuel contains an inert diluent such as nitrogen or carbon dioxide the lean limit can still be calculated by not including the inert in Eq. (12-1). For example the lean limit for methane in air is 5 percent. If methane is diluted with an inert and this mixture is used as a fuel the lean limit for this mixture would be

$$LFL_m = \frac{1.0}{x_1/5 + 0} = \frac{5}{x_1}$$

where $x_1$ is the mole fraction of $CH_4$ in the fuel mixture. The Le Châtelier formula works quite well if the fuel mixture contains hydrogen, carbon monoxide, or ordinary hydrocarbons. It does not work well for unusual compounds such as carbon disulfide, or when inhibiters are present.

One empirical observation[7] is that for many hydrocarbon–air mixtures, the lean limit volume percent of fuel multiplied by its heating value in kilojoules per mole is approximately $4.34 \times 10^3$. This means, of course, that the flame temperature has approximately the same value at extinction for all hydrocarbon fuels. This result is similar to that obtained for diffusion flames (see Table 12-1).

The flammability limit is thought to be caused by flame extinction due to a combination of heat loss from the flame, flame stretch, and/or spontaneous flame instabilities. Furthermore, all of these effects are complicated by the fact that the flame temperature of limit flames is so low that the competition between the hydrogen atom chain-branching and chain-breaking reactions mentioned in Secs. 6-10 and 8-5 occurs at a location very close to the hot reaction zone.

Unfortunately, there is no adequate theory for flammability limits. In essence, all the extant theories have so simplified the problem that the actual physical processes that occur at extinction are no longer modeled properly. As an example, take the heat-loss theory of Spalding.[8] He modeled the flame as a one-dimensional flame and removed heat from the hot gases while retaining the one-dimensionality of the flame. He removed heat at a fixed rate per unit length of gas column downstream of the flame front and found that there is a maximum value for this quantity above which one can no longer find a solution to the equations that were formulated. This critical maximum quantity of heat removal per unit length was then considered to be the condition for flame extinction. However, when large amounts of heat are removed from this flame the burning velocity drops and the flame thickness increases markedly. Under these conditions, even though the heat abstracted per unit length along the flame goes through a maximum, the total heat that is abstracted from the flame increases monotonically. Furthermore, since one can calculate the structure of a flame using nonsteady one-dimensional flame equations and full kinetics for compositions considerably leaner than the observed lean limit composition, it appears that one-dimensional theories for flame extinction, even with heat loss, are not adequate.

Experimentally it is found that different mechanisms of extinction are operative in different geometries. Tsuji and Yamaoka,[9] have recently studied the extinction of two opposed premixed flames using the type IV burner of Fig. 7-6. In this experiment they pass the same fuel–air mixture through the porous cylindrical holder and up the duct. When ignition is effected two opposed premixed flames appear as shown in Fig. 12-6. These have the property that they are separated by a stagnation-point region which is exactly adiabatic. As the compositions of the two mixtures approach either the lean- or the rich-limit

7. F. T. Bodurtha, *Industrial Explosion Prevention and Protection*, McGraw-Hill (1980).

8. D. B. Spalding, *Proc. R. Soc. London*, **A240**:83 (1957).

9. H. Tsuji and I. Yamaoka, "Structure and Extinction of Near Limit Flames in a Stagnation Flow," *Nineteenth Symposium (International) on Combustion*, The Combustion Institute, Pittsburgh, Pa., p. 1533, (1983).

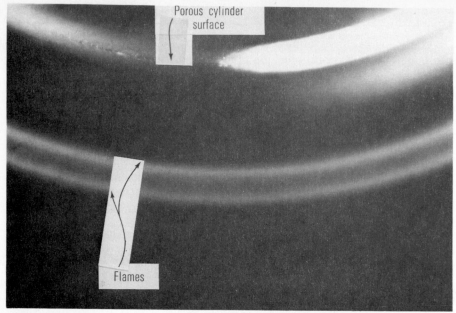

(a)

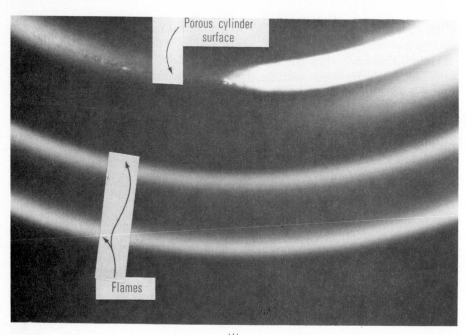

(b)

**Figure 12-6** Visible light photograph of counterflow, premixed, twin flames established in the forward stagnation region of a porous cylinder. (a) Lean methane–air flames near the limit $\Phi = 0.53$. (b) Rich methane–air flames near the limit $\Phi = 1.58$. Notice how close the lean flames approach each other when compared to the rich flames. *(Courtesy Prof. H. Tsuji, University of Tokyo, from Ref. 8, with permission.)*

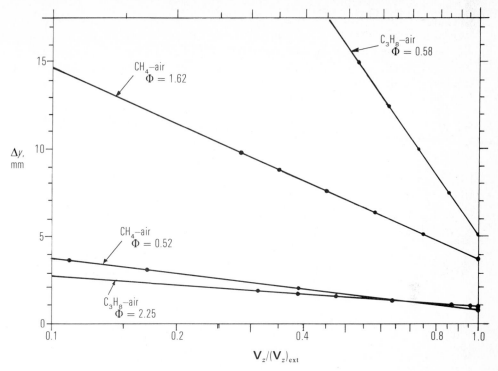

**Figure 12-7** The separation distance between the two luminous zones of opposed premixed flames. *(Adapted from Ref. 8, with permission.)*

composition the flames approach each other. Both of the flames are stretched because of the nature of stagnation-point flow and eventually at some critical lean or rich composition the flames blow out. Experimentally it is observed that near the limit rich methane or lean propane flames stand quite some distance apart while lean methane or rich propane flames approach each other very closely. This is shown in Fig. 12-7, which is a plot of the distance between the two luminous zones at the centerline of the flow as a function of the relative blowing rate divided by the blowing rate at extinction.

Their study of the temperature profiles and composition of the gases at the stagnation point for a flame pair that is near extinction shows that the temperature is high and the chemical reactions are going to completion for the rich methane and lean propane flames and that the temperature is low and the chemical reactions are not going to completion for the lean methane and rich propane flames. This is shown in Fig. 12-8. This means that the mechanism of extinction is strongly dependent on the effective Lewis number of the deficient species. If the Lewis number is less than unity, as it is for a lean methane or rich propane flame, the flame extinguishes because the rate of stretch is so large that the chemical reactions cannot go to completion. However if it is greater than unity, as it is in a rich methane or lean propane flame, extinguishment is not caused by the chemical reactions being incomplete but instead must be caused by a true stretch mechanism. We note that at the extinction limit for a rich methane or lean propane flame, stretch will cause a local reduction of the deficient species, i.e., at the centerline the local propane concentration will drop because of stretch in a propane–air flame and the local oxygen concentration will drop because of stretch in a methane–air flame.

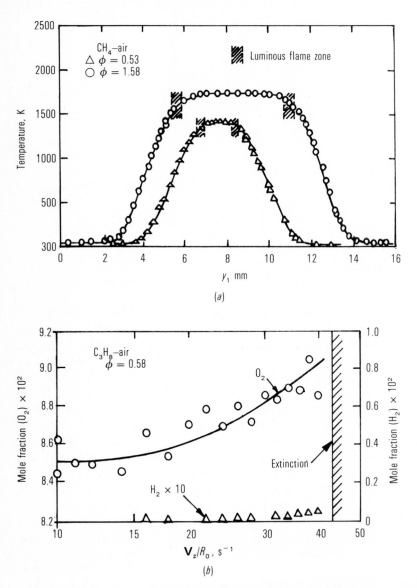

**Figure 12-8** Temperature profiles and centerline concentrations for flames near extinction. *(From Ref. 8, with permission.)*

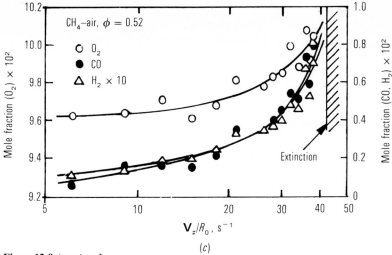

**Figure 12-8** (*continued*)

Experiments using a standard flammability tube at one g and zero g have shown that the upward propagating flame and the zero g flames have a flame-cap shape that is controlled primarily by inviscid flow behavior ahead of the flame (see Fig. 10-8) and that the rate of propagation through the tube is determined by bouyancy forces.[10] Thus the length of the flame skirt for an upward propagating flame is related primarily to the speed at which the flame propagates through the tube, because the normal burning velocity is not affected by buoyancy.

In an upward propagating flame extinguishment occurs first at the center of the flame where it is held (see Sec. 10-4) and then propagates over the remainder of the flame.[11] A calculation based on the observed flame shape using potential flow theory indicates that stretch is indeed maximum at the holding point of this flame and therefore that extinguishment should start there.

The downward propagating flame in a standard flammability tube is markedly affected by gravity. Photographs of this flame taken with an image intensifier are shown in Fig. 12-9. The flame is almost flat but has a cellular structure, and it does not propagate uniformly down the tube but sort of oscillates as it propagates. At incipient extinguishment the flame recedes from the walls, and heat loss to the walls cools the gas in the neighborhood of the walls. At this time differential buoyancy of the central hot column and the surrounding cooler

10. R. A. Strehlow and D. Reuss, "Flammability Limits in a Standard Tube," *Combustion Experiments in a Zero Gravity Laboratory*, ed. T. H. Cochran, *Prog. Astronaut. Aeronaut.*, AIAA, **73**: 61–90 (1981).

11. J. Jarosinski, R. A. Strehlow, and A. Azarbarzin, "The Mechanism of Lean Limit Extinguishment of an Upward and Downward Propagating Flame in a Standard Flammability Tube," *Nineteenth Symposium (International) on Combustion*, The Combustion Institute, Pittsburgh, Pa., p. 1549 (1983).

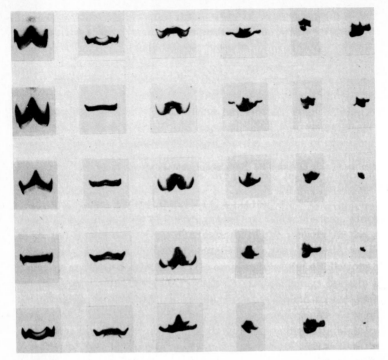

**Figure 12-9** Extinction of a downward propagating lean limit methane–air flame in a standard 51 mm flammability tube. Propagation from the top open end toward the bottom closed end. Time increases from top to bottom and then from left to right. Fifty-five frames per second. Except for the last three frames, every third frame is shown. *(Adapted from A. Azarbarzin, A Study of Flame Extinction in a Vertical Flammability Tube. M.S. Thesis, University of Illinois at Urbana-Champaign, 50 pp. (1981).)*

annular space causes the cooler product gases to travel ahead of the flame and the small flame kernel that is left actually rises just prior to complete extinction. Thus extinction of a downward propagating flame is actually caused by a very complex sequence of events.

Chan and T'ien[12] have observed that a wick-held diffusion flame in a large vessel whose oxygen is slowly being depleted eventually takes the shape of a small blue hemispherical cap over the wick. At incipient extinguishment they observed that the flame's edge oscillates in and out symmetrically around the centerline with no apparent motion of the central region. Then, after a few oscillations a final oscillation causes complete extinguishment of the flame.

12. W. Y. Chan and J. S. T'ien, *Combust. Sci. Tech.*, **18**: 139–143 (1978).

# 12-4 MINIMUM IGNITION ENERGY AND QUENCHING DISTANCE

The minimum ignition energy is the smallest quantity of energy that must be added to a system to start flame propagation. Its value is quite dependent on the local rate and method of heat addition, and on the geometry of the heat source. Quenching distance is defined as the largest channel dimension that will just stop a flame from propagating through the channel if there is no large pressure drop along the channel. Values of the quenching distance are dependent on experimental geometry.

Combustible mixtures may be exploded "homogeneously" in "vessel" experiments like those discussed in Chap. 6, or they may be ignited by rapid heat addition at a localized region in the gas. In the latter case, successful ignition leads to the propagation of either a flame or a detonation. As the total quantity of energy is reduced for any one ignition technique, direct detonation ceases to occur, if it did at all (see the next chapter for a discussion of detonation initiation), and flames are observed to propagate from the source. Further reduction of the total quantity of energy that is added to the fluid continues to produce ordinary flame propagation until the minimum ignition energy for that source configuration is reached. At this point only a small quantity of the combustible gas is converted to products and the energy is dissipated harmlessly by thermal conduction.

A multitude of different energy sources and source configurations have been used to study flame ignition experimentally. However, an ordinary capacitor discharge spark has consistently yielded the lowest ignition energy for any specific combustible mixture. For an experimental determination of an ignition energy, high-performance (usually air-gap) condensers are used so that the majority of the stored energy will appear in the spark gap. The stored energy may be calculated using the equation $E = CV^2/2$, where $C$ is the capacity of the condenser and $V$ the voltage just before the spark is passed through the gas. The spark must always be produced by a spontaneous breakdown of the gap because an electronic firing circuit or a trigger electrode would either obviate the measurement of spark energy or grossly change the geometry of the ignition source. It has been found experimentally that for this type of spontaneous spark up to 95 percent of the stored energy appears in the hot kernel of gas in less than $10^{-5}$ s. The losses are thought to be due primarily to heat conduction to the electrodes. Since the total stored energy can be varied by changing either the capacity or the voltage, the electrode spacing (which is proportional to the voltage at breakdown) may be varied as an independent parameter. Two problems now arise: If the electrode spacing is too small, the electrodes will interfere with the propagation of the incipient flame and the apparent ignition energy will increase. If, however, the spacing is too large, the source geometry will become essentially cylindrical and the ignition energy will again increase because the area of the incipient flame is greatly increased in this geometry. This condition is shown schematically in Fig. 12-10. The fact that the increase

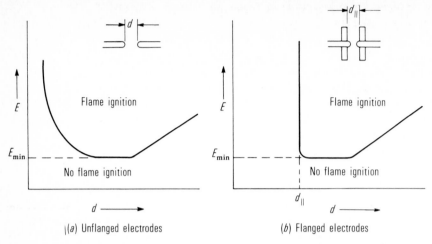

**Figure 12-10** Quenching distance and minimum-ignition-energy measurement using a capacitor spark.

in minimum ignition energy is due to quenching at small electrode spacings may be confirmed by using electrodes flanged with electrically insulating material. Figure 12-10 shows the effect of electrode flanging on the ignition energy at small separations. Note that below a certain spacing, as the spacing is reduced the spark energy required to ignite the bulk of the gas sample rises very much more rapidly with flanged tips. This critical flange spacing is defined as the flat-plate quenching distance of the flame. Therefore this simple experiment may be used to determine the flat-plate quenching distance and the minimum ignition energy for spherical geometry, and may also be used to evaluate whether the ignition source is essentially spherical or cylindrical. The latter piece of information is obtained by determining the effect of electrode separation on $E_{min}$ at large separations.

Quenching distances may also be measured by quickly stopping the flow through a tube of the desired geometry when a flame is seated on the exit of the tube. If the flame flashes down the tube, the minimum dimensions are above the quenching distance for that mixture. Quenching distances that have been measured between two flat plates using this technique agree quite well with the quenching distances measured using the flanged electrodes in the spark-ignition experiment described above. Parallel-plate quenching distances and minimum ignition energies determined by the spark method for propane, oxygen, and nitrogen mixtures at various initial pressures are given in Fig. 12-11. It should be noted that, in general, minimum ignition energies are extremely small—in some cases, corresponding to the passage of a barely audible spark through the mixture. In fact, it is well known that static electric sparks can cause the ignition of many explosive mixtures.

The relation between the characteristic quenching distance and the geometry of the quenching tube has been developed theoretically and confirmed experimentally in a number of instances. The theory can be based on the assumption either that (1) wall capture of reactive species, or (2) heat transfer to the wall controls the quenching. Both these approaches yield the same theoretical relations relative to geometry effects, because of the similarity between diffusion and thermal conduction. Quenching measurements for a variety of geometries show quite good relative agreement ($\pm 10\%$) with the theoretical predictions for a number of hydrocarbon systems. The theory predicts that the circular tube quenching diameter should be related to the parallel plate quenching distance by the equation

$$d_{\parallel} = 0.65 d_0$$

Experimentally, one may show that the quenching distance is simply related to the pressure. This is because it is related to the preheat zone thickness of the flame. Also it is related to the minimum ignition energy of the system for virtually all hydrocarbon–air flames, through the equation $E_{min} = 0.06 d_{\parallel}^2$ (see Fig. 12-12) where $E_{min}$ has units of millijoules and $d$ is measured in millimeters.

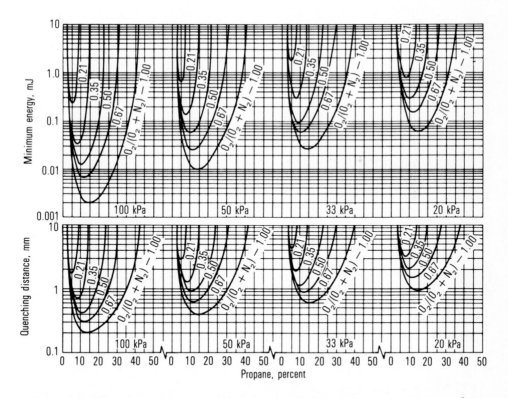

**Figure 12-11** Effect of pressure, nitrogen dilution, and equivalence ratio on propane–oxygen flat-plate quenching distances and spark minimum-ignition energies. *(With permission from B. Lewis and G. von Elbe, Combustion Flames and Explosions of Gases, 2d ed., Academic, New York, 1961.)*

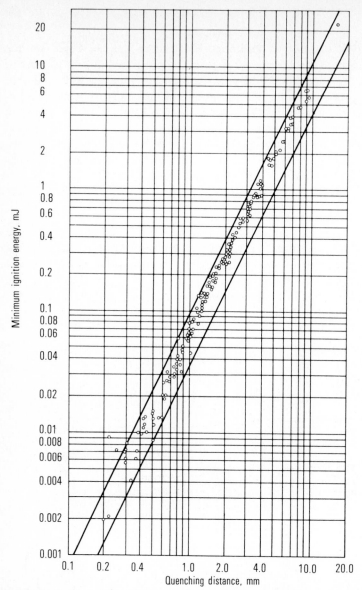

**Figure 12-12** The relation between flat-plate quenching and spark minimum-ignition energies for a number of hydrocarbon–air mixtures. *(With permission from B. Lewis and G. von Elbe, Combustion Flames and Explosions of Gases, 1st ed., Academic, New York, 1951.)*

---

**Figure 12-13** Effect of change in ignition energy upon flame propagation. Times are in microseconds. *(Courtesy H. Lowell Olsen, Applied Physics Laboratory, Silver Spring, Md. Originally in AGARD, Selected Combustion Problems, II, Butterworth, London, 1956.)* *(With permission.)*

Read down:

|  | Gas | $\frac{1}{2}CV^2$ |
|---|---|---|
|  | Air only | 1.33 mJ |
| Subcritical energy | Propane–Air | 1.14 mJ |
| Supercritical energy | Propane–Air | 1.23 mJ |

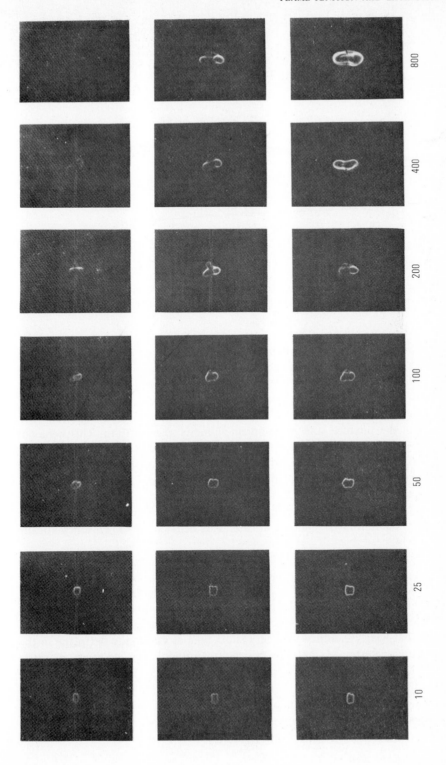

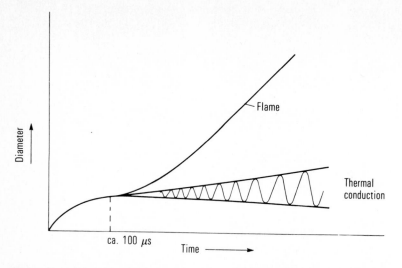

**Figure 12-14** Schematic diagram of the decay or growth behavior of an ignition kernel (incipient flame ball) near the threshold value of energy for ignition of a propagating flame.

We will now look at the details of incipient growth from a small spark. Figure 12-13 contains repetitive schlieren photographs illustrating the appearance of the hot kernel of gas for an inert mixture and for a reactive mixture with both a subcritical and supercritical energy addition. In the reactive case, the spark kernel initially contains temperatures well above the ignition temperature of the mixture, and an incipient flame is seen to form. This behavior should be contrasted with the inert case where the initially high gradient simply flattens with time and eventually disappears. However, as the two reactive cases show, the formation of an incipient flame does not ensure subsequent propagation. At some point, well after spark passage (ca. 100 $\mu$s in Fig. 12-13) the incipient flame becomes a true flame in the supercritical case and starts to decay by thermal conduction in the subcritical case. This behavior is shown schematically in Fig. 12-14.

Ignition and quenching are difficult to discuss theoretically and complete theories are not as yet available. However, Rosen[13] has justified Lewis and von Elbe's[14] empirical relation for a plane source (i.e., one-dimensional) minimum ignition energy by applying the nonsteady equations to a simple first-order $A \to B$ flame with Le = 1, similar to the steady flame discussed in Sec. 8-3. He assumed that heat was added in a very short time at an infinite flat plane in the mixture, and determined that the critical quantity of heat addition to cause

13. G. Rosen, *J. Chem. Phys.*, **30**:298 (1959).

14. B. Lewis and G. von Elbe, *Combustion Flames and Explosion of Gases*, 2d ed., Academic, New York, 1961.

flame propagation is related to the steady-flame properties by the equation

$$E_{min} \simeq \frac{\kappa_u q}{(C_p/M)S_u} = \frac{\kappa_u}{S_u}(T_b - T_u) \tag{12-2}$$

where $E_{min}$ has units of joules per square meter and $q = (C_p/M)(T_b - T_u)$, the heat of reaction at constant pressure in joules per kilogram. We note that $q$ is defined as an enthalpy, while $E_{min}$ is an energy. However, this equation is correct because the energy added to the system appears as an enthalpy in the system, since the process is taking place at constant pressure.

We note that when we substitute Eq. (8-7) from elementary flame theory into Eq. (12-2) we obtain the relationship

$$E_{min} \simeq \eta_0 \rho_u c_p (T_b - T_u) \tag{12-3}$$

In Eq. (12-3) one should really use a variable density and not $\rho_u$. This would make the right-hand side of this equation smaller. This, plus the experiments on ignition delay described above show, in at least an approximate way, that one must add enough energy to the system to form a preheat zone before propagation will occur.

In addition, Aly and Hermance[15] have presented a two-dimensional theory for parallel plate or circular tube quenching using either overall first- or second-order kinetics with an Arrhenius temperature dependence to simulate a stoichiometric propane–air flame. They also allowed the Lewis number to range from one to infinity and neglected transverse diffusion of reactants. They justify this because, as Fig. 12-15c shows, there is quite a thick cool region near the wall where chemistry is not occurring. Figures 12-15a and b show the centerline longitudinal and the transverse structure of such a flame.

The authors also investigated the effect of varying the Lewis number. These results are shown in Fig. 12-16. Note from Fig. 12-16a that the "adiabatic" flame speed increases as the Lewis number goes from $1 \to \infty$, just as it did for the Friedman and Burke flame with simple Arrhenius kinetics. Also note that the theoretically determined quenching distance drops and that the Peclet number defined as

$$Pe = \frac{\rho_u c_{pu} S_u d_{\parallel}}{\kappa_u} \tag{12-4}$$

is independent of the Lewis number. In this equation $d_{\parallel}$ is the parallel plate quenching distance. This means that in this two-dimensional flame, quenching is determined by the competition of the rate of flame propagation (or rate of heat generation) and rate of heat loss to the walls by thermal conduction. Also note that the maximum temperature of the centerline flame at incipient extinction is about 2100 K, that is, about 300 kelvins less than the "adiabatic" flame temperature in a 50 mm tube.

15. S. L. Aly and C. E. Hermance, *Combust. Flame*, **40**: 173–185 (1981).

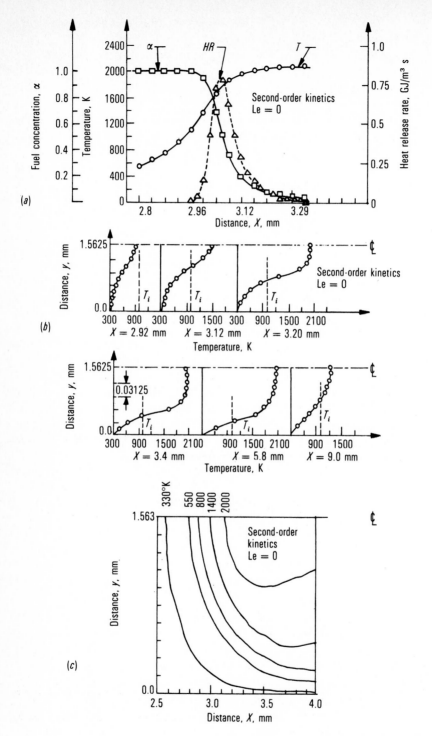

**Figure 12-15** Longitudinal (*a*), transverse (*b*), and isothermic (*c*) structures of the flame at the quenching limit; second-order kinetics. (*Ref. 14, with permission.*)

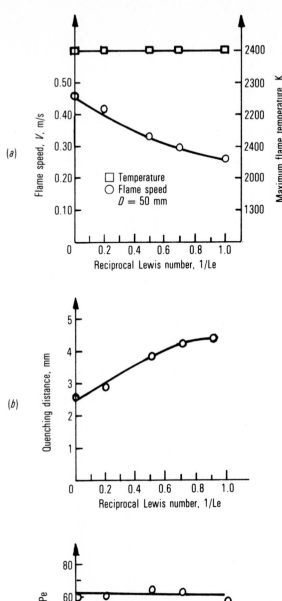

(a)

(b)

(c)

**Figure 12-16** Temperature and flame speed (a), quenching distance (b), and Peclet number (c) dependence on Lewis number; first-order kinetics, all calculated for a 50 mm diameter tube. (*Ref. 14, with permission.*)

A different problem arises during spherical ignition. The incipient flame is always a stretched flame in the Karlovitz sense (see Sec. 10-3) because its area increases at an ever-increasing rate while its radius of curvature decreases. Using the same assumptions as were used in Sec. 10-3 one finds that the non-dimensional stretch or the Karlovitz number for such a flame is given by the expression

$$K = 4 \frac{\eta_0}{d} \frac{\rho_u}{\rho_b} \qquad (12\text{-}5)$$

As a result of this stretch, preferential diffusion of the lighter species, either fuel or oxygen, toward the ignition region causes two unique effects to occur. In the first place preferential diffusion alters the equivalence ratio in the incipient preheat zone by enriching the concentration of the lighter species there. This causes a shift in the equivalence ratio for the lowest minimum ignition energy as shown in Fig. 12-17. This is a plot of minimum ignition energy vs. composi-

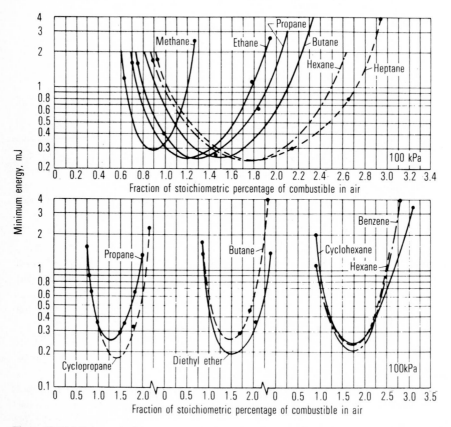

**Figure 12-17** Minimum ignition energy vs. equivalence ratio for various fuel–air mixtures showing the effects of preferential diffusion. (*With permission from B. Lewis and G. von Elbe, Combustion Flames and Explosions in Gases, 2d ed., Academic, New York, 1961.*)

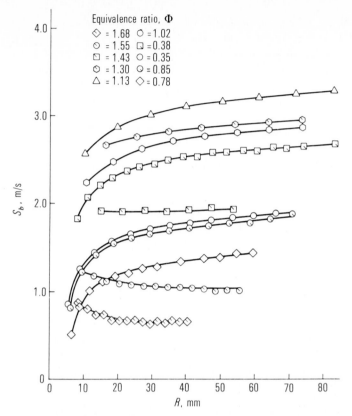

**Figure 12-18** Space velocity $S_b$ vs. radius for a laminar flame ball in propane–air mixtures.

tion for various fuels in air. Since from the elementary theory one would expect the minimum ignition energy to occur at the maximum burning velocity of these mixtures, these curves should all show a minimum in the neighborhood of $\Phi = 1.1$ (the approximate equivalence ratio for the maximum laminar burning velocity for all these systems). Notice that for methane, which is lighter than oxygen, $\Phi_{min}$ is 0.8, while for those fuels whose diffusivity is less than oxygen $\Phi_{min}$ is always greater than one and increases as the molecular weight of the fuel increases. Thus, the larger the difference in relative diffusivity, the larger the deviation of $\Phi_{min}$ from stoichiometric.

Another way that preferential diffusion plays an important role in early flame behavior is its effect on the burning velocity of an outwardly propagating flame. For example as shown in Fig. 12-18, in propane–air mixtures (i.e., for a heavy fuel and light oxidizer mixture), rich mixtures exhibit a higher space velocity at small radii while lean mixtures show a lower space velocity at small radii. In all cases the space velocity of the flame relaxes to the theoretically expected laminar space velocity after a few centimeters of flame travel. The effect is somewhat masked by two additional effects which cannot be easily discussed quantitatively: (1) the slow relaxation of the burned products to the equilibrium composition adds a slow initial increase in velocity to the experimental curves, and (2) any excess ignition energy at the spark source can cause a higher propagation velocity of the early flame. However, the preferential diffusion effect is the only effect which shifts sign as the equivalence ratio passes through unity and is therefore quite definitely operating in this system.

Recently Frankel and Sivashinsky[16] have verified, using a large activation asymptotic analysis with the Lewis number defined for the deficient species, that rich and lean propane–air flames should exhibit the behavior shown in Fig. 12-18.

## 12-5 HOT SURFACE OR HOT GAS IGNITION: THE AUTOIGNITION TEMPERATURE AND THE MINIMUM EXPERIMENTAL SAFE GAP

There have been many attempts to study the ignition process using either a heated gas stream as an igniter, passing the combustible mixture over a heated surface, or injecting heated spheres into the mixture. These are all quite complex aerodynamically and therefore will not be discussed in this text. There are however two standardized ignition techniques which are important from the point of view of combustion safety that will be discussed here. These are the measurement of the autoignition temperature and the minimum experimental safe gap.

For gases, the autoignition temperature is the lowest vessel temperature at which any mixture of that fuel and air will explode. For liquid fuels the autoignition temperature is the temperature of a specified flask that contains air below which a flash of light will not be seen after a specified amount of liquid fuel is dropped into the flask. The autoignition temperature yields an imperfect indication of the maximum surface temperatures that can be tolerated in an environment where that liquid or gas may be handled in bulk.

The minimum experimental safe gap is also a safety-related combustion property. It is based on the fact that it is very expensive either to purge electrical switch boxes or motor housings or to attempt to make them completely air tight. Therefore when electrical devices are to be used in the presence of combustible vapors or gases the electrical boxes must be constructed to meet two requirements. These are: (1) the box should be strong enough to be able to stand an internal explosion without rupturing and (2) openings to the outside should be small enough so that escaping combustion products will not cause ignition of the surrounding flammable vapors or gases.

Studies of the ignition of a combustible mixture by hot product gases escaping from a vessel through a long slot have shown that for a sufficiently wide

16. M. L. Frankel and G. I. Sivashinsky, "On Effects Due to Thermal Expansion and Lewis Number in Spherical Flame Propagation," *Combust. Sci. Tech.*, **31**:131 (1983).

slot width, a flame simply propagates through the slot into the surrounding combustible mixture. However when the slot width becomes somewhat less than the ordinary parallel plate quenching distance, a flame no longer propagates through the slot but the hot gases that escape cause a slightly delayed ignition of the surrounding combustible mixture. Schlieren observations of this process have shown that ignition occurs some distance from the slot in the turbulent mixing region and that initially the ignition is a volumetric "explosion" which subsequently ignites a flame front which propagates away from the "exploded" turbulent mixing region. It has also been found that there is a critical slot width below which ignition of the surrounding gas no longer occurs. This critical width is called minimum experimental safe gap (MESG). For most hydrocarbons it is approximately $d_{\parallel}/2$.

## 12-6 FLASH POINT

Liquid fuels all exhibit an equilibrium vapor pressure which is dependent on the temperature of the fuel. As the temperature of the fuel is raised from some low value to some high value the equilibrium vapor pressure increases monotonically. If a mixture of this equilibrium fuel vapor with air contains less fuel vapor than required for lean limit combustion the liquid is said to be below its lower flash-point temperature. Above this temperature, as the liquid temperature is increased its equilibrium vapor pressure increases to the point where the vapor–air mixture contains so much fuel vapor that it is above the upper flammable limit for that fuel. Under these circumstances the liquid is said to be above its upper flash-point temperature.

In actual practice, the lower and upper flash-point temperatures are measured in any of a number of standardized pieces of apparatus. By far the most accurate measurements are obtained using equilibrium or closed-cup techniques. In these techniques the liquid is placed in a loosely closed container in a thermostat bath and is held at some temperature until the vapor space in the container holds fuel vapor which is in equilibrium with the liquid at that temperature. Then a part of the top of the container is opened and a flame from a torch is swept across the opening. If the flame "flashes" into the vapor space the thermostat temperature is between the lower and upper flash points. The test is then repeated at different temperatures until both the lower and upper flash points are found. A flame is used as a strong igniter that is easily available. Note that since the flame is applied to the top of the container the measured flash point corresponds to the limit for downward propagation of the flame.

There are other "approved" techniques for measuring the flash point. One set of these uses an "open cup." Here the fuel is heated in a cup which is open at the top and the high vapor density of the fuel is expected to stop the fuel vapors from escaping before the flame is applied. This is obviously less satisfactory than the closed-cup "equilibrium" technique described above.

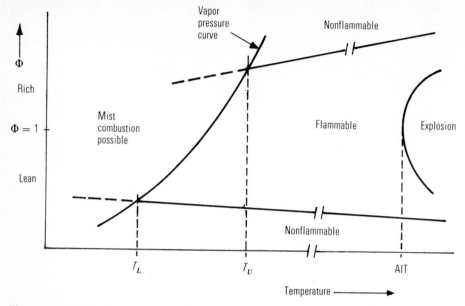

**Figure 12-19** Effect of temperature on limits of flammability of a pure liquid fuel in air. $T_L$ = lean (or lower) flash point; $T_U$ = rich (or upper) flash point; AIT = auto ignition temperature. *(Adapted from W. E. Baker, P. A. Cox, P. E. Westine, J. Kulesz and R. A. Strehlow, Explosion Hazards and Evaluation, Elsevier, Amsterdam (1983) with permission.)*

The measurement of a flash point for pure liquids is relatively straightforward. However special problems arise when the liquid contains a mixture of fuels and/or fuels and inhibiters. In these cases the vapor composition will be different from the liquid composition. Therefore too many successive tests with the same sample in the cup will lead to erroneous results because the composition of the liquid in the cup will change with time due to fractional distillation. Problems also arise when one is dealing with a highly viscous liquid or with a liquid (such as a paint) that "skins" in air.

The relationship between the flash points and the lower and upper flammability limits for a pure liquid are shown schematically in Fig. 12-19. The AIT or autoignition temperature that is also shown schematically has no relationship to the flash point of the liquid.

## PROBLEMS

**12-1** Calculate the lean flammability limit for a fuel that contains 50% $CH_4$, 40% $C_4H_{10}$, and 10% $N_2$.

**12-2** Calculate the flame temperature of stoichiometric methane–air–diluent mixtures at their flammability-limit concentration for the diluents He, $N_2$, $H_2O$, and $CO_2$. Compare your answers to the data of Table 12-1.

**12-3** For the lean methane–air opposed flames of Fig. 12-8 determine if the water–gas reaction $CO + H_2O \rightleftarrows CO_2 + H_2$ is in equilibrium in the hot gas region between the two flames. First calculate the unstressed (i.e., normal) flame temperature (with no CO or $H_2O$ present) and use this for the blowing rate of 5 s$^{-1}$. Then calculate the composition near extinction at a blowing rate of ca. 35–40 s$^{-1}$.

*Hint:* A carbon and hydrogen balance can be used to determine the $CO_2$ and $H_2O$ concentrations assuming that preferential diffusion is absent at the centerline.

**12-4** Work Prob. 8-3 to determine the preheat zone thickness of a propane–air flame. Compare your answer to the quenching distance of Fig. 12-11.

**12-5** Calculate the minimum ignition energy for the flame of Prob. 12-4 using Eq. (12-3) and compare it to the value given in Fig. 12-11.

**12-6** Determine the Peclet number for quenching of the flame of Prob. 12-4.

**12-7** Derive Eq. (12-5).

**12-8** In the eighteenth and nineteenth century whisky was tested for its proof by determining if a lit match would ignite gunpowder saturated with it. If the flash point of the whisky was below room temperature the gunpowder would not ignite and the whisky was considered to be less than 100 proof. Later this was standardized to mean 50 volume percent $C_2H_5OH$ in water. If the vapor pressure of ethyl alcohol is 7.9 kPa at 25°C how accurate is this standardization?

*Hint:* Assume that the vapor pressure of the alcohol in an alcohol–water mixture varies linearly with mole fraction (it does not).


# BIBLIOGRAPHY

Baker, W. E., P. A. Cox, P. S. Westine, J. J. Kulesz, and R. A. Strehlow, *Explosion Hazards and Evaluation*, Elsevier, New York (1983).

Barnett, H. C., and R. R. Hibbard, *Basic Considerations in the Combustion of Hydrocarbon Fuels with Air*, NACA Report 1300, U.S. Government Printing Office, Washington D.C. (1959).

Jost, W., *Explosion and Combustion Processes in Gases*, McGraw-Hill, New York (1946).

Lovachev, L. D., "Flammability Limits—A Review," *Combust. Sci. Tech.*, **20**: 209–224 (1980).

Mullins, B. P., and S. S. Penner, *Explosions, Detonations, Flammability and Ignition*, Pergamon, New York (1959).

Lewis, B., and G. von Elbe, *Combustion, Flames and Explosions of Gases*, 2nd ed. Academic, New York (1961).

Potter, A. E., Jr., "Flame Quenching," *Progress in Combustion and Fuel Technology Volume I*, ed. J. Ducarme, M. Gerstein, and A. H. Lefebvre, Pergamon, New York 145–182 (1960).

# THIRTEEN

## DETONATION INITIATION AND FAILURE

### 13-1 INTRODUCTION

In this chapter we will discuss the nonsteady processes which cause a detonation to either start or stop propagating. To describe these processes we will use the terms initiation and failure, to distinguish them from the processes which cause the ignition or extinction of flames. We make this distinction because the mechanisms which cause the initiation and failure of detonations are so markedly different from those that cause the ignition or extinction of flames.

### 13-2 DIRECT INITIATION BY PLANAR WAVES

Initiation behind the reflected shock wave in a conventional shock tube produces the simplest one-dimensional initiation that has been observed experimentally. This process has been analyzed rather thoroughly by Strehlow and coworkers[1] and their results are summarized here. In mixtures containing a high percentage of argon, the shock-reflection process is quite ideal because the reflected shock is not grossly distorted by its interaction with the boundary layer (as it is in gas mixtures having a higher heat capacity). In this ideal case, the region between the reflected shock and the end wall of the shock tube is relatively quiescent and has quite uniform properties initially. Furthermore,

---

1. R. A. Strehlow and A. Cohen, *Phys. Fluids*, **5**:97 (1962); R. B. Gilbert and R. A. Strehlow, *AIAA J.*, **4**:1777 (1966).

because the reflected shock must be reasonably strong to produce high temperatures, the temperature jump across the reflected shock will be sufficiently large to cause a short delay to adiabatic explosion behind the reflected shock, whereas the concurrent delay behind the incident shock is quite long. Thus one may easily find a range of incident Mach numbers over which the first explosion process always occurs behind the reflected shock. This range of Mach numbers is bounded on the lower end by the weakest shock strength which will just cause initiation behind the reflected shock during the available run time, and on the upper end by an incident shock strength which is sufficiently high to cause initiation before reflection. In this short Mach-number range, initiation is observed behind the reflected shock.

Four types of initiation have been observed behind the reflected shock in these experiments. The first type is spurious and is due to the presence of bumps or crevasses on the tube walls or in the end plate. If large enough, these irregularities will destroy the one-dimensional nature of the flow and lead to local shock reenforcement and non-one-dimensional initiation. The second and third types of initiation are strictly one-dimensional processes and strikingly illustrate the nonsteady wave nature of initiation. Figure 13-1 contains schlieren and schlieren-interferometric photographs illustrating these two forms of initiation. Figure 13-1a illustrates one type of initiation, Figs. 13-1b and c the other. In both cases the initiation starts as a simple wave phenomenon traveling away from the end wall behind the reflected shock. This wave overtakes the reflected shock some distance from the end wall to produce a detonation in the incident gas flow. The nature of this strictly one-dimensional initiation process has been rather extensively analyzed by assuming that the development of a detonation under these conditions is primarily controlled by the presence of a "reaction wave" triggered by reflected shock passage. This nonsteady "reaction wave" is fundamentally different from the "flames" discussed in Chap. 8 because its velocity is dictated by (1) the reflected shock velocity, and (2) a subtle interaction of local gas dynamics and reaction kinetics.

Two separate theoretical approaches have been used to model flows of the type illustrated in Fig. 13-1. In the first a constant-gamma gas containing a thermally neutral chain-branching reaction with an activation energy and oxygen concentration dependence similar to that of the high-temperature hydrogen–oxygen induction zone reaction was assumed to control the length of the induction zone along each particle path. This induction region was assumed to be followed by a region of heat addition in which the effect of the recombination reactions could be modeled by the equation

$$\frac{q}{a_{rs}^2} = \frac{q_{CJ}}{a_{rs}^2} \left[ 1 - \exp\left(-t/\tau_{rec}\right) \right]$$

where $\tau_{rec}$ is a characteristic time and $t$ is measured from the end of the induction zone for each fluid particle. This model is a reasonable representation of the recombination kinetics for the hydrogen–oxygen system. This model was

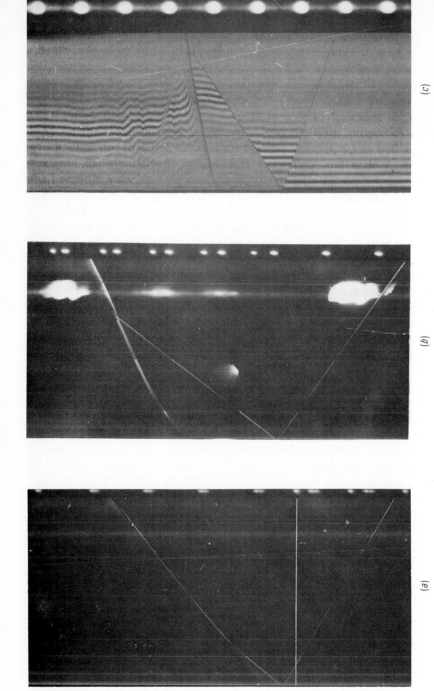

(c)

(b)

(a)

**Figure 13-1** (x, t) photographs obtained behind the reflected shock in a conventional shock tube containing $2H_2 + O_2 + 70\%$ argon mixtures. Time increases upward, back wall to the right, timing dots are 100 $\mu s$ apart on left. (a) Schlieren photograph, (b) Schlieren photograph, (c) Schlieren interferometric photograph.

398

programmed for an $(x, t)$ method of characteristics calculation in a constant-gamma gas with the boundary conditions $u = 0$ at the end wall and $u = 0$ immediately downstream of the reflected shock. In this model, each fluid element of the quiescent gas downstream of the reflected shock initially starts "exploding" at a constant time after shock passage. Thus the system first produces a reaction wave which is traveling away from the end wall at the reflected shock velocity. The initial velocity of the reaction wave (i.e., the locus of the beginning of heat addition in the $x, t$ plane) will therefore be highly subsonic relative to the gas. However, an adiabatic explosion occurring in the quiescent region at the end wall results in a local pressure rise and causes a compression wave to propagate ahead of the reaction wave. This process has been found to be very strongly self-accelerating because the compression process both heats and raises the density of a gas element which has not yet completed its induction reaction. Thus the induction period of an element of gas which is located some distance from the end wall is decreased by this process. This causes heat addition to occur earlier, and the compression wave is accelerated. This model has been very favorably compared to observations of initiation in the hydrogen–oxygen system. These results are summarized in Fig. 13-2. We note from Fig. 13-2a that the calculated reaction-wave–shock-wave pattern overlays the experimental reaction-wave–shock-wave pattern exactly (the reaction wave and reaction shock wave are unresolvable on the photograph). We also see that the calculated particle paths asymptotically approach the experimental contact surface motion in the figure. In Fig. 13-2b we have plotted the calculated $(P, v)$ behavior of various individual particle paths which start at the condition $rs$, at different distances from the end wall of the tube. These show that an element of gas at the end wall undergoes a simple adiabatic explosion (in a vessel whose volume increases slightly during the process), while all other particles first undergo a shock transition (shown either as a dashed line or as an absence of a line) followed by a further isentropic compression process until heat addition occurs. At this point there is a discontinuous change of slope to a slope corresponding to the heat-addition region of the flow along that particle path. We note that for particles near the end wall the pressure still rises continuously during heat addition but that farther from the end wall the particles undergo an expansion process during heat addition. As is shown, this process could only be followed out to the dimensionless distance $\xi_0 = 11$ in the calculation. The CJ detonation behavior for the region behind the reflected shock is plotted on Fig. 13-2b for comparison with the behavior of these nonsteady path process lines. We note (without any real proof) that the process appears to be systematically approaching the heat-addition behavior dictated by steady CJ conditions behind the reflected shock. The point BWRS in this diagram has been calculated for the isentropic expansion of the gases leaving the traveling detonation at the point CJRS to the end-wall velocity through a rightward-facing expansion fan. We note further that the final pressure along each streamline in the nonsteady calculation appears to be approaching this pressure but that the final volume of each point will be slightly different for each path process. This results from the fact that entropy addition along the different streamlines is different due to processing by a variable-strength shock and to the assumption that $q_{CJ}$ = const for each particle (which leads to a different chemical entropy change along each particle path). The different nature of the two processes shown in Fig. 13-1 may be explained by noting that the line S-CJ-BW in Fig. 13-2$b$ is the locus of a CJ detonation occurring behind the incident shock in the flow. Thus if the acceleration process produces a local pressure above that of point $S$, the interaction of the reflected shock and the reaction shock will initially produce an overdriven detonation wave in the incident gases as shown in Figs. 13-1$b$ and $c$. However, if the acceleration process has not progressed for a sufficient time to yield a pressure above the pressure at point $S$ at the time of intersection, the compression wave will simply cause the reflected shock to accelerate to a CJ detonation wave as shown in Fig. 13-1$a$.

A few more comments about this model and its agreement with experiment may be made. First we note that the model implicitly assumes that transport

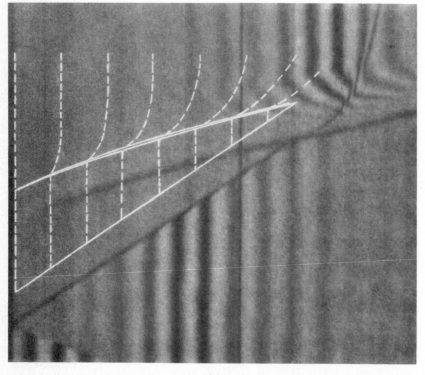

($a$)

**Figure 13-2** ($a$) Comparison of an ($x, t$) experimental photograph with a theoretical wave pattern calculated for the initial conditions. Theoretical pattern is displaced upward about one-third of the delay time at the back wall on the left. ($b$) A ($P, V$) diagram for the process shown in ($a$). Each $\xi_0$ line is the ($P, v$) behavior of a particle path. $\xi_0$ is the dimensionless distance of the particle from the back wall at the moment it is contacted by the reflected shock.

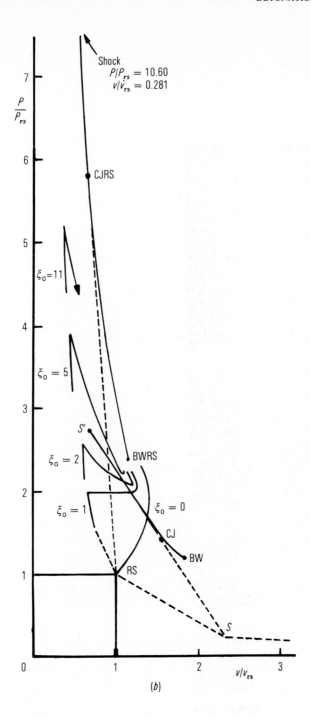

**Figure 13-2** (*continued*)

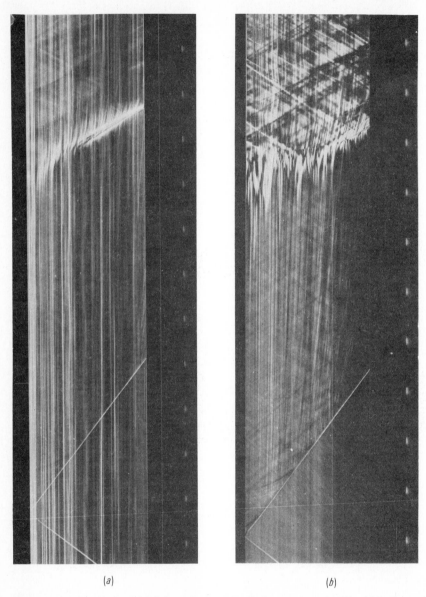

(a)                                                              (b)

**Figure 13-3** Reflected shock-initiation photographs similar to those in Fig. 13-1, illustrating the occurrence of hot-spot initiation at low temperature and high pressure. Timing dots on the right are 100 $\mu$s apart and test section is 200 mm long. Time increasing upward, back wall to the left. Mixture is 0.08 $H_2$ + 0.02 $O_2$ + 0.90 Ar. (a) $P_0$ = 2.4 kPa, $T_{rs}$ = 972 K. (b) $P_0$ = 3.3 kPa, $T_{rs}$ = 1020 K. *(Courtesy of A. Cohen, Ballistic Research Laboratories, Aberdeen Proving Ground, Maryland.)*

phenomena are absent or not important. Thus the "reaction wave" of this model has no similarity to the "flames" of Chap. 8. Second, we have assumed that the local kinetics at any point in the flow may be modeled by assuming that the kinetics of an ordinary static system may be applied at each point in the flow. At present the validity of this assumption is justified only by the results. There is no a priori reason for its use except that it is the simplest assumption that one can make to model these nonsteady flows. Third, the model predicts that the reaction shock wave will always start as a sound wave at the end wall and that its acceleration starts at zero and goes through a maximum before again returning to zero. This behavior is primarily the result of the assumption that the rate of heat addition is finite along each particle path, and that heat addition starts along a subsonic reaction wave. Fourth, it is sometimes observed experimentally that a weak compression wave precedes the shock wave for some distance. The model discussed here does not predict this behavior because it assumes that there is a discontinuous change in the rate of heat addition at the start of the heat-addition process. If one were to remove this restriction the model would undoubtedly yield a relatively broad compression region ahead of the reaction shock in the neighborhood of the end wall. Lastly, we note that sub-CJ nonsteady shock waves are both observed experimentally and predicted theoretically for this case of homogeneous one-dimensional initiation. Thus we see that sub-CJ shock waves followed by an exothermic reaction are physically realizable as nonsteady processes.

The fourth type of reflected shock-initiation behavior that has been observed in the hydrogen–oxygen system is illustrated in Fig. 13-3. In this case, even though the system is apparently homogeneous, hot spots of weak ignition occur and these each propagate spherically growing flames. When either these flames or their pressure waves coalesce they produce a local detonation and the initiation process is no longer one-dimensional. This type of initiation has been observed to occur in hydrogen–oxygen systems at high pressure and low temperature where the reaction

$$H + O_2 + M \rightarrow HO_2 + M$$

becomes important during the chain-branching portion of the reaction. We note from Fig. 6-19 that this is exactly the $(P, T)$ region where the delay to explosion becomes extremely sensitive to the initial conditions and therefore slight fluctuations in the properties of the mixture that is exploding can cause large differences in the explosion delay of localized regions.

An extension of the reaction theory of initiation to the process of initiation behind an accelerating incident shock has led to the observation that these same principles may be used to predict the locus of the explosion line behind an accelerating shock wave. The accelerating shock has been produced by passing an incident step shock wave into a slowly converging channel so that the phenomenon may be approximately modeled as a one-dimensional process. In this case the locus of the explosion line may be determined by finding the fraction of

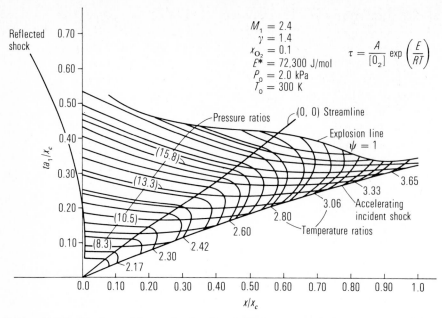

**Figure 13-4** Isopressure and isotemperature lines behind an accelerating shock wave produced by passing a step shock into a slowly convergent channel. The locus of the explosion line for high-temperature hydrogen–oxygen induction-delay time is shown. (*Adapted from R. A. Strehlow, A. J. Crooker, and R. E. Cusey, Combust. Flame, vol. 11, p. 339 (1967). With permission.*)

explosion delay time, $\psi$, consumed at any point along a particle path by integrating the equation

$$\psi = \int_{t_s}^{t} \frac{dt}{\tau([O_2], T)} \tag{13-1}$$

where $\tau([O_2], T)$ is the local value of the explosion delay time and $t_s$ the shock passage time along the particle path. A typical calculation of the flow properties and explosion delay locus ($\psi = 1$) is shown in Fig. 13-4. In this calculation heat was not added at the explosion line. Therefore its minimum in true time is the only explosion detail which may be calculated by this technique. Note that the isopressure and isotemperature lines in this plot are parallel in the region to the left of the streamline running through the point (0, 0). This region is an isentropic flow region (because the reflected shock is weak), and the local pressure history therefore conforms to the local temperature history. However, to the right of the entrance streamline the temperature rises markedly because of the large entropy gradient produced by shock acceleration. Thus, as shown in this figure, the locus of the explosion line obtained by integrating Eq. (13-1) along a particle path from the shock wave to the point where $\psi = 1$ exhibits a dip in the $(x, t)$ plane, even though the total delay time along neighboring particle paths is decreasing monotonically as the shock strength increases. This behavior has been verified experimentally.

**Figure 13-5** A smoked-foil record taken during the initiation experiment of Fig. 13-1c. Back wall is to left. Discontinuity in writing pattern about 90 mm from back wall is point at which the reflected shock and reaction shock merged to produce a detonation wave. The original smoked-foil is 100 mm wide.

One interesting observation that has been made during initiation experiments is that transverse waves appear spontaneously even during the most nicely controlled one-dimensional initiations. Figure 13-5 is a smoked-foil record taken on the bottom wall of the shock tube during the initiation experiment of Fig. 13-1c. Note that the structure appears some distance from the point of initiation at the end wall, and that the transverse waves are weak at their appearance point and then grow in amplitude as they propagate away from the end wall. This is evident from the refraction of the waves at their intersections. We also note that the transverse waves can appear before the reaction shock intersects the reflected shock. This means that the occurrence of transverse waves is not limited to CJ detonation, but that they can appear spontaneously on a shock which is simply followed by an exothermic reaction.

## 13-3 DIRECT INITIATION BY BLAST WAVES

The rapid deposition of sufficient energy in a small volume or along a line can lead to the direct initiation of spherical or cylindrical detonation, respectively. Lee and his coworkers[2] have done most of the recent work on spark initiation. Bach et al.[3] first reported that there is a critical power density for initiation by spark which is defined as the energy deposited divided by the spark kernel volume and the time of deposition. This quantity is invariant when different spark sources are used, even though the total spark energy ranges over a factor of $10^3$. The spark energy was varied by such a large factor by using 20 nano-

2. J. H. S. Lee and I. O. Moen, *Prog. Energy Combust. Sci.*, **6**:359–389 (1980).

3. G. G. Bach, R. Knystautas, and J. H. S. Lee, *Thirteenth Symposium (International) on Combustion*, The Combustion Institute, Pittsburgh, Pa., p. 1097 (1971).

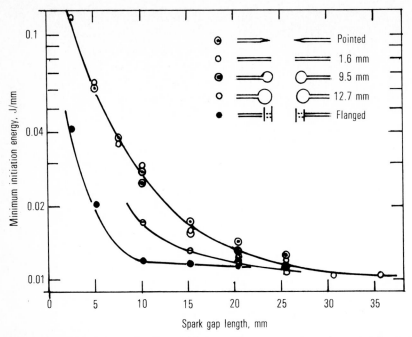

**Figure 13-6** Critical energy for direct initiation with respect to electrode spacing for four electrode geometries in stoichiometric oxyacetylene mixture at 13.2 kPa initial pressure. *(Adapted from Ref. 5, with permission.)*

second laser sparks, 1 to 10 microsecond conventional capacitance sparks, and 100 microsecond exploding wires. Matsui and Lee[4] next showed that electrode shape had a profound effect on the initiation energy per millimeter of gap length for line sparks, particularly for short gaps. This is shown in Fig. 13-6. Finally, Knystautas and Lee[5] showed that the first half-cycle of a ringing-spark discharge was the only phase of the discharge that determined the critical energy for direct initiation. In that paper, they also determined that for very short sparks there is a critical minimum energy for initiation even though for sparks of longer duration there is a limiting power. This is shown in Fig. 13-7.

These experimental observations show that spark initiation of detonation occurs when the spark generates a sufficiently strong shock that has a sufficient duration to heat the exothermic mixture to above its autoignition temperature and hold it there for a time that is longer than the induction time. This has been verified theoretically by Abouseif and Toong.[6] This is the reason why, for long-duration sparks, the power density is important. Power density determines shock strength when the shock duration is long and, as Fig. 13-7 shows, the

4. H. Matsui and J. H. Lee, *Combust. Flame*, **27**:217–220 (1976).
5. R. Knystautas and J. H. S. Lee, *Combust. Flame*, **27**:221–228 (1976).
6. G. E. Abouseif and T. Y. Toong, *Combust. Flame*, **45**:39–46 (1982).

power for initiation increases with total energy, because the kernel size also increases with energy. Also note that there is a threshold critical energy below which initiation will not occur. This minimum energy reflects the fact that a critical minimum shock duration at a critical minimum strength is required for initiation. Finally, the fact that only the first half-cycle of the power pulse from the spark is important to initiation reinforces the view that it is the initial shock wave generated by the spark that causes direct initiation.

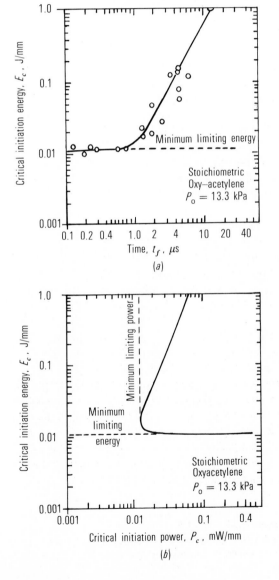

**Figure 13-7** (*a*) The variation of critical initiation energy with time of energy deposition. *(With permission from Ref. 5.)* (*b*) The dependence of critical initiation energy on the peak average power of the source. Note that there is a minimum initiation energy also. *(With permission from Ref. 5.)*

Recently Lee and coworkers[7] have shown that the critical energy for direct initiation of detonation is directly related to the characteristic transverse spacing in a well-developed propagating detonation. Specifically they find that, for a large variety of fuel–oxidizer mixtures,

$$E_c = \frac{287.6}{(\gamma_0 + 1)} P_0 M_{CJ}^2 Z^3$$

where $E_c$ is the critical energy for direct initiation, J, $P_0$ is the initial pressure, Pa, $\gamma_0$ is the heat capacity ratio of the initial fuel-oxidizer mixture, and $M_{CJ}$ and $Z$ are the Chapman–Jouguet detonation Mach number and cell spacing, m, respectively.

## 13-4 DIRECT INITIATION BY RADICALS (SWACER)

Direct initiation of detonation under conditions where the autoignition temperature is never exceeded has been observed and studied by Lee et al.[8] They irradiated acetylene–oxygen, hydrogen–oxygen and hydrogen–chlorine mixtures in a chamber with ultraviolet light, using quartz windows. The ultraviolet light is absorbed either by the oxygen or chlorine molecule and causes these molecules to dissociate to form the free radicals O and Cl, respectively. These radicals, if in sufficiently high concentration, will trigger chain reactions and thereby cause a local "explosion" in the system. Three different explosion behaviors are observed. If the light intensity is low, the UV light penetrates only a very thin layer near the window and the radicals in this layer ignite a flame or deflagration wave which propagates away from the wall. For very high intensity irradiation, absorption is rather uniform throughout the volume, and the uniform high radical concentration causes a constant-volume explosion of the vessel contents. In this case, there are no waves of any consequence in the system during the explosion.

In the case of intermediate levels of radiation, a detonation is observed to form immediately and propagate away from the irradiated window. This behavior can be explained in the following way. The irradiation produces a high concentration of radicals at the wall and more importantly a gradient of radical concentration away from the wall. When this gradient has the proper shape, the localized explosion of the layer nearest the wall generates a pressure wave which pressurizes the neighboring layer and shortens the explosion delay time in that layer. This augmentation continues until a fully developed CJ detonation wave is formed. The authors of Ref. 8 call this Shock Wave Amplification by Coherent Energy Release or the SWACER effect. They make an analogy to the

7. J. H. S. Lee, R. Knystautas, and C. M. Guirao, "The Link between Cell Energy Critical Tube Diameter, Initiation Energy, and Detonatability Limits," *Fuel Air Explosions*, J. H. S. Lee and C. M. Guirao, eds., *SM Study 16*, University of Waterloo Press, Waterloo, Canada, pp. 157–188 (1982).
8. J. H. S. Lee, R. Knystautas, and N. Yoshikawa, Acta Astronautica, 5:971–982 (1978).

behavior of a laser because the simple explosion of one layer would not cause a significant reaction in the next layer if it had not been preconditioned to be ready to explode.

The significance of this mechanism of direct initiation is that the process is not triggered by the presence of a strong shock wave but instead offers a chemical mechanism for *generating* an accelerating shock wave which quickly reaches the CJ detonation velocity.


## 13-5 DETONATION LIMITS AND FAILURE

It has long been observed that any premixed fuel–oxidizer mixture has a range of composition within which a self-sustaining detonation will propagate through a long straight tube. It has also been observed that there is a pressure level, below which a self-sustaining detonation will not propagate in a normally detonatable mixture. This behavior is analogous to the flammability limit behavior of flames discussed in the previous chapter. The analogy ends at this level, however, because the mechanisms of failure and extinction are as different as the mechanisms of initiation and ignition.

A near-limit detonation in a straight tube with a circular cross section propagates either as the single-spin detonation of Fig. 9.2 or as a galloping detonation. A galloping detonation is one which degenerates into a shock followed by a separated reaction zone, but periodically reignites to form an overdriven detonation. The length of a typical cycle is about 60 tube diameters. Galloping detonations occur slightly outside the single-spin range and have not been studied extensively.

The fact that a limit composition detonation propagates as a single-spin detonation in a long straight tube of circular cross section indicates that at the limit the tube diameter and the characteristic cell size of that particular fuel–oxidizer mixture should be simply related. Lee and coworkers have shown that in all systems studied, the critical tube diameter for failure is equal to $Z/2$ where $Z$ is the characteristic cell size discussed in Secs. 9-3 and 9-4.

Direct transmission of a detonation into a large open volume from a long straight pipe or through an orifice has been found to be possible only if the pipe or orifice is larger than some critical size, which is dependent on the cross-sectional shape of the pipe or orifice. If the section is circular, the critical diameter for transmission is $13Z$.[9] If it is square the critical diameter is $11Z$. The critical width for a long slot is $3Z$.[10] It is interesting to note that a pipe and orifice have the same effectiveness relative to transmission. This simply means that it is only the rate of decay of the lead shock that determines whether or

9. R. Knystautas, J. H. S. Lee, and C. M. Guirao, *Combust. Flame,* **48**: 63–84 (1982).

10. J. H. S. Lee, R. Knystautas, C. Guirao, W. B. Benedick, and J. E. Shepard, "Hydrogen–Air Detonation," a paper presented at the Second International Workshop on the Impact of Hydrogen on Water Reactor Safety, at Albuquerque, N.Mex. on October 3–7, 1982.

not transmission occurs and that the flow behind the coupled shock-reaction zone of the detonation is not important to transmission. It has also been observed that two or more closely spaced orifices will allow the transmission of detonation even though the orifices are below the critical sizes listed above. This is because, for this geometry, the emerging and expanding shock fronts from neighboring orifices will interact to produce hotter regions where reignition can occur.

## PROBLEMS

**13-1** Figures 13-1*b* and *c* show the wave behavior for planar detonation initiation. Construct the appropriate diagram in the *P–u* plane and show that the intersection of the newly formed detonation with the reflected shock produces an overdriven detonation traveling into the incident gas and a rarefaction wave propagating towards the end wall. Assume that the detonation in the reflected shock region is CJ (it is not). Do no calculations—just construct the (*P*, *u*) diagram qualitatively.

**13-2** A chemical plant is handling a gaseous propane–butane mixture at a pressure of 80 kPa in a pipeline system of 1 m diameter. Since the pressure is less than atmospheric there is the possibility of an air leak into the pipeline system and subsequent ignition in the piping. Can a detonation propagate through the piping under these conditions? Will such a detonation be transmitted into a large process vessel? If detonation can be transmitted what size orifice should be used to prevent transmission?

**13-3** Determine the lean and rich detonation limits of a hydrogen–air mixture in straight pipes of 10 mm diameter and 0.1 m diameter.

**13-4** Calculate the energy required to directly initiate a detonation in a stoichiometric ethylene–air mixture and in a stoichiometric methane–air mixture.

## BIBLIOGRAPHY

Fickett, W., and W. C. Davis, *Detonation*, University of California Press, Berkeley (1979).

Lee, J. H. S., and C. M. Guirao, eds., *Fuel–Air Explosions, SM Study 16*, University of Waterloo Press, Waterloo (1982).

Oppenheim, A. K., *Introduction to Gasdynamics of Explosions*, International Centre for Mechanical Sciences, Courses and Lectures No. 48 CISM, Undine (1972).

Shchelkin, K. I., and Y. K. Troshin, *Gas Dynamics of Combustion*, Mono, Baltimore (1965).

Voitsekhovskii, B. V., V. V. Mitrofanov, and M. E. Topchian, *Structure of the Detonation Front in Gases*, Izd-vo Sibirsk. Otdel. Akad. Nauk., SSSR, Novosibirsk (1963).

# FOURTEEN

## COMBUSTION AERODYNAMICS—
## NONSTEADY FLOWS

## 14-1 INTRODUCTION

We have already discussed, in Chaps. 10 and 11, the aerodynamics of steady flows with combustion, both laminar and turbulent. We now wish to discuss the aerodynamics of nonsteady flows with combustion, both laminar and turbulent. These flames are transient and are important because they represent a class of combustion phenomena that are normally unwanted or dangerous. Combustion-induced instabilities, which are usually simply unwanted, have been encountered during the development of many high-intensity continuous-combustion engines, such as rocket engines and afterburners for jet engines. They will be discussed first. Other nonsteady combustion phenomena, such as transient flame propagation, will be discussed next. Flame acceleration and transition to detonation, which can be dangerous because they represent processes that are operative during accidental explosions, will be discussed last.

## 14-2 ACOUSTIC INSTABILITY

In Sec. 5-9 we investigated the nature of the acoustic oscillations that could be generated and sustained in a pipe with open and closed ends. We did this both for the simple case where no energy is being added to the system and for the case where, even though there is no net energy addition, there is a fluctuating

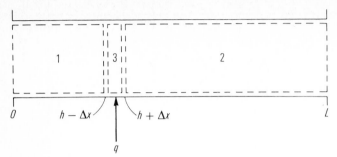

**Figure 14-1** Tube geometry for acoustic model. Ends (at 0 and $L$) may be on either open, as shown, or closed.

and homogeneous flow of energy in and out of the system, which is proportional to the magnitude of the fluctuations of the density in the system and has some phase relation to those fluctuations. In that analysis, we found that if the coupling constant $k$ is positive, the amplitude of the oscillations will grow exponentially with time [see Eq. (5-78)]; i.e., the system is linearly unstable to any perturbation. In Chap. 9, this principle was applied to the structure of a one-dimensional ZND detonation to assess the inherent stability of that structure, and it was found to be linearly unstable at all sufficiently high frequencies.

One of the assumptions used in deriving Eq. (5-78) makes the application of this equation to combustion-driven instabilities questionable. In most combustion systems that can exhibit combustion instability, heat release is usually concentrated in only one region rather than being distributed homogeneously throughout the system, as was assumed in the derivation. We therefore approach the problem again, using a different model which is more closely related to physical reality than the homogeneous model discussed above.[1] The result of this new analysis is a formal mathematical proof of the criterion for instability which Rayleigh first stated without proof in 1878.[2]

Consider a tube of length $L$ with either open or closed ends containing a homogeneous heater in the region $h - \Delta x$ to $h + \Delta x$, as shown in Fig. 14-1. We assume as before that the acoustic approximations hold, that there is no mean flow, and that $\int_{cycle} q \, dx = 0$ at the heater. We also assume that $2\Delta x \ll L$.

For this analysis we divide the tube into three control volumes as labeled in Fig. 14-1 and calculate the rate of change, with time, of the energy stored as acoustic energy in volumes 1 and 2 so that we may relate this to a fluctuating heat input at the heater through a set of compatible boundary conditions at the interfaces 1–3 and 2–3. We use the theorem of Gauss, that states that

$$\int_A \phi_i n_i \, dA = \int_V \frac{\partial \phi_i}{\partial x_i} \, dV$$

1. Boa-Teh Chu, "Stability of Systems Containing a Heat Source—The Rayleigh Criterion," NACA RM 56D27 (1956).
2. J. W. S. Rayleigh, *Nature*, **18**: 319 (1878).

and we apply this to the quantity $Pu$ for one-dimensional flow occurring in the control volumes 1 and 2 for a tube of unit cross-sectional area. For the first control volume we obtain

$$(P_1 + \delta P_1)_{h-\Delta x}(u_1 + \delta u_1)_{h-\Delta x} - (P_1 + \delta P_1)_0(u_1 + \delta u_1)_0$$

$$= \int_0^{h-\Delta x} \frac{\partial}{\partial x} [(P_1 + \delta P_1)(u_1 + \delta u_1 + \delta u_1)] \, dx$$

Since $P_1 = $ constant and $u_1 = 0$ by definition, this equation may be written as

$$(\delta P_1)_{h-\Delta x}(\delta u_1)_{h-\Delta x} - (\delta P_1)_0(\delta u_1)_0$$

$$= \int_0^{h-\Delta x} \left[ \delta P_1 \frac{\partial}{\partial x}(\delta u_1) + \delta u_1 \frac{\partial}{\partial x}(\delta P_1) \right] dx \qquad (14\text{-}1)$$

However, for an open end $(\delta P_1)_0 = 0$ while for a closed end $(\delta u_1)_0 = 0$. Therefore, the second term on the left-hand side of Eq. (14-1) becomes zero for either of these cases. Furthermore, the coordinate derivatives under the integral on the right-hand side of Eq. (14-1) may be replaced by time derivatives using Eqs. (5-68) and (5-69) to yield the equation

$$(\delta P_1)_{h-\Delta x} \cdot (\delta u_1)_{h-\Delta x} = -\frac{\partial}{\partial t} \int_0^{h-\Delta x} \left( \frac{1}{2} \rho_1 a_1^2 s_1^2 + \frac{1}{2} \rho_1 u_1^2 \right) dx \qquad (14\text{-}2)$$

Here we have defined $s = \delta P_1/\gamma P_1 = \delta \rho_1/\rho_1$, since $q = 0$ in this control volume. In a similar manner Gauss' theorem when applied to the second control volume yields the equation

$$(\delta P_2)_{h+\Delta x} \cdot (\delta u_2)_{h+\Delta x} = \frac{\partial}{\partial t} \int_{h+\Delta x}^L \left( \frac{1}{2} \rho_2 a_2^2 s_2^2 + \frac{1}{2} \rho_2 u_2^2 \right) dx \qquad (14\text{-}3)$$

The integrands in Eqs. (14-2) and (14-3) represent the total acoustic energy contained in these two control volumes at any instant of time. The second term under the integral represents the kinetic energy per cubic meter, while the first term is the internal or potential energy of the gas relative to the energy of a fluid element at rest at a pressure $P_0$ and a denisty $\rho_0$. To prove this we consider an elemental volume of fluid $\Delta V$ whose volume at $s = 0$ is $\Delta V_0$ where $\Delta V = \Delta V_0(1 - s)$. The potential energy (P.E.) is defined as either the work which must be performed on the system to reversibly reach the desired state from the reference state or the work that the fluid element does on its surroundings when it is reversibly returned to the reference state. For our case this definition may be summarized with the equation

$$\text{P.E.} = \int_{\Delta V}^{\Delta V_0} \delta P \, d(\Delta V)$$

However, $d(\Delta V) = -\Delta V_0 \, ds$ and $\delta P = \rho_0 a_0^2 s$. Therefore

$$\text{P.E.} = -\Delta V_0 \int_s^0 \rho_0 a_0^2 s \, ds = \frac{\rho_0 a_0^2 s^2}{2} \Delta V_0$$

Thus in the framework of acoustic theory the total potential energy of the gas in a control volume is

$$\text{P.E.} = \int_V \frac{\rho_0 a_0^2 s^2}{2} \, dV$$

and Eqs. (14-2) and (14-3) may be added to obtain

$$(\delta P_2)_{h+\Delta x} \cdot (\delta u_2)_{h+\Delta x} - (\delta P_1)_{h-\Delta x} \cdot (\delta u_1)_{h-\Delta x} = \frac{\partial}{\partial t}(E) \qquad (14\text{-}4)$$

where $E$ is the total energy in the acoustic modes of the tube.

We now turn our attention to the third control volume, where the heater is located. We assume that $2\Delta x$ is sufficiently small so that $\delta P_2 = \delta P_1$ at any instant of time (i.e., the region has uniform static properties throughout) and we write the specifying equation for this region as

$$\frac{\partial(\delta\rho)}{\partial t} = \frac{1}{a^2} \frac{\partial(\delta P)}{\partial t} - \frac{\gamma - 1}{a^2} q$$

This equation may be integrated over the control volume to yield

$$\int_{h-\Delta x}^{h+\Delta x} \left[ \frac{a^2}{\gamma - 1} \frac{\partial(\delta\rho)}{\partial t} - \frac{1}{\gamma - 1} \frac{\partial(\delta P)}{\partial t} \right] dx = - \int_{h-\Delta x}^{h+\Delta x} \rho_0 q \, dx = -\delta q$$

where we now define $\delta q$ to be independent of $\Delta x$. This will allow us to reduce the extent of the heating region without varying the magnitude of the heat-addition term. Substituting Eq. (5-68) we obtain

$$\int_{h-\Delta x}^{h+\Delta x} \left[ \frac{a^2 \rho}{\gamma - 1} \frac{\partial u}{\partial x} - \frac{1}{\gamma - 1} \frac{\partial(\delta P)}{\partial t} \right] dx = -\delta q \qquad (14\text{-}5)$$

(it should be noted that $\delta\rho/\rho_0 \neq s$ in this region because of the heat addition). Equation 14-5 may be rewritten to yield

$$\frac{a^2}{\gamma - 1} [(\delta u_2)_{h+\Delta x} - (\delta u_1)_{h-\Delta x}] + \int_{h-\Delta x}^{h+\Delta x} \frac{1}{\gamma - 1} \frac{\partial(\delta P)}{\partial t} \, dx = \delta q \qquad (14\text{-}6)$$

Combining Eqs. 14-4 and 14-6 and recalling that

$$(\delta P_1)_{h-\Delta x} = (\delta P_2)_{h+\Delta x} = \delta P$$

yields the equation

$$\frac{\partial}{\partial t}(E) = \delta P[(\delta u_2)_{h+\Delta x} - (\delta u_1)_{h-\Delta x}]$$

$$= \frac{\delta q \delta P(\gamma - 1)}{a^2 \rho} - \delta P \int_{h-\Delta x}^{h+\Delta x} \frac{\gamma P_0}{\gamma - 1} \frac{\partial s}{\partial t} \, dx \qquad (14\text{-}7)$$

We now assume that $\Delta x \rightarrow 0$ and integrate Eq. 14-7 over a single cycle of a pure mode of the tube to obtain the equation

$$\Delta E_{\text{cycle}} = \frac{(\gamma - 1)}{\gamma P_0} \int_{\text{cycle}} \delta P \cdot \delta q \, dt \qquad (14\text{-}8)$$

Equation 14-8 states that the local-energy addition or abstraction must be positively correlated with a naturally available pressure fluctuation in the system to produce a spontaneous amplification of that particular pressure fluctuation. While this theory was developed for an acoustic fluctuation, its application is not limited to acoustic-mode amplification. Examples of nonacoustic modes which are amplified by this mechanism are given in Secs. 15-5 and 15-6.

The criterion developed above has been found to be generally applicable to the discussion of combustion instabilities in a variety of systems. We note, however, that it places no limitation on the nature of the coupling which leads to positive correlation (i.e., to amplification). Thus while this theory may be used to discover if a particular coupling mechanism is viable in a system, it cannot be used to rule out all other possible types of coupling that may occur in the system. In other words, even though the theory cannot predict the stability of a real system it can be used to discuss possible mechanisms of coupling in systems which are known to be unstable.

We note that in real "steady combustors," such as rockets or jet engines, heat is continually being released and the background conditions appear steady because the hot gases that are produced by the combustion process are continuously flowing out of the chamber. This background flow can be included in any analysis. However in most cases the velocity of this flow is low and neglecting it is a reasonable approximation because, even though it causes resonant frequencies to shift somewhat, it does not markedly affect the rate of growth of the instability. We will neglect mean flow effects in our discussions of combustion instability. The interested reader is referred to the bibliography for further information on the subject.

In order to show how Rayleigh's criterion, Eq. (14-8), allows one to evaluate a particular case of acoustic instability, we will investigate the behavior of one simple system, the Rijke tube. In 1859 Rijke[3] discovered that an open vertical tube length $L$ containing a heated gauze at a position $L/4$ from the bottom spontaneously sounds an extremely loud tone. If the gauze is heated by an external burner he found that the tone becomes audible approximately one second after removal of the burner, and that after this time lag the amplitude of the acoustic vibration grows to a maximum and then decays. He also observed that an electrically heated gauze produces a continuous tone whose amplitude decreases slowly as the surrounding tube is warmed by the heater. Furthermore, he found that the maximum sound level that is reached occurs when the gauze is located at a position $L/4$ from the bottom end of the tube.

3. P. L. Rijke, *Phil. Mag.*, **17**:419 (1859).

Elevation
in tube

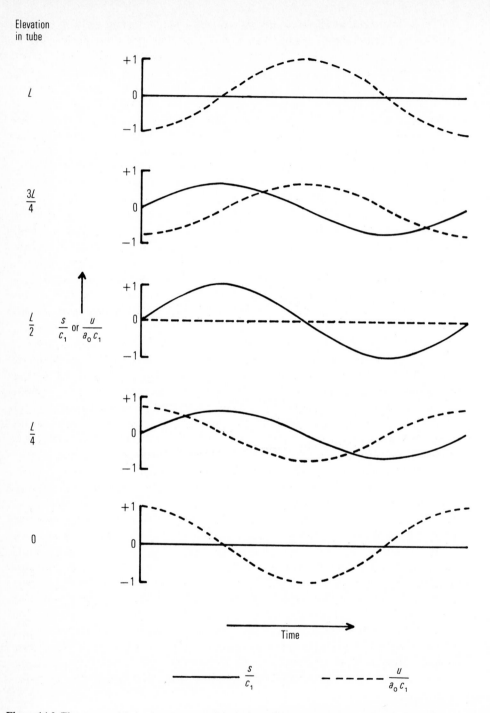

**Figure 14-2** The temporal behavior of the condensation and flow velocity in a tube that is open at both ends. Note how the two fields are coupled and how the coupling depends on location. This figure is replotted from Fig. 5-10.

In order to model the configuration of the Rijke tube, assume that the tube of Fig. 14-1 is rotated to a vertical position so that the end labeled zero is down. Also assume that region 3 in that figure represents the location of the heated gauze. Recall, from Sec. 5-9, that the acoustic pressure and acoustic velocity fields are coupled in a unique way for any standing mode of the tube. This coupled $(u, s)$ field is given by Eqs. (5-73) and (5-74) for all resonant frequencies of the tube and is graphed in Fig. 5-10 for the fundamental or first organ-pipe mode. The wavelength of this mode is $2L$. Figure 14-2 contains plots of the temporal behavior of the coupled $(u, s)$ fields for those positions in the tube that can best be used to explain the instability behavior of the Rijke tube.

We now note that heat can be transmitted from the wires of the heated gauze to the gas by conduction and convection. We also note that the presence of a heated gauze in a vertical tube will cause an upward flow through the tube and that convective heat transfer is a function of the velocity of that flow. We see, however, that if the rate of heat transfer is simply proportional to the flow velocity, as it is for a steady flow, there will be no net amplification for any heater position because, as Fig. 14-2 shows, there is no position in the tube which will yield the required positive correlation. However, it has been shown that heat transfer to a fluctuating flow from a transverse fine wire lags the flow velocity with a phase angle which approaches $\pi/2$ at reasonably high frequencies. As Fig. 14-2 shows, such a lag will cause a positive correlation in the fundamental mode if the heater is placed in the bottom quadrant of the tube. Figure 14-2 also shows that a cooled gauze placed in the upper quadrant with a forced upward convection should cause the tube to sound. This second effect was first observed experimentally by Bosscha and Riesse[4] in 1859.

It has also been observed that a premixed gas flame seated on a gauze flame holder in the tube will cause the tube to sound if the flame holder's position is the same as that for the heated gauze. Here there is a marked temperature difference across the gauze, but the theory may be applied by simply writing the equations for the upper and lower portions separately and connecting them with compatible boundary conditions at the flame's location. A careful investigation by Bailey[5] of both this and the Rijke tube effect has been reported. In his work the frequency of the first tone to sound was determined as a function of heater position and compared to a rather detailed theoretical analysis which included system losses. Bailey assumed that the coupling for the flame case was caused by a perturbation of the flame's burning velocity due to its nearness to the gauze. The agreement between experiment and theory was found to be quite good.

Other coupling mechanisms which cause instability behavior have been observed. One of the more interesting examples is the instability that may occur when a diffusion flame is located in a pipe as shown in Fig. 14-3. This resonance was first observed by Higgins[6] in 1777, using a hydrogen diffusion flame.

4. H. Bosscha and P. Riesse, *Pogg. Ann.*, **108**: 653 (1859).
5. J. J. Bailey, *J. Appl. Mech.*, **24**: 333 (1957).
6. B. Higgins, *Nicholson's Journal*, **1**, 130 (1802).

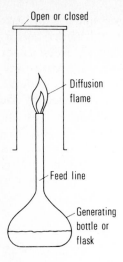

Open or closed

Diffusion flame

Feed line

Generating bottle or flask

**Figure 14-3** Diffusion-flame-driven resonance apparatus (Higgins, 1777).

In this case the coupling is with a resonant oscillation of the flow in the fuel feed line such that the size of the diffusion flame is caused to change with the proper phase relationship. Here the active coupling agent is pressure and the oscillation may be suppressed by simply changing the length and therefore the resonant frequency of either the feed line or the surrounding pipe. Kaskan[7] has observed and described another instability mechanism which requires that a flat flame propagating down a tube extend into the acoustic boundary layer before amplification is effected. This is undoubtedly the mechanism which causes tone production when a hydrogen flame flashes down a test tube after the flame has been ignited at the open end.

Solid-propellant and liquid-propellant rockets also show instabilities which may be related to Rayleigh's criterion. The discussion of these systems will be deferred to Chap. 15.

## 14-3 TAYLOR–MARKSTEIN INSTABILITY

Taylor[8] showed that a contact discontinuity between two fluids of different density is linearly unstable to acceleration in a direction perpendicular to its orientation if the acceleration is toward the lighter fluid. In other words, that type of contact surface will only remain flat if it is accelerated in such a way that the lighter fluid is pushing the heavier fluid. In a premixed combustible system, the transient flame that separates two fluids of markedly different densities, namely the high-density combustible mixture and the low-density product

7. W. E. Kaskan, *Fourth Symposium (International) on Combustion*, The Combustion Institute, Pittsburgh, Pa., p. 575 (1953).

8. G. I. Taylor, *Proc. R. Soc.*, **A201**: 192 (1950).

gases, while not a contact surface in the strict sense, responds to impulsive accelerations as though it were.

Figure 14-4 illustrates this flame instability for the case of a weak shock interacting with a spherically growing flame ball. Note that the shock is propagating downward and that the upper section of flame surface quickly becomes unstable while the lower surface remains stable until it is contacted by the highly accelerated upper surface. Markstein[9] first identified the cause of the interaction in the 1950s, even though earlier photographs show the effect. This effect is important when flames propagate in enclosures, because the increased flame area produced by the instability causes a more rapid rate of pressure rise. Further examples will be presented in the next two sections.

## 14-4 SIMPLE NONSTEADY FLAMES

By far the simplest nonsteady flames that can be generated are the growing spherical flames produced by central ignition inside a soap bubble, a free cloud, or a spherical vessel. In the first case the soap bubble expands and finally is destroyed without markedly influencing the pressure and the spherically growing flame burns at constant pressure (see Sec. 8-2). In a centrally ignited spherical vessel the flame also remains spherical until it contacts the walls of the vessel and extinguishes. We note, however, that if the bubble or vessel is too large, or if the normal burning velocity is too low, gravity can cause the flame to rise appreciably and distort, ultimately forming a torus-shaped flame whose central axis is vertical. The behavior of such a flame is shown in Fig. 14-5. This flame was ignited near the bottom of a spherical vessel of 0.2 m diameter. The normal burning velocity of this flame, estimated from the size of the flame at $t = 0.204$ s and the calculated expansion ratio, is 44 mm/s. Note the buoyant rise of the flame and the formation of a cusp on its underside. At $t = 0.476$ s the top of the rising flame contacts the top of the vessel and extinguishes. The extinguishment process continues until the final parcel burns out at the bottom of the vessel.

A centrally ignited spherical flame in a spherical vessel is aerodynamically simple if the burning velocity is high enough and the vessel is small enough, because the gas motion generated by the flame is everywhere radial and therefore normal to the wall. Thus, there is no possibility of a boundary-layer–flame interaction.

Lewis and von Elbe[10] have presented an extensive analysis of this system and have reached the following conclusions:

1. Throughout the propagation process the local flame may be assumed to be burning as though in a constant-pressure environment. This assumption is

9. G. H. Markstein, "A Shock Tube Study of Flame Front-Pressure Wave Interactions," *Sixth Symposium (International) on Combustion*, The Combustion Institute, Pittsburg, Pa., p. 387 (1957).

10. B. Lewis and G. von Elbe, *Combustion Flames and Explosions of Gases*, 2d ed., Academic, 1961, pp. 367–381.

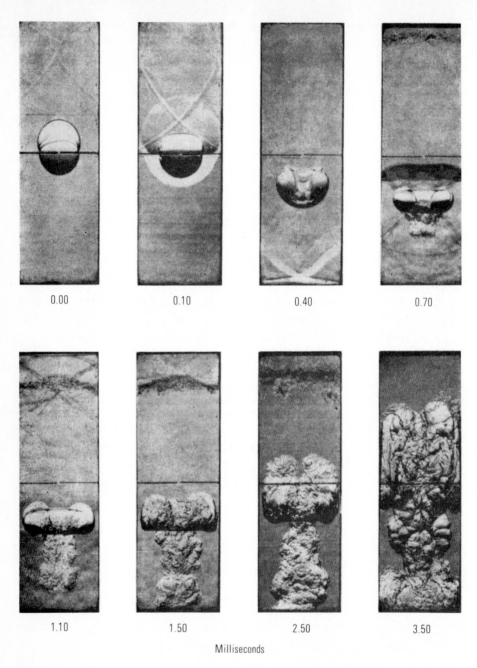

**Figure 14-4** Interaction of a shock wave and a flame of initially roughly spherical shape. Pressure ratio of incident shock wave 1.3; stoichiometric butane–air mixture ignited at center of combustion chamber, 8.70 millisec before origin of the time scale. *(Courtesy G. H. Markstein, Factory Mutual Research, Norwood, Mass.; originally in Combustion and Propulsion, Third AGARD Colloquium, p. 153, Pergamon, 1958, with permission.)*

valid, since pressure changes due to the formation of hot products are relatively slow during the combustion of a gas layer of thickness equivalent to the preheat-zone thickness of the flame.

2. Because of spherical geometry the early burning takes place at essentially constant pressure (when the flame radius is one-half of the bomb radius, the hot gas ball occupies only one-eighth of the bomb's volume).

3. The early pressure rise is a cubic function of time.

4. Burning velocity is not constant during the entire propagation process. This is the result of adiabatic compression which significantly raises the temperature and pressure of the outer shell of gas before it is reached by the combustion wave.

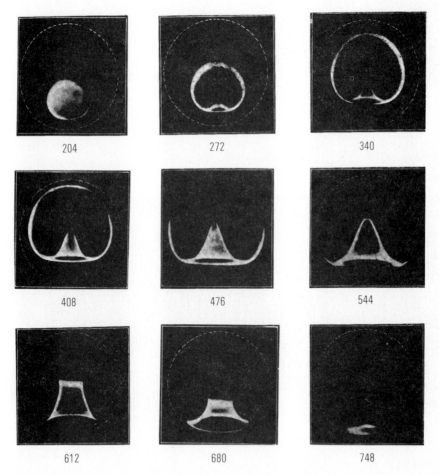

**Figure 14-5** A series of photographs of a very slow burning flame in a 13.5% CO–air mixture saturated with water vapor at 18 C. This flame is markedly affected by buoyancy. The numbers under each frame are the time after ignition in milliseconds. *(O. C. de C. Ellis, Fuel in Practice and Science, vol. 7, pp. 245–252 (1928), with permission.)*

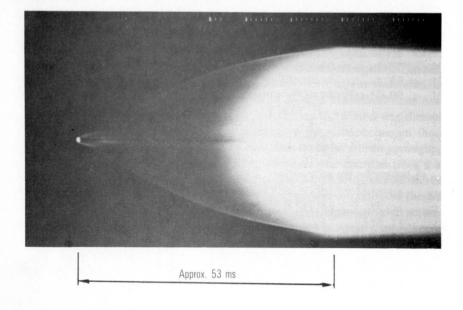

Approx. 53 ms

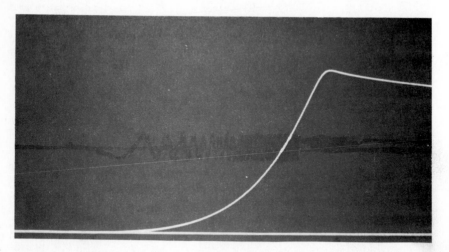

**Figure 14-6** A streak photograph of a centrally ignited flame in a spherical vessel and a pressure record taken at the same time. Stoichiometric propane–air mixture at $P = 101.3$ kPa and $T = 293$ K. *(Courtesy Dr. A. M. Garforth, Johannesburg, South Africa; Originally appeared in Combust. Flame, vol. 26, pp. 343–352 (1976).) (Adapted, with permission.)*

5. The combustion process produces a gas sample which is significantly hotter at the center of the vessel. We note that the thermodynamic path for a sample at the center is different than that for a sample in the outer shell because the center sample is first burned irreversibly and then compressed isentropically, whereas the outer shell sample is first compressed isentropically and then converted to products by the irreversible combustion process. Thus the specific entropy at the center will be significantly higher than that near the wall, and temperature will be higher in the center of the bomb after combustion is complete. This result has been verified experimentally.

The above observations mean that one should be able to determine the normal burning velocity of a flame, as a function of pressure along an isentrope, from careful measurements of the $(x, t)$ trace of the flame and the pressure-time behavior in the vessel. Techniques to do this have been developed at a number of laboratories. Usually, the gas ahead of the flame is assumed to be compressed isentropically as an inert gas, while the gas behind the flame is assumed to be compressed along a number of equilibrium isentropes, each corresponding to a specific spherical shell of initial gas. Each of these thin shells is programmed to burn sequentially at constant pressure and the resulting individual burned shells and the remaining gas ahead of the flame are each compressed isentropically by the same fractional amount until their total volume again equals the vessel volume. The pressure and the location of the flame are then determined at the end of that particular incremental burn. This information and the experimental data are used to determine the normal burning velocity at that pressure and temperature.

The results of such an experiment and its theoretical interpretation are shown in Figs. 14-6 and 14-7. It should be noted that this measurement of burning velocity has direct application to spark-ignition internal combustion engines because, in that type of engine, the combustible mixture ahead of the flame is also compressed isentropically during flame propagation.

As the vessel shape or location of the ignition point is changed the behavior of the flame becomes more complex. In 1928 Ellis[11] published a remarkable series of photographs that show the effect of such simple changes on flame behavior. Figure 14-5, which was discussed earlier, shows the effect of off-center ignition as well as buoyancy. Figure 14-8 shows the effect of increasing the length-to-diameter ($L/D$) ratio from small to relatively large values. These photos were taken using a slotted wheel and the light from the flame. Notice how the effect of Taylor-Markstein (T–M) instability becomes more pronounced as the length-to-diameter ratio gets larger. In all cases the downward-propagating portion of the initially hemispherical flame is strongly decelerated when it reaches the side walls. This deceleration causes the flame to become T–M unstable and the central region becomes cusp shaped. The effect is most

11. O. C. de C. Ellis, *Fuel Sci. Pract.*, 7:195–205, 245–252, 300–304, 336–344, 408–415, 449–454, 502–508, 526–534 (1928).

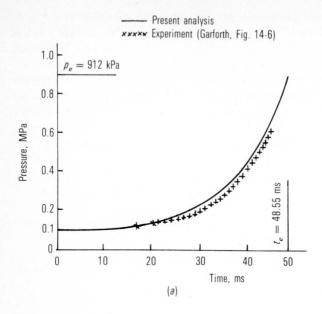

(a)

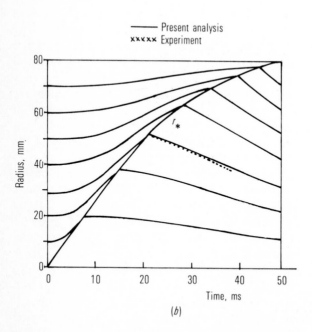

(b)

**Figure 14-7** Analysis of Fig. 14-6. Note the residual temperature gradient shown in (c). In (d) the quantity $U_f$ is the cold gas flow velocity just ahead of the flame. (*Adapted from T. Takeno and T. Iijima, Nonsteady Flame Propagation in Closed Vessels, in "Combustion in Reactive Systems," Prog. Astronaut. Aeronaut., vol. 76, pp. 578–595 (1981), with permission.*)

striking in tubes with larger $L/D$, because they contain more unburned gas when the deceleration occurs. Thus there is more time for the instability mechanism to be operative.

In Sec. 8-2 soap-bubble flames were discussed because they have been used to measure the normal burning velocity of a number of premixed flames. Detached, freely expanding flames can also be produced in a laminar stream when ignition is effected in a region devoid of flame holders and the stream velocity is sufficient to cause the entire flame to translate downstream as it propagates to encompass an ever-increasing volume of gas. Observations of these outwardly propagating flames in a predominantly uniform laminar stream

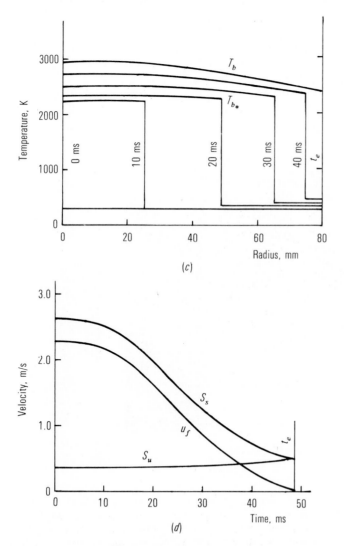

(c)

(d)

**Figure 14-7** (*continued*)

show some interesting effects triggered by aerodynamic interactions. The two flame balls shown in Fig. 14-9 were both obtained in a laminar flow of about 7 m/s by passing a spark between two neighboring 25 $\mu$m tungsten wires placed mutually perpendicular to each other and to the flow direction. In this case the wires are sufficiently fine to prevent attachment and are also fine enough to yield a laminar wake. Figure 14-9 $a$ illustrates the large effect of this very weak wake on the flame-front shape in a situation where cellular instabilities are not present. The presence of the wake in the flow has caused a deep depression in the upper surface of the flame ball and a corresponding extension in the lower portions of the flame ball, and the effect is most pronounced in the region where the wakes of both the wires are present. Figure 14-9$b$ was obtained in an extremely lean hydrogen–air mixture which tends to propagate with a highly developed cellular structure. The very pronounced lower tip of this flame is in reality a simple example of the interaction of aerodynamic and flame-curvature effects in an inherently unstable flame system. In this case the strong curvature of the tip coupled with the extremely high preferential diffusion of the light hydrogen into the preheat region ahead of the tip yields a sufficient increase of the local burning velocity to cause the tip to dominate the flame shape through-

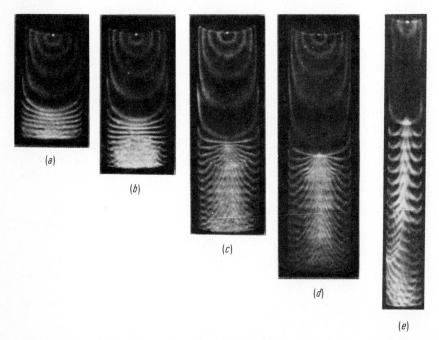

(a)

(b)

(c)

(d)

(e)

**Figure 14-8** The effect of $L/D$ on the growth of Taylor–Markstein instability. Mixture: 10 parts $CO$ + 1 part $O_2$, saturated with water vapor at 15°C. Cylindrical vertical tube closed at both ends. (a)–(d) 50 mm diameter, (e) 25 mm diameter. Lengths are: (a) 95 mm, (b) 120 mm, (c) 170 mm, (d) 195 mm, and (e) 230 mm. Ignition is at the top center. *(From O. C. de C. Ellis, Fuel Sci. Pract., vol. 7, p. 502 (1928), with permission.)*

out the flame-ball life. The smooth conical section in this flame is an oblique flame generated by this tip, and this oblique surface is seen to spontaneously develop cellular instabilities only in the region near the main flame ball. Thus this flame illustrates an enhancement of the burning velocity in a strong curvature region, and in addition shows the transient slow development of cellular instability on the initially smooth conical section generated by the tip motion.

In Sec. 8-6 cellular flames were discussed and hydrodynamic instabilities were mentioned. Hydrodynamic instabilities are found to cause gross flame-wrinkling when the flame is a spherically expanding flame either in the open or in a spherical vessel. The scale of the wrinkling is normally much larger than the usual cellular size and the effect is not dependent on equivalence ratio, as it is for cellular structure. Also, it is not observed in small-scale experiments. In experiments in which a hemispherically expanding flame propagated from the center of 5 and 10 m radius thin plastic hemispherical bags, Lind and Whitson[12] observed that the flame had a roughness with a scale that was much larger than the cellular scale (see Fig. 14-10). Also, the measured space velocities were 1.6 to 1.8 times the space velocities calculated from the laboratory-determined normal burning velocities and the theoretical expansion ratio. Furthermore, this ratio was the same for all five fuels studied, even though their normal burning velocities varied by a factor of four. Michelson and Sivashinsky[13] have found this type of instability theoretically, using asymptotic analysis. Figure 14-11 is a computer generated flame shape for a portion of a spherically expanding flame. They found that the flame front showed no tendency to stabilize and that the flame surface is in a continual state of agitation. They also found that the net average space velocity of their flame was about 1.7 times the normal smooth flame space velocity.

We now wish to turn our attention to freely propagating premixed flames in a low-intensity isotropic turbulent flow. In Chap. 11 we noted that attached flames burning in a turbulent approach flow are always transient flames because they are held in a laminar region and relax to produce a wrinkled structure as they enter the turbulent flow. Since each oblique flame sheet has a finite lifetime, the flame never reaches equilibrium with the flow turbulence. A freely expanding flame in an isotropic turbulent flow also has a finite lifetime. However, in such a flame, each element of the flame has the same "age" at any instant of time, because the entire flame originates from a single spark. Thus such a flame can possibly be used to determine if an "equilibrium isotropic turbulent burning velocity" similar to the well-established "eigenvalue" laminar burning velocity can be observed.

We define an "equilibrium isotropic turbulent burning velocity" in the following way. Consider a semi-infinite isotropic turbulent field of intensity $I$ and

12. C. D. Lind and J. Whitson, "Explosion Hazards Associated with Spills of Large Quantities of Hazardous Materials," Phase III Report No. CG-D-85-77, Department of Transportation, U.S. Coast Guard Final Report, ADA 047585 (1977).

13. D. M. Michelson and G. I. Sivashinsky, *Combust. Flame*, **48**: 211–217 (1982).

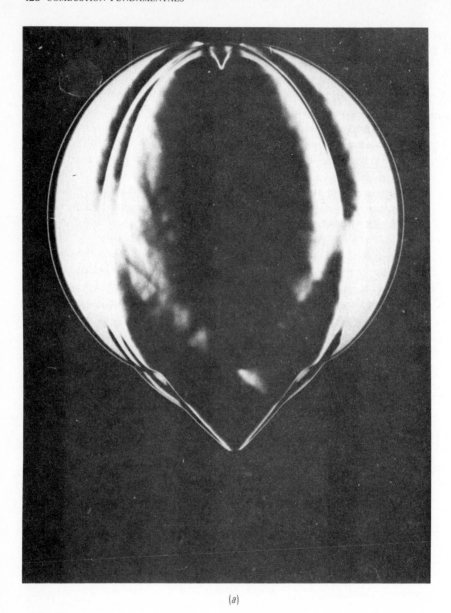

(a)

**Figure 14-9** Free flame balls obtained in a low-speed laminar flow by spark ignition. (a) Lean butane–air flame which is inherently stable. Note the surface features caused by the wake of the crossed ignition wires upstream in the flow. (b) Lean hydrogen–air flame which is inherently unstable. Note the extremely rough flame surface caused by spontaneous cellular instability and the

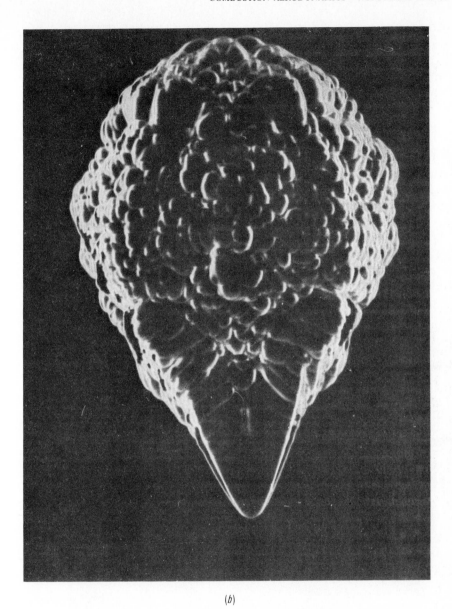

(b)

very pronounced lower tip caused by the enhancement of the burning velocity at the point where the wake of the crossed wires in the upstream flow produces an initially low-radius-of-curvature section in the flame.

scale $\lambda_g$ flowing in the $x$ direction at a velocity $U$ and continuously supporting a flame zone of infinite lateral extent lying in the neighborhood of $x = 0$. If we find that we can define a value of $U = S_T$ at which such a flame zone exists as a time-average steady phenomenon, we will say that we have defined the turbulent-burning velocity of the system. If such a flame velocity is definable (i.e., if such a flame truly exists) we should be able to determine the dependence of $S_T$ on the initial mixture composition and the static properties of the flow (which allow one to define the laminar-burning velocity, $S_u$) as well as its dependence on the approach-flows turbulent intensity $I$ and scale $\lambda_g$ at $x = $ some small negative value.

The most readily obtainable free flame is the outwardly propagating "spherical" flame which may be obtained by spark ignition, using a set of thin crossed wires placed in the turbulent flow. This type of free flame is most useful for comparison to the theoretical model described above. However, there is no a priori reason to expect even this spherical flame to exist as a fully developed "equilibrium" turbulent flame immediately after its inception. In fact, high-speed photographs of flame balls taken in relatively low-intensity turbulent

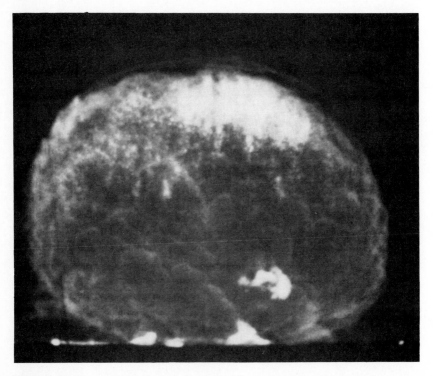

**Figure 14-10** Flame near the end of propagation in a ten meter radius hemispherical bag. Fuel is methane at 10% concentration in air. Note the large-scale roughness of the flame surface. In these experiments the bag always tore loose from the pad at an early time and the propagation occurred at essentially constant pressure. *(Courtesy Dr. Doug Lind: Originally in Ref. 12.) (With permission.)*

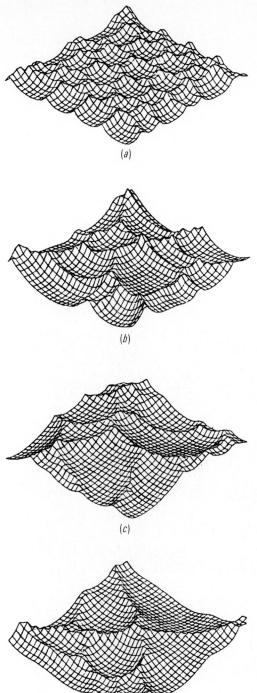

(a)

(b)

(c)

(d)

**Figure 14-11** Successive stages in the evolution of a spherically expanding flame front showing hydrodynamic instability. Propagation is down and time is increasing from (a) to (d). *(From Ref. 13, with permission.)*

**431**

flows show that during the initial phase of flame-ball growth the flame appears as an ordinary laminar flame ball similar to that shown in Fig. 12-13. The flame then slowly relaxes until it contains a roughened or turbulent-flame surface as illustrated in Fig. 14-12. Measurements of the turbulent burning velocity $S_T$, made by measuring the average local $(S_b)_T$ as a function of flame-ball radius

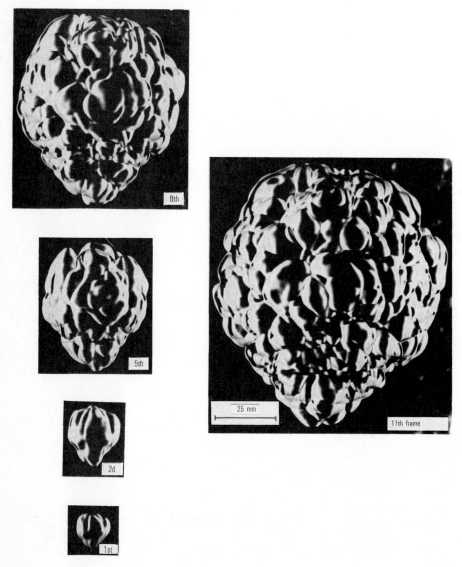

**Figure 14-12** Flame ball burning in an isotropic turbulent stream. Note that surface roughness increases with time but that characteristic roughness spacing or size is remarkably constant. Framing rate = 1/300 s.

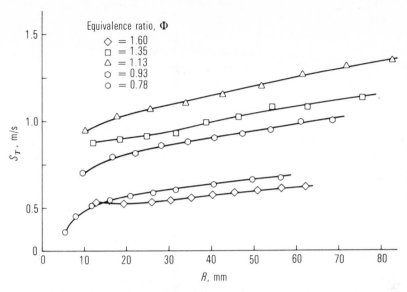

Equivalence ratio, $\Phi$

$\diamond = 1.60$
$\square = 1.35$
$\triangle = 1.13$
$\circ = 0.93$
$\circ = 0.78$

**Figure 14-13** $S_T$ vs. radius of a turbulent flame ball.

and dividing this by the calculated isenthalpic expansion ratio yields the interesting results that $S_T$ is increasing with the flame-ball radius over the entire experimental range. Figure 14-13 is such a plot for a series of equivalence ratios in a propane–air mixture. The implication of these results is quite clear. A turbulent approach flow presents a continuous source of velocity perturbations to the advancing flame front, and an equilibrium turbulent flame, as defined, can exist only if the flame front has had sufficient time to reach a relaxation equilibrium with these perturbations, if it is indeed capable of relaxing to an equilibrium state. Thus an important parameter in turbulent-burning studies should be the characteristic relaxation time to each "element" of the flame front for a known approach turbulence. Here an element is again defined as a section of the flame front propagating normal to itself into the flow. In determining the data of Fig. 14-13 the flame was observed for a travel distance of 80 mm. This corresponds to a travel of 40 mm relative to the gas ahead of the flame, and thus the observation that this flame is still accelerating is consistent with the results of Petersen and Emmons,[14] who observed that for a simple driven perturbation the flame required a considerable time to reach an equilibrium cusped perturbation shape. However, since Boyer et al.[15] have been able to stabilize a "flat" turbulent flame in a slowly expanding duct, it appears as though there may be a unique turbulent flame velocity for low-intensity approach-flow turbulence even though the relaxation of the flame to this equilibrium condition may

14. H. E. Peterson and H. W. Emmons, *Phys. Fluids*, **4**:456 (1961).

15. L. Boyer, P. Clavin, and F. Sabathier, "Dynamical Behavior of Premixed Turbulent Flame Front." *Eighteenth Symposium (International) on Combustion*, The Combustion Institute, Pittsburgh, Pa., p. 1041 (1981).

take of the order of seconds. This is in agreement with the observations of Boyer et al. and of Clavin and Williams,[16] presented in Sec. 11-2.

## 14-5 FLAME ACCELERATION PROCESSES—
## DEFLAGRATION TO DETONATION TRANSITION

There are three primary mechanisms that cause flame acceleration when a transient flame propagates as a premixed combustible gas mixture in an enclosure. The first of these is the Taylor–Markstein mechanism discussed in Sec. 14-3. In the other two mechanisms, the acceleration process is a direct result of the fact that the unburned mixture ahead of the flame is set into motion by the expansion of the hot gases produced by the flame. In the second mechanism, gas motion ahead of the flame causes a turbulent boundary layer to form on a surface of the enclosure. When the flame reaches the turbulent boundary layer, it rapidly spreads into it and the effective rate of heat release increases markedly. In the third mechanism, the inpulsive flow generated by the flame causes the separated flow region of every obstacle to shed a vortex. As in the second mechanism, when the flame enters a vortex, the effective rate of heat release increases rapidly. In this section we will discuss three different experiments that illustrate how these processes produce flame accelerations that can ultimately cause transition to detonation.

The first experiment that we will consider is the propagation of a flame through a long smooth pipe, after ignition by a spark at the center of a closed end. The pipe will be assumed to be long enough so that the other end does not influence the flow that we are discussing. There are two unique limit behaviors for this system and these are dependent primarily on the normal burning velocity of the mixture and the diameter of the pipe. If the burning velocity is low and the pipe has a relatively small diameter, after some initial transient behavior, the flame will propagate through the rest of the pipe at a constant speed, which is determined by shape of the flame and the normal burning velocity. The usual shape of the flame has been described in Sec. 10-5. The reason for this behavior is twofold. Firstly, the initial acceleration processes associated with ignition are not strong enough to generate a turbulent boundary layer ahead of the flame. Secondly, after propagating some distance, heat losses to the wall are sufficient to return the product gases to room temperature. When this happens, gas motion ahead of the flame ceases and no new boundary layer is generated.

The other unique limit behavior of transient flame propagation in a long pipe occurs when the flame's normal burning velocity is high and the pipe is large enough to make heat loss to the wall unimportant. In this case, the flame propagation process goes through a number of distinct phases. Figure 14-14 is an $(x, t)$ schlieren photograph that shows the first three phases of flame develop-

16. P. Clavin and F. A. Williams, *J. Fluid Mech.*, **116**:251 (1982).

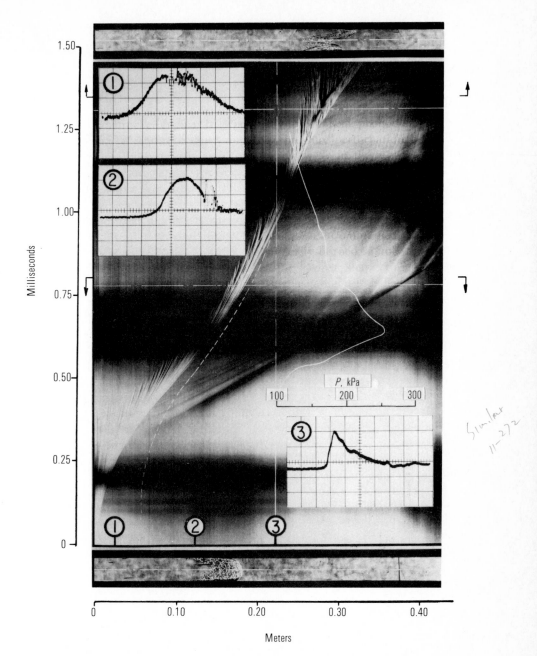

**Figure 14-14** Flame propagation from the closed end of a long rectangular duct. Note that the flame is becoming highly turbulent near the end of the record and that compression waves ahead of the flame are coalescing to produce shock waves. Pressure records taken from stations 1, 2, and 3 are shown as well as two pulsed laser photographs of the flame and shock wave development. *(Courtesy Prof. A. K. Oppenheim, University of California, Berkeley; photograph originally in paper presented at the OAR Research Applications Conference, 1966).*

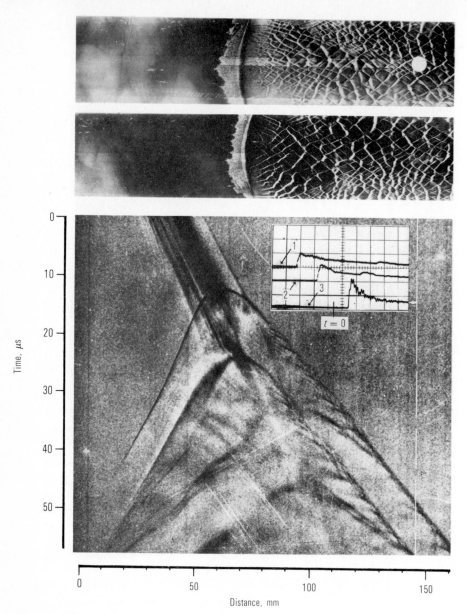

**Figure 14-15** Schlieren $(x,t)$ photograph, smoke track, and simultaneously obtained pressure records of the later stages of the detonation initiation process in a smooth tube. Time increasing downward, flow to the right in a $0.5H_2 + 0.5O_2$ mixture initially at 11.7 kPa. Oscilloscope record: 50 $\mu$s/div, and 140 kPa/div. Pressure gauges 1 and 2 were mounted out of the field of view to the left and show that the initial shock is accelerating. Gauge 3 was located at the white dot in the upper smoked foil. Note the very much higher detonation pressure recorded by this gauge. Also note the immediate appearance of transverse structure when the detonation is formed. *(Courtesy Prof. A. K. Oppenheim, University of California, Berkeley, with permission, from Proc. Roy. Soc., vol. A295, pp. 13–26 (1966).)*

ment. Near the closed end, before reaching station 1, the flame is a growing spherical flame. After reaching the wall the flame becomes Taylor–Markstein unstable. This causes a mild acceleration of the flame between stations 1 and 2. This accelerating flame generates a compression wave which steepens to form a propagating shock wave on the right-hand side of the photograph. The speed of propagation of the flame system then slows down in the region between stations 2 and 3. The shape of the flame at this time is shown in the lower full-frame photograph. Just to the right of station 3 the flame first encounters the turbulent boundary layer and starts to accelerate again. The shape of this developing boundary layer flame is shown in the upper full-frame photograph. This newly accelerating flame generates another compression-wave system which is considerably stronger than the first. This second compression wave also steepens to become a shock wave.

At some later time these two shock systems merge and the boundary layer becomes thicker and more turbulent. When this happens the flame propagates through the wall boundary layer and takes the shape of an elongated cone. This behavior causes the shock to continue to strengthen. Eventually the shock becomes strong enough to heat the explosive mixture to above autoignition temperature. Then, after some induction delay similar to that described for the accelerating shock case of Sec. 13-2 (see Fig. 13-4), a homogeneous constant-volume explosion occurs some distance behind the shock and causes a detonation to propagate forward and a retonation to propagate back down the tube. This behavior is shown in Fig. 14-15. This figure is a double schlieren streak photograph of the interaction process in a rectangular tube. One slit was placed near the bottom of the tube while the other was placed near the top. Also, smoked foils were placed on the top and bottom of the tube. The wave system that has a rounded top in the middle of this streak photograph is an explosion in a distributed region in which the effective propagation velocity approaches infinity. Notice that smoked-foil writing begins the instant the explosion wave overtakes the lead shock wave in the tube. Thus the mechanism of the last stage of initiation is exactly the same in this case as it is for the planar initiations of Sec. 13-2.

The acceleration of a flame in a tube is very dependent on the condition of the tube wall. The above experiments were performed in a straight tube with a smooth interior surface. The presence of sharp bends or obstacles will cause a much more rapid acceleration of the flame because the impulsive flow ahead of the flame generates separated flow regions behind each obstacle or bend. Each of these regions contains a separation vortex. When the flame enters these vortices it is strongly accelerated. Experiments have shown that repeated obstacles can cause very rapid acceleration of the rate of heat release because the interaction produced downstream of the first obstacle causes the flow ahead of it to accelerate, and this causes the next interaction to be more violent. This behavior is illustrated in Fig. 14-16, which shows the effect of a repeated obstacle on the flame speed for the hydrogen–air system.[17] The obstacle in this case was a 6 mm diameter wire attached to the inner surface of the first 3 m of the tube as a spiral with a pitch of 50 mm. The effective blockage ratio for this spiral is 0.44.

We note from Fig. 14-16 that below 13 percent hydrogen the accelerations are not very violent and the flame tends to return to a low constant velocity after passing through the obstacle region. At 15 percent hydrogen the effect of the obstacle on flame propagation is much more severe but the flame velocity again drops after it leaves the obstacle region. Evidently the normal burning velocity is still sufficiently low and heat loss to the tube walls stops any later acceleration processes from occurring. Over the range of 17 to 25% $H_2$ the

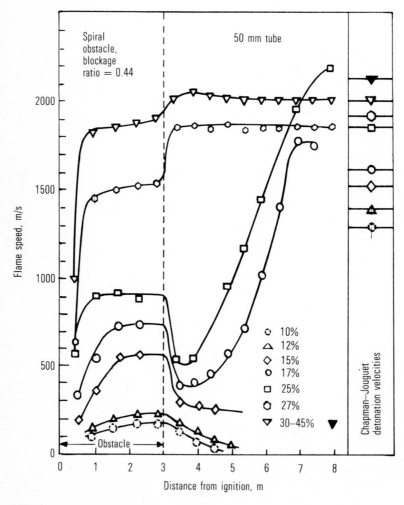

**Figure 14-16** Flame speed profiles for $H_2$–air mixtures in a 50 mm diameter tube with a spiral obstacle. Chapman–Jouguet detonation velocities for the test mixtures are plotted on the right-hand side of this graph. (*Adapted from Ref. 17, with permission.*)

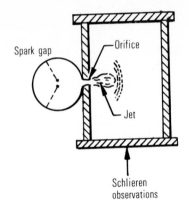

**Figure 14-17** Test arrangement used by the authors of Ref. 18. The main chamber was about 0.3 m on a side. The test gas was an equimolar $C_2H_2-O_2$ mixture at 20 kPa initial pressure. *(With permission from W. E. Baker, P. A. Cox, P. E. Westine, J. Kulesz, and R. A. Strehlow, "Explosion Hazards and Evaluation," Elsevier, Amsterdam (1983).)*

acceleration becomes increasingly rapid, and a very high steady state propagation velocity is reached in the 3 m length that contains the obstacles. After the flame leaves the obstacle region the flame speed drops markedly but then quickly accelerates to velocities that are higher than the CJ detonation velocity. Because of the transient nature of the acceleration processes, the observation of local velocities above CJ is not surprising.

The difference in behavior between 25% $H_2$ and 27% $H_2$ is striking. At 27% $H_2$ the flame quickly accelerates to form a CJ detonation. This occurs within the first meter of travel. The velocity in the region that contains the spiral is somewhat less than the classic CJ velocity because the large blockage ratio acts to attenuate the wave. However, once the wave leaves the obstacle field it quickly reaches a velocity close to CJ.

In the experiments of Ref. 17, it is not known whether the initiation of detonation occurred by the autoignition mechanism described earlier in this section or if it occurred by the SWACER mechanism described in Sec. 13-4. Knystautas and coworkers[18] have observed SWACER initiation during flame propagation. Their experimental apparatus is shown schematically in Fig. 14-17. They filled both test chambers with the test mixture and ignited a flame at the spark gap. The channel between the two vessels contained a variety of orifices of different shapes. Multiple orifices were also used. The sequence of events in this experiment are as follows. Ignition will cause the pressure to rise in the donor chamber and this causes a turbulent jet (or jets) of unreacted gas to form in the second larger receptor chamber. After a short period of time the flame reaches the orifices and since these are always larger than the quenching distance, turbulent product gases and the flame front mix into the originally cold

17. J. H. S. Lee, C. Chan, and T. Knystautas, Hydrogen–Air Deflagrations: Recent Results. A paper presented at the Second International Workshop on the Impact of Hydrogen on Water Reactor Safety, Albuquerque, N. Mex., on Oct. 3–7, 1982.

18. R. Knystautas, J. H. S. Lee, I. O. Moen, and H. G. Wagner, "Direct Initiation of Spherical Detonation by a Hot Turbulent Gas Jet," *Seventeenth Symposium (International) on Combustion,* The Combustion Institute, Pittsburgh, Pa., p. 1235 (1979).

jet. The authors found that this can simply cause combustion of the gases in the second chamber. However, they also found that if there are multiple orifices of the proper size and spacing the turbulent mixing region suddenly initiates a Chapman–Jouguet detonation which propagates through the remainder of the receptor vessel. This behavior is shown in Fig. 14-18, which is taken from Ref. 18.

We note from Fig. 14-18 that after a very short period of time the shock front is traveling at the CJ velocity of 3160 m/s. In fact, this time is so short that it is not possible for a signal from the jet to reach the far wall and return prior to the initiation of the detonation wave. This means that the initiation

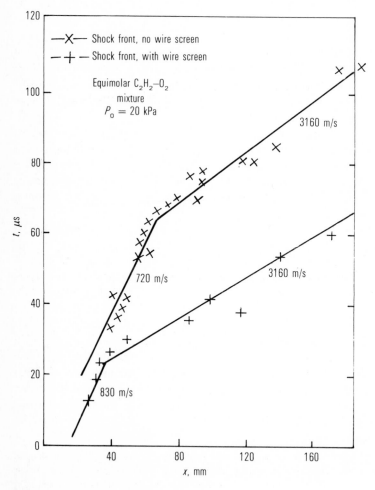

**Figure 14-18** The shock front position with time after reactive jet penetration into the cold turbulent jet. *(Taken from Ref. 18, with permission.)*

must be shock-free initially and that the SWACER mechanism first introduced in Sec. 13-4 must be operative. However, the exact details of the process are not known. For example, it could be that multiple flame-folding causes a tremendous increase in the rate of heat release—or it could be that the flame is actually momentarily extinguished by shear, and cold-gas–hot-gas mixing produces high radical concentrations in a cold mixture—and that this leads to a radical initiated explosion similar to that discussed in Sec. 6-6. In either case this explosion then makes a rapid transition to detonation. There is no doubt about two features of this type of initiation. In the first place the reactive volume must have some minimum critical size so that heat release can cause the local pressure to rise before wave propagation allows the volume to increase significantly. Thus there must be a critical Damköhler number based on the time it will take the pressure to be equilibrated and the time of the explosive reaction, below which transition will not occur.

It is also rather obvious that the region where transition occurs must be properly conditioned to exhibit this type of SWACER initiation. This must be true because if one were to burst a sphere containing a gas with the same pressure and velocity of sound as the hot product gas in the surrounding combustible mixture, the shock-wave strength would be far below that needed to cause autoignition in the surroundings even at the contact surface of the sphere gases. Thus, simple compression alone is not sufficient to cause initiation in the manner that is observed.

## PROBLEMS

**14-1** Figure 14-2 and the discussion in the text show that for a vertical tube containing a heated gauze, the first mode is excited when the gauze is placed $L/4$ from the tube bottom. Find the positions in the tube where the gauze would have to be placed to preferentially excite the second, third and sixth harmonics of the tube. Assume a $\pi/2$ phase lag in the heat transfer. In general how would the optimum position shift if the phase lag were somewhat less than $\pi/2$?

**14-2** Assume that a stoichiometric methane–air mixture is centrally ignited in a spherical bomb. Use the constants for the working fluid heat-addition model listed in Table 5-2 to calculate the difference between the final temperature and density for the first and last elements burned. Assume that the final pressure is the constant-volume explosion pressure given in Table 5-2, that combustion takes place at constant pressure, and that either the cold or hot gases are compressed isentropically. Determine if the actual final pressure will be higher than, equal to, or lower than this value.

**14-3** Show that the spherical flame of Prob. 14-2 initially causes the pressure in the bomb to rise as a cubic of time.

**14-4** Assume that a flame with a normal burning velocity of 0.45 m/s is ignited at the center of the top closed end of a long cylindrical pipe 50 mm in diameter. Assume that the flame remains a section of a growing sphere (initially a hemisphere) even after it becomes large enough to contact the walls of the pipe. Calculate the acceleration of the flame front in room coordinates up until its radius is three times the radius of the pipe. Assume an infinitely thin flame and use the heat-addition working-fluid model for stoichiometric propane–air from Table 5-2.

# BIBLIOGRAPHY

**Acoustic instability**

Putnam, A. A., *Combustion-Driven Oscillations in Industry*, Elsevier, New York (1971).

**Transition to detonation**

Lee, J. H. S., "Initiation of Gaseous Detonation," *Annu. Review Phys. Chem.*, **28**:75–104 (1977).
———— and I. O. Moen, "The Mechanism of Transition from Deflagration to Detonation in Vapor Cloud Explosions," *Prog. Energy Combust. Sci.*, **6**:359–389 (1980).

# FIFTEEN

## CONDENSED PHASE COMBUSTION

### 15-1 INTRODUCTION

There are many solids and liquids, both pure compounds and mixtures of compounds, that support highly exothermic decomposition reactions that produce copious quantities of gas. However, not all such highly exothermic compounds or mixtures detonate. Some simply burn. Therefore, four classes of exothermic compounds and mixtures are usually defined relative to their innate stability. These are:

1. Extremely unstable. There are a number of compounds and mixtures that can be prepared which are so unstable that they are never handled in bulk or stored in any quantity. Examples are nitrogen trichloride, nitrogen triodide, and a mixture of ammonium perchlorate and yellow phosphorus. These all have been found to decompose violently without confinement (detonate) at any time with no apparent initiation source.
2. Primary explosives. Primary explosives are the most sensitive explosive substances in commercial use. Indications are that they propagate a CJ (or high-order) detonation very soon after being initiated by friction. In this group are the metal styphnates, fulminates, and azides. In general these compounds detonate when heated, before melting.
3. Secondary explosives. These substances must usually be traversed by a strong shock wave before detonation appears. In general, they melt before detonating and in most cases of initiation by friction during confinement they burn before detonating. This group contains the substances trinitrotoluene (TNT), ammonium nitrate and pentaerythritol tetranitrate (PETN).

**443**

4. Propellants. These substances, even though highly exothermic, have not exhibited a detonative decomposition even when struck by a strong shock wave. Commercial solid rocket propellants usually fall in this group.

It must be emphasized that the boundaries between these four groups are not, in general, clearly defined. They are based primarily on accumulated experience. In general one should always be cautious when handling any exothermic compound or mixture because they all have the potential of releasing their energy rapidly when not expected.

## 15-2 DETONATION PROPAGATION, STRUCTURE, INITIATION, AND FAILURE

The one-dimensional theory for the propagation of detonation in solids and liquids is based on the same equations of steady motion developed for gas-phase detonation, i.e., Eqs. (5-28) to (5-30). Both the CJ velocity and CJ pressure are known to be functions of the initial density of the high explosive. The extremely high initial density of solid and liquid explosives leads to detonation pressures which lie in the range of 1 to 10 GPa. These extremely high pressures cause two major complications in the application of the elementary theory. In the first place, the product gases no longer obey the ideal gas law even at the high temperatures encountered; second, the type of confinement required to produce one-dimensional flow is impossible to obtain except in the central area of a wave front whose breadth is very much larger than the reaction-zone thickness. The first complication means that calculated detonation velocities will not exhibit as good an agreement with experimental values as for the gas-phase case. Note that the only way to produce the temperatures and pressures encountered in condensed-phase detonations is to have a condensed-phase detonation. Thus, there is no independent experimental method available for directly measuring equation-of-state parameters to allow a truly independent calculation of detonation velocity. Furthermore, theoretical estimates are not sufficiently accurate to choose an unambiguous mathematical form for the equation of state. The second complication means that relatively large-diameter charges must be used to determine equation of state parameters.

We also note that pressure cannot be measured directly during the detonation of a high explosive. CJ detonation pressures are well above the yield strength of any known material and therefore direct reading pressure gauges cannot be used. Pressures can be inferred by measuring the CJ velocity and by measuring the shock velocity and plate velocity of a plate of inert material attached to the detonation either at some oblique angle or exactly normal to the direction of travel of the detonation wave. By using different plate thicknesses and materials with different shock impedances one can determine a CJ pressure and also obtain a $(P, V)$ isentrope for the explosive products. Fickett and Davis[1] discuss various modern forms of the equation of state for detonation products. They feel that the best empirical equation of state is the constant $\gamma$ equation of state

$$e - e_0 = \frac{Pv}{(\gamma - 1)} - q$$

which, with the assumption that $P_0 = 0$ (good for high explosives) yields the Hugoniot

$$P\left(\frac{v}{v_0} - \frac{\gamma - 1}{\gamma + 1}\right) = 2\frac{\gamma - 1}{\gamma + 1}\frac{q}{v_0}$$

and expressions for the CJ pressure and velocity,

$$P_{CJ} = 2(\gamma - 1)\rho_0 q$$

and

$$V_{CJ}^2 = 2(\gamma^2 - 1)q$$

Using various techniques an expansion isentrope of the detonation products can be determined

$$Pv^\gamma = P_{CJ} v_{CJ}^\gamma$$

and with sufficient data $q$ and $\gamma$ can be determined. The value of $\gamma$ usually lies near 3, in contrast to its range of 1 to $5/3$ for an ideal gas.

This simple constant-gamma equation of state has one major drawback. It predicts that $P_{CJ}$ and $V_{CJ}$ are independent of the initial loading density $\rho_0$. Experimentally it is observed that both are functions of the initial density. This behavior can be handled in a simple manner by defining a reference isentrope, ref, for a specific initial density of loading and expanding the equation of state using a first-order Taylor expansion. This yields

$$E(P, v) = E_{ref} + v\frac{P - P_{ref}}{\Gamma}$$

where

$$\Gamma = \frac{v}{(\partial E/\partial P)_v}$$

and

$$P_{ref} = P_{CJ}\left(\frac{v_{CJ}}{v}\right)^\gamma$$

$$E_{ref} = P_{ref}\frac{v}{\gamma - 1}$$

1. W. Fickett and W. C. Davis, *Detonation*, University of California Press, Berkeley, California, (1979).

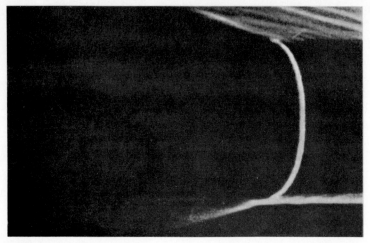

**Figure 15-1** Shape of a detonation front in a small-diameter unconfined charge. Propagation is from left to right. *(With permission from J. Berger and J. Viard, Physique des Explosifs Solides, Dunod, Paris, 1962.)*

where $\Gamma$ is called the Gruniesen coefficient. The remaining constants in the equations are determined by fitting experimental results to data obtained with explosives pressed to different densities.

There are more sophisticated equations of state based on the principles of statistical thermodynamics and even full chemical equilibrium of the product gases. Assumptions such as the general shape of the molecular interaction potential must be made for these equations of state. Then the parameters must be determined by fitting known properties of the detonation wave and the isentrope, including the effects of initial density.

The very high pressures generated by a detonation wave means that even heavy metal walls can impose only an "inertial" confinement and true one-dimensionality can only be obtained in the central region of large-diameter charges. Figure 15-1 is a photograph of the shape of a detonation front for weak confinement of a cylindrical charge. Inadequate confinement causes the detonation velocity to drop appreciably and eventually can lead to extinguishment. It has been found experimentally that for unconfined charges there is a critical minimum-charge diameter and that charges with a diameter less than this critical value will not spontaneously propagate detonation.

Most commercial secondary explosives have critical diameters in the range of 3 to 25 mm. However, ammonium nitrate has a minimum charge diameter of approximately 0.2 m, and the nitrate-sulphate $2NH_4NO_3 \cdot (NH_4)_2SO_4$ which was used commercially as a fertilizer after World War I has an even larger minimum-charge diameter. This fact resulted in an unfortunate accident soon after World War I. At that time it was thought that the nitrate-sulphate would be incapable of sustaining detonation (since the usual small sample impact tests showed it to be insensitive) and the material was handled in bulk like any other industrial chemical. This substance cakes rather easily, especially in humid atmospheres, and in order to recover the salt from storage a congealed pile of 4.5 million kilograms was dynamited on September 21, 1921, in the industrial town of Oppau, Germany (pop. 7500). Instead of simply fracturing, the pile detonated as a unit. The disaster was the largest man-made explosion prior to the advent of the atom bomb. The explosion killed an estimated 1100 people, left a hole 60 m deep by 120 m in diameter, and caused severe damage to a radius of 6 km. It is best to assume that all exothermic compounds are dangerous.

The occurrence of transverse structure in condensed-phase detonation has been confirmed but its role in the propagation has not been settled as yet. Figure 15-2 is a photograph of transverse structure in a homogeneous nitromethane–acetone mixture. In most high explosives the reaction zone is so thin that any investigation of the effect of transverse structure on the chemistry would be exceedingly difficult when compared to investigations of that type in the gas phase. Also, most commercial explosives are granular substances (if pressed) or fine-grain crystalline substances (if cast), and it is known that this type of inhomogeneity introduces a structural roughness in the detonation front which would overshadow any inherent transverse structure. In short, we know that condensed-phase detonation propagates as a self-sustaining phenomenon under the proper conditions, and we also know that the one-dimensional Chapman–Jouguet theory does not justify the observed stability of the wave front. Therefore we suspect that three-dimensional structure is important for propagation, but at the present time we do not have sufficient information to corroborate these suppositions.

Initiation of detonation in a condensed medium may be accomplished by friction in the more sensitive substances; Bowden[2] has shown that the friction of grit particles will ignite a particular explosive only if the melting point of the grit material is above a certain temperature. Since the temperature rise associated with frictional heating cannot exceed the melting temperature of the substance involved, this implies that there is a minimum hot-spot temperature required for initiation of primary explosives. In this connection it is interesting to note that most primary explosives detonate *before* melting.

Most secondary explosives cannot be initiated to detonation by friction alone and require the presence of a strong shock wave before initiation occurs. In practice, the strong shock is usually supplied by the detonation of a donor charge, and for testing purposes this shock is attenuated by passage through an appropriate thickness of inert material. A typical experimental setup is shown in Fig. 15-3 and a set of high-speed photographs of the initiation process are shown in Fig. 15-4. As can be seen from Fig. 15-4, the initiation process is not

2. F. P. Bowden and A. Yoffe, *Initiation and Growth of Explosion*, Cambridge University Press, London, 1952.

strictly one-dimensional and there is a considerable spreading of the high-explosive charge by the entering shock wave before initiation occurs. Experiments using different geometries have shown that initiation usually occurs first at the center of the charge some distance behind the entering shock and then propagates to overtake the entering shock wave and produce a high-order detonation in the undisturbed charge. Note that the process is quite similar hydrodynamically to reflected shock initiation in the gas phase.

The reader is referred to the bibliography for more information on the behavior of high explosives.

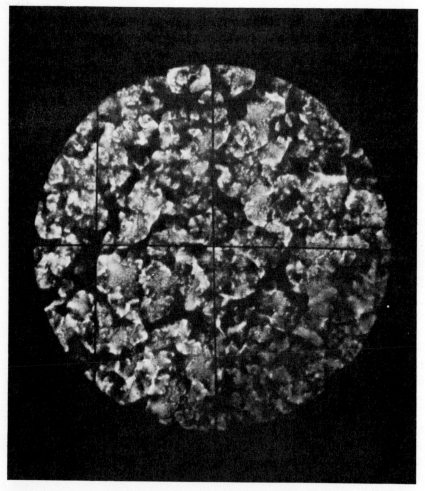

**Figure 15-2** End-on view of the structure of an approaching detonation wave in a 80%/20% nitromethane–acetone mixture. Tube diameter is 19.05 mm. Photograph was taken just before the detonation reached the window. *(Courtesy of Dr. W. C. Davis, Los Alamos National Laboratory, Los Alamos, N.Mex.)*

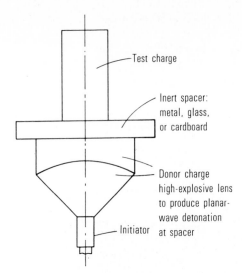

Test charge

Inert spacer:
metal, glass,
or cardboard

Donor charge
high-explosive lens
to produce planar-
wave detonation

Initiator   at spacer

**Figure 15-3** Typical setup to study detonation initiation in a test charge of explosive.

## 15-3 SOLID ROCKET PROPELLANT COMBUSTION

Solid-propellant rockets have played an important role in modern propulsion technology. While the steady-state theory of the rocket as a thrust producer has been adequately discussed in a number of texts, the combustion process which leads to this thrust production is usually not discussed in detail. We will therefore circumvent the external aspects of the rocket by looking in detail only at the processes that occur in the combustion chamber. Thus we will not discuss the effect of either recombination or of nonaxial flow in the nozzle. These can be handled adequately by techniques based on the reactive-flow equations introduced in Chap. 5 and the assumption of steady ($M = 1$) flow through the throat (see Sec. 5-6 for an example).

We will also not concern ourselves with the techniques of calculating the specific impulse (newtons of thrust per kilogram of propellant) for a particular propellant system. This calculation is ordinarily based on the assumption of full chemical equilibrium in the flow through the throat, and the techniques for this type of calculation have been discussed adequately in other texts and introduced in Chaps. 4 and 5 of this text. We will therefore assume, that for any system under consideration, the equilibrium composition of the combustion products, the isobaric flame temperature, and the specific impulse are available as a function of the chamber pressure for steady operation. With this information available, the rocket designer must choose an operating pressure, throat area, and nozzle shape to yield the desired thrust. The choice of these variables in turn fixes the total mass flow through the nozzle for steady operation and therefore establishes the steady rate of production of combustion products in the chamber.

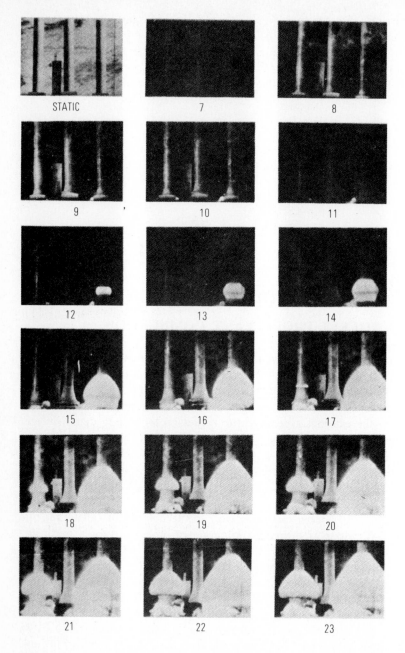

**Figure 15-4** Initiation of three sticks of high explosive by a planar-wave initiator similar to that shown in Fig. 15-3. Note that (1) the right-hand grain ignites first and detonates both forward and back toward the initiation plate, (2) the left-hand grain ignites later and propagates a detonation in the forward direction only, (3) the center charge does not ignite, and (4) all charges are extensively crushed by the initiating shock before initiation occurs. *(With permission from M. A. Cook, The Science of High Explosives, ACS Monograph No. 39, New York: Reinhold, 1958.)*

In the case of solid fuels, the rate of production of combustion products is primarily determined by the exposed surface area of the solid-propellant grain. This is because at a fixed pressure the regression rate of the surface of a solid-propellant grain is everywhere constant in a direction normal to the surface. Furthermore, the pressure-dependence of this regression rate for steady operation may usually be approximated by an equation of the form

$$r = B_p P^n \tag{15-1}$$

where $B_p$ and $n$ are constants and $n < 1$. Thus if a surface area $s$ is exposed in the rocket chamber, the mass rate of production of product gases is given by the equation

$$\dot{m}_p = sr\rho_s = s\rho_s B_p P^n \tag{15-2}$$

where $s$ is usually a slowly varying function of time. However, the equilibrium mass-flow rate through a nozzle is, to a first approximation, simply proportional to the pressure

$$\dot{m}_e = B_e A P \tag{15-3}$$

where $B_e$ is a constant dependent on the nozzle efficiency and the equilibrium combustion properties in the gas phase, and $A$ is the throat area. Steady operation is obtained when $\dot{m}_p = \dot{m}_e$. Equations (15-2) and (15-3) may therefore be combined to yield the equation

$$s = A \mathcal{K}(P) \tag{15-4}$$

where

$$\mathcal{K}(P) = \frac{B_e P^{(1-n)}}{B_p \rho_s}$$

Here $\mathcal{K}(P)$ is the ratio of the propellant-surface area to nozzle-throat area that will just produce the equilibrium combustion-chamber pressure $P$ during steady operation. Note that Eq. (15-4) predicts a stable operating condition only if $n < 1$, since if $n > 1$ an incremental increase in pressure $\Delta P$ will cause $\Delta \dot{m}_p > \Delta \dot{m}_e$ and the pressure will continue to increase spontaneously with time, whereas if $n < 1$ an incremental increase in $\Delta P$ will cause $\Delta \dot{m}_e > \Delta \dot{m}_p$ and the pressure will spontaneously seek an equilibrium level. Most solid propellants have values of $n$ in the range 0.4 to 0.85. There are a few combustible substances which have a value of $n$ greater than unity.

Equations (15-1) to (15-4) indicate that the relationship between throat area, exposed propellant-surface area, and chamber pressure will be a dynamic relationship. In fact, in a solid-propellant rocket the initial surface shape and the value of $\mathcal{K}(P)$ determines the $s(t)$ relationship and, therefore, the specific $P(t)$ relationship for the rocket's operation. However, the exact power-law dependence of Eq. (15-1) (with $n =$ const) is not, in general, required to allow the prediction of a $P(t)$ relationship for any specific initial surface shape and throat

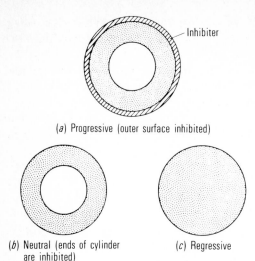

(a) Progressive (outer surface inhibited)

(b) Neutral (ends of cylinder are inhibited)

(c) Regressive

**Figure 15-5** Effect of surface shape on the burning characteristics of a solid-propellant grain in a rocket engine. Stippled areas represent propellant.

area. All that is needed is an experimentally determined plot of $\mathcal{K}(P)$ versus the chamber pressure. Numerical integration may then be used to determine $s(t)$ for a specific initial shape. In general, three types of instantaneous behavior of the grain surface area may be defined:

$$\frac{ds}{dt} > 0 \qquad \text{progressive}$$

$$\frac{ds}{dt} = 0 \qquad \text{neutral}$$

$$\frac{ds}{dt} < 0 \qquad \text{regressive}$$

Typical examples of surfaces which exhibit these three behaviors are shown in Fig. 15-5.

There are two major classifications of solid propellants in use today—*composite* and *double-base*. Composite propellants contain a crystalline inorganic oxidizer dispersed as a powder through a matrix of an organic polymer which ordinarily acts primarily as a fuel. Typical oxidizers are ammonium perchlorate, potassium perchlorate, or ammonium nitrate. Typical polymers are polyurethane, asphalt, or rubber type compounds. These propellants are formulated by mixing the oxidizer with the uncured polymer, casting the mix in a mold of the desired shape, and then curing the polymer either with thermal or chemical techniques. Double-base rocket propellants are related to the nitroglycerin–nitrocellulose formulations used in the manufacture of smokeless powder for guns. This material may be formed into grains either by extrusion or by casting. Extrusions of either a heated mix or an acetone-gelled mix is used to produce the smaller grains. Grains extruded with the acetone process may not be larger than about 10 mm in diameter because they are cured by allowing the acetone to diffuse to the surface and evaporate. The size of hot extruded grains is limited by the size of available presses.

The casting process is used to produce the larger double-based grains. These castings are manufactured by first extruding small ($<2$ mm diameter) pellets of double-base propellant of low nitroglycerin content by the usual acetone or hot-extrusion process. The pellets (casting powder) are then placed in a mold and the mold evacuated prior to immersion of the entire pellet bed in liquid nitroglycerin. The filled mold is then cured at a high temperature. During this curing process the individual grains of casting powder absorb the interstitial nitroglycerin and swell to contact each other and produce a relatively uniform single grain of double-base propellant.

In many applications these basic formulations are modified for specific purposes. For example, powdered metals may be used as additives, or an exothermic binder may be used in a composite-propellant formulation. Modifications in composition are continually being made to alter the properties of existing propellants in an attempt to increase structural integrity or specific impulse, modify the burning rate, and decrease the propellant's tendency to burn unstably in a rocket engine.

Once a sample propellant is formulated, its burning rate must be determined experimentally. This is usually done in either a strand burner, window bomb, or an actual small experimental rocket engine. In the strand burner a stick of propellant is coated on its sides with a relatively inert material (inhibiter) so that the combustion wave will propagate as a one-dimensional wave throughout the entire length of the stick. Thin fuse wires are then placed in small holes drilled across the stick at suitable intervals and the prepared sample is placed in a high-pressure bomb of relatively large volume. The fuse wires are connected to an external timing device and the bomb pressurized to the desired pressure level. The sample is ignited at one end (by a heated wire, for example), and the arrival of the combustion wave at each fuse wire is timed by noting the instant at which the circuit resistance increases markedly. In the window bomb, inhibitors may not be used because they tend to obscure the burning surface. In this case a flow of inert gas counter to the direction of regression has been used to retain a one-dimensional combustion front. A choked nozzle is usually used to maintain constant pressure in a window bomb.

Window-bomb experiments are primarily used for observing the details of the flame structure during combustion. A small experimental rocket engine may also be used to determine propellant burning rate if one knows the nozzle's throat area, the propellant surface area as a function of the fraction burned, and has available an experimental pressure-time curve. This method is usually less accurate than the strand or window-bomb techniques because it is difficult to estimate heat-transfer losses and the effects of nonideal flow in a small motor. However, the rocket motor technique is a reasonably good technique for determining the pressure exponent $n$ because for most propellants $1 - n \ll 1$. Thus, a small change in the surface area of the propellant will cause a large change in the equilibrium operating pressure of the rocket engine. Regression rates for most solid propellants lie in the range 5 to 75 mm/s.

The burning zone of a solid propellant is somewhat more complex than the equivalent region in a premixed gas flame. Furthermore, even though it is customary to discuss a one-dimensional solid-propellant combustion wave, it is doubtful if one can ever hope to observe strictly one-dimensional combustion during the burning of a solid-propellant grain. The inherent three-dimensional structure that appears in the burning zone of a solid propellant does not arise from the type of transverse instability which leads to non-one-dimensional detonation fronts but mainly appears because of inhomogeneities in the structure of the propellant grain itself. It is observed that this burning roughness or lack of one-dimensionality becomes more severe as one changes from a simple extruded double-base to a composite propellant. It has also been observed that the presence of metal additives decreases the one-dimensionality of the burning process. We will, however, follow convention and discuss the structure of a solid-propellant combustion wave as though it were a strictly one-dimensional process. Figure 15-6 is a schematic of the structure in a coordinate system that is steady relative to the burning surface. There are four rather distinct and different regions involved in the combustion process. These are:

### 15-3-1 The Solid Phase Heat-Transfer Region

In this region the propellant is being heated by conduction from the hot surface but is not reacting. Thus the temperature gradient in this region is governed by the usual steady heat-conduction equation with no source terms

$$\frac{d}{dx}\left( \kappa_s \frac{dT}{dx} - m_0 c_s T \right) = 0 \qquad (15\text{-}5)$$

where $m_0 =$ mass flux, $c_s =$ heat capacity, and $\kappa_s$ is the heat conductivity of the solid. The boundary conditions for this region are $T = T_0$ at $x = -\infty$ and $T = T_{us}$ at $x = x_{us}$.

### 15-3-2 The Gasification Layer

In this region vaporization of the propellant takes place with some reaction. The exact nature of this region is at present unknown and in the case of composites is most unsatisfactorily treated as a one-dimensional flow region. The region must be included in the model, however, because gasification cannot be assumed to be either a thermally neutral or an infinitely rapid process, and because the rate of gasification primarily controls the rate of surface regression. Due to the lack of detailed information, one usually assumes that the rate of gasification in this layer is related to the average temperature of the layer through an Arrhenius temperature-dependence. The activation energy for gasification then becomes an important constant in the model. The relationship between $T_a$ at $x = 0$ and $T_{us}$ must be obtained by assuming an exothermicity or endothermicity for the gasification reaction and by knowing a surface-regression rate. More experimental data is needed before a truly adequate model of this zone may be constructed.

### 15-3-3 The Gas-Phase Induction Zone Region

The products of the gasification process next traverse a gas-phase induction zone without undergoing extensive reaction. For the pure double-base propellants this zone has been shown to contain a large quantity of nitric oxide and have a temperature equal to about one-half the flame temperature of the propellant. In composite propellants this zone is far from being one-dimensional because it is the region where the fuel and oxidizer mix through diffusional and turbulent mixing processes. Thus, the length of this induction zone is governed by either a thermal or chain-branching delay mechanism or by a mixing-delay mechanism. It is a region in which heat transfer is of primary importance. A steady-flow model for this region yields an equation similar to (15-5) with the inclusion of a heat-source term.

### 15-3-4 The Hot-Reaction Zone Region

In this region the final chemical reactions occur and the temperature of the propellant gases approach the adiabatic flame temperature. The boundary between this zone and the induction zone (at $x_{ind}$ in Fig. 15-6) is never sharp

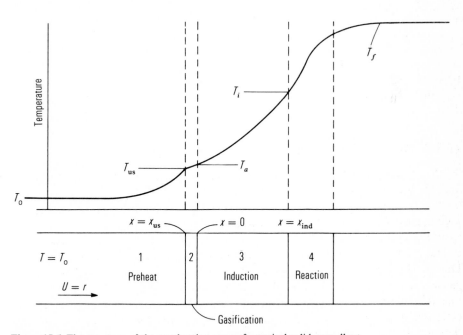

**Figure 15-6** The structure of the combustion zone of a typical solid propellant.

but may be located for convenience by requiring that the reaction rate at $x_{ind}$ be a specified fraction of the maximum reaction rate of region 4. This is analogous to the definition used in Chap. 8 for a one-dimensional gas flame.

Thus we see that the structure of a solid propellant's combustion zone differs from the structure of a gas-phase combustion zone mainly by the addition of a heat-transfer zone in the solid phase and of a solid-phase gasification zone. These differences are significant and they play an important role in the transient response of the burning rate to pressure fluctuations. This will be discussed in Sec. 15-5. The model as described has never been used to predict a burning rate for a solid propellant. In fact, the process is so complex that it is doubtful whether a burning rate will ever be calculated from first principles.

Solid propellants will usually support combustion in an inert atmosphere down to pressures of about 40 kPa. In the case of double-base propellants, combustion at these low pressures is incomplete, the flame zone is absent and the propellant is said to "fizz-burn." Under these conditions the nitric oxide decomposition reactions do not go to completion, and only about half of the available energy appears as thermal energy in the product gases.

## 15-4 LIQUID AND HYBRID ROCKET PROPELLANT COMBUSTION

In a liquid-propellant rocket engine the fuel and oxidizer are usually introduced into the chamber independently. It is simply too dangerous to premix the two or use a single exothermic compound as a propellant. In general, there are two types of fuel–oxidizer combinations in use today: the hypergolic, which burn on contact, or the nonhypergolic, which form relatively inert mixtures. Nonhypergolic systems require an independent ignition system and, in some cases, continuous ignition.

The design of efficient and reliable liquid-propellant rocket engines is mainly dependent on the existence of a proper arrangement for spray mixing in the chamber. The determination of this arrangement is primarily experimental at the present time and it appears that it will remain so in the future. There has been a considerable amount of work on spray-nozzle operation with and without combustion, but a discussion of this facet of combustor operation is outside the scope of this text.

Even with optimum mixing nozzle design, it takes some time for combustion to be complete in a liquid propellant rocket. Each chemical system has been found to have its own characteristic time. In general a short burning time means a more efficient engine because a large combustion chamber means a heavier chamber and less payload for that particular specific impulse.

Hybrid engines use a combination of solid and liquid propellants in their operation. One of the difficulties with the use of a composite solid propellant on an engine is that up to 80 percent of inorganic oxidizer must usually be included in the formulation to obtain a maximum specific impulse. However, such high concentrations of crystalline additives adversely affect the physical properties of the grain. One advantage of hybrid operation is that a fuel-rich grain with good physical properties may be burned with the injection of liquid oxidizer from the head end of the rocket. This increases the specific impulse of the entire system and has the added advantage of allowing more freedom in thrust programing for the rocket designer.

# 15-5 SOLID-PROPELLANT COMBUSTION INSTABILITY IN ROCKETS

A rocket engine is very susceptible to unstable operation. It is, after all, an enclosure in which heat is released at a very rapid rate, and it therefore conforms to the general requirements for self-excited instability discussed in Sec. 14-2. Solid- and liquid-propellant rocket instabilities are sufficiently different that each rocket type will be discussed in a separate section.

Solid-propellant rocket motors have been observed to suffer from four relatively distinct types of instabilities: (1) sporadic instabilities; (2) low-pressure instabilities (chuffing); and acoustic instabilities, both (3) linear and (4) nonlinear. Sporadic instabilities are not inherent to the system. They are caused by various types of structural failure of the rocket grain such as the opening of a fissure, the uncovery of an included bubble, or the loss of a large piece of propellant. These processes all cause the effective area of the burning surface to change rapidly and therefore cause the equilibrium operating pressure of the rocket to seek a new level rapidly. In addition, the loss of a sizable piece of propellant may cause a momentary nozzle blockage and lead to a destructive overpressure in the rocket engine. All of these difficulties may be remedied by choosing a solid propellant with the proper physical properties and by imposing adequate quality control during the manufacture of the grain.

Chuffing or low-pressure instability was at one time thought to be caused primarily by poor ignition; it is, in the sense that poor ignition of the grain will lead to low-pressure operation because the entire surface is not burning, which in turn will lead to the instability. It is now known that this type of instability is caused by the coupling of the response time of the solid-phase heat-conduction zone (region 1 of Fig. 15-6) and the characteristic exhaust time of the rocket engine. The solid-phase heat-conduction zone responds most slowly to a change in the operating pressure in the engine. Furthermore, a mathematical analysis of the effect of pressure change on regression rate indicates that when the propellant is subjected to a step pressure change, the burning rate does not respond instantaneously, and when it does respond it overshoots the new equilibrium burning rate momentarily. This behavior is shown schematically in Fig. 15-7. The time delay to the maximum overshoot has been found to be directly related to the thickness of the preheat zone in the propellant prior to the transient behavior. On the other hand, all rocket engines have a characteristic exhaust time that is proportional to the value of $L^* = V/A$ for the chamber,

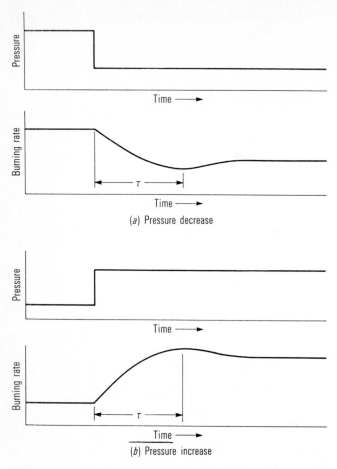

Pressure

Time ⟶

Burning rate

Time ⟶

(a) Pressure decrease

Pressure

Time ⟶

Burning rate

τ

Time ⟶

(b) Pressure increase

**Figure 15-7** The effect of a pressure step on the burning rate of a solid propellant (schematic).

where $V$ is the instantaneous free volume of the chamber and $A$ the throat area. $L^*$ is related to the characteristic exhaust time through the velocity of sound of the propellant gases at the throat. Since $L^*$ is essentially independent of the chamber pressure, the rocket's characteristic response time due to nozzle outflow is itself independent of the chamber pressure. The time delay to over-shoot, $\tau$, shown in Fig. 15-7 is, however, proportional to the preheat zone thickness which in turn is a function of the regression rate. Thus this overshoot time increases markedly as the operating pressure of the rocket decreases. At the pressure where the overshoot time and the exhaust time are equal, the two will couple to produce an instability in which the entire chamber pressure is oscil-lating as a unit. A typical photograph of the growth of the pressure oscillation just prior to extinguishment is shown in Fig. 15-8a. It has been observed that this oscillation always leads to extinguishment of the grain. A correlation of extinction pressure for a number of grain shapes burning in a small rocket

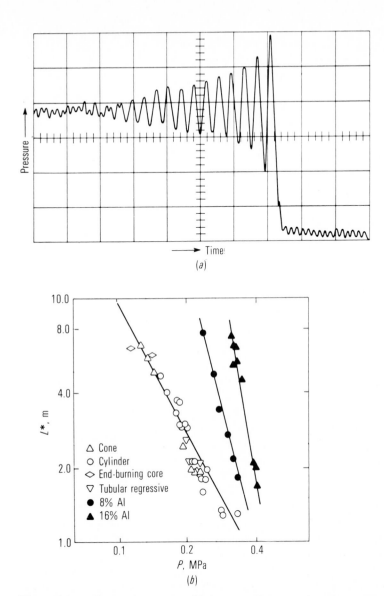

Time
(a)

$L^*$, m

△  Cone
○  Cylinder
◇  End-burning core
▽  Tubular regressive
●  8% Al
▲  16% Al

$P$, MPa
(b)

**Figure 15-8** (a) The low-frequency oscillation preceding $L^*$ extinguishment, 70 kPa/major division on the ordinate and 0.25 s/major division on the abscissa. *(With permission from N. W. Ryan, Tenth Symposium (International) on Combustion, The Combustion Institute, Pittsburgh, Pa., 1965, p. 1082.)* (b) The pressure-$L^*$-dependence of extinguishment for a number of regressive burning propellant shapes and for two different aluminium additions. Open symbols are for 0% Al addition. *(Adapted from F. A. Anderson, R. A. Strehlow, and L. D. Strand, AIAAJ., vol. 1, p. 2269 (1963), with permission.)*

engine is shown in Fig. 15-8b. The slope of the limit line for the case of an unaluminized propellant is $-1.80$ and the linearized theory for the instability predicts that it should be $-2n$, where $n$ is the pressure exponent (equal to 0.86 for this propellant). The steeper slopes that are observed for aluminized propellants have not been adequately explained as yet.

In all the firings shown in Fig. 15-8b the grain was burning regressively and stably for at least 1 s before extinguishment. $L^*$ was measured by removing the unburned portion of the grain after the firing and calculating the residual chamber volume at extinguishment. The grain could be recovered in these cases only because the rockets were fired onto a large vacuum chamber, and once the grain was extinguished by the instability it could not reignite because the chamber pressure itself fell below the minimum combustion pressure of the propellant. However, it has been observed that when the chamber is operated in an external environment whose pressure is above the propellant's minimum burning pressure, the presence of hot product gases in the chamber and the existence of a residual preheat zone in the propellant can cause the extinguished grain to reignite and burn. This behavior is called *chuffing*. Chuffing has been observed to occur repeatedly at low chamber pressures until either all the propellant is consumed or, if the grain is progressive, until the equilibrium burning pressure exceeds the $L^*$ limit for the propellant and stable combustion occurs. The theory for the reignition phase of the chuffing phenomenon is not complete, but it appears that reignition occurs by rapid decomposition of the solid phase in a thin zone, and that gas evolution during this process can be sufficiently rapid to momentarily pressurize the chamber to pressure levels above its equilibrium operating pressure.

This problem may be eliminated by designing the rocket engine to operate well above the $L^*$ limit of that particular propellant, and by ensuring that the entire burning surface is ignited rapidly during the ignition phase. This problem is usually handled adequately by placing the igniter at the head end of the grain so that the igniter gases and the initial propellant gases sweep over the entire unignited surface during the early phases of the ignition process.

Acoustic instability is so labeled because in its initial stages the growth mechanism is a linearly coupled mechanism involving an acoustic mode of the chamber. Acoustic instability is important because it can (and usually does) lead to very large amplitude pressure fluctuations in the rocket engine. It is not uncommon to find a $\pm 3$ MPa fluctuation in a chamber whose average steady operating pressure is 4.5 MPa. This large a fluctuating pressure can cause the average chamber pressure to change markedly, thereby changing the thrust level. It also may lead to a rapid temperature rise in the propellant grain, since the propellant is a viscoelastic material and will absorb energy if compressed at high amplitudes and high frequencies. In addition, these internal motor oscillations adversely affect any instrument package on the rocket and cause a weight penalty if one must design the chamber to contain the overpressures.

Acoustic instability occurs in a solid-propellant rocket engine when any of the possible acoustic modes of the chamber have been excited. The available modes and their frequencies may be found by solving a wave equation similar to Eq. (5-75) with $k = 0$. In this case one looks for a solution that is separable in space coordinates and time and is harmonic in time. For example, a rocket engine whose chamber is a right circular cylinder could possibly excite any of the frequencies of the chamber's longitudinal organ-pipe modes with a fundamental of wavelength $4L$, where $L$ is the chamber length, and in addition, could excite the chamber's transverse, radial, and rotating modes. Cross-modes and coupled modes are also possible in this simple geometry.

The fundamental and higher modes of the more complex internal geometries such as that of an internal burning star-shaped grain are more difficult to determine by purely mathematical techniques. Nevertheless, in general, one must assume that any mode over a broad frequency range is available for excitation. This behavior differs somewhat from the behavior of the Rijke tube as discussed in Chap. 14. In the Rijke tube discussion we noted that instability of a particular standing mode of the tube (only longitudinal modes were considered) was dependent through a particular coupling mechanism on the heater location in the tube and on the direction of the drift velocity. In the case of a solid-propellant rocket, however, the region of active heat addition is always relatively close to the surface of the grain and, therefore, to the boundary of the "acoustic" chamber. This is particularly true for high-pressure operation in a large motor.

As McClure, Hart, and Bird[3] first pointed out, this localization of the possible source of instability allows one to greatly simplify the problem because the bulk of the chamber volume may be treated as a resonant cavity which is filled with inert gas, and the response of the regressing surface to an acoustic level fluctuation may be examined independently as the source of acoustic energy. In general, a particular acoustic mode of the chamber will produce a local oscillation behavior of the gas near the surface which can contain both a fluctuating pressure component (usually thought of as acting normal to the surface) and a fluctuating velocity component (usually assumed to be acting along the surface). Figure 15-9 illustrates this behavior schematically for a specific longitudinal mode in a cylindrical chamber. Note that an acoustic-level signal on the surface at $x = L/3$ consists entirely of a pure velocity perturbation, while at the points $x = 0$ and $x = 2L/3$ it contains only a fluctuating pressure component. Points between these two locations are affected by a coupled transverse velocity–pressure perturbation whose nature changes with the location. In the framework of a linearized coupling theory one may assume, as a first approximation, that a purely pressure-coupled mode is acting independently of the velocity perturbations, even though both may be present.

The theory which discusses the amplification of a purely one-dimensional pressure perturbation by the burning surface is now highly developed. In short,

3. F. T. McClure, R. W. Hart, and J. F. Bird, *Solid Propellant Rocket Research*, M. Summerfield (ed.), Academic, New York, 1960.

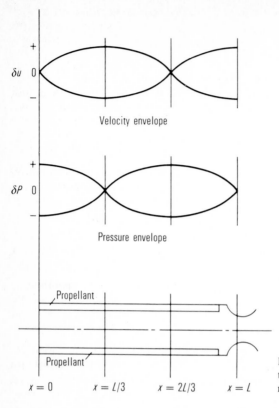

Figure 15-9 Pressure and velocity perturbations of the second longitudinal mode of a simple rocket engine.

the coupled equations for the one-dimensional model shown in Fig. 15-6 have been linearized to determine the reflection coefficient of the surface as a function of the frequency for an acoustic–pressure perturbation. A typical response curve is shown in Fig. 15-10. This curve shows that the reflection coefficient of an actively burning surface is greater than unity for a large range of frequencies, and therefore that this propellant surface could be expected to spontaneously amplify incident acoustic-level signals in a broad frequency range. The important result here is not the quantitative shape of the curve shown in Fig. 15-10 since this is dependent in a very complex way on the numerical values of a number of constants which are very difficult to determine. It is, instead, the qualitative proof that a regressing propellant surface can linearly amplify an incident acoustic signal over an extremely broad frequency range. In fact, because of the difficulty involved in determining the proper constants to use in the model, at the present time the reflection coefficient is determined by using an experimental technique. Figure 15-11 shows a drawing of a T burner which is used for this determination. This chamber is operated with a port which may be either a sonic nozzle or a larger-aperture nozzle connected to a large-volume, high-pressure chamber. Active propellant surfaces are placed at either end or both ends of the burner and a high-frequency pressure transducer is used to

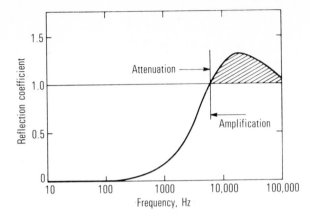

**Figure 15-10** The reflection of a burning solid-propellant surface as a function of the imposed frequency. This curve was calculated by McClure et al. using a linearized theory. Note the broad frequency range in which the amplification per cycle is positive (reflection coefficient > 1). *(Adapted with permission from F. T. McClure, R. W. Hart, and J. F. Bird, Solid Propellant Rocket Research (M. Summerfield, ed.), Academic, New York, 1960.)*

monitor the rate of growth of the first longitudinal mode of the chamber (wavelength = 2L). Frequency is changed by changing the chamber length L from run to run. A typical pressure record and its analysis is shown in Fig. 15-12. Note that the fluctuating pressure amplitude is growing exponentially at the beginning of oscillation. This is typical for a linearly self-excited instability. The experimental technique is somewhat complicated by the fact that the chamber has losses which must be determined independently to obtain a true value for the reflection coefficient of the active propellant surface. However, the technique has verified the theoretical result and may be used quite effectively for studying the relative response of different propellants to an acoustic perturbation.

The behavior of a velocity-coupled perturbation has not been examined as fully as the purely pressure-coupled amplification behavior discussed above. However, there is ample evidence that velocity-coupled modes do occur in solid-propellant rocket engines. In this case, the presence of the velocity fluctuation must preferentially excite a pressure fluctuation to cause self-amplification (Rayleigh's criterion, Sec. 14-2). Therefore, the mode cannot actively couple with the chamber oscillation as a linear phenomenon at the location of the pressure node for that particular mode (i.e., at $x = L/3$ in Fig. 15-9). It must instead either couple in regions where both fluctuations are present or must couple by

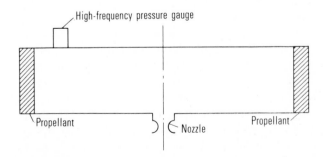

**Figure 15-11** The T burner used to elevate the instability behavior of a propellant to a pure pressure perturbation.

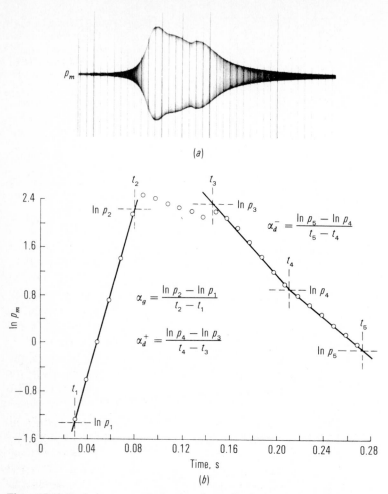

**Figure 15-12** Typical test record showing the oscillating pressure envelope and an analysis of its amplitude behavior. *(Adapted from M. D. Horton, AIAAJ., vol. 2, p. 1112 (1964), with permission.)*

an essentially nonlinear mechanism. Such a mechanism is shown schematically in Fig. 15-13. This figure illustrates a weak shock wave passing over the burning surface. Because of the shock compression and the extreme temperature sensitivity of the induction zone, the rate of heat release behind the shock wave will be enhanced, and this could lead to an active though nonlinear coupling mechanism. Even though we are still discussing the excitation of a standing mode of the chamber, in this case local excitation occurs by means of a traveling wave, and the usual acoustic theory of coupling is no longer a valid approximation.

Experimental observations of this type of coupling may be made in a modified T burner in which the nozzle is located in a position so as to preferentially excite the second mode ($L$ = wavelength). The propellant sample under study is placed in the wall of the chamber at a location well removed from the nozzle

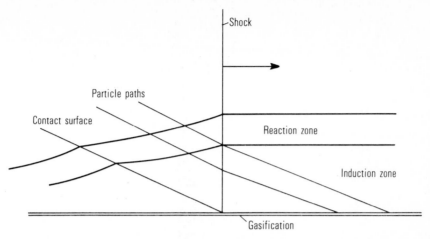

**Figure 15-13** Shock passing over the burning zone of a solid propellant in a shock-wave-burning-surface-oriented system (steady for a step shock wave). Note the movement of the reaction zone, the curvature of the particle paths at the shock, and the contact surface. The shock is not flat in the actual case because a gradient of sound velocity exists in the neighborhood of the surface. (Schematic.)

location. The nozzle's location is important to the mode type because a nozzle will greatly attenuate local pressure perturbations. It therefore favors an acoustic mode which has a pressure node at its location. A complete understanding of this type of coupling is not yet available.[4]

It has been observed that the presence of fine, high-melting-point particles in the grain, particularly combustible particles, usually enhances the stability behavior of the propellant. This is particularly true for aluminum powder, and therefore a number of practical propellants have been formulated with high concentrations of aluminum powder to suppress unstable burning. The reason why aluminum is effective in suppressing instability is not adequately understood at present. However, two possibilities exist. In the first place, immediately above the propellant surface the aluminum concentration on a mass basis may be considerably above its average concentration because it appears in the gas phase at the solid-phase density and velocity. This is caused by the fact that it takes time for the propellant gases to accelerate the particles. This effect is shown schematically in Fig. 15-14 for a specific steady regression rate. Secondly, there is evidence that combustion of the aluminum in the gas phase above the surface has a stabilizing effect on the induction-zone thickness which in turn reduces the acoustic-reflection coefficient sufficiently to produce stable oper-

4. J. N. Levine and J. D. Baum, "Modeling of Nonlinear Combustion Instability in Solid Propellant Rocket Motors," *Nineteenth Symposium (International) on Combustion*, The Combustion Institute, Pittsburgh, Pa., pp. 769–776 (1983).

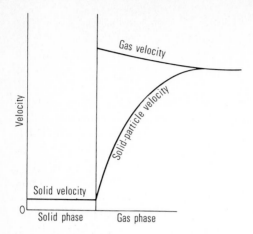

**Figure 15-14** Velocity of the gas and solid particles above a regressing surface. Surface velocity = 0.

ation. The presence of a fine powder above the surface would also alter the response of the flow to the weak transverse shock wave illustrated in Fig. 15-13. There is considerable literature available on this subject. The interested reader is referred to the bibliography.

## 15-6 LIQUID-ROCKET-ENGINE INSTABILITIES

The reader will note that the title of this section is different from that of the previous section. This difference arises because the instability behavior of solid-propellant engines has been quite definitely related to the behavior of the thin gas-phase region in the neighborhood of the burning surface, which constitutes the active heat-release region for the solid propellant. In the case of liquid rockets, however, the propellant is *pumped* into the chamber in an unmixed state and the active heat-release zone is somewhere in the interior of the chamber. Thus the coupling mechanisms which cause instability are much more strongly related to the transient behavior of the overall system in the case of a liquid engine.

There are three major types of instabilities in liquid-propellant engines. Two of these are fundamentally different from any of the solid-propellant rocket instabilities discussed above. The three major types are (1) pump-line coupling, (2) entropy instabilities (low frequency), and (3) acoustic instabilities.

Pump-line coupling problems can lead to a low-frequency instability in the liquid-propellant rocket engine. This type of instability is reminiscent of the instability first observed by Higgins, and in a liquid-propellant engine it is caused by a coupling of the characteristic relaxation time of the fuel feed rate and the exhaust time of the chamber. High-frequency excitation is seldom observed with this type of interaction because it is primarily controlled by the relatively slow inertial response of a column of liquid in the feed lines to a pressure fluctuation in the chamber. This instability may be remedied by operating the injectors at higher pressures, by changing the length of the injector lines, or by using positive-displacement pumps in the feed lines. It is not really a serious problem at the present time.

Entropy instability is a low-frequency instability which is unique to the liquid-propellant rocket engine. It is primarily a nonlinear phenomenon and as such it appears to require a finite-amplitude pressure perturbation before it is excited. This instability travels as a longitudinal wave in the motor—from the nozzle to the injector plate at the velocity of sound and from the injector plate to the nozzle at the flow velocity. The instability occurs when a finite pressure pulse from the nozzle so alters the mixing and burning process near the injector region that the local mixture burns incompletely and then passes through the nozzle as a nonequilibrium mixture. The oscillation is maintained if the conditions are such that the passage of this incompletely burned sample through the nozzle generates a new pressure pulse of the proper shape to again cause incomplete combustion at the injectors. This instability is called an *entropy instability* because the fluid motion must carry an entropy inhomogeneity through the nozzle to maintain the instability. It has sometimes been incorrectly called an "entropy-wave" instability, even though entropy cannot be propagated as a wave phenomenon. It has been found that entropy instabilities disappear if the fuel and oxidizer are injected in the proper manner. However, the problem has not been solved at the fundamental level, since the "proper" manner of injection must still be arrived at by trial-and-error techniques, and the design of injectors and injector configurations is still an empirical art.

Acoustic instabilities in liquid-propellant rockets, as in solid-propellant rockets, occur because the enclosure can support a variety of standing-wave modes, each with a characteristic acoustic frequency, and because coupling mechanisms are available which lead to slight amplification of the signal level during each acoustic cycle. However, it is not always obvious in the case of liquid-propellant rockets that the coupling mechanisms are linear mechanisms, and fundamental studies of liquid-propellant rocket instabilities have still not produced even a realistic categorization of all the possible coupling modes that may be available to the system. The major difficulty is that a large fraction of the volume of the engine always contains unburned and partially unmixed fuel and oxidizer, and the system is thus susceptible to the occurrence of catastrophic and self-energized pressure waves. Local detonation waves have been observed in certain circumstances, and other high-amplitude oscillations are observed repeatedly. Even though a great body of empirical information on different stable designs has accumulated in the past 20 years, there is still always the possibility of observing unstable operation on the first firing of a newly designed engine or on the first firing of an engine with a new fuel–oxidizer system. The reader is referred to the bibliography for more information on this subject.

## PROBLEMS

**15-1** Suppose a propellant of density 1600 kg/m$^3$ has a value of the pressure exponent $n = 0.5$ in Eq. (10-1) and a regression rate of 20 mm/s at 10.0 MPa. Assume the flame temperature is 2500 K, $\gamma = 1.2$, and the molecular weight of the product gases is constant = 0.024 kg/mole. Calculate the surface area required to maintain pressures of 2.0, 5.0, 10.0, and 20.0 MPa if the nozzle has a 15 mm diameter throat.

**15-2** For the propellant and rocket of Prob. 15-1 calculate a theoretical pressure–time curve for the progressive and regressive cylinders shown in Fig. 10-1. Assume that the ends of the grains are inhibited and the grain is initially burning at 10 MPa and has an initial length to diameter ratio of 10 : 1 for the burning surfaces. Further assume for the progressive grain that the outer diameter of the grain is twice the inner diameter.

**15-3** Hydrazine is a monopropellant which is quite frequently used as a gas generator for small rockets to provide thrust for orbit corrections, etc. If liquid hydrazine has a heat of vaporization of $-95.19$ kJ/mole calculate the flame temperature. Assume that $2H_2 + N_2$ are the products of decomposition.

**15-4** Liquid ozone can support a detonation. Its heat of vaporization is 10 kJ/mole and its sp. gravity at 100 K is 1.71. Calculate the Chapman–Jouguet detonation pressure and temperature, assuming the ideal gas law and that gaseous $O_2$ is the only product. Note that the ideal gas law is not valid at these extremely high pressures.

## BIBLIOGRAPHY

Cohen, N. S., "Review of Composite Propellant Burn Rate Modeling," *AIAA J.*, **18**, 277–293 (1980).

Cook, M. A., *The Science of High Explosives*, ACS Monograph No. 139, Reinhold, New York (1958).

Harrje, D. T., and F. H. Reardon, eds. *Liquid Propellant Rocket Combustion Instability*, NASA SP-194, (1972).

Krier, H., and M. Summerfield, eds., *Interior Ballistics of Guns*, AIAA, New York, **66** (1979).

Kuo, K., and G. Coates, "Review of Dynamic Burning of Solid Propellants in Gun and Rocket Propulsion Systems," *Sixteenth Symposium (International) on Combustion*, The Combustion Institute, Pittsburgh, Pa., pp. 1177–1190 (1977).

Lengelle, G., J. Deuterque, C. Verdier, A. Bizot, and J. F. Trubert, "Combustion Mechanisms of Double Based Solid Propellants," *Seventeenth Symposium (International) on Combustion*, The Combustion Institute, Pittsburgh, Pa., pp. 1443–1450 (1979).

Mader, C. L., *Numerical Modeling of Detonations*, University of California Press, Berkeley (1979).

Penner, S. S., *Chemical Rocket Propulsion and Combustion Research*, Gordon and Breach (1962).

Penner, S. S., and F. A. Williams, eds., *Detonation and Two-Phase Flow*, Progress in Astronautics and Rocketry, **6**, Academic, New York (1962).

# SIXTEEN

## POLLUTION AND COMBUSTION

### 16-1 INTRODUCTION

Ever since man first learned how to build fires, he has had to contend in one way or another with the pollutants that are produced by the combustion process. Before the advent of the industrial revolution almost two centuries ago the main problem was wood or coal smoke produced by incomplete combustion in open fireplaces. Wood combustion produces primarily soot, carbon monoxide, and unburned and partially oxygenated hydrocarbons. Coal combustion, which came much later, added fly ash, the sulfur oxides, and the nitrogen oxides to the list of pollutants produced by combustion.

In some ways the pollution problem is local, in others it is global in scope. The first large-scale problem that occurred because of combustion pollution was the trapping of combustion products by air inversion in or around large population or industrial centers for three or more cloudy days. Under these conditions the high sulfur content trapped in the local region causes sickness and excess deaths. The second large-scale problem is that of smog suffered by large urban centers with bright sunlight. The third major effect is the acid-rain effect caused by long-distance transport of sulfur and nitrogen acid aerosols by winds above the earth's boundary layer. In this chapter we will discuss not only the source of pollutants but also their fate in the earth's atmosphere.

### 16-2 CARBON MONOXIDE AND HYDROCARBONS

Tables 16-1 and 16-2 show that the major source of carbon monoxide and hydrocarbons from combustion is the motor vehicle. This is because it is difficult to construct an internal combustion engine that burns the fuel completely.

**Table 16-1  Annual emissions of carbon monoxide (estimates for 1970)**

| Source category | Emissions, millions of tons per year | Percent of total |
|---|---|---|
| Fuel combustion | | |
| Transportation | | |
|   Gasoline motor vehicles | 95.8 | 64.3 |
|   Diesel, aircraft, trains, vessels | 5.6 | 3.8 |
|   Off-highway vehicles | 9.5 | 6.4 |
| Total transportation | 110.9 | 74.5 |
| Stationary sources | | |
|   Coal | 0.5 | 0.3 |
|   Fuel oil | 0.1 | 0.1 |
|   Natural gas | 0.1 | 0.1 |
|   Wood | 0.1 | 0.1 |
| Total stationary source | 0.8 | 0.6 |
| Total fuel combustion | 111.7 | 75.1 |
| Industrial processes | 11.4 | 7.7 |
| Agricultural burning | 13.8 | 9.3 |
| Solid waste disposal | 7.2 | 4.9 |
| Miscellaneous* | 4.5 | 3.0 |
| Total | 148.6 | 100.0 |

* Includes coal-refuse burning, structural fires, and accidental and intentional forest fires.

*Source:* J. H. Cavender, D. S. Kircher, and A. J. Hoffman, *Nationwide Air Pollutant Emission Trends 1940–1970.* EPA Publication No. AP-115, January 1973. Reproduced with permission from D. A. Lynn, *Air Pollution—Threat and Response,* Addison-Wesley, Reading, Mass. (1976).

Stationary sources produce only a very small amount of these pollutants because it is relatively easy to effect complete combustion in these devices.

Hydrocarbons and carbon monoxide are oxidized in the atmosphere by the same set of reactions that occur during high-temperature combustion processes. It has been estimated that the normal concentration of OH radicals is about $10^2$ to $10^3$ molecules per cubic millimeter (i.e., a mole fraction of about $10^{-14}$ to $10^{-13}$) and that this is a sufficient amount to oxidize all the hydrocarbons and carbon monoxide to carbon dioxide. Indeed it is well known that the concentration of carbon dioxide, now at 330 ppm, is increasing at a rate of about 7 ppm/year. This slow increase is thought to be due to the combustion of hydrocarbon fuels by man. The half-life of carbon monoxide in the atmosphere is about 0.3 years and the majority of that present comes from the oxidation of methane produced by anaerobic bacteria in swamps and paddies.

Recall that the major man made source of carbon monoxide and unburned hydrocarbons is incomplete combustion. It has been observed that irrespective of how complete the combustion process is the ratio of CO to hydrocarbons emitted is about 10 : 1 and that the CO : odorant ratio is more like 150 : 1. This means that if you can smell wood smoke or automobile exhaust gases you are also breathing a gas which has more CO than the EPA (Environmental Protection Agency) standards allow.

## 16-3 SULFUR OXIDES

Table 16-3 shows that the major sources of atmospheric sulfur are coal combustion and industrial smelting. There is very little sulfur in fuel oils, gasoline, or commercially available combustible gases because the sulfur in these fuels is corrosive and it is removed for economic reasons. Thus combustion of these fuels adds little sulfur to the atmosphere. The chemistry of the formation of sulfur oxides is rather straightforward. Essentially all the sulfur in the fuel is oxidized to $SO_2$ or $SO_3$ during the combustion process. In most combustion processes only a small portion ends up as $SO_3$.

The half-life of this pollutant in the atmosphere is only about 6 to 10 days. During that period of time it is oxidized to form sulfuric acid and precipitates

## Table 16-2 Annual emissions of hydrocarbons (estimates for 1970)

| Source category | Emissions, millions of tons per year | Percent of total |
|---|---|---|
| Fuel combustion | | |
| Transportation | | |
| Gasoline motor vehicles | 16.6 | 47.6 |
| Diesel, aircraft, trains, vessels | 0.9 | 2.6 |
| Off-highway vehicles | 2.0 | 5.7 |
| Total transportation | 19.5 | 55.9 |
| Stationary sources | | |
| Coal | 0.2 | 0.6 |
| Fuel oil | 0.1 | 0.3 |
| Natural gas | 0.3 | 0.8 |
| Wood | — | — |
| Total stationary source | 0.6 | 1.7 |
| Total fuel combustion | 20.1 | 57.6 |
| Industrial processes | 9.5 | 27.2 |
| Agricultural burning | 2.8 | 8.0 |
| Solid waste disposal | 2.0 | 5.7 |
| Miscellaneous* | 0.5 | 1.5 |
| Total | 34.9 | 100.0 |

\* Includes coal-refuse burning, structural fires, and accidental and intentional forest fires.

*Source:* J. H. Cavender, D. S. Kircher, and A. J. Hoffman, *Nationwide Air Pollutant Emission Trends 1940–1970*, EPA Publication No. AP-115, January 1973. Reproduced with permission from D. A. Lynn, *Air Pollution—Threat and Response*, Addison-Wesley, Reading, Mass. (1976).

**Table 16-3 Annual emissions of sulfur oxides (estimates for 1970)**

| Source category | Emissions, millions of tons per year | Percent of total |
|---|---|---|
| Fuel combustion | | |
| Transportation | | |
|    Gasoline motor vehicles | 0.2 | 0.6 |
|    Diesel, aircraft, trains, vessels | 0.6 | 1.8 |
|    Off-highway vehicles | 0.2 | 0.6 |
|   Total transportation | 1.0 | 3.0 |
| Stationary sources | | |
|    Coal | 22.2 | 65.4 |
|    Fuel oil | 4.2 | 12.4 |
|    Natural gas | — | — |
|    Wood | 0.1 | 0.3 |
|   Total stationary source | 26.5 | 78.1 |
| Total fuel combustion | 27.5 | 81.1 |
| Industrial processes | 6.0 | 17.7 |
| Agricultural burning | 0.1 | 0.3 |
| Solid waste disposal | 0.1 | 0.3 |
| Miscellaneous* | 0.2 | 0.6 |
| Total | 33.9 | 100.0 |

* Includes coal-refuse burning, structural fires, and accidental and intentional forest fires.

*Source:* J. H. Cavender, D. S. Kircher, and A. J. Hoffman, *Nationwide Air Pollutant Emission Trends 1940–1970*, EPA Publication No. AP-115, January 1973. Reproduced with permission from D. A. Lynn, *Air Pollution—Threat and Response*, Addison-Wesley, Reading, Mass. (1976).

as "acid rain." Acid rains with pH's below that of vinegar have been reported on occasion. The problem is particularly severe in localities that contain mainly granitic rocks because limestone tends to neutralize the rain and prevent further damage. Thus the New England states, southeastern Canada, and Norway all suffer from the generation of sulfur oxides from sites as far away as Illinois and Ohio on the U.S. continent and France and Germany on the European continent. Also the major limestone sculptures in Europe have suffered irreparable damage in the past 40 years from acid rain. It must be pointed out that the nitric oxides, which will be discussed in the next section, also contribute to the acid-rain problem because they are ultimately oxidized to nitric acid.

Sulfur oxides production by combustion cannot be controlled. If one knows the percent sulfur in the fuel and the fuel–air ratio one can directly calculate the ppm of $SO_x$ in the stack gases. Absorption of this pollutant can be accomplished by injecting pulverized limestone within the right region of the furnace and then collecting the $CaSO_4$ that is formed or by scrubbing the exhaust gases with an alkaline suspension of lime in water. One can also meet emission standards by blending high-sulfur coal with low-sulfur coal to reach the required stack gas ppm set by the EPA.

## 16-4 NITROGEN OXIDES

The two oxides of nitrogen that are considered to be pollutants are nitric oxide, NO, and nitrogen dioxide, $NO_2$. A third oxide, nitrous oxide, $N_2O$, is extremely stable in the earth's atmosphere and is not considered to be a pollutant. Its half-life in the troposphere is estimated to be about 100 years, about the length of time it takes for it to diffuse to the stratosphere and be destroyed there by a series of photochemically triggered reactions.

Table 16-4 shows that $NO_x$ production is primarily from combustion processes and that the rate of production is rather evenly divided between transportation and stationary sources. The two higher oxides of nitrogen are the only gaseous pollutants that have interesting chemistry and there has been a great deal of research on both their formation in flames and their chemistry in the atmosphere. They are formed in the higher-temperature regions of flames either by direct oxidation of fuel-bound nitrogen or by the Zel'dovich chain

$$N + O_2 \rightarrow NO + O$$

$$O + N_2 \rightarrow NO + N$$

**Table 16-4  Annual emissions of nitrogen oxides (estimates for 1970)**

| Source category | Emissions, millions of tons per year | Percent of total |
|---|---|---|
| Fuel combustion | | |
| Transportation | | |
| Gasoline motor vehicles | 7.8 | 34.2 |
| Diesel, aircraft, trains, vessels | 2.0 | 8.8 |
| Off-highway vehicles | 1.9 | 8.3 |
| Total transportation | 11.7 | 51.3 |
| Stationary sources | | |
| Coal | 3.9 | 17.1 |
| Fuel oil | 1.3 | 5.7 |
| Natural gas | 4.7 | 20.6 |
| Wood | 0.1 | 0.4 |
| Total stationary source | 10.0 | 43.8 |
| Total fuel combustion | 21.7 | 95.1 |
| Industrial processes | 0.2 | 0.9 |
| Agricultural burning | 0.3 | 1.3 |
| Solid waste disposal | 0.4 | 1.8 |
| Miscellaneous* | 0.2 | 0.9 |
| Total | 22.8 | 100.0 |

\* Includes coal-refuse burning, structural fires, and accidental and intentional forest fires.

*Source:* J. H. Cavender, D. S. Kircher, and A. J. Hoffman, *Nationwide Air Pollutant Emission Trends 1940–1970*, EPA Publication No. AP-115, January 1973. Reproduced with permission from D. A. Lynn. *Air Pollution—Threat and Response*, Addison-Wesley, Reading, Mass. (1976).

with the less important reaction

$$OH + N \rightarrow NO + H$$

also occurring to some extent.

Ordinarily $NO_x$ is oxidized in the atmosphere to produce nitric acid and this is washed out of the atmosphere in rain. Thus the half-life of $NO_x$ in the atmosphere is about 6 to 10 days and its concentration is highly variable from location to location. However, if there are sufficient concentrations of $NO_x$ and reactive hydrocarbons (not methane) in the atmosphere in the presence of strong sunlight, photochemical smog can form. When these three ingredients are present, the concentrations of ozone and nitrogen dioxide rise and a large variety of partially oxygenated and very irritating hydrocarbons are also formed. The chemistry of smog formation has been studied extensively. Even though the hydrocarbon oxidation reactions are very complex the basic smog formation process is a relatively simple chain called the Ford–Endow[1] chain. The central driving reaction in this chain reaction is the production of oxygen atoms by the photolysis of the brown-colored nitrogen dioxide molecule

$$NO_2 + hv \xrightarrow{\phi k_a} NO + O \tag{16-1}$$

where $\phi$, the quantum efficiency, is one when the wavelength of light is below 366 nm and drops to zero at 435 nm. This reaction is followed by two fast reactions. These are

$$O + O_2 + M \xrightarrow{k_3} O_3 + M \tag{16-2}$$

and

$$NO + O_3 \xrightarrow{k_{31}} NO_2 + O_2 \tag{16-3}$$

where the unique subscripts on the $k$'s refer to those originally assigned by Ford and Endow. These reactions are represented by the open arrows in Fig. 16-1. These three reactions are very rapid and the concentration ratios of the species come to equilibrium, as for the case of the radicals during the induction period of an explosion. In this case, however, we are dealing with concentrations in the parts-per-million (ppm) or parts-per-billion (ppb) range, and the net change of their concentration due to other very much slower competing reactions determines what the equilibrium concentrations are.

We note that it takes hours to develop smog conditions once sunlight irradiates air that contains ppm concentrations of $NO_x$ and reactive hydrocarbons. Thus we use the Lindeman steady state approximation to write

$$0 = \frac{d[NO]}{dt} = \phi k_a[NO_2] - k_{31}[NO][O_3] \tag{16-4}$$

---

1. H. W. Ford and N. Endow, *J. Chem. Phys.*, **27**:1156 (1957).

This yields an expression for the relative concentration of the active species in the smog cycle

$$\frac{[NO][O_3]}{[NO_2]} = \frac{\phi k_a}{k_{31}}$$  (16-5)

The constant $\phi k_a/k_{31}$ is approximately $10^{-2}$ ppm for sunlight irradiating a cloudless area at $45°$ incidence angle, assuming the concentrations of the species are measured in ppm, as they usually are in air pollution work.

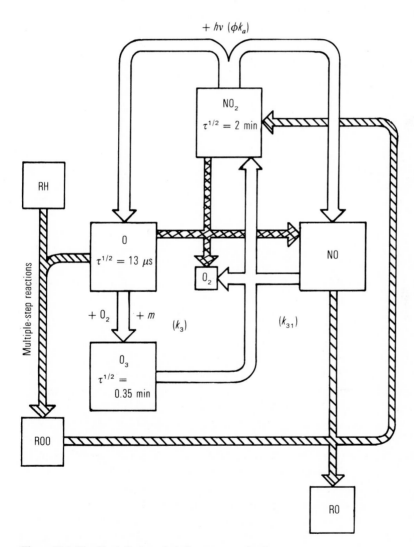

**Figure 16-1** The Ford–Endow chain for smog production.

Now, if the air does not contain an appreciable quantity of reactive hydrocarbons, the relatively slow reduction of $NO_2$ by the oxygen atom through the reaction

$$O + NO_2 \rightarrow NO + O_2 \tag{16-6}$$

reduces the $NO_2$ concentration. This in turn slows the photolysis reaction and, in order to maintain the equilibrium required by Eq. (16-5), both the ozone and nitrogen dioxide concentrations become small and virtually all the $NO_x$ exists as NO. This slow reaction is shown by the one set of cross-hatched arrows in the central region of Fig. 16-1.

When reactive hydrocarbons are present in the atmosphere, conditions change drastically. This is because the oxygen atom can attack reactive hydrocarbons and through multiple steps form reactive peroxides (generic formula, ROO) as shown in Fig. 16-1. These peroxides react with nitric oxide and reduce its concentration. These two generic reactions are shown as slashed arrows in Fig. 16-1. Now the equilibrium requirements of Eq. (16-5) mean that the concentrations of $NO_2$ and $O_3$ increase together. Notice that, as was stated earlier, it may take 2 hours for smog to reach equilibrium in sunlight and the half-life of the oxygen atom is only 13 $\mu s$ while NO, $NO_2$, and $O_3$ are being cycled through the chain scheme in times that are of the order of minutes. It should be noted that the active ingredient in this smog cycle, $NO_2$, is produced by this cycle. Also note that the cycle stops immediately when the sunlight is shut off. However, the products of the smog reactions, primarily the oxidized hydrocarbons, will persist for some time.

As was mentioned earlier, a number of irritating and corrosive hydrocarbons are produced during a photochemical smog incident. In particular it has been found that olefinic hydrocarbons that contain four or more carbon atoms produce peroxyacetyl nitrates. These have been given the acronym PAN and have the following structure.

$$
\begin{array}{c}
\quad\quad O \quad\quad\quad\quad O \\
\quad\quad \| \quad\quad\quad\quad \| \\
R-C-O-O-N \\
\quad\quad\quad\quad\quad\quad\quad \| \\
\quad\quad\quad\quad\quad\quad\quad O
\end{array}
$$

These compounds are lachrymators even at ppm concentrations. In addition, aldehydes, ketones, other peroxy compounds, and organic nitrites and nitrates are formed. Most of these are strongly oxidizing compounds and along with ozone and nitrogen dioxide are responsible for plant damage.

Los Angeles was the first location to have a noticeable smog problem. There are two reasons for this. In the first place it is a natural valley on the coast and wind conditions create an inversion layer at about 750 m elevation which traps the air in the valley, thus allowing pollutant concentrations to become large. In the second place, after World War II an extensive freeway system was installed in the sprawling metropolitan area. Public transportation became almost nonexistent and private automobiles without pollution-control devices were used almost exclusively for transportation. These, along with uncontrolled stationary sources and the millions of private backyard incinerators, produced relatively large quantities of both $NO_x$ and unburned hydrocarbons. At the present time smog can be observed over many western cities such as Denver and Phoenix. The only real solution to the problem is to reduce the $NO_x$ and unburned hydrocarbon emissions in localities that are prone to photochemical smog production.

## 16-5 PARTICULATES AND SOOT

Particulates are generally considered to be oxidized inorganic materials from the fuel. As shown in Table 16-5 they primarily arise from the combustion of coal which contains an ash content, and from industrial processes. Some of the ash is formed as large particles and quickly settles out of the product gases. However, any particles that are less than about 10 $\mu$m in diameter will remain airborne for a long time. They are usually called fly ash. These can be removed

**Table 16-5 Annual emissions of particulate matter (estimates for 1970)**

| Source category | Emissions, millions of tons per year | Percent of total |
|---|---|---|
| Fuel combustion | | |
| Transportation | | |
| Gasoline motor vehicles | 0.3 | 1.1 |
| Diesel, aircraft, trains, vessels | 0.3 | 1.2 |
| Off-highway vehicles | 0.1 | 0.4 |
| Total transportation | 0.7 | 2.7 |
| Stationary sources | | |
| Coal | 5.6 | 21.5 |
| Fuel oil | 0.4 | 1.5 |
| Natural gas | 0.2 | 0.8 |
| Wood | 0.6 | 2.3 |
| Total stationary source | 6.8 | 26.1 |
| Total fuel combustion | 7.5 | 28.8 |
| Industrial processes | 13.3 | 51.0 |
| Agricultural burning | 2.4 | 9.2 |
| Solid waste disposal | 1.4 | 5.3 |
| Miscellaneous* | 1.5 | 5.7 |
| Total | 26.1 | 100.0 |

* Includes coal-refuse burning, structural fires, and accidental and intentional forest fires.

*Source:* J. H. Cavender, D. S. Kircher, and A. J. Hoffman, *Nationwide Air Pollutant Emission Trends 1940–1970*, EPA Publication No. AP-115, January 1973. Reproduced with permission from D. A. Lynn, *Air Pollution—Threat and Response*, Addison-Wesley, Reading, Mass. (1976).

from the product gases either by using a bag house, which operates essentially like a household vacuum cleaner, or by using an electrostatic precipitator. This makes use of the fact that particles will invariably be charged. In this device the particle-laden stream is passed through a strong electric field, the charged particles attach themselves to the wall and are eventually removed by shaking the device or by backflow.

Soot forms in the fuel-rich regions of diffusion flames and in fuel-rich premixed flames. Fresh soot has the empirical formula $C_8H$, that is, it contains about 12 molar percent hydrogen. Thus soot is *not* carbon. In premixed flames the critical C/O ratio for the formation of soot has been found to be about 0.5. We note that thermodynamic equilibrium considerations lead to the conclusion that soot should only form when C/O $\geq$ 1. This means, of course, that soot is formed by nonequilibrium kinetic processes such as those that allow $C_2H_2$ and $C_2H_4$ to form in a methane flame.

A number of distinct stages or processes that lead to the formation of soot can be identified.[2] These are (1) nucleation, (2) growth to form spherical particles, and (3) aggregation of these to form chain-like structures that are soot.

It is now thought that the nucleation process involves ion reactions and that the ion species are formed primarily by the reaction

$$CH + O \rightarrow CHO^+ + e^-$$

followed by reactions of the type

$$CHO^+ + C_2H_2 \rightarrow C_2H_3^+ + CO$$

which then lead to rapid ionic polymerization. These ions grow in size until they become incipient particles with atomic masses of about $10^4$. They then form crystallites and finally spherical particles whose masses range from $10^5$ to $10^7$. These soot spheres have diameters in the 10 to 50 nm range. After they are formed they aggregate to form chains. In a premixed sooting flame this all happens in about 10 to 50 ms.

Different hydrocarbons have different sooting tendencies. In general the aromatics have the highest tendency to soot and the paraffinic hydrocarbons the least, with the olefins lying in between.

Soot can present serious problems, e.g., in Diesel engines. However it is also desirable in many applications. In large water-wall boilers for power generation almost half of the energy transferred to the water from the combustion process is transferred by radiation from the sooting diffusion flame in the boiler. This soot is eventually completely oxidized to $CO_2$ but its presence in the boiler markedly enhances heat-transfer to the wall.

The reader is referred to the bibliography for reviews of the subject of soot formation and destruction.

2. H. F. Calcote, *Combust. Flame*, **42**: 215–242 (1982).

# PROBLEMS

**16-1** Given that the half-life of CO in the atmosphere is about 0.3 years and that the major path for its destruction is the homogeneous gas-phase reaction $CO + OH \rightarrow CO_2 + H$ with a rate constant $k_1 = 10^{12.88} \, T^{1.3} \exp(-3,222/RT)$, calculate the background concentration of OH radicals in the atmosphere.

**16-2** Illinois coal has a high sulfur content and it is usually mixed with Wyoming coal to meet stack-emission requirements in a large utility boiler. If IL-H and WY-G coals listed in App. C1 are mixed to produce a 25/75 blend, what would be the sulfur oxides content of the wet-stack gases at 20% excess air (in ppm)?

**16-3** Use the calculated equilibrium compositions of a methane–air flame to determine the excess air that would be required to produce stack gases that contain 10 ppm CO and 100 ppm CO. What would the nitric oxide levels be at these compositions?

# BIBLIOGRAPHY

**General**

Bowman, C. T., "Kinetics of Pollutant Formation and Destruction in Combustion," *Prog. Energy Combust. Sci.*, **1**: 33–45 (1975).

Caretto, L. S., "Mathematical Modeling of Pollutant Formation," *Prog. Energy Combust. Sci.*, **1**: 47–71 (1975).

Chameides, W. L., and D. D. Davis, "Chemistry in the Troposphere," *Chem. Eng. News*, **76**: 38–52, Oct. 4, 1982.

Perkins, H. C., *Air Pollution*, McGraw-Hill, New York (1974).

Sawyer, R., "The Formation and Destruction of Pollutants in Combustion Processes: Clearing the Air on the Role of Combustion Research," *Eighteenth Symposium (International) on Combustion*, The Combustion Institute, Pittsburgh, Pa., pp. 1–22 (1981).

**Nitrogen oxides**

Hayhurst, A. N., and I. M. Vince, "Nitric Oxide Formation from $N_2$ in Flames: The Importance of 'Prompt' NO," *Prog. Energy Combust. Sci.*, **6**: 35–51 (1980).

Rosenburg, H. S., L. M. Curran, A. V. Slack, J. Ando, and J. H. Oxley, "Post Combustion Methods for Control of $NO_x$ Emissions," *Prog. Energy Combust. Sci.*, **6**: 287–302 (1980).

Sarofim, A. M., and R. C. Flagan, "$NO_x$ Control for Stationary Combustion Sources," *Prog. Energy Combust. Sci.*, **2**: 1–25 (1976).

**Pollutant production and control**

Henein, N. A., "Analysis of Pollutant Formation and Control and Fuel Economy in Diesel Engines," *Prog. Energy Combust. Sci.*, **1**: 165–207 (1975).

Heywood, J. B., "Pollutant Formation and Control in Spark-Ignition Engines," *Prog. Energy Combust. Sci.*, **1**: 135–164 (1975).

Jones, R. E., "Gas Turbine Engine Emissions—Problems, Progress and Future," *Prog. Energy Combust. Sci.*, **4**: 73–113 (1978).

Levy, A., "Unresolved Problems in $SO_x$, $NO_x$ and Soot Control in Combustion," *Nineteenth Symposium (International) on Combustion*, The Combustion Institute, Pittsburgh, Pa., pp. 1223–1242 (1983).

Mellor, A. M., "Gas Turbine Engine Pollution," *Prog. Energy Combust. Sci.*, **1**: 111–133 (1975).

Wendt, J. O. L., "Fundamental Coal Combustion Mechanisms and Pollutant Formation in Furnaces," *Prog. Energy Combust. Sci.*, **6**: 201–222 (1980).

**Soot**

Calcote, H. F., "Mechanisms of Soot Nucleation in Flames—A Critical Review," *Combust. Flame,* **42**: 215–242 (1981).

Haynes, B. S., and H. G. Wagner, "Soot Formation," *Prog. Energy Combust. Sci.,* **7**: 229–273 (1981).

Howard, J. B., and W. J. Kausch, Jr.: "Soot Control by Fuel Additives," *Prog. Energy Combust. Sci.,* **6**: 263–276 (1980).

Longwell, J. P., "The Formation of Polycyclic Hydrocarbons by Combustion," *Nineteenth Symposium (International) on Combustion,* The Combustion Institute, Pittsburgh, Pa., pp. 1339–1351 (1983).

Smith, O. I., "Fundamentals of Soot Formation in Flames with Application to Diesel Engine Particulate Emissions," *Prog. Energy Combust. Sci.,* **7**: 275–291 (1981).

## Appendix A1 Lennard–Jones Parameters‡

| Species | Molecular weight | From viscosity data§ | | From second virial coefficients data¶ | | |
|---|---|---|---|---|---|---|
| | | $\varepsilon/k$ K | $\sigma_0$ nm | $\varepsilon/k$ K | $\sigma_0$ nm | $b_0$ $m^3$/mole $\times 10^6$ |
| Air | 0.028964 | 78.6 | 0.3711 | 99.2 | 0.3522 | 55.11 |
| $O_2$ | 0.031999 | 106.7 | 0.3467 | 117.5 | 0.358 | 57.75 |
| O | 0.016000 | 106.7 | 0.3050 | — | — | — |
| $N_2$ | 0.028013 | 71.4 | 0.3798 | 95.05 | 0.3698 | 63.78 |
| N | 0.014007 | 71.4 | 0.3298 | — | — | — |
| $H_2$ | 0.002016 | 59.7 | 0.2827 | 29.2 | 0.287 | 29.76 |
| H | 0.001008 | 37.0 | 0.2708 | — | — | — |
| $H_2O$ | 0.018016 | 809.1 | 0.2641 | — | — | — |
| OH | 0.017008 | 79.8 | 0.3147 | — | — | — |
| $CO_2$ | 0.044010 | 195.2 | 0.3941 | 189.0 | 0.4486 | 113.9 |
| CO | 0.028011 | 91.7 | 0.3690 | 100.2 | 0.3763 | 67.22 |
| NO | 0.030008 | 116.7 | 0.3492 | 131.0 | 0.317 | 40.0 |
| $N_2O$ | 0.044016 | 232.4 | 0.3828 | 189.0 | 0.459 | 122.0 |
| $NH_3$ | 0.017031 | 558.3 | 0.2900 | — | — | — |
| NH | 0.015015 | 65.3 | 0.3312 | — | — | — |
| $CH_4$ | 0.016043 | 148.6 | 0.3758 | 148.2 | 0.3817 | 70.16 |
| CH | 0.013009 | 68.6 | 0.3370 | — | — | — |
| $C_2H_2$ | 0.026038 | 231.8 | 0.4033 | — | — | — |
| $C_2H_4$ | 0.028054 | 224.7 | 0.4163 | 199.2 | 0.4523 | 116.7 |
| CN | 0.026018 | 75.0 | 0.3856 | — | — | — |
| HCN | 0.027026 | 569.1 | 0.3630 | — | — | — |
| $F_2$ | 0.037998 | 112.6 | 0.3357 | — | — | — |
| F | 0.018999 | 112.6 | 0.2986 | — | — | — |
| $Cl_2$ | 0.070906 | 316.0 | 0.4217 | — | — | — |
| Cl | 0.035453 | 130.8 | 0.3613 | — | — | — |
| $Br_2$ | 0.159832 | 507.9 | 0.4296 | — | — | — |
| Br | 0.079916 | 236.6 | 0.3672 | — | — | — |
| HF | 0.020006 | 330.0 | 0.3148 | — | — | — |
| HCl | 0.036465 | 344.7 | 0.3339 | — | — | — |
| HBr | 0.080924 | 449.0 | 0.3353 | — | — | — |
| He | 0.004003 | 10.22 | 0.2551 | 10.22 | 0.2556 | 21.07 |
| Ne | 0.020179 | 32.8 | 0.2820 | 35.60 | 0.2749 | 26.21 |
| Ar | 0.039948 | 93.3 | 0.3542 | 119.8 | 0.3405 | 49.80 |

‡ It is suggested that the viscosity data be used exclusively for transport coefficient calculations and that the second virial coefficient be used exclusively for equation of state calculations.

§ Data taken from R. A. Svehla, NASA Tech. Report R-132, Lewis Research Center, Cleveland, Ohio (1962).

¶ Data and tables in App. A2 taken from J. O. Hirshfelder, C. F. Curtis, and R. B. Bird. *Molecular Theory of Gases and Liquids*, Wiley, New York (1954) (with permission).

# Appendix A2 Collision Integrals and Reduced Virial Coefficients

| $kT/\varepsilon$ | $\Omega_{\mu/\kappa}$ | $\Omega_D$ | $C\dagger$ | $B^*$ | $C^*$ |
|---|---|---|---|---|---|
| 0.30 | 2.785 | 2.662 | 0.847 | -27.880581 | |
| 0.35 | 2.628 | 2.476 | 0.839 | -18.754895 | |
| 0.40 | 2.492 | 2.318 | 0.833 | -13.798835 | |
| 0.45 | 2.368 | 2.184 | 0.828 | -10.754975 | |
| | | | | | |
| 0.50 | 2.257 | 2.066 | 0.825 | -8.720205 | |
| 0.55 | 2.156 | 1.966 | 0.823 | -7.2740858 | |
| 0.60 | 2.065 | 1.877 | 0.822 | -6.1979708 | |
| 0.65 | 1.982 | 1.798 | 0.823 | -5.3681918 | |
| 0.70 | 1.908 | 1.729 | 0.823 | -4.7100370 | -3.37664 |
| | | | | | |
| 0.75 | 1.841 | 1.667 | 0.825 | -4.1759283 | -1.79197 |
| 0.80 | 1.780 | 1.612 | 0.826 | -3.7342254 | -0.84953 |
| 0.85 | 1.725 | 1.562 | 0.829 | -3.3631193 | -0.27657 |
| 0.90 | 1.675 | 1.517 | 0.831 | -3.0471143 | 0.07650 |
| 0.95 | 1.629 | 1.476 | 0.834 | -2.7749102 | 0.29509 |
| | | | | | |
| 1.00 | 1.587 | 1.439 | 0.837 | -2.5380814 | 0.42966 |
| 1.05 | 1.549 | 1.406 | 0.839 | -2.3302208 | 0.51080 |
| 1.10 | 1.514 | 1.375 | 0.841 | -2.1463742 | 0.55762 |
| 1.15 | 1.482 | 1.346 | 0.844 | -1.9826492 | 0.58223 |
| 1.20 | 1.452 | 1.320 | 0.848 | -1.8359492 | 0.59240 |
| | | | | | |
| 1.25 | 1.424 | 1.296 | 0.850 | -1.7037784 | 0.59326 |
| 1.30 | 1.399 | 1.273 | 0.853 | -1.5841047 | 0.58815 |
| 1.35 | 1.375 | 1.253 | 0.856 | -1.4752571 | 0.57933 |
| 1.40 | 1.353 | 1.233 | 0.859 | -1.3758479 | 0.56831 |
| 1.45 | 1.333 | 1.215 | 0.861 | -1.2847160 | 0.55611 |
| | | | | | |
| 1.50 | 1.314 | 1.198 | 0.863 | -1.2008832 | 0.54339 |
| 1.55 | 1.296 | 1.182 | 0.865 | -1.1235183 | 0.53059 |
| 1.60 | 1.279 | 1.167 | 0.868 | -1.0519115 | 0.51803 |
| 1.65 | 1.264 | 1.153 | 0.871 | -0.98545337 | 0.50587 |
| 1.70 | 1.248 | 1.140 | 0.872 | -0.92361639 | 0.49425 |
| | | | | | |
| 1.75 | 1.234 | 1.128 | 0.874 | -0.86594279 | 0.48320 |
| 1.80 | 1.221 | 1.116 | 0.876 | -0.81203328 | 0.47277 |
| 1.85 | 1.209 | 1.105 | 0.878 | -0.76153734 | 0.46296 |
| 1.90 | 1.197 | 1.094 | 0.880 | -0.71414733 | 0.45376 |
| 1.95 | 1.186 | 1.084 | 0.882 | -0.66959030 | 0.44515 |
| | | | | | |
| 2.00 | 1.175 | 1.075 | 0.884 | -0.62762535 | 0.43710 |
| 2.10 | 1.156 | 1.057 | 0.887 | -0.55063308 | 0.42260 |
| 2.20 | 1.138 | 1.041 | 0.890 | -0.48170997 | 0.40999 |
| 2.30 | 1.122 | 1.026 | 0.893 | -0.41967761 | 0.39900 |
| 2.40 | 1.107 | 1.012 | 0.896 | -0.36357566 | 0.38943 |

| $kT/\varepsilon$ | $\Omega_{\mu/\kappa}$ | $\Omega_D$ | $C\dagger$ | $B^*$ | $C^*$ |
|---|---|---|---|---|---|
| 2.50 | 1.093 | 0.9996 | 0.899 | -0.31261340 | 0.38108 |
| 2.60 | 1.081 | 0.9878 | 0.902 | -0.26613345 | 0.37378 |
| 2.70 | 1.069 | 0.9770 | 0.904 | -0.22358626 | 0.36737 |
| 2.80 | 1.058 | 0.9672 | 0.906 | -0.18450728 | 0.36173 |
| 2.90 | 1.048 | 0.9576 | 0.909 | -0.14850215 | 0.35675 |
| 3.00 | 1.039 | 0.9490 | 0.910 | -0.11523390 | 0.35234 |
| 3.10 | 1.030 | 0.9406 | 0.912 | -0.08441245 | 0.34842 |
| 3.20 | 1.022 | 0.9328 | 0.913 | -0.05578696 | 0.34491 |
| 3.30 | 1.014 | 0.9256 | 0.915 | -0.02913997 | 0.34177 |
| 3.40 | 1.007 | 0.9186 | 0.917 | -0.00428086 | 0.33894 |
| 3.50 | 0.9999 | 0.9120 | 0.918 | 0.01895684 | 0.33638 |
| 3.60 | 0.9932 | 0.9058 | 0.919 | 0.04072012 | 0.33407 |
| 3.70 | 0.9870 | 0.8998 | 0.921 | 0.06113882 | 0.33196 |
| 3.80 | 0.9811 | 0.8942 | 0.922 | 0.08032793 | 0.33002 |
| 3.90 | 0.9755 | 0.8888 | 0.923 | 0.09839014 | 0.32825 |
| 4.00 | 0.9700 | 0.8836 | 0.924 | 0.11541691 | 0.32662 |
| 4.10 | 0.9649 | 0.8788 | 0.925 | 0.13149021 | 0.32510 |
| 4.20 | 0.9600 | 0.8740 | 0.926 | 0.14668372 | 0.32369 |
| 4.30 | 0.9553 | 0.8694 | 0.927 | 0.16106381 | 0.32238 |
| 4.40 | 0.9507 | 0.8652 | 0.928 | 0.17469039 | 0.32115 |
| 4.50 | 0.9464 | 0.8610 | 0.928 | 0.18761774 | 0.32000 |
| 4.60 | 0.9422 | 0.8568 | 0.929 | 0.19989511 | 0.31891 |
| 4.70 | 0.9382 | 0.8530 | 0.930 | 0.21156728 | 0.31788 |
| 4.80 | 0.9343 | 0.8492 | 0.931 | 0.22267507 | 0.31690 |
| 4.90 | 0.9305 | 0.8456 | 0.931 | 0.23325577 | 0.31596 |
| 5.0 | 0.9269 | 0.8422 | 0.932 | 0.24334351 | 0.31508 |
| 6.0 | 0.8963 | 0.8124 | 0.936 | 0.32290437 | 0.30771 |
| 7.0 | 0.8727 | 0.7896 | 0.940 | 0.37608846 | 0.30166 |
| 8.0 | 0.8538 | 0.7712 | 0.941 | 0.41343396 | 0.29618 |
| 9.0 | 0.8379 | 0.7556 | 0.943 | 0.44059784 | 0.29103 |
| 10.0 | 0.8242 | 0.7424 | 0.945 | 0.46087529 | 0.28610 |
| 20.0 | 0.7432 | 0.6640 | 0.948 | 0.52537420 | 0.24643 |
| 30.0 | 0.7005 | 0.6232 | 0.948 | 0.52692546 | 0.21954 |
| 40.0 | 0.6718 | 0.5960 | 0.948 | 0.51857502 | 0.20012 |
| 50.0 | 0.6504 | 0.5756 | 0.948 | 0.50836143 | 0.18529 |
| 60.0 | 0.6335 | 0.5596 | 0.948 | 0.49821261 | 0.17347 |
| 70.0 | 0.6194 | 0.5464 | 0.948 | 0.48865069 | 0.16376 |
| 80.0 | 0.6076 | 0.5352 | 0.948 | 0.47979009 | 0.15560 |
| 90.0 | 0.5973 | 0.5256 | 0.948 | 0.47161504 | 0.14860 |
| 100.0 | 0.5882 | 0.5170 | 0.948 | 0.46406948 | 0.14251 |
| 200.0 | 0.5320 | 0.4644 | 0.948 | 0.41143168 | 0.10679 |
| 300.0 | 0.5016 | 0.4360 | 0.948 | 0.38012787 | 0.08943 |
| 400.0 | 0.4811 | 0.4170 | 0.948 | 0.35835117 | 0.07862 |

# APPENDIX B

**Thermodynamic Properties of Selected Species in the Ideal Gas State (Except for Graphite and Bromine) (1 cal = 4.184 J).**
**List of Species[1] in Appendices B1 and B2**

| Substance | Formula | Page numbers for molecular weight $\Delta H_f^\circ$ $C_p$, H–H$_{298}$, $S$ | $\log_{10} K_p$ |
|---|---|---|---|
| Real Air[2] | — | 486 | —[2] |
| 21% Oxygen Air[2] | — | 486 | —[2] |
| Oxygen (SSE)[3, 4] | $O_2$ | 487 | —[5] |
| Oxygen, Monatomic[4] | O | 487 | 502 |
| Nitrogen (SSE)[4] | $N_2$ | 488 | —[5] |
| Nitrogen, Monatomic[4] | N | 488 | 510 |
| Hydrogen (SSE)[4] | $H_2$ | 489 | —[5] |
| Hydrogen, Monatomic[4] | H | 489 | 508 |
| Water Vapor | $H_2O$ | 490 | 508 |
| Hydroxyl[4] | OH | 490 | 508 |
| Carbon dioxide | $CO_2$ | 491 | 508 |
| Carbon monoxide | CO | 491 | 508 |
| Nitric oxide | NO | 492 | 508 |
| Nitrogen dioxide | $NO_2$ | 492 | 508 |
| Nitrous oxide | $N_2O$ | 493 | 509 |
| Ammonia[4] | $NH_3$ | 493 | 509 |
| Amidogen[4] | $NH_2$ | 494 | 509 |
| Imidogen[4] | NH | 494 | 509 |

| Substance | Formula | Page numbers for molecular weight $\Delta H_f^\circ$ $C_p$, H–H$_{298}$, S | Page numbers for $\log_{10} K_p$ |
|---|---|---|---|
| Methane | CH$_4$ | 495 | 509 |
| Methyl | CH$_3$ | 495 | 509 |
| Methylene[6] | CH$_2$ | 496 | 509 |
| Methylidyne | CH | 496 | 509 |
| Acetylene | C$_2$H$_2$ | 497 | 510 |
| Ethylene | C$_2$H$_4$ | 497 | 510 |
| Graphite[4] (SSE) | C(s) | 498 | —[5] |
| Cyano | CN | 498 | 510 |
| Hydrogen Cyanide | HCN | 499 | 510 |
| Isocyanic acid[7] | HNCO | 499 | 510 |
| Formyl[7] | HCO | 500 | 510 |
| Formaldehyde | CH$_2$O | 500 | 510 |
| Hydroperoxyl | HO$_2$ | 501 | 511 |
| Nitroxyl hydride | HNO | 501 | 511 |
| Fluorine (SSE)[4] | F$_2$ | 502 | —[5] |
| Fluorine, Monatomic | F | 502 | 511 |
| Chlorine (SSE) | Cl$_2$ | 503 | —[5] |
| Chlorine, Monatomic[7] | Cl | 503 | 511 |
| Bromine (SSE) | Br$_2$ | 504 | —[5] |
| Bromine, Monatomic[6] | Br | 504 | 511 |
| Hydrogen Fluoride[4] | HF | 505 | 511 |
| Hydrogen Chloride | HCl | 505 | 511 |
| Hydrogen Bromide | HBr | 506 | 511 |
| Helium[4] | He | 506 | —[5] |
| Neon[4] | Ne | 507 | —[5] |
| Argon[4] | Ar | 507 | —[5] |

1. Unless otherwise marked the data for these species were obtained from D. R. Stull and H. Prophet, JANAF Thermodynamic Tables, second edition, NSRDS-NBS 37, National Bureau of Standards, June 1971. These tables may be obtained from the Superintendent of Documents, U.S. Government Printing Office, Washington, D.C. (Cat. No. C 13.48:37) for $9.75. Horizontal lines in certain tables indicate a change in state of the species.

2. Calculated from the tabular values for the appropriate species. Values of log $k_p$ are not appropriate for these mixtures.

3. (SSE) means standard state element.

4. Data for these species were taken from M. W. Chase, Jr., J. L. Curnutt, J. R. Downey, Jr., R. A. McDonald, A. N. Syverud, and E. A. Valenzuela, JANAF Thermochemical Tables. 1982 Supplement, *J. Phys. Chem. Ref. Data*, **11**:695–940 with permission.

5. log $k_p$ for these species is defined to be 0.000.

6. Data for these species were taken from M. W. Chase, Jr., J. L. Curnutt, H. Prophet, R. A. McDonald, and A. N. Syverud, JANAF Thermochemical Tables, 1975 Supplement, *J. Phys. Chem. Ref. Data*, **4**:1–75 (1975) with permission.

7. Data for these species were taken from M. W. Chase, Jr., J. L. Curnutt, A. T. Hu, H. Prophet, A. N. Syverud, and L. C. Walker, JANAF Thermochemical Tables, 1974 Supplement, *J. Phys. Chem. Ref. Data*, **3**:311–480 (1974) with permission.

# Appendix B1

| T, K | Real air (App. C4) $(\Delta H_f^\circ)_{298.15} = -0.13$ kJ/mole M.W. = 0.028964 | | | 21% $O_2$ air $(\Delta H_f^\circ)_{298.15} = 0.00$ kJ/mole M.W. = 0.028850 | | |
|---|---|---|---|---|---|---|
| | $C_p^\circ$, J/mole $K$ | $H^\circ - H_{298}^\circ$, kJ/mole | $S^\circ$, J/mole $K$ | $C_p^\circ$, J/mole $K$ | $H^\circ - H_{298}^\circ$, kJ/mole | $S^\circ$, J/mole $K$ |
| 100 | 29.02 | -5.76 | 162.28 | 29.10 | -5.77 | 162.54 |
| 200 | 29.03 | -2.85 | 182.40 | 29.11 | -2.86 | 182.71 |
| 298 | 29.10 | 0.00 | 194.00 | 29.18 | 0.00 | 194.34 |
| 300 | 29.10 | .05 | 194.17 | 29.18 | .05 | 194.52 |
| 400 | 29.33 | 2.97 | 202.58 | 29.41 | 2.98 | 202.94 |
| 500 | 29.82 | 5.93 | 209.18 | 29.90 | 5.95 | 209.56 |
| 600 | 30.44 | 8.95 | 214.66 | 30.53 | 8.97 | 215.07 |
| 700 | 31.13 | 12.02 | 219.41 | 31.22 | 12.05 | 219.82 |
| 800 | 31.82 | 15.17 | 223.61 | 31.92 | 15.21 | 224.04 |
| 900 | 32.47 | 18.38 | 227.40 | 32.57 | 18.43 | 227.84 |
| 1000 | 33.05 | 21.66 | 230.85 | 33.15 | 21.72 | 231.30 |
| 1100 | 33.56 | 25.00 | 234.02 | 33.67 | 25.07 | 234.48 |
| 1200 | 34.02 | 28.37 | 236.97 | 34.13 | 28.46 | 237.43 |
| 1300 | 34.41 | 31.79 | 239.71 | 34.53 | 31.89 | 240.18 |
| 1400 | 34.76 | 35.25 | 242.27 | 34.89 | 35.36 | 242.75 |
| 1500 | 35.07 | 38.75 | 244.68 | 35.20 | 38.87 | 245.17 |
| 1600 | 35.35 | 42.27 | 246.95 | 35.48 | 42.40 | 247.45 |
| 1700 | 35.60 | 45.81 | 249.10 | 35.73 | 45.96 | 249.61 |
| 1800 | 35.82 | 49.39 | 251.14 | 35.95 | 49.54 | 251.66 |
| 1900 | 36.02 | 52.98 | 253.08 | 36.15 | 53.15 | 253.60 |
| 2000 | 36.21 | 56.59 | 254.94 | 36.34 | 56.77 | 255.47 |
| 2100 | 36.38 | 60.22 | 256.71 | 36.51 | 60.42 | 257.24 |
| 2200 | 36.53 | 63.86 | 258.40 | 36.67 | 64.08 | 258.95 |
| 2300 | 36.68 | 67.53 | 260.03 | 36.82 | 67.75 | 260.58 |
| 2400 | 36.82 | 71.20 | 261.60 | 36.96 | 71.44 | 262.15 |
| 2500 | 36.94 | 74.89 | 263.10 | 37.09 | 75.14 | 263.66 |
| 2600 | 37.07 | 78.59 | 264.55 | 37.21 | 78.86 | 265.12 |
| 2700 | 37.18 | 82.30 | 265.95 | 37.32 | 82.58 | 266.52 |
| 2800 | 37.28 | 86.03 | 267.31 | 37.43 | 86.32 | 267.88 |
| 2900 | 37.38 | 89.76 | 268.62 | 37.53 | 90.07 | 269.20 |
| 3000 | 37.48 | 93.50 | 269.88 | 37.62 | 93.83 | 270.47 |
| 3100 | 37.57 | 97.25 | 271.11 | 37.71 | 97.60 | 271.71 |
| 3200 | 37.66 | 101.02 | 272.31 | 37.80 | 101.37 | 272.91 |
| 3300 | 37.74 | 104.78 | 273.47 | 37.89 | 105.15 | 274.07 |
| 3400 | 37.81 | 108.56 | 274.60 | 37.96 | 108.95 | 275.20 |
| 3500 | 37.89 | 112.35 | 275.69 | 38.03 | 112.75 | 276.30 |

| | Oxygen ($O_2$) (Standard state element) $(\Delta H_f^\circ)_{298.15} = 0.00$ kJ/mole Log $K_p = 0.00$ M.W. = 0.031999 | | | Monatomic oxygen (O) $(\Delta H_f^\circ)_{289.15} = 249.17$ kJ/mole M.W. = 0.015999 | | |
|---|---|---|---|---|---|---|
| $T$, K | $C_p^\circ$, J/mole $K$ | $H^\circ - H_{298}^\circ$, kJ/mole | $S^\circ$, J/mole $K$ | $C_p^\circ$, J/mole $K$ | $H^\circ - H_{298}^\circ$, kJ/mole | $S^\circ$, J/mole $K$ |
| 100 | 29.10 | −5.78 | 173.20 | 23.70 | −4.52 | 135.84 |
| 200 | 29.12 | −2.87 | 193.38 | 22.73 | −2.19 | 152.04 |
| 298 | 29.38 | 0.00 | 205.04 | 21.91 | 0.00 | 160.95 |
| 300 | 29.38 | .05 | 205.18 | 21.90 | .04 | 161.08 |
| 400 | 30.00 | 3.03 | 213.76 | 21.48 | 2.21 | 167.32 |
| 500 | 31.09 | 6.08 | 220.58 | 21.26 | 4.34 | 172.09 |
| 600 | 32.09 | 9.24 | 226.34 | 21.13 | 6.46 | 175.95 |
| 700 | 32.98 | 12.50 | 231.36 | 21.04 | 8.57 | 179.20 |
| 800 | 33.73 | 15.84 | 235.81 | 20.98 | 10.67 | 182.01 |
| 900 | 34.35 | 19.24 | 239.82 | 20.95 | 12.77 | 184.48 |
| 1000 | 34.87 | 22.70 | 243.47 | 20.92 | 14.86 | 186.68 |
| 1100 | 35.30 | 26.21 | 246.81 | 20.89 | 16.95 | 188.67 |
| 1200 | 35.67 | 29.76 | 249.90 | 20.88 | 19.04 | 190.49 |
| 1300 | 35.99 | 33.34 | 252.77 | 20.87 | 21.13 | 192.16 |
| 1400 | 36.28 | 36.96 | 255.45 | 20.85 | 23.21 | 193.71 |
| 1500 | 36.54 | 40.60 | 257.96 | 20.84 | 25.30 | 195.15 |
| 1600 | 36.80 | 44.27 | 260.32 | 20.84 | 27.38 | 196.49 |
| 1700 | 37.04 | 47.96 | 262.56 | 20.83 | 29.46 | 197.75 |
| 1800 | 37.28 | 51.67 | 264.69 | 20.83 | 31.55 | 198.95 |
| 1900 | 37.51 | 55.41 | 266.71 | 20.83 | 33.63 | 200.07 |
| 2000 | 37.74 | 59.17 | 268.64 | 20.83 | 35.71 | 201.14 |
| 2100 | 37.97 | 62.96 | 270.49 | 20.83 | 37.79 | 202.15 |
| 2200 | 38.20 | 66.77 | 272.26 | 20.83 | 39.88 | 203.12 |
| 2300 | 38.42 | 70.60 | 273.96 | 20.84 | 41.96 | 204.05 |
| 2400 | 38.64 | 74.45 | 275.60 | 20.84 | 44.04 | 204.94 |
| 2500 | 38.86 | 78.33 | 277.18 | 20.85 | 46.13 | 205.79 |
| 2600 | 39.07 | 82.22 | 278.71 | 20.86 | 48.22 | 206.61 |
| 2700 | 39.28 | 86.14 | 280.19 | 20.88 | 50.30 | 207.39 |
| 2800 | 39.48 | 90.08 | 281.62 | 20.89 | 52.39 | 208.15 |
| 2900 | 39.67 | 94.04 | 283.01 | 20.92 | 54.48 | 208.89 |
| 3000 | 39.87 | 98.01 | 284.36 | 20.94 | 56.58 | 209.59 |
| 3100 | 40.05 | 102.01 | 285.67 | 20.96 | 58.67 | 210.28 |
| 3200 | 40.22 | 106.02 | 286.94 | 20.99 | 60.77 | 210.95 |
| 3300 | 40.40 | 110.05 | 288.18 | 21.02 | 62.87 | 211.59 |
| 3400 | 40.56 | 114.10 | 289.39 | 21.05 | 64.97 | 212.22 |
| 3500 | 40.71 | 118.16 | 290.57 | 21.09 | 67.08 | 212.83 |

| | Nitrogen ($N_2$) (Standard state element) $(\Delta H_f^\circ)_{298.15} = 0.00$ kJ/mole Log $K_p = 0.00$ M.W. = 0.028013 | | | Monatomic nitrogen (N) $(\Delta H_f^\circ)_{298.15} = 472.69$ kJ/mole M.W. = 0.014007 | | |
|---|---|---|---|---|---|---|
| $T$, $K$ | $C_p^\circ$, J/mole $K$ | $H^\circ - H_{298}^\circ$, kJ/mole | $S^\circ$, J/mole $K$ | $C_p^\circ$, J/mole $K$ | $H^\circ - H_{298}^\circ$, kJ/mole | $S^\circ$, J/mole $K$ |
| 100 | 29.10 | -5.77 | 159.70 | 20.79 | -4.12 | 130.48 |
| 200 | 29.11 | -2.86 | 179.87 | 20.79 | -2.04 | 144.89 |
| 298 | 29.12 | 0.00 | 191.50 | 20.79 | 0.00 | 153.19 |
| 300 | 29.12 | .05 | 191.68 | 20.79 | .04 | 153.32 |
| 400 | 29.25 | 2.97 | 200.07 | 20.79 | 2.12 | 159.30 |
| 500 | 29.58 | 5.91 | 206.63 | 20.79 | 4.20 | 163.94 |
| 600 | 30.11 | 8.90 | 212.07 | 20.79 | 6.28 | 167.73 |
| 700 | 30.75 | 11.94 | 216.76 | 20.79 | 8.35 | 170.93 |
| 800 | 31.43 | 15.05 | 220.91 | 20.79 | 10.43 | 173.71 |
| 900 | 32.09 | 18.22 | 224.65 | 20.79 | 12.51 | 176.15 |
| 1000 | 32.70 | 21.46 | 228.06 | 20.79 | 14.59 | 178.35 |
| 1100 | 33.24 | 24.76 | 231.20 | 20.79 | 16.67 | 180.33 |
| 1200 | 33.72 | 28.11 | 234.12 | 20.79 | 18.74 | 182.13 |
| 1300 | 34.15 | 31.50 | 236.84 | 20.79 | 20.82 | 183.80 |
| 1400 | 34.52 | 34.94 | 239.38 | 20.79 | 22.90 | 185.34 |
| 1500 | 34.84 | 38.40 | 241.77 | 20.79 | 24.98 | 186.77 |
| 1600 | 35.13 | 41.90 | 244.03 | 20.79 | 27.06 | 188.12 |
| 1700 | 35.38 | 45.43 | 246.17 | 20.79 | 29.14 | 189.38 |
| 1800 | 35.60 | 48.98 | 248.19 | 20.79 | 31.22 | 190.56 |
| 1900 | 35.79 | 52.55 | 250.12 | 20.79 | 33.30 | 191.69 |
| 2000 | 35.97 | 56.14 | 251.96 | 20.79 | 35.38 | 192.75 |
| 2100 | 36.12 | 59.74 | 253.72 | 20.79 | 37.46 | 193.77 |
| 2200 | 36.27 | 63.36 | 255.41 | 20.80 | 39.53 | 194.74 |
| 2300 | 36.40 | 66.99 | 257.02 | 20.80 | 41.61 | 195.66 |
| 2400 | 36.51 | 70.64 | 258.58 | 20.82 | 43.69 | 196.54 |
| 2500 | 36.61 | 74.30 | 260.06 | 20.83 | 45.78 | 197.40 |
| 2600 | 36.71 | 77.96 | 261.50 | 20.84 | 47.86 | 198.21 |
| 2700 | 36.80 | 81.64 | 262.89 | 20.87 | 49.94 | 199.00 |
| 2800 | 36.88 | 85.32 | 264.23 | 20.89 | 52.03 | 199.76 |
| 2900 | 36.96 | 89.01 | 265.53 | 20.92 | 54.12 | 200.49 |
| 3000 | 37.03 | 92.71 | 266.78 | 20.96 | 56.22 | 201.20 |
| 3100 | 37.10 | 96.42 | 268.00 | 21.01 | 58.32 | 201.89 |
| 3200 | 37.16 | 100.14 | 269.18 | 21.06 | 60.42 | 202.56 |
| 3300 | 37.22 | 103.85 | 270.32 | 21.13 | 62.53 | 203.21 |
| 3400 | 37.27 | 107.57 | 271.43 | 21.20 | 64.65 | 203.84 |
| 3500 | 37.32 | 111.31 | 272.51 | 21.28 | 66.77 | 204.46 |

| | Hydrogen (H$_2$) (Standard state element) $(\Delta H_f^\circ)_{298.15} = 0.00$ kJ/mole Log $K_p = 0.00$ M.W. $= 0.002016$ | | | Monatomic hydrogen (H) $(\Delta H_f^\circ)_{298.15} = 218.00$ kJ/mole M.W. $= 0.001008$ | | |
|---|---|---|---|---|---|---|
| $T$, $K$ | $C_p^\circ$, J/mole $K$ | $H^\circ - H_{298}^\circ$, kJ/mole | $S^\circ$, J/mole $K$ | $C_p^\circ$, J/mole $K$ | $H^\circ - H_{298}^\circ$, kJ/mole | $S^\circ$, J/mole $K$ |
| 100 | 28.15 | −5.47 | 100.62 | 20.79 | −4.12 | 91.90 |
| 200 | 27.45 | −2.77 | 119.30 | 20.79 | −2.04 | 106.31 |
| 298 | 28.84 | 0.00 | 130.57 | 20.79 | 0.00 | 114.61 |
| 300 | 28.85 | .05 | 130.75 | 20.79 | .04 | 114.73 |
| 400 | 29.18 | 2.96 | 139.11 | 20.79 | 2.12 | 120.72 |
| 500 | 29.26 | 5.88 | 145.63 | 20.79 | 4.20 | 125.35 |
| 600 | 29.33 | 8.81 | 150.97 | 20.79 | 6.28 | 129.14 |
| 700 | 29.44 | 11.75 | 155.49 | 20.79 | 8.35 | 132.35 |
| 800 | 29.62 | 14.70 | 159.44 | 20.79 | 10.43 | 135.12 |
| 900 | 29.88 | 17.68 | 162.94 | 20.79 | 12.51 | 137.57 |
| 1000 | 30.20 | 20.68 | 166.10 | 20.79 | 14.59 | 139.76 |
| 1100 | 30.58 | 23.72 | 169.00 | 20.79 | 16.67 | 141.74 |
| 1200 | 30.99 | 26.80 | 171.68 | 20.79 | 18.74 | 143.55 |
| 1300 | 31.42 | 29.92 | 174.18 | 20.79 | 20.82 | 145.21 |
| 1400 | 31.86 | 33.08 | 176.52 | 20.79 | 22.90 | 146.75 |
| 1500 | 32.30 | 36.29 | 178.74 | 20.79 | 24.98 | 148.19 |
| 1600 | 32.72 | 39.54 | 180.83 | 20.79 | 27.06 | 149.53 |
| 1700 | 33.14 | 42.84 | 182.83 | 20.79 | 29.14 | 150.79 |
| 1800 | 33.54 | 46.17 | 184.74 | 20.79 | 31.22 | 151.98 |
| 1900 | 33.92 | 49.54 | 186.56 | 20.79 | 33.30 | 153.10 |
| 2000 | 34.28 | 52.95 | 188.31 | 20.79 | 35.38 | 154.17 |
| 2100 | 34.62 | 56.40 | 189.99 | 20.79 | 37.46 | 155.18 |
| 2200 | 34.95 | 59.88 | 191.61 | 20.79 | 39.53 | 156.15 |
| 2300 | 35.26 | 63.39 | 193.17 | 20.79 | 41.61 | 157.08 |
| 2400 | 35.56 | 66.93 | 194.68 | 20.79 | 43.69 | 157.96 |
| 2500 | 35.84 | 70.50 | 196.13 | 20.79 | 45.77 | 158.81 |
| 2600 | 36.11 | 74.09 | 197.54 | 20.79 | 47.85 | 159.62 |
| 2700 | 36.37 | 77.72 | 198.91 | 20.79 | 49.92 | 160.41 |
| 2800 | 36.62 | 81.37 | 200.24 | 20.79 | 52.00 | 161.16 |
| 2900 | 36.86 | 85.04 | 201.53 | 20.79 | 54.08 | 161.89 |
| 3000 | 37.09 | 88.74 | 202.78 | 20.79 | 56.16 | 162.60 |
| 3100 | 37.31 | 92.46 | 204.00 | 20.79 | 58.24 | 163.28 |
| 3200 | 37.53 | 96.20 | 205.19 | 20.79 | 60.32 | 163.94 |
| 3300 | 37.74 | 99.96 | 206.35 | 20.79 | 62.40 | 164.58 |
| 3400 | 37.94 | 103.75 | 207.48 | 20.79 | 64.48 | 165.20 |
| 3500 | 38.15 | 107.55 | 208.58 | 20.79 | 66.55 | 165.80 |

| | Water vapor (H$_2$O) $(\Delta H_f^\circ)_{298.15} = -241.83$ kJ/mole M.W. = 0.018016 | | | Hydroxyl (OH) $(\Delta H_f^\circ)_{298.15} = 38.99$ kJ/mole M.W. = 0.17008 | | |
|---|---|---|---|---|---|---|
| $T$, K | $C_p^\circ$, J/mole $K$ | $H^\circ - H_{298}^\circ$, kJ/mole | $S^\circ$, J/mole $K$ | $C_p^\circ$, J/mole $K$ | $H^\circ - H_{298}^\circ$, kJ/mole | $S^\circ$, J/mole $K$ |
| 100 | 33.31 | -6.61 | 152.28 | 32.63 | -6.14 | 149.48 |
| 200 | 33.34 | -3.28 | 175.38 | 30.78 | -2.97 | 171.48 |
| 298 | 33.58 | 0.00 | 188.72 | 29.99 | 0.00 | 183.60 |
| 300 | 33.58 | .06 | 188.93 | 29.98 | .05 | 183.79 |
| 400 | 34.25 | 3.45 | 198.67 | 29.65 | 3.03 | 192.36 |
| 500 | 35.21 | 6.92 | 206.41 | 29.52 | 5.99 | 198.96 |
| 600 | 36.30 | 10.50 | 212.93 | 29.53 | 8.94 | 204.34 |
| 700 | 37.46 | 14.18 | 218.61 | 29.66 | 11.90 | 208.90 |
| 800 | 38.69 | 17.99 | 223.69 | 29.92 | 14.88 | 212.87 |
| 900 | 39.94 | 21.92 | 228.32 | 30.26 | 17.89 | 216.42 |
| 1000 | 41.22 | 25.98 | 232.60 | 30.68 | .20.94 | 219.63 |
| 1100 | 42.48 | 30.17 | 236.58 | 31.12 | 24.02 | 222.57 |
| 1200 | 43.70 | 34.48 | 240.33 | 31.59 | 27.16 | 225.30 |
| 1300 | 44.87 | 38.90 | 243.88 | 32.05 | 30.34 | 227.84 |
| 1400 | 45.97 | 43.45 | 247.24 | 32.49 | 33.57 | 230.24 |
| 1500 | 47.00 | 48.10 | 250.45 | 32.92 | 36.84 | 232.49 |
| 1600 | 47.96 | 52.84 | 253.51 | 33.32 | 40.15 | 234.63 |
| 1700 | 48.84 | 57.68 | 256.45 | 33.69 | 43.50 | 236.66 |
| 1800 | 49.66 | 62.61 | 259.26 | 34.05 | 46.89 | 238.60 |
| 1900 | 50.41 | 67.61 | 261.97 | 34.37 | 50.31 | 240.45 |
| 2000 | 51.10 | 72.69 · | 264.57 | 34.67 | 53.76 | 242.22 |
| 2100 | 51.74 | 77.83 | 267.08 | 34.95 | 57.25 | 243.91 |
| 2200 | 52.32 | 83.04 | 269.50 | 35.21 | 60.75 | 245.55 |
| 2300 | 52.86 | 88.29 | 271.84 | 35.45 | 64.28 | 247.12 |
| 2400 | 53.36 | 93.60 | 274.10 | 35.67 | 67.84 | 248.63 |
| 2500 | 53.82 | 98.96 | 276.29 | 35.88 | 71.42 | 250.09 |
| 2600 | 54.25 | 104.37 | 278.41 | 36.07 | 75.01 | 251.50 |
| 2700 | 54.64 | 109.81 | 280.46 | 36.25 | 78.63 | 252.87 |
| 2800 | 55.00 | 115.29 | 282.45 | 36.43 | 82.27 | 254.19 |
| 2900 | 55.35 | 120.81 | 284.39 | 36.58 | 85.92 | 255.47 |
| 3000 | 55.66 | 126.36 | 286.27 | 36.74 | 89.58 | 256.71 |
| 3100 | 55.96 | 131.94 | 288.10 | 36.88 | 93.27 | 257.92 |
| 3200 | 56.24 | 137.55 | 289.88 | 37.01 | 96.96 | 259.09 |
| 3300 | 56.50 | 143.19 | 291.62 | 37.14 | 100.67 | 260.24 |
| 3400 | 56.74 | 148.85 | 293.31 | 37.26 | 104.39 | 261.35 |
| 3500 | 56.97 | 154.54 | 294.96 | 37.38 | 108.12 | 262.43 |

| T, K | Carbon dioxide (CO$_2$) ($\Delta H_f^\circ$)$_{298.15}$ = −393.52 kJ/mole M.W. = 0.044010 $C_p^\circ$, J/mole K | $H^\circ - H_{298}^\circ$, kJ/mole | $S^\circ$, J/mole K | Carbon monoxide (CO) ($\Delta H_f^\circ$)$_{298.15}$ = −110.53 kJ/mole M.W. = 0.028011 $C_p^\circ$, J/mole K | $H^\circ - H_{298}^\circ$, kJ/mole | $S^\circ$, J/mole K |
|---|---|---|---|---|---|---|
| 100 | 29.21 | −6.46 | 178.90 | 29.10 | −5.77 | 165.74 |
| 200 | 32.36 | −3.41 | 199.87 | 29.11 | −2.87 | 185.92 |
| 298 | 37.13 | 0.00 | 213.69 | 29.14 | 0.00 | 197.54 |
| 300 | 37.22 | .07 | 213.92 | 29.14 | .05 | 197.72 |
| 400 | 41.33 | 4.01 | 225.22 | 29.34 | 2.97 | 206.12 |
| 500 | 44.63 | 8.31 | 234.81 | 29.79 | 5.93 | 212.72 |
| 600 | 47.32 | 12.92 | 243.20 | 30.44 | 8.94 | 218.20 |
| 700 | 49.56 | 17.76 | 250.66 | 31.17 | 12.02 | 222.95 |
| 800 | 51.43 | 22.82 | 257.41 | 31.90 | 15.18 | 227.16 |
| 900 | 53.00 | 28.04 | 263.56 | 32.58 | 18.40 | 230.96 |
| 1000 | 54.31 | 33.41 | 269.22 | 33.18 | 21.69 | 234.42 |
| 1100 | 55.41 | 38.89 | 274.45 | 33.71 | 25.03 | 237.61 |
| 1200 | 56.34 | 44.48 | 279.31 | 34.17 | 28.43 | 240.56 |
| 1300 | 57.14 | 50.16 | 283.85 | 34.57 | 31.87 | 243.32 |
| 1400 | 57.80 | 55.91 | 288.11 | 34.92 | 35.34 | 245.89 |
| 1500 | 58.38 | 61.71 | 292.11 | 35.22 | 38.85 | 248.31 |
| 1600 | 58.89 | 67.58 | 295.90 | 35.48 | 42.38 | 250.59 |
| 1700 | 59.32 | 73.49 | 299.48 | 35.71 | 45.94 | 252.75 |
| 1800 | 59.70 | 79.44 | 302.88 | 35.91 | 49.52 | 254.80 |
| 1900 | 60.05 | 85.43 | 306.12 | 36.09 | 53.12 | 256.74 |
| 2000 | 60.35 | 91.45 | 309.21 | 36.25 | 56.74 | 258.60 |
| 2100 | 60.62 | 97.50 | 312.16 | 36.39 | 60.38 | 260.37 |
| 2200 | 60.86 | 103.57 | 314.99 | 36.52 | 64.02 | 262.06 |
| 2300 | 61.09 | 109.67 | 317.70 | 36.64 | 67.68 | 263.69 |
| 2400 | 61.29 | 115.79 | 320.30 | 36.74 | 71.35 | 265.25 |
| 2500 | 61.47 | 121.93 | 322.81 | 36.84 | 75.02 | 266.76 |
| 2600 | 61.65 | 128.08 | 325.22 | 36.92 | 78.71 | 268.20 |
| 2700 | 61.80 | 134.26 | 327.55 | 37.00 | 82.41 | 269.60 |
| 2800 | 61.95 | 140.44 | 329.80 | 37.08 | 86.12 | 270.94 |
| 2900 | 62.09 | 146.65 | 331.98 | 37.15 | 89.83 | 272.25 |
| 3000 | 62.23 | 152.86 | 334.08 | 37.22 | 93.54 | 273.51 |
| 3100 | 62.35 | 159.09 | 336.13 | 37.28 | 97.27 | 274.73 |
| 3200 | 62.47 | 165.33 | 338.11 | 37.34 | 101.00 | 275.91 |
| 3300 | 62.58 | 171.59 | 340.03 | 37.39 | 104.73 | 277.06 |
| 3400 | 62.68 | 177.85 | 341.90 | 37.44 | 108.48 | 278.18 |
| 3500 | 62.79 | 184.12 | 343.72 | 37.49 | 112.22 | 279.27 |

| | Nitric oxide (NO) $(\Delta H_f^\circ)_{298.15} = 90.29$ kJ/mole M.W. = 0.030008 | | | Nitrogen dioxide ($NO_2$) $(\Delta H_f^\circ)_{298.15} = 33.10$ kJ/mole M.W. = 0.046007 | | |
|---|---|---|---|---|---|---|
| $T$, K | $C_p^\circ$, J/mole $K$ | $H^\circ - H_{298}^\circ$, kJ/mole | $S^\circ$, J/mole $K$ | $C_p^\circ$, J/mole $K$ | $H^\circ - H_{298}^\circ$, kJ/mole | $S^\circ$, J/mole $K$ |
| 100 | 32.30 | −6.07 | 176.92 | 33.28 | −6.86 | 202.45 |
| 200 | 30.42 | −2.95 | 198.64 | 34.38 | −3.49 | 225.74 |
| 298 | 29.84 | 0.00 | 210.65 | 36.97 | 0.00 | 239.92 |
| 300 | 29.84 | .05 | 210.84 | 37.03 | .07 | 240.15 |
| 400 | 29.94 | 3.04 | 219.43 | 40.17 | 3.93 | 251.23 |
| 500 | 30.49 | 6.06 | 226.16 | 43.18 | 8.10 | 260.53 |
| 600 | 31.24 | 9.15 | 231.78 | 45.84 | 12.56 | 268.65 |
| 700 | 32.03 | 12.31 | 236.66 | 47.99 | 17.25 | 275.88 |
| 800 | 32.77 | 15.55 | 240.98 | 49.71 | 22.14 | 282.40 |
| 900 | 33.42 | 18.86 | 244.88 | 51.08 | 27.18 | 288.34 |
| 1000 | 33.99 | 22.23 | 248.43 | 52.17 | 32.34 | 293.78 |
| 1100 | 34.47 | 25.65 | 251.70 | 53.04 | 37.61 | 298.80 |
| 1200 | 34.88 | 29.12 | 254.71 | 53.75 | 42.95 | 303.44 |
| 1300 | 35.23 | 32.63 | 257.52 | 54.33 | 48.35 | 307.77 |
| 1400 | 35.53 | 36.17 | 260.14 | 54.81 | 53.81 | 311.81 |
| 1500 | 35.78 | 39.73 | 262.60 | 55.20 | 59.31 | 315.61 |
| 1600 | 36.00 | 43.32 | 264.92 | 55.53 | 64.85 | 319.18 |
| 1700 | 36.20 | 46.93 | 267.11 | 55.81 | 70.42 | 322.56 |
| 1800 | 36.37 | 50.56 | 269.18 | 56.06 | 76.01 | 325.75 |
| 1900 | 36.51 | 54.20 | 271.15 | 56.26 | 81.63 | 328.79 |
| 2000 | 36.65 | 57.86 | 273.03 | 56.44 | 87.26 | 331.68 |
| 2100 | 36.77 | 61.53 | 274.82 | 56.60 | 92.91 | 334.44 |
| 2200 | 36.87 | 65.22 | 276.53 | 56.74 | 98.58 | 337.08 |
| 2300 | 36.97 | 68.91 | 278.17 | 56.85 | 104.26 | 339.60 |
| 2400 | 37.06 | 72.61 | 279.75 | 56.96 | 109.95 | 342.02 |
| 2500 | 37.14 | 76.32 | 281.26 | 57.05 | 115.65 | 344.35 |
| 2600 | 37.22 | 80.04 | 282.72 | 57.14 | 121.36 | 346.59 |
| 2700 | 37.29 | 83.76 | 284.13 | 57.21 | 127.08 | 348.74 |
| 2800 | 37.35 | 87.49 | 285.48 | 57.28 | 132.80 | 350.83 |
| 2900 | 37.41 | 91.23 | 286.79 | 57.34 | 138.54 | 352.84 |
| 3000 | 37.47 | 94.98 | 288.06 | 57.40 | 144.27 | 354.78 |
| 3100 | 37.52 | 98.73 | 289.29 | 57.45 | 150.01 | 356.67 |
| 3200 | 37.57 | 102.48 | 290.48 | 57.49 | 155.76 | 358.49 |
| 3300 | 37.62 | 106.24 | 291.64 | 57.53 | 161.51 | 360.26 |
| 3400 | 37.66 | 110.00 | 292.77 | 57.57 | 167.27 | 361.98 |
| 3500 | 37.71 | 113.77 | 293.86 | 57.61 | 173.03 | 363.65 |

| T, K | Nitrous oxide (N$_2$O) $(\Delta H_f^\circ)_{298.15} = 82.05$ kJ/mole M.W. = 0.044016 | | | Ammonia (NH$_3$) $(\Delta H_f^\circ)_{298.15} = -45.90$ kJ/mole M.W. = 0.017031 | | |
|---|---|---|---|---|---|---|
| | $C_p^\circ$, J/mole K | $H^\circ - H_{298}^\circ$, kJ/mole | $S^\circ$, J/mole K | $C_p^\circ$, J/mole K | $H^\circ - H_{298}^\circ$, kJ/mole | $S^\circ$, J/mole K |
| 100 | 29.35 | -6.67 | 184.05 | 33.28 | -6.74 | 155.73 |
| 200 | 33.61 | -3.55 | 205.47 | 33.76 | -3.39 | 178.88 |
| 298 | 38.62 | 0.00 | 219.85 | 35.65 | 0.00 | 192.66 |
| 300 | 38.70 | .07 | 220.09 | 35.70 | .07 | 192.89 |
| 400 | 42.68 | 4.15 | 231.79 | 38.71 | 3.78 | 203.55 |
| 500 | 45.83 | 8.58 | 241.67 | 42.05 | 7.82 | 212.55 |
| 600 | 48.39 | 13.30 | 250.26 | 45.29 | 12.19 | 220.51 |
| 700 | 50.50 | 18.24 | 257.88 | 48.35 | 16.87 | 227.72 |
| 800 | 52.24 | 23.38 | 264.74 | 51.23 | 21.85 | 234.37 |
| 900 | 53.68 | 28.68 | 270.98 | 53.95 | 27.11 | 240.56 |
| 1000 | 54.86 | 34.11 | 276.70 | 56.49 | 32.64 | 246.37 |
| 1100 | 55.85 | 39.65 | 281.98 | 58.86 | 38.40 | 251.87 |
| 1200 | 56.67 | 45.28 | 286.88 | 61.05 | 44.40 | 257.09 |
| 1300 | 57.35 | 50.98 | 291.44 | 63.06 | 50.61 | 262.06 |
| 1400 | 57.93 | 56.74 | 295.71 | 64.89 | 57.01 | 266.80 |
| 1500 | 58.41 | 62.56 | 299.72 | 66.56 | 63.58 | 271.33 |
| 1600 | 58.83 | 68.42 | 303.51 | 68.08 | 70.32 | 275.68 |
| 1700 | 59.18 | 74.32 | 307.08 | 69.45 | 77.19 | 279.85 |
| 1800 | 59.49 | 80.26 | 310.47 | 70.69 | 84.20 | 283.85 |
| 1900 | 59.76 | 86.22 | 313.70 | 71.82 | 91.33 | 287.70 |
| 2000 | 59.99 | 92.21 | 316.77 | 72.84 | 98.56 | 291.42 |
| 2100 | 60.19 | 98.22 | 319.70 | 73.75 | 105.89 | 294.99 |
| 2200 | 60.37 | 104.24 | 322.51 | 74.58 | 113.31 | 298.44 |
| 2300 | 60.53 | 1.10.29 | 325.19 | 75.33 | 120.80 | 301.78 |
| 2400 | 60.66 | 116.35 | 327.77 | 76.01 | 128.37 | 305.00 |
| 2500 | 60.79 | 122.42 | 330.25 | 76.63 | 136.01 | 308.11 |
| 2600 | 60.90 | 128.50 | 332.64 | 77.17 | 143.70 | 311.13 |
| 2700 | 61.00 | 134.60 | 334.94 | 77.67 | 151.44 | 314.05 |
| 2800 | 61.09 | 140.70 | 337.16 | 78.13 | 159.23 | 316.88 |
| 2900 | 61.17 | 146.82 | 339.30 | 78.53 | 167.06 | 319.63 |
| 3000 | 61.25 | 152.94 | 341.38 | 78.90 | 174.93 | 322.30 |
| 3100 | 61.32 | 159.07 | 343.39 | 79.23 | 182.84 | 324.89 |
| 3200 | 61.38 | 165.20 | 345.33 | 79.52 | 190.78 | 327.41 |
| 3300 | 61.44 | 171.34 | 347.23 | 79.78 | 198.74 | 329.86 |
| 3400 | 61.49 | 177.49 | 349.06 | 80.01 | 206.74 | 332.25 |
| 3500 | 61.53 | 183.64 | 350.84 | 80.22 | 214.74 | 334.57 |

## Appendix B1—(continued)

| T, K | Amidogen (NH$_2$) $(\Delta H_f^\circ)_{298.15} = 190.37$ kJ/mole M.W. = 0.016023 $C_p^\circ$, J/mole K | $H^\circ - H_{298}^\circ$, kJ/mole | $S^\circ$, J/mole K | Imidogen (NH) $(\Delta H_f^\circ)_{298.15} = 376.56$ kJ/mole M.W. = 0.015015 $C_p^\circ$, J/mole K | $H^\circ - H_{298}^\circ$, kJ/mole | $S^\circ$, J/mole K |
|---|---|---|---|---|---|---|
| 100 | 33.26 | -6.60 | 158.22 | 29.12 | -5.77 | 149.32 |
| 200 | 33.28 | -3.28 | 181.27 | 29.13 | -2.86 | 169.51 |
| 298 | 33.57 | 0.00 | 194.60 | 29.15 | 0.00 | 181.14 |
| 300 | 33.58 | .06 | 194.81 | 29.15 | .05 | 181.32 |
| 400 | 34.40 | 3.46 | 204.57 | 29.18 | 2.97 | 189.71 |
| 500 | 35.53 | 6.95 | 212.36 | 29.26 | 5.89 | 196.23 |
| 600 | 36.84 | 10.57 | 218.96 | 29.46 | 8.82 | 201.58 |
| 700 | 38.25 | 14.33 | 224.74 | 29.79 | 11.79 | 206.15 |
| 800 | 39.71 | 18.23 | 229.94 | 30.22 | 14.79 | 210.15 |
| 900 | 41.18 | 22.27 | 234.71 | 30.72 | 17.83 | 213.74 |
| 1000 | 42.60 | 26.46 | 239.12 | 31.26 | 20.93 | 217.00 |
| 1100 | 43.94 | 30.79 | 243.25 | 31.80 | 24.09 | 220.01 |
| 1200 | 45.20 | 35.24 | 247.12 | 32.33 | 27.29 | 222.80 |
| 1300 | 46.35 | 39.82 | 250.78 | 32.82 | 30.55 | 225.40 |
| 1400 | 47.42 | 44.51 | 254.26 | 33.30 | 33.86 | 227.85 |
| 1500 | 48.39 | 49.30 | 257.57 | 33.73 | 37.21 | 230.17 |
| 1600 | 49.29 | 54.19 | 260.72 | 34.13 | 40.60 | 232.35 |
| 1700 | 50.13 | 59.16 | 263.73 | 34.51 | 44.03 | 234.44 |
| 1800 | 50.91 | 64.21 | 266.62 | 34.85 | 47.50 | 236.42 |
| 1900 | 51.65 | 69.34 | 269.39 | 35.17 | 51.00 | 238.31 |
| 2000 | 52.35 | 74.54 | 272.06 | 35.47 | 54.53 | 240.12 |
| 2100 | 53.02 | 79.81 | 274.63 | 35.76 | 58.09 | 241.86 |
| 2200 | 53.66 | 85.14 | 277.11 | 36.02 | 61.68 | 243.53 |
| 2300 | 54.29 | 90.54 | 279.51 | 36.28 | 65.30 | 245.14 |
| 2400 | 54.89 | 96.00 | 281.83 | 36.52 | 68.94 | 246.68 |
| 2500 | 55.48 | 101.52 | 284.09 | 36.75 | 72.60 | 248.18 |
| 2600 | 56.05 | 107.09 | 286.27 | 36.97 | 76.29 | 249.63 |
| 2700 | 56.61 | 112.73 | 288.40 | 37.20 | 80.00 | 251.03 |
| 2800 | 57.15 | 118.42 | 290.47 | 37.41 | 83.73 | 252.38 |
| 2900 | 57.66 | 124.16 | 292.48 | 37.61 | 87.48 | 253.70 |
| 3000 | 58.16 | 129.95 | 294.44 | 37.82 | 91.25 | 254.98 |
| 3100 | 58.64 | 135.79 | 296.36 | 38.02 | 95.04 | 256.22 |
| 3200 | 59.10 | 141.67 | 298.23 | 38.22 | 98.86 | 257.43 |
| 3300 | 59.54 | 147.61 | 300.06 | 38.42 | 102.69 | 258.61 |
| 3400 | 59.96 | 153.58 | 301.84 | 38.61 | 106.54 | 259.76 |
| 3500 | 60.36 | 159.60 | 303.58 | 38.80 | 110.41 | 260.88 |

| | Methane ($CH_4$) $(\Delta H_f^\circ)_{298.15} = -74.87$ kJ/mole M.W. = 0.016043 | | | Methyl ($CH_3$) $(\Delta H_f^\circ)_{298.15} = 145.69$ kJ/mole M.W. = 0.015035 | | |
|---|---|---|---|---|---|---|
| $T$, K | $C_p^\circ$, J/mole K | $H^\circ - H_{298}^\circ$, kJ/mole | $S^\circ$, J/mole K | $C_p^\circ$, J/mole K | $H^\circ - H_{298}^\circ$, kJ/mole | $S^\circ$, J/mole K |
| 100 | 33.26 | -6.70 | 149.39 | 33.40 | -6.81 | 155.79 |
| 200 | 33.48 | -3.37 | 172.47 | 35.64 | -3.46 | 179.00 |
| 298 | 35.64 | 0.00 | 186.15 | 0.00 | 0.00 | 0.00 |
| 300 | 35.71 | .07 | 186.37 | 38.75 | .07 | 193.26 |
| 400 | 40.50 | 3.86 | 197.25 | 42.04 | 3.92 | 204.34 |
| 500 | 46.34 | 8.20 | 206.91 | 45.25 | 8.16 | 213.78 |
| 600 | 52.23 | 13.13 | 215.88 | 48.29 | 12.79 | 222.20 |
| 700 | 57.79 | 18.64 | 224.35 | 51.17 | 17.78 | 229.89 |
| 800 | 62.93 | 24.67 | 232.41 | 53.92 | 23.12 | 237.02 |
| 900 | 67.60 | 31.20 | 240.10 | 56.53 | 28.78 | 243.68 |
| 1000 | 71.80 | 38.18 | 247.45 | 58.95 | 34.73 | 249.94 |
| 1100 | 75.53 | 45.55 | 254.47 | 61.19 | 40.93 | 255.85 |
| 1200 | 78.83 | 53.27 | 261.18 | 63.22 | 47.36 | 261.45 |
| 1300 | 81.75 | 61.30 | 267.61 | 65.05 | 53.99 | 266.75 |
| 1400 | 84.31 | 69.61 | 273.76 | 66.69 | 60.79 | 271.79 |
| 1500 | 86.56 | 78.15 | 279.66 | 68.15 | 67.74 | 276.59 |
| 1600 | 88.54 | 86.91 | 285.31 | 69.46 | 74.83 | 281.16 |
| 1700 | 90.29 | 95.86 | 290.73 | 70.63 | 82.03 | 285.52 |
| 1800 | 91.83 | 104.96 | 295.93 | 71.67 | 89.32 | 289.69 |
| 1900 | 93.19 | 114.21 | 300.94 | 72.59 | 96.71 | 293.69 |
| 2000 | 94.40 | 123.60 | 305.75 | 73.42 | 104.18 | 297.52 |
| 2100 | 95.48 | 133.09 | 310.38 | 74.17 | 111.72 | 301.19 |
| 2200 | 96.44 | 142.69 | 314.85 | 74.83 | 119.32 | 304.73 |
| 2300 | 97.30 | 152.37 | 319.15 | 75.43 | 126.97 | 308.13 |
| 2400 | 98.08 | 162.14 | 323.31 | 75.97 | 134.67 | 311.41 |
| 2500 | 98.78 | 171.99 | 327.33 | 76.46 | 142.41 | 314.57 |
| 2600 | 99.40 | 181.90 | 331.21 | 76.90 | 150.20 | 317.62 |
| 2700 | 99.97 | 191.87 | 334.98 | 77.30 | 158.02 | 320.57 |
| 2800 | 100.49 | 201.89 | 338.62 | 77.67 | 165.87 | 323.43 |
| 2900 | 100.96 | 211.96 | 342.16 | 78.00 | 173.76 | 326.20 |
| 3000 | 101.39 | 222.08 | 345.59 | 78.31 | 181.67 | 328.88 |
| 3100 | 101.78 | 232.24 | 348.92 | 78.59 | 189.60 | 331.48 |
| 3200 | 102.14 | 242.44 | 352.15 | 78.84 | 197.56 | 334.00 |
| 3300 | 102.48 | 252.67 | 355.31 | 79.08 | 205.53 | 336.46 |
| 3400 | 102.78 | 262.93 | 358.37 | 79.30 | 213.53 | 338.85 |
| 3500 | 103.06 | 273.22 | 361.35 | 79.50 | 221.54 | 341.17 |

| T, K | Methylene ($CH_2$) $(\Delta H_f^\circ)_{298.15} = 386.39$ kJ/mole M.W. = 0.014027 | | | Methylidyne (CH) $(\Delta H_f^\circ)_{298.15} = 594.13$ kJ/mole M.W. = 0.013019 | | |
|---|---|---|---|---|---|---|
| | $C_p^\circ$, J/mole $K$ | $H^\circ - H_{298}^\circ$, kJ/mole | $S^\circ$, J/mole $K$ | $C_p^\circ$, J/mole $K$ | $H^\circ - H_{298}^\circ$, kJ/mole | $S^\circ$, J/mole $K$ |
| 100 | 33.26 | -6.67 | 157.18 | 29.26 | -5.78 | 151.04 |
| 200 | 33.50 | -3.33 | 180.26 | 29.17 | -2.86 | 171.28 |
| 298 | 34.60 | 0.00 | 193.82 | 29.17 | 0.00 | 182.93 |
| 300 | 34.63 | .06 | 194.03 | 29.17 | .05 | 183.11 |
| 400 | 36.14 | 3.60 | 204.20 | 29.22 | 2.97 | 191.51 |
| 500 | 37.66 | 7.29 | 212.43 | 29.40 | 5.90 | 198.04 |
| 600 | 39.20 | 11.13 | 219.43 | 29.76 | 8.86 | 203.43 |
| 700 | 40.81 | 15.13 | 225.59 | 30.29 | 11.86 | 208.06 |
| 800 | 42.44 | 19.30 | 231.15 | 30.97 | 14.92 | 212.15 |
| 900 | 44.05 | 23.62 | 236.24 | 31.74 | 18.06 | 215.84 |
| 1000 | 45.57 | 28.10 | 240.96 | 32.56 | 21.27 | 219.22 |
| 1100 | 46.98 | 32.73 | 245.37 | 33.41 | 24.57 | 222.36 |
| 1200 | 48.25 | 37.50 | 249.52 | 34.25 | 27.95 | 225.31 |
| 1300 | 49.38 | 42.38 | 253.42 | 35.06 | 31.42 | 228.08 |
| 1400 | 50.38 | 47.37 | 257.12 | 35.83 | 34.97 | 230.71 |
| 1500 | 51.26 | 52.45 | 260.63 | 36.55 | 38.58 | 233.20 |
| 1600 | 52.04 | 57.62 | 263.96 | 37.21 | 42.27 | 235.58 |
| 1700 | 52.72 | 62.86 | 267.14 | 37.81 | 46.02 | 237.86 |
| 1800 | 53.33 | 68.16 | 270.17 | 38.35 | 49.83 | 240.04 |
| 1900 | 53.86 | 73.52 | 273.06 | 38.83 | 53.69 | 242.12 |
| 2000 | 54.34 | 78.93 | 275.84 | 39.25 | 57.60 | 244.12 |
| 2100 | 54.76 | 84.38 | 278.50 | 39.63 | 61.54 | 246.05 |
| 2200 | 55.14 | 89.88 | 281.06 | 39.96 | 65.52 | 247.90 |
| 2300 | 55.48 | 95.41 | 283.52 | 40.25 | 69.53 | 249.68 |
| 2400 | 55.80 | 100.98 | 285.88 | 40.49 | 73.57 | 251.40 |
| 2500 | 56.08 | 106.57 | 288.17 | 40.71 | 77.63 | 253.06 |
| 2600 | 56.34 | 112.19 | 290.37 | 40.89 | 81.71 | 254.66 |
| 2700 | 56.58 | 117.84 | 292.50 | 41.05 | 85.81 | 256.21 |
| 2800 | 56.79 | 123.50 | 294.57 | 41.19 | 89.92 | 257.70 |
| 2900 | 56.99 | 129.19 | 296.56 | 41.30 | 94.04 | 259.15 |
| 3000 | 57.18 | 134.90 | 298.50 | 41.40 | 98.18 | 260.55 |
| 3100 | 57.35 | 140.63 | 300.38 | 41.49 | 102.32 | 261.91 |
| 3200 | 57.50 | 146.37 | 302.20 | 41.56 | 106.47 | 263.23 |
| 3300 | 57.65 | 152.13 | 303.97 | 41.63 | 110.63 | 264.51 |
| 3400 | 57.79 | 157.90 | 305.70 | 41.68 | 114.80 | 265.75 |
| 3500 | 57.91 | 163.69 | 307.37 | 41.74 | 118.97 | 266.96 |

| | Acetylene (C$_2$H$_2$) $(\Delta H_f^\circ)_{298.15} = 226.73$ kJ/mole M.W. = 0.026038 | | | Ethylene (C$_2$H$_4$) $(\Delta H_f^\circ)_{298.15} = 52.47$ kJ/mole M.W. = 0.028054 | | |
|---|---|---|---|---|---|---|
| $T$, K | $C_p^\circ$, J/mole $K$ | $H^\circ - H_{298}^\circ$, kJ/mole | $S^\circ$, J/mole $K$ | $C_p^\circ$, J/mole $K$ | $H^\circ - H_{298}^\circ$, kJ/mole | $S^\circ$, J/mole $K$ |
| 100 | 29.35 | -7.10 | 163.18 | 33.27 | -7.19 | 180.44 |
| 200 | 35.58 | -3.92 | 184.99 | 35.36 | -3.80 | 203.85 |
| 298 | 44.10 | 0.00 | 200.85 | 42.89 | 0.00 | 219.22 |
| 300 | 44.23 | .08 | 201.12 | 43.06 | .08 | 219.49 |
| 400 | 50.48 | 4.83 | 214.75 | 53.05 | 4.88 | 233.24 |
| 500 | 54.87 | 10.12 | 226.52 | 62.48 | 10.67 | 246.11 |
| 600 | 58.29 | 15.78 | 236.83 | 70.66 | 17.33 | 258.24 |
| 700 | 61.15 | 21.75 | 246.04 | 77.71 | 24.76 | 269.68 |
| 800 | 63.76 | 28.00 | 254.38 | 83.84 | 32.85 | 280.47 |
| 900 | 66.11 | 34.50 | 262.02 | 89.20 | 41.51 | 290.65 |
| 1000 | 68.27 | 41.22 | 269.10 | 93.90 | 50.66 | 300.18 |
| 1100 | 70.25 | 48.15 | 275.70 | 98.02 | 60.27 | 309.45 |
| 1200 | 72.05 | 55.26 | 281.90 | 101.63 | 70.25 | 318.13 |
| 1300 | 73.69 | 62.55 | 287.73 | 104.78 | 80.58 | 326.40 |
| 1400 | 75.18 | 69.99 | 293.24 | 107.55 | 91.20 | 334.27 |
| 1500 | 76.53 | 77.58 | 298.48 | 109.98 | 102.08 | 341.77 |
| 1600 | 77.75 | 85.30 | 303.46 | 112.11 | 113.18 | 348.94 |
| 1700 | 78.85 | 93.13 | 308.21 | 113.98 | 124.49 | 355.79 |
| 1800 | 79.85 | 101.06 | 312.74 | 115.63 | 135.97 | 362.36 |
| 1900 | 80.76 | 109.09 | 317.08 | 117.09 | 147.61 | 368.65 |
| 2000 | 81.60 | 117.21 | 321.25 | 118.39 | 159.39 | 374.69 |
| 2100 | 82.36 | 125.41 | 325.25 | 119.54 | 171.28 | 380.49 |
| 2200 | 83.06 | 133.68 | 329.10 | 120.57 | 183.29 | 386.08 |
| 2300 | 83.70 | 142.02 | 332.80 | 121.49 | 195.39 | 391.46 |
| 2400 | 84.31 | 150.42 | 336.38 | 122.32 | 207.58 | 396.65 |
| 2500 | 84.86 | 158.88 | 339.83 | 123.07 | 219.85 | 401.66 |
| 2600 | 85.37 | 167.39 | 343.17 | 123.74 | 232.20 | 406.50 |
| 2700 | 85.85 | 175.96 | 346.40 | 124.35 | 244.60 | 411.18 |
| 2800 | 86.29 | 184.56 | 349.53 | 124.90 | 257.06 | 415.71 |
| 2900 | 86.72 | 193.21 | 352.56 | 125.41 | 269.58 | 420.10 |
| 3000 | 87.11 | 201.91 | 355.51 | 125.87 | 282.14 | 424.36 |
| 3100 | 87.49 | 210.64 | 358.38 | 126.29 | 294.75 | 428.50 |
| 3200 | 87.85 | 219.40 | 361.16 | 126.67 | 307.40 | 432.51 |
| 3300 | 88.19 | 228.20 | 363.87 | 127.03 | 320.08 | 436.41 |
| 3400 | 88.51 | 237.04 | 366.50 | 127.35 | 332.80 | 440.21 |
| 3500 | 88.81 | 245.91 | 369.07 | 127.65 | 345.56 | 443.91 |

## Appendix B1—(*continued*)

| | Graphite (C) (Standard state element) $(\Delta H_f^\circ)_{298.15} = 0.00$ kJ/mole Log $K_p = 0.00$ M.W. $= 0.012011$ | | | Cyano (CN) $(\Delta H_f^\circ)_{298.15} = 435.14$ kJ/mole M.W. $= 0.026018$ | | |
|---|---|---|---|---|---|---|
| $T,$ $K$ | $C_p^\circ,$ J/mole $K$ | $H^\circ - H_{298}^\circ,$ kJ/mole | $S^\circ,$ J/mole $K$ | $C_p^\circ,$ J/mole $K$ | $H^\circ - H_{298}^\circ,$ kJ/mole | $S^\circ,$ J/mole $K$ |
| 100 | 1.67 | -.99 | .95 | 29.10 | -5.77 | 170.72 |
| 200 | 5.00 | -.67 | 3.08 | 29.11 | -2.86 | 190.90 |
| 298 | 8.52 | 0.00 | 5.74 | 29.15 | 0.00 | 202.53 |
| 300 | 8.58 | .02 | 5.79 | 29.16 | .05 | 202.71 |
| 400 | 11.82 | 1.04 | 8.72 | 29.41 | 2.98 | 211.12 |
| 500 | 14.62 | 2.36 | 11.66 | 29.94 | 5.95 | 217.74 |
| 600 | 16.84 | 3.94 | 14.54 | 30.65 | 8.97 | 223.26 |
| 700 | 18.54 | 5.72 | 17.26 | 31.42 | 12.08 | 228.04 |
| 800 | 19.83 | 7.64 | 19.83 | 32.17 | 15.26 | 232.29 |
| 900 | 20.82 | 9.67 | 22.22 | 32.85 | 18.51 | 236.12 |
| 1000 | 21.61 | 11.79 | 24.46 | 33.46 | 21.82 | 239.61 |
| 1100 | 22.24 | 13.99 | 26.55 | 33.99 | 25.20 | 242.82 |
| 1200 | 22.77 | 16.24 | 28.51 | 34.46 | 28.62 | 245.80 |
| 1300 | 23.20 | 18.54 | 30.35 | 34.88 | 32.09 | 248.58 |
| 1400 | 23.58 | 20.88 | 32.08 | 35.28 | 35.60 | 251.18 |
| 1500 | 23.90 | 23.25 | 33.72 | 35.66 | 39.14 | 253.62 |
| 1600 | 24.19 | 25.66 | 35.27 | 36.03 | 42.73 | 255.94 |
| 1700 | 24.45 | 28.09 | 36.74 | 36.41 | 46.35 | 258.13 |
| 1800 | 24.68 | 30.55 | 38.15 | 36.79 | 50.01 | 260.22 |
| 1900 | 24.89 | 33.02 | 39.49 | 37.20 | 53.71 | 262.22 |
| 2000 | 25.09 | 35.53 | 40.77 | 37.61 | 57.45 | 264.14 |
| 2100 | 25.28 | 38.05 | 42.00 | 38.03 | 61.23 | 265.99 |
| 2200 | 25.45 | 40.58 | 43.18 | 38.47 | 65.06 | 267.77 |
| 2300 | 25.62 | 43.13 | 44.32 | 38.92 | 68.93 | 269.49 |
| 2400 | 25.77 | 45.71 | 45.41 | 39.38 | 72.84 | 271.15 |
| 2500 | 25.92 | 48.29 | 46.46 | 39.83 | 76.80 | 272.77 |
| 2600 | 26.07 | 50.89 | 47.48 | 40.28 | 80.81 | 274.34 |
| 2700 | 26.21 | 53.50 | 48.47 | 40.73 | 84.86 | 275.87 |
| 2800 | 26.35 | 56.13 | 49.43 | 41.17 | 88.95 | 277.36 |
| 2900 | 26.48 | 58.77 | 50.35 | 41.60 | 93.09 | 278.81 |
| 3000 | 26.61 | 61.43 | 51.25 | 42.01 | 97.27 | 280.23 |
| 3100 | 26.74 | 64.09 | 52.13 | 42.40 | 101.49 | 281.61 |
| 3200 | 26.86 | 66.78 | 52.98 | 42.78 | 105.75 | 282.96 |
| 3300 | 26.99 | 69.47 | 53.81 | 43.14 | 110.05 | 284.29 |
| 3400 | 27.10 | 72.17 | 54.61 | 43.47 | 114.38 | 285.58 |
| 3500 | 27.23 | 74.89 | 55.40 | 43.78 | 118.74 | 286.84 |

| T, K | Hydrogen cyanide (HCN) $(\Delta H^\circ_f)_{298.15} = 135.14$ kJ/mole M.W. $= 0.027026$ | | | Isocyanic acid (HNCO) $(\Delta H^\circ_f)_{298.15} = -101.67$ kJ/mole M.W. $= 0.043026$ | | |
|---|---|---|---|---|---|---|
| | $C^\circ_p$, J/mole K | $H^\circ - H^\circ_{298}$, kJ/mole | $S^\circ$, J/mole K | $C^\circ_p$, J/mole K | $H^\circ - H^\circ_{298}$, kJ/mole | $S^\circ$, J/mole K |
| 100 | 29.17 | -6.33 | 167.50 | 33.47 | -7.63 | 197.21 |
| 200 | 31.72 | -3.32 | 188.27 | 38.27 | -4.10 | 221.55 |
| 298 | 35.86 | 0.00 | 201.72 | 45.04 | 0.00 | 238.12 |
| 300 | 35.93 | .07 | 201.94 | 45.16 | .08 | 238.40 |
| 400 | 39.23 | 3.83 | 212.75 | 50.73 | 4.89 | 252.19 |
| 500 | 41.73 | 7.89 | 221.79 | 55.01 | 10.19 | 263.99 |
| 600 | 43.81 | 12.16 | 229.58 | 58.41 | 15.86 | 274.32 |
| 700 | 45.64 | 16.64 | 236.48 | 61.23 | 21.85 | 283.55 |
| 800 | 47.32 | 21.29 | 242.68 | 63.63 | 28.10 | 291.88 |
| 900 | 48.84 | 26.10 | 248.34 | 65.70 | 34.56 | 299.50 |
| 1000 | 50.23 | 31.05 | 253.56 | 67.51 | 41.22 | 306.52 |
| 1100 | 51.48 | 36.14 | 258.41 | 69.10 | 48.06 | 313.03 |
| 1200 | 52.61 | 41.34 | 262.94 | 70.50 | 55.04 | 319.11 |
| 1300 | 53.62 | 46.66 | 267.19 | 71.72 | 62.15 | 324.80 |
| 1400 | 54.52 | 52.06 | 271.19 | 72.80 | 69.38 | 330.15 |
| 1500 | 55.33 | 57.56 | 274.99 | 73.75 | 76.71 | 335.21 |
| 1600 | 56.05 | 63.12 | 278.58 | 74.59 | 84.13 | 340.00 |
| 1700 | 56.70 | 68.76 | 282.00 | 75.33 | 91.62 | 344.54 |
| 1800 | 57.27 | 74.46 | 285.25 | 75.99 | 99.19 | 348.86 |
| 1900 | 57.79 | 80.22 | 288.37 | 76.58 | 106.82 | 352.99 |
| 2000 | 58.26 | 86.02 | 291.34 | 77.09 | 114.50 | 356.93 |
| 2100 | 58.69 | 91.87 | 294.19 | 77.56 | 122.24 | 360.70 |
| 2200 | 59.07 | 97.75 | 296.93 | 77.98 | 130.01 | 364.32 |
| 2300 | 59.42 | 103.68 | 299.57 | 78.35 | 137.83 | 367.79 |
| 2400 | 59.74 | 109.64 | 302.10 | 78.69 | 145.68 | 371.14 |
| 2500 | 60.03 | 115.63 | 304.55 | 78.99 | 153.57 | 374.36 |
| 2600 | 60.30 | 121.65 | 306.91 | 79.27 | 161.48 | 377.46 |
| 2700 | 60.54 | 127.69 | 309.19 | 79.52 | 169.42 | 380.46 |
| 2800 | 60.76 | 133.75 | 311.39 | 79.74 | 177.38 | 383.35 |
| 2900 | 60.97 | 139.84 | 313.53 | 79.95 | 185.37 | 386.15 |
| 3000 | 61.17 | 145.95 | 315.60 | 80.14 | 193.37 | 388.87 |
| 3100 | 61.35 | 152.07 | 317.61 | 80.32 | 201.39 | 391.50 |
| 3200 | 61.51 | 158.21 | 319.56 | 80.48 | 209.43 | 394.05 |
| 3300 | 61.67 | 164.37 | 321.46 | 80.62 | 217.49 | 396.53 |
| 3400 | 61.81 | 170.55 | 323.30 | 80.76 | 225.56 | 398.94 |
| 3500 | 61.95 | 176.74 | 325.09 | 80.88 | 233.64 | 401.28 |

| | Formyl (HCO) $(\Delta H^\circ_f)_{298.15} = 43.51$ kJ/mole M.W. $= 0.029019$ | | | Formaldehyde (CH$_2$O) $(\Delta H^\circ_f)_{298.15} = -115.90$ kJ/mole M.W. $= 0.030027$ | | |
|---|---|---|---|---|---|---|
| $T$, $K$ | $C^\circ_p$, J/mole $K$ | $H^\circ - H^\circ_{298}$, kJ/mole | $S^\circ$, J/mole $K$ | $C^\circ_p$, J/mole $K$ | $H^\circ - H^\circ_{298}$, kJ/mole | $S^\circ$, J/mole $K$ |
| 100 | 33.26 | -6.67 | 187.91 | 33.26 | -6.69 | 181.92 |
| 200 | 33.47 | -3.33 | 210.99 | 33.50 | -3.36 | 205.00 |
| 298 | 34.60 | 0.00 | 224.53 | 35.40 | 0.00 | 218.66 |
| 300 | 34.63 | .06 | 224.75 | 35.46 | .07 | 218.88 |
| 400 | 36.53 | 3.62 | 234.96 | 39.27 | 3.79 | 229.57 |
| 500 | 38.73 | 7.38 | 243.35 | 43.76 | 7.94 | 238.81 |
| 600 | 40.97 | 11.36 | 250.60 | 48.22 | 12.54 | 247.19 |
| 700 | 43.07 | 15.57 | 257.08 | 52.32 | 17.57 | 254.94 |
| 800 | 44.96 | 19.97 | 262.96 | 55.98 | 22.99 | 262.17 |
| 900 | 46.62 | 24.55 | 268.35 | 59.20 | 28.75 | 268.95 |
| 1000 | 48.05 | 29.29 | 273.34 | 61.99 | 34.82 | 275.33 |
| 1100 | 49.28 | 34.16 | 277.98 | 64.41 | 41.14 | 281.36 |
| 1200 | 50.33 | 39.14 | 282.32 | 66.50 | 47.69 | 287.06 |
| 1300 | 51.22 | 44.22 | 286.38 | 68.29 | 54.43 | 292.45 |
| 1400 | 51.99 | 49.38 | 290.20 | 69.84 | 61.34 | 297.57 |
| 1500 | 52.65 | 54.61 | 293.81 | 71.18 | 68.39 | 302.44 |
| 1600 | 53.22 | 59.91 | 297.23 | 72.35 | 75.57 | 307.07 |
| 1700 | 53.72 | 65.25 | 300.47 | 73.36 | 82.86 | 311.48 |
| 1800 | 54.16 | 70.65 | 303.55 | 74.25 | 90.24 | 315.70 |
| 1900 | 54.55 | 76.09 | 306.49 | 75.03 | 97.70 | 319.74 |
| 2000 | 54.89 | 81.56 | 309.30 | 75.71 | 105.24 | 323.60 |
| 2100 | 55.20 | 87.06 | 311.99 | 76.32 | 112.84 | 327.31 |
| 2200 | 55.49 | 92.60 | 314.56 | 76.86 | 120.50 | 330.87 |
| 2300 | 55.74 | 98.16 | 317.03 | 77.33 | 128.21 | 334.30 |
| 2400 | 55.98 | 103.75 | 319.41 | 77.76 | 135.97 | 337.60 |
| 2500 | 56.19 | 109.35 | 321.70 | 78.14 | 143.76 | 340.79 |
| 2600 | 56.39 | 114.98 | 323.91 | 78.49 | 151.59 | 343.86 |
| 2700 | 56.58 | 120.63 | 326.04 | 78.80 | 159.46 | 346.82 |
| 2800 | 56.75 | 126.30 | 328.10 | 79.09 | 167.35 | 349.70 |
| 2900 | 56.91 | 131.98 | 330.10 | 79.34 | 175.28 | 352.48 |
| 3000 | 57.07 | 137.68 | 332.03 | 79.58 | 183.22 | 355.17 |
| 3100 | 57.21 | 143.39 | 333.90 | 79.79 | 191.19 | 357.78 |
| 3200 | 57.35 | 149.12 | 335.72 | 79.98 | 199.18 | 360.32 |
| 3300 | 57.48 | 154.87 | 337.49 | 80.16 | 207.18 | 362.78 |
| 3400 | 57.60 | 160.62 | 339.21 | 80.32 | 215.21 | 365.18 |
| 3500 | 57.72 | 166.39 | 340.87 | 80.48 | 223.25 | 367.51 |

| T,<br>K | Hydroperoxyl (HO$_2$)<br>$(\Delta H_f^\circ)_{298.15}$ = 20.92 kJ/mole<br>M.W. = 0.033007 | | | Nitroxyl hydride (HNO)<br>$(\Delta H_f^\circ)_{298.15}$ = 99.58 kJ/mole<br>M.W. = 0.031015 | | |
|---|---|---|---|---|---|---|
| | $C_p^\circ$,<br>J/mole K | $H^\circ - H_{298}^\circ$,<br>kJ/mole | $S^\circ$,<br>J/mole K | $C_p^\circ$,<br>J/mole K | $H^\circ - H_{298}^\circ$,<br>kJ/mole | $S^\circ$,<br>J/mole K |
| 100 | 33.26 | -6.68 | 190.87 | 33.26 | -6.66 | 184.00 |
| 200 | 33.48 | -3.34 | 213.95 | 33.45 | -3.33 | 207.08 |
| 298 | 34.89 | 0.00 | 227.54 | 34.64 | 0.00 | 220.62 |
| 300 | 35.04 | .06 | 227.75 | 34.67 | .06 | 220.84 |
| 400 | 37.27 | 3.67 | 238.11 | 36.77 | 3.63 | 231.08 |
| 500 | 39.66 | 7.52 | 246.69 | 39.09 | 7.43 | 239.54 |
| 600 | 41.76 | 11.59 | 254.11 | 41.29 | 11.44 | 246.86 |
| 700 | 43.53 | 15.86 | 260.68 | 43.28 | 15.68 | 253.38 |
| 800 | 45.06 | 20.29 | 266.60 | 45.04 | 20.10 | 259.27 |
| 900 | 46.39 | 24.87 | 271.99 | 46.58 | 24.68 | 264.67 |
| 1000 | 47.55 | 29.56 | 276.93 | 47.92 | 29.41 | 269.65 |
| 1100 | 48.58 | 34.37 | 281.52 | 49.08 | 34.25 | 274.27 |
| 1200 | 49.50 | 39.28 | 285.78 | 50.08 | 39.22 | 278.59 |
| 1300 | 50.31 | 44.27 | 289.78 | 50.95 | 44.27 | 282.63 |
| 1400 | 51.03 | 49.33 | 293.53 | 51.70 | 49.40 | 286.44 |
| 1500 | 51.67 | 54.47 | 297.08 | 52.35 | 54.61 | 290.03 |
| 1600 | 52.24 | 59.67 | 300.43 | 52.91 | 59.87 | 293.42 |
| 1700 | 52.74 | 64.91 | 303.61 | 53.40 | 65.18 | 296.65 |
| 1800 | 53.20 | 70.22 | 306.64 | 53.84 | 70.55 | 299.71 |
| 1900 | 53.60 | 75.55 | 309.53 | 54.22 | 75.95 | 302.63 |
| 2000 | 53.95 | 80.93 | 312.29 | 54.55 | 81.39 | 305.42 |
| 2100 | 54.27 | 86.35 | 314.93 | 54.84 | 86.86 | 308.09 |
| 2200 | 54.56 | 91.78 | 317.46 | 55.11 | 92.36 | 310.65 |
| 2300 | 54.83 | 97.26 | 319.89 | 55.34 | 97.88 | 313.11 |
| 2400 | 55.06 | 102.75 | 322.23 | 55.55 | 103.42 | 315.47 |
| 2500 | 55.27 | 108.27 | 324.48 | 55.74 | 108.99 | 317.73 |
| 2600 | 55.46 | 113.80 | 326.65 | 55.91 | 114.57 | 319.93 |
| 2700 | 55.64 | 119.36 | 328.75 | 56.06 | 120.17 | 322.04 |
| 2800 | 55.80 | 124.93 | 330.77 | 56.20 | 125.78 | 324.08 |
| 2900 | 55.94 | 130.52 | 332.73 | 56.33 | 131.41 | 326.05 |
| 3000 | 56.08 | 136.12 | 334.63 | 56.44 | 137.05 | 327.96 |
| 3100 | 56.20 | 141.73 | 336.47 | 56.55 | 142.70 | 329.82 |
| 3200 | 56.31 | 147.36 | 338.26 | 56.64 | 148.36 | 331.62 |
| 3300 | 56.42 | 153.00 | 340.00 | 56.73 | 154.03 | 333.36 |
| 3400 | 56.51 | 158.64 | 341.68 | 56.81 | 159.70 | 335.05 |
| 3500 | 56.60 | 164.30 | 343.32 | 56.89 | 165.39 | 336.70 |

| | Fluorine ($F_2$) (Standard state element) $(\Delta H_f^\circ)_{298.15} = 0.00$ kJ/mole Log $K_p = 0.00$ M.W. = 0.037998 | | | Monatomic fluorine (F) $(\Delta H_f^\circ)_{298.15} = 78.91$ kJ/mole M.W. = 0.018999 | | |
|---|---|---|---|---|---|---|
| $T$, $K$ | $C_p^\circ$, J/mole $K$ | $H^\circ-H_{298}^\circ$, kJ/mole | $S^\circ$, J/mole $K$ | $C_p^\circ$, J/mole $K$ | $H^\circ-H_{298}^\circ$, kJ/mole | $S^\circ$, J/mole $K$ |
| 100 | 29.11 | -5.92 | 170.27 | 21.20 | -4.43 | 134.37 |
| 200 | 29.69 | -2.99 | 190.55 | 22.61 | -2.23 | 149.56 |
| 298 | 31.30 | 0.00 | 202.69 | 22.75 | 0.00 | 158.64 |
| 300 | 31.33 | .06 | 202.88 | 22.74 | .04 | 158.79 |
| 400 | 32.98 | 3.28 | 212.13 | 22.43 | 2.30 | 165.29 |
| 500 | 34.24 | 6.64 | 219.63 | 22.10 | 4.53 | 170.26 |
| 600 | 35.14 | 10.11 | 225.96 | 21.83 | 6.72 | 174.26 |
| 700 | 35.79 | 13.66 | 231.43 | 21.63 | 8.90 | 177.61 |
| 800 | 36.28 | 17.27 | 236.24 | 21.48 | 11.05 | 180.49 |
| 900 | 36.65 | 20.91 | 240.53 | 21.36 | 13.19 | 183.01 |
| 1000 | 36.94 | 24.59 | 244.41 | 21.27 | 15.33 | 185.25 |
| 1100 | 37.18 | 28.30 | 247.94 | 21.20 | 17.45 | 187.28 |
| 1200 | 37.38 | 32.03 | 251.19 | 21.14 | 19.56 | 189.12 |
| 1300 | 37.56 | 35.77 | 254.19 | 21.09 | 21.67 | 190.81 |
| 1400 | 37.71 | 39.54 | 256.97 | 21.05 | 23.78 | 192.37 |
| 1500 | 37.84 | 43.32 | 259.58 | 21.02 | 25.89 | 193.82 |
| 1600 | 37.97 | 47.11 | 262.03 | 21.00 | 27.99 | 195.18 |
| 1700 | 38.08 | 50.91 | 264.33 | 20.97 | 30.09 | 196.46 |
| 1800 | 38.18 | 54.72 | 266.51 | 20.96 | 32.18 | 197.65 |
| 1900 | 38.28 | 58.54 | 268.58 | 20.94 | 34.28 | 198.79 |
| 2000 | 38.38 | 62.38 | 270.55 | 20.92 | 36.37 | 199.86 |
| 2100 | 38.46 | 66.22 | 272.42 | 20.92 | 38.46 | 200.88 |
| 2200 | 38.55 | 70.07 | 274.21 | 20.90 | 40.55 | 201.85 |
| 2300 | 38.64 | 73.93 | 275.93 | 20.89 | 42.64 | 202.78 |
| 2400 | 38.71 | 77.80 | 277.57 | 20.89 | 44.73 | 203.67 |
| 2500 | 38.79 | 81.67 | 279.15 | 20.88 | 46.82 | 204.52 |
| 2600 | 38.87 | 85.55 | 280.68 | 20.87 | 48.91 | 205.34 |
| 2700 | 38.94 | 89.45 | 282.14 | 20.87 | 50.99 | 206.13 |
| 2800 | 39.02 | 93.35 | 283.56 | 20.86 | 53.08 | 206.89 |
| 2900 | 39.09 | 97.25 | 284.93 | 20.86 | 55.17 | 207.62 |
| 3000 | 39.17 | 101.16 | 286.26 | 20.85 | 57.25 | 208.33 |
| 3100 | 39.24 | 105.08 | 287.55 | 20.85 | 59.34 | 209.01 |
| 3200 | 39.30 | 109.01 | 288.79 | 20.84 | 61.42 | 209.67 |
| 3300 | 39.38 | 112.94 | 290.00 | 20.84 | 63.50 | 210.31 |
| 3400 | 39.44 | 116.88 | 291.18 | 20.84 | 65.59 | 210.94 |
| 3500 | 39.51 | 120.83 | 292.32 | 20.84 | 67.67 | 211.54 |

| | Chlorine (Cl$_2$) (Standard state element) $(\Delta H_f^\circ)_{298.15} = 0.00$ kJ/mole Log $K_p = 0.00$ M.W. = 0.070906 | | | Monatomic chlorine (Cl) $(\Delta H_f^\circ)_{298.15} = 121.29$ kJ/mole M.W. = 0.035453 | | |
|---|---|---|---|---|---|---|
| $T$, K | $C_p^\circ$, J/mole K | $H^\circ - H_{298}^\circ$, kJ/mole | $S^\circ$, J/mole K | $C_p^\circ$, J/mole K | $H^\circ - H_{298}^\circ$, kJ/mole | $S^\circ$, J/mole K |
| 100 | 29.29 | −6.27 | 188.91 | 20.79 | −4.19 | 142.06 |
| 200 | 31.70 | −3.23 | 209.85 | 21.08 | −2.10 | 156.52 |
| 298 | 33.94 | 0.00 | 222.96 | 21.84 | 0.00 | 165.08 |
| 300 | 33.97 | .06 | 223.17 | 21.85 | .04 | 165.21 |
| 400 | 35.30 | 3.54 | 233.15 | 22.47 | 2.26 | 171.59 |
| 500 | 36.08 | 7.10 | 241.12 | 22.74 | 4.52 | 176.64 |
| 600 | 36.57 | 10.74 | 247.74 | 22.78 | 6.80 | 180.79 |
| 700 | 36.91 | 14.41 | 253.40 | 22.69 | 9.08 | 184.30 |
| 800 | 37.15 | 18.12 | 258.35 | 22.55 | 11.34 | 187.32 |
| 900 | 37.33 | 21.84 | 262.74 | 22.39 | 13.59 | 189.97 |
| 1000 | 37.47 | 25.59 | 266.68 | 22.23 | 15.82 | 192.32 |
| 1100 | 37.59 | 29.34 | 270.25 | 22.09 | 18.03 | 194.43 |
| 1200 | 37.70 | 33.10 | 273.53 | 21.96 | 20.23 | 196.35 |
| 1300 | 37.79 | 36.88 | 276.55 | 21.84 | 22.41 | 198.10 |
| 1400 | 37.87 | 40.66 | 279.35 | 21.74 | 24.60 | 199.71 |
| 1500 | 37.94 | 44.45 | 281.97 | 21.65 | 26.77 | 201.21 |
| 1600 | 38.02 | 48.25 | 284.42 | 21.57 | 28.93 | 202.61 |
| 1700 | 38.08 | 52.05 | 286.73 | 21.51 | 31.09 | 203.91 |
| 1800 | 38.15 | 55.86 | 288.91 | 21.44 | 33.23 | 205.14 |
| 1900 | 38.21 | 59.68 | 290.97 | 21.39 | 35.38 | 206.30 |
| 2000 | 38.28 | 63.51 | 292.93 | 21.34 | 37.51 | 207.39 |
| 2100 | 38.35 | 67.34 | 294.80 | 21.30 | 39.64 | 208.43 |
| 2200 | 38.43 | 71.18 | 296.59 | 21.26 | 41.77 | 209.42 |
| 2300 | 38.51 | 75.02 | 298.30 | 21.23 | 43.89 | 210.37 |
| 2400 | 38.59 | 78.88 | 299.94 | 21.20 | 46.02 | 211.27 |
| 2500 | 38.68 | 82.74 | 301.52 | 21.17 | 48.13 | 212.13 |
| 2600 | 38.78 | 86.61 | 303.03 | 21.14 | 50.25 | 212.96 |
| 2700 | 38.88 | 90.50 | 304.50 | 21.12 | 52.36 | 213.76 |
| 2800 | 38.99 | 94.39 | 305.92 | 21.10 | 54.48 | 214.53 |
| 2900 | 39.10 | 98.29 | 307.29 | 21.08 | 56.58 | 215.27 |
| 3000 | 39.22 | 102.21 | 308.61 | 21.06 | 58.69 | 215.98 |
| 3100 | 39.34 | 106.14 | 309.90 | 21.05 | 60.79 | 216.67 |
| 3200 | 39.46 | 110.08 | 311.15 | 21.03 | 62.90 | 217.34 |
| 3300 | 39.58 | 114.03 | 312.37 | 21.02 | 65.00 | 217.99 |
| 3400 | 39.71 | 118.00 | 313.55 | 21.01 | 67.10 | 218.61 |
| 3500 | 39.82 | 121.98 | 314.70 | 21.00 | 69.20 | 219.22 |

|  | Bromine (Br$_2$) (Standard state element) $(\Delta H_f^\circ)_{298.15} = 0.00$ kJ/mole Log $K_p = 0.00$ M.W. = 0.159832 | | | Monatomic Bromine (Br) $(\Delta H_f^\circ)_{298.15} = 111.86$ kJ/mole M.W. = 0.079916 | | |
|---|---|---|---|---|---|---|
| $T$, $K$ | $C_p^\circ$, J/mole $K$ | $H^\circ - H_{298}^\circ$, kJ/mole | $S^\circ$, J/mole $K$ | $C_p^\circ$, J/mole $K$ | $H^\circ - H_{298}^\circ$, kJ/mole | $S^\circ$, J/mole $K$ |
| 100 | 43.59 | -21.72 | 53.85 | 20.79 | -4.12 | 152.20 |
| 200 | 53.77 | -16.82 | 87.44 | 20.79 | -2.04 | 166.61 |
| 298 | 75.69 | 0.00 | 152.23 | 20.79 | 0.00 | 174.90 |
| 300 | 75.63 | .14 | 152.70 | 20.79 | .04 | 175.03 |
| 400 | 36.71 | 34.61 | 256.07 | 20.79 | 2.12 | 181.01 |
| 500 | 37.06 | 38.31 | 264.31 | 20.80 | 4.20 | 185.65 |
| 600 | 37.27 | 42.02 | 271.09 | 20.83 | 6.28 | 189.45 |
| 700 | 37.42 | 45.76 | 276.84 | 20.91 | 8.36 | 192.66 |
| 800 | 37.53 | 49.51 | 281.85 | 21.03 | 10.46 | 195.46 |
| 900 | 37.62 | 53.27 | 286.27 | 21.18 | 12.57 | 197.95 |
| 1000 | 37.70 | 57.03 | 290.24 | 21.36 | 14.70 | 200.19 |
| 1100 | 37.77 | 60.81 | 293.84 | 21.56 | 16.84 | 202.23 |
| 1200 | 37.83 | 64.58 | 297.13 | 21.75 | 19.01 | 204.12 |
| 1300 | 37.89 | 68.37 | 300.16 | 21.94 | 21.20 | 205.87 |
| 1400 | 37.94 | 72.16 | 302.97 | 22.11 | 23.40 | 207.50 |
| 1500 | 38.00 | 75.96 | 305.59 | 22.26 | 25.61 | 209.03 |
| 1600 | 38.05 | 79.76 | 308.04 | 22.39 | 27.85 | 210.47 |
| 1700 | 38.10 | 83.57 | 310.35 | 22.50 | 30.09 | 211.83 |
| 1800 | 38.15 | 87.38 | 312.53 | 22.59 | 32.35 | 213.12 |
| 1900 | 38.19 | 91.20 | 314.59 | 22.66 | 34.61 | 214.34 |
| 2000 | 38.25 | 95.02 | 316.55 | 22.71 | 36.88 | 215.51 |
| 2100 | 38.29 | 98.85 | 318.42 | 22.75 | 39.15 | 216.61 |
| 2200 | 38.34 | 102.68 | 320.20 | 22.77 | 41.43 | 217.67 |
| 2300 | 38.38 | 106.52 | 321.91 | 22.79 | 43.70 | 218.69 |
| 2400 | 38.43 | 110.36 | 323.54 | 22.79 | 45.98 | 219.66 |
| 2500 | 38.48 | 114.20 | 325.11 | 22.78 | 48.26 | 220.58 |
| 2600 | 38.52 | 118.05 | 326.62 | 22.77 | 50.54 | 221.48 |
| 2700 | 38.56 | 121.91 | 328.08 | 22.75 | 52.81 | 222.34 |
| 2800 | 38.61 | 125.77 | 329.48 | 22.73 | 55.09 | 223.17 |
| 2900 | 38.65 | 129.63 | 330.84 | 22.70 | 57.36 | 223.96 |
| 3000 | 38.70 | 133.49 | 332.15 | 22.67 | 59.63 | 224.73 |
| 3100 | 38.74 | 137.37 | 333.42 | 22.64 | 61.89 | 225.48 |
| 3200 | 38.79 | 141.24 | 334.65 | 22.60 | 64.15 | 226.19 |
| 3300 | 38.83 | 145.13 | 335.84 | 22.56 | 66.41 | 226.89 |
| 3400 | 38.87 | 149.01 | 337.00 | 22.53 | 68.67 | 227.56 |
| 3500 | 38.92 | 152.90 | 338.13 | 22.49 | 70.92 | 228.21 |

| | Hydrogen fluoride (HF) $(\Delta H_f^\circ)_{298.15} = -272.55$ kJ/mole M.W. = 0.020006 | | | Hydrogen chloride (HCl) $(\Delta H_f^\circ)_{298.15} = -92.31$ kJ/mole M.W. = 0.036465 | | |
|---|---|---|---|---|---|---|
| $T$, K | $C_p^\circ$, J/mole K | $H^\circ - H_{298}^\circ$, kJ/mole | $S^\circ$, J/mole K | $C_p^\circ$, J/mole K | $H^\circ - H_{298}^\circ$, kJ/mole | $S^\circ$, J/mole K |
| 100 | 29.13 | -5.77 | 141.85 | 29.12 | -5.77 | 154.98 |
| 200 | 29.13 | -2.86 | 162.04 | 29.12 | -2.86 | 175.16 |
| 298 | 29.14 | 0.00 | 173.67 | 29.14 | 0.00 | 186.79 |
| 300 | 29.14 | .05 | 173.85 | 29.14 | .05 | 186.97 |
| 400 | 29.15 | 2.97 | 182.23 | 29.18 | 2.97 | 195.36 |
| 500 | 29.17 | 5.88 | 188.74 | 29.30 | 5.89 | 201.89 |
| 600 | 29.23 | 8.80 | 194.07 | 29.58 | 8.84 | 207.25 |
| 700 | 29.35 | 11.73 | 198.58 | 29.99 | 11.81 | 211.84 |
| 800 | 29.55 | 14.68 | 202.51 | 30.50 | 14.84 | 215.87 |
| 900 | 29.83 | 17.64 | 206.01 | 31.06 | 17.91 | 219.50 |
| 1000 | 30.17 | 20.64 | 209.17 | 31.63 | 21.05 | 222.80 |
| 1100 | 30.56 | 23.68 | 212.06 | 32.19 | 24.24 | 225.84 |
| 1200 | 30.97 | 26.76 | 214.74 | 32.71 | 27.48 | 228.66 |
| 1300 | 31.40 | 29.87 | 217.23 | 33.20 | 30.78 | 231.30 |
| 1400 | 31.82 | 33.04 | 219.57 | 33.65 | 34.12 | 233.78 |
| 1500 | 32.24 | 36.24 | 221.78 | 34.06 | 37.51 | 236.12 |
| 1600 | 32.64 | 39.48 | 223.88 | 34.43 | 40.93 | 238.32 |
| 1700 | 33.02 | 42.76 | 225.86 | 34.77 | 44.39 | 240.43 |
| 1800 | 33.38 | 46.09 | 227.76 | 35.07 | 47.89 | 242.42 |
| 1900 | 33.71 | 49.44 | 229.58 | 35.35 | 51.41 | 244.32 |
| 2000 | 34.03 | 52.83 | 231.32 | 35.60 | 54.96 | 246.14 |
| 2100 | 34.33 | 56.25 | 232.98 | 35.83 | 58.53 | 247.89 |
| 2200 | 34.60 | 59.69 | 234.58 | 36.04 | 62.12 | 249.56 |
| 2300 | 34.86 | 63.17 | 236.13 | 36.23 | 65.73 | 251.17 |
| 2400 | 35.10 | 66.66 | 237.62 | 36.41 | 69.37 | 252.71 |
| 2500 | 35.32 | 70.18 | 239.06 | 36.58 | 73.01 | 254.20 |
| 2600 | 35.53 | 73.73 | 240.45 | 36.73 | 76.68 | 255.64 |
| 2700 | 35.73 | 77.29 | 241.79 | 36.87 | 80.36 | 257.03 |
| 2800 | 35.92 | 80.87 | 243.09 | 37.00 | 84.06 | 258.37 |
| 2900 | 36.09 | 84.47 | 244.36 | 37.13 | 87.76 | 259.67 |
| 3000 | 36.25 | 88.09 | 245.58 | 37.25 | 91.48 | 260.93 |
| 3100 | 36.40 | 91.72 | 246.77 | 37.35 | 95.21 | 262.15 |
| 3200 | 36.55 | 95.37 | 247.93 | 37.46 | 98.95 | 263.34 |
| 3300 | 36.69 | 99.03 | 249.06 | 37.56 | 102.70 | 264.50 |
| 3400 | 36.82 | 102.71 | 250.16 | 37.65 | 106.46 | 265.62 |
| 3500 | 36.94 | 106.39 | 251.22 | 37.74 | 110.23 | 266.71 |

| | Hydrogen bromide (HBr) $(\Delta H^\circ_f)_{298.15} = -36.44$ kJ/mole M.W. = 0.080924 | | | Helium (He) $(\Delta H^\circ_f)_{298.15} = 0.00$ kJ/mole M.W. = 0.004003 | | |
|---|---|---|---|---|---|---|
| $T$, $K$ | $C^\circ_p$, J/mole $K$ | $H^\circ - H^\circ_{298}$, kJ/mole | $S^\circ$, J/mole $K$ | $C^\circ_p$, J/mole $K$ | $H^\circ - H^\circ_{298}$, kJ/mole | $S^\circ$, J/mole $K$ |
| 100 | 29.12 | −5.77 | 166.78 | 20.79 | −4.12 | 103.34 |
| 200 | 29.12 | −2.86 | 186.96 | 20.79 | −2.04 | 117.74 |
| 298 | 29.14 | 0.00 | 198.59 | 20.79 | 0.00 | 126.04 |
| 300 | 29.14 | .05 | 198.77 | 20.79 | .04 | 126.17 |
| 400 | 29.22 | 2.97 | 207.17 | 20.79 | 2.12 | 132.15 |
| 500 | 29.46 | 5.90 | 213.71 | 20.79 | 4.20 | 136.79 |
| 600 | 29.87 | 8.87 | 219.11 | 20.79 | 6.28 | 140.58 |
| 700 | 30.43 | 11.88 | 223.76 | 20.79 | 8.35 | 143.78 |
| 800 | 31.06 | 14.96 | 227.86 | 20.79 | 10.43 | 146.56 |
| 900 | 31.70 | 18.10 | 231.56 | 20.79 | 12.51 | 149.01 |
| 1000 | 32.32 | 21.30 | 234.93 | 20.79 | 14.59 | 151.20 |
| 1100 | 32.90 | 24.56 | 238.04 | 20.79 | 16.67 | 153.18 |
| 1200 | 33.43 | 27.87 | 240.92 | 20.79 | 18.74 | 154.99 |
| 1300 | 33.90 | 31.24 | 243.62 | 20.79 | 20.82 | 156.65 |
| 1400 | 34.33 | 34.65 | 246.14 | 20.79 | 22.90 | 158.19 |
| 1500 | 34.71 | 38.10 | 248.53 | 20.79 | 24.98 | 159.62 |
| 1600 | 35.05 | 41.59 | 250.78 | 20.79 | 27.06 | 160.97 |
| 1700 | 35.36 | 45.11 | 252.91 | 20.79 | 29.14 | 162.23 |
| 1800 | 35.64 | 48.66 | 254.94 | 20.79 | 31.22 | 163.41 |
| 1900 | 35.89 | 52.24 | 256.88 | 20.79 | 33.30 | 164.54 |
| 2000 | 36.11 | 55.84 | 258.72 | 20.79 | 35.38 | 165.61 |
| 2100 | 36.31 | 59.46 | 260.49 | 20.79 | 37.46 | 166.62 |
| 2200 | 36.50 | 63.10 | 262.18 | 20.79 | 39.53 | 167.59 |
| 2300 | 36.67 | 66.76 | 263.81 | 20.79 | 41.61 | 168.51 |
| 2400 | 36.83 | 70.44 | 265.37 | 20.79 | 43.69 | 169.39 |
| 2500 | 36.98 | 74.13 | 266.88 | 20.79 | 45.77 | 170.24 |
| 2600 | 37.11 | 77.83 | 268.33 | 20.79 | 47.85 | 171.06 |
| 2700 | 37.24 | 81.55 | 269.73 | 20.79 | 49.92 | 171.84 |
| 2800 | 37.36 | 85.28 | 271.09 | 20.79 | 52.00 | 172.60 |
| 2900 | 37.47 | 89.02 | 272.40 | 20.79 | 54.08 | 173.33 |
| 3000 | 37.57 | 92.77 | 273.68 | 20.79 | 56.16 | 174.03 |
| 3100 | 37.67 | 96.53 | 274.91 | 20.79 | 58.24 | 174.72 |
| 3200 | 37.77 | 100.31 | 276.11 | 20.79 | 60.32 | 175.38 |
| 3300 | 37.86 | 104.09 | 277.27 | 20.79 | 62.40 | 176.01 |
| 3400 | 37.94 | 107.88 | 278.40 | 20.79 | 64.48 | 176.64 |
| 3500 | 38.02 | 111.68 | 279.50 | 20.79 | 66.55 | 177.24 |

| T, K | Neon (Ne) $(\Delta H^\circ_f)_{298.15} = 0.00$ kJ/mole M.W. $= 0.020179$ | | | Argon (Ar) $(\Delta H^\circ_f)_{298.15} = 0.00$ kJ/mole M.W. $= 0.039948$ | | |
|---|---|---|---|---|---|---|
| | $C^\circ_p$, J/mole K | $H^\circ - H^\circ_{298}$, kJ/mole | $S^\circ$, J/mole K | $C^\circ_p$, J/mole K | $H^\circ - H^\circ_{298}$, kJ/mole | $S^\circ$, J/mole K |
| 100 | 20.79 | -4.12 | 123.51 | 20.79 | -4.12 | 132.03 |
| 200 | 20.79 | -2.04 | 137.92 | 20.79 | -2.04 | 146.44 |
| 298 | 20.79 | 0.00 | 146.22 | 20.79 | 0.00 | 154.74 |
| 300 | 20.79 | .04 | 146.35 | 20.79 | .04 | 154.87 |
| 400 | 20.79 | 2.12 | 152.33 | 20.79 | 2.12 | 160.85 |
| 500 | 20.79 | 4.20 | 156.96 | 20.79 | 4.20 | 165.48 |
| 600 | 20.79 | 6.28 | 160.75 | 20.79 | 6.28 | 169.27 |
| 700 | 20.79 | 8.35 | 163.96 | 20.79 | 8.35 | 172.48 |
| 800 | 20.79 | 10.43 | 166.73 | 20.79 | 10.43 | 175.25 |
| 900 | 20.79 | 12.51 | 169.18 | 20.79 | 12.51 | 177.70 |
| 1000 | 20.79 | 14.59 | 171.37 | 20.79 | 14.59 | 179.89 |
| 1100 | 20.79 | 16.67 | 173.35 | 20.79 | 16.67 | 181.87 |
| 1200 | 20.79 | 18.74 | 175.16 | 20.79 | 18.74 | 183.68 |
| 1300 | 20.79 | 20.82 | 176.82 | 20.79 | 20.82 | 185.34 |
| 1400 | 20.79 | 22.90 | 178.37 | 20.79 | 22.90 | 186.89 |
| 1500 | 20.79 | 24.98 | 179.80 | 20.79 | 24.98 | 188.32 |
| 1600 | 20.79 | 27.06 | 181.14 | 20.79 | 27.06 | 189.66 |
| 1700 | 20.79 | 29.14 | 182.40 | 20.79 | 29.14 | 190.92 |
| 1800 | 20.79 | 31.22 | 183.59 | 20.79 | 31.22 | 192.11 |
| 1900 | 20.79 | 33.30 | 184.72 | 20.79 | 33.30 | 193.23 |
| 2000 | 20.79 | 35.38 | 185.78 | 20.79 | 35.38 | 194.30 |
| 2100 | 20.79 | 37.46 | 186.79 | 20.79 | 37.46 | 195.31 |
| 2200 | 20.79 | 39.53 | 187.76 | 20.79 | 39.53 | 196.28 |
| 2300 | 20.79 | 41.61 | 188.69 | 20.79 | 41.61 | 197.20 |
| 2400 | 20.79 | 43.69 | 189.57 | 20.79 | 43.69 | 198.09 |
| 2500 | 20.79 | 45.77 | 190.42 | 20.79 | 45.77 | 198.94 |
| 2600 | 20.79 | 47.85 | 191.23 | 20.79 | 47.85 | 199.75 |
| 2700 | 20.79 | 49.92 | 192.02 | 20.79 | 49.92 | 200.53 |
| 2800 | 20.79 | 52.00 | 192.77 | 20.79 | 52.00 | 201.29 |
| 2900 | 20.79 | 54.08 | 193.50 | 20.79 | 54.08 | 202.02 |
| 3000 | 20.79 | 56.16 | 194.21 | 20.79 | 56.16 | 202.73 |
| 3100 | 20.79 | 58.24 | 194.89 | 20.79 | 58.24 | 203.41 |
| 3200 | 20.79 | 60.32 | 195.55 | 20.79 | 60.32 | 204.07 |
| 3300 | 20.79 | 62.40 | 196.19 | 20.79 | 62.40 | 204.71 |
| 3400 | 20.79 | 64.48 | 196.81 | 20.79 | 64.48 | 205.33 |
| 3500 | 20.79 | 66.55 | 197.41 | 20.79 | 66.55 | 205.93 |

# Appendix B2 JANAF values of $\text{Log}_{10} K_{pf}$ for the species listed in App. B1

| $T, K$ | O | H | $H_2O$ | OH | $CO_2$ | CO | NO | $NO_2$ |
|---|---|---|---|---|---|---|---|---|
| 600 | −18.574 | −16.336 | 18.633 | −2.568 | 34.405 | 14.318 | −7.210 | −6.111 |
| 700 | −15.449 | −13.599 | 15.583 | −2.085 | 29.506 | 12.946 | −6.086 | −5.714 |
| 800 | −13.101 | −11.539 | 13.289 | −1.724 | 25.830 | 11.914 | −5.243 | −5.417 |
| 900 | −11.272 | −9.934 | 11.498 | −1.444 | 22.970 | 11.108 | −4.587 | −5.185 |
| 1000 | −9.807 | −8.646 | 10.062 | −1.222 | 20.680 | 10.459 | −4.062 | −5.000 |
| 1100 | −8.606 | −7.589 | 8.883 | −1.041 | 18.806 | 9.926 | −3.633 | −4.848 |
| 1200 | −7.604 | −6.707 | 7.899 | −.890 | 17.243 | 9.479 | −3.275 | −4.721 |
| 1300 | −6.755 | −5.958 | 7.064 | −.764 | 15.920 | 9.099 | −2.972 | −4.612 |
| 1400 | −6.027 | −5.315 | 6.347 | −.656 | 14.785 | 8.771 | −2.712 | −4.519 |
| 1500 | −5.395 | −4.756 | 5.725 | −.563 | 13.801 | 8.485 | −2.487 | −4.438 |
| 1600 | −4.842 | −4.266 | 5.180 | −.482 | 12.940 | 8.234 | −2.290 | −4.367 |
| 1700 | −4.353 | −3.833 | 4.699 | −.410 | 12.180 | 8.011 | −2.116 | −4.304 |
| 1800 | −3.918 | −3.448 | 4.270 | −.347 | 11.504 | 7.811 | −1.962 | −4.248 |
| 1900 | −3.529 | −3.102 | 3.886 | −.291 | 10.898 | 7.631 | −1.823 | −4.198 |
| 2000 | −3.178 | −2.790 | 3.540 | −.240 | 10.353 | 7.469 | −1.699 | −4.152 |
| 2100 | −2.860 | −2.508 | 3.227 | −.195 | 9.860 | 7.321 | −1.586 | −4.111 |
| 2200 | −2.571 | −2.251 | 2.942 | −.153 | 9.411 | 7.185 | −1.484 | −4.074 |
| 2300 | −2.307 | −2.016 | 2.682 | −.116 | 9.001 | 7.061 | −1.391 | −4.040 |
| 2400 | −2.065 | −1.800 | 2.443 | −.082 | 8.625 | 6.946 | −1.305 | −4.008 |
| 2500 | −1.842 | −1.601 | 2.224 | −.050 | 8.280 | 6.840 | −1.227 | −3.979 |
| 2600 | −1.636 | −1.417 | 2.021 | −.021 | 7.960 | 6.741 | −1.154 | −3.953 |
| 2700 | −1.446 | −1.247 | 1.833 | .005 | 7.664 | 6.649 | −1.087 | −3.928 |
| 2800 | −1.268 | −1.089 | 1.658 | .030 | 7.388 | 6.563 | −1.025 | −3.905 |
| 2900 | −1.103 | −.941 | 1.495 | .053 | 7.132 | 6.483 | −.967 | −3.884 |
| 3000 | −.949 | −.803 | 1.343 | .074 | 6.892 | 6.407 | −.913 | −3.864 |
| 3100 | −.805 | −.674 | 1.201 | .094 | 6.668 | 6.336 | −.863 | −3.846 |
| 3200 | −.670 | −.553 | 1.067 | .112 | 6.458 | 6.269 | −.815 | −3.828 |
| 3300 | −.543 | −.439 | .942 | .129 | 6.260 | 6.206 | −.771 | −3.812 |
| 3400 | −.423 | −.332 | .824 | .145 | 6.074 | 6.145 | −.729 | −3.797 |
| 3500 | −.310 | −.231 | .712 | .160 | 5.898 | 6.088 | −.690 | −3.783 |
| 3600 | −.204 | −.135 | .607 | .174 | 5.732 | 6.034 | −.653 | −3.770 |
| 3700 | −.103 | −.044 | .507 | .188 | 5.574 | 5.982 | −.618 | −3.757 |
| 3800 | −.007 | .042 | .413 | .200 | 5.425 | 5.933 | −.585 | −3.746 |
| 3900 | .084 | .123 | .323 | .212 | 5.283 | 5.886 | −.554 | −3.734 |
| 4000 | .170 | .201 | .238 | .223 | 5.149 | 5.841 | −.524 | −3.724 |

**Appendix B2**—(*continued*)

| T, K | N₂O | NH₃ | NH₂ | NH | CH₄ | CH₃ | CH₂ | CH |
|------|------|------|------|------|------|------|------|------|
| 600 | -11.040 | -1.377 | -18.326 | -31.732 | 2.001 | -13.212 | -30.678 | -45.842 |
| 700 | -10.021 | -2.023 | -15.996 | -27.049 | .951 | -11.458 | -25.898 | -38.448 |
| 800 | -9.253 | -2.518 | -14.255 | -23.537 | .146 | -10.152 | -22.319 | -32.905 |
| 900 | -8.654 | -2.910 | -12.905 | -20.806 | -.493 | -9.145 | -19.540 | -28.597 |
| 1000 | -8.171 | -3.228 | -11.827 | -18.621 | -1.011 | -8.344 | -17.321 | -25.152 |
| 1100 | -7.774 | -3.490 | -10.948 | -16.834 | -1.440 | -7.693 | -15.508 | -22.336 |
| 1200 | -7.442 | -3.710 | -10.216 | -15.345 | -1.801 | -7.153 | -14.000 | -19.991 |
| 1300 | -7.158 | -3.897 | -9.598 | -14.084 | -2.107 | -6.698 | -12.726 | -18.008 |
| 1400 | -6.914 | -4.058 | -9.069 | -13.004 | -2.372 | -6.309 | -11.635 | -16.310 |
| 1500 | -6.701 | -4.197 | -8.610 | -12.068 | -2.602 | -5.974 | -10.691 | -14.838 |
| 1600 | -6.514 | -4.319 | -8.210 | -11.249 | -2.803 | -5.681 | -9.866 | -13.551 |
| 1700 | -6.347 | -4.426 | -7.856 | -10.526 | -2.981 | -5.423 | -9.139 | -12.417 |
| 1800 | -6.065 | -4.521 | -7.542 | -9.883 | -3.139 | -5.195 | -8.493 | -11.409 |
| 1900 |  | -4.605 | -7.261 | -9.308 | -3.281 | -4.991 | -7.916 | -10.507 |
| 2000 | -5.943 | -4.681 | -7.009 | -8.790 | -3.408 | -4.808 | -7.397 | -9.696 |
| 2100 | -5.833 | -4.749 | -6.780 | -8.322 | -3.523 | -4.642 | -6.929 | -8.963 |
| 2200 | -5.732 | -4.810 | -6.572 | -7.896 | -3.627 | -4.492 | -6.503 | -8.296 |
| 2300 | -5.639 | -4.866 | -6.382 | -7.507 | -3.722 | -4.355 | -6.115 | -7.687 |
| 2400 | -5.554 | -4.916 | -6.208 | -7.151 | -3.809 | -4.230 | -5.760 | -7.130 |
| 2500 | -5.475 | -4.963 | -6.048 | -6.823 | -3.889 | -4.115 | -5.433 | -6.617 |
| 2600 | -5.401 | -5.005 | -5.899 | -6.520 | -3.962 | -4.009 | -5.133 | -6.144 |
| 2700 | -5.333 | -5.044 | -5.762 | -6.240 | -4.030 | -3.911 | -4.854 | -5.706 |
| 2800 | -5.270 | -5.079 | -5.635 | -5.979 | -4.093 | -3.820 | -4.596 | -5.300 |
| 2900 | -5.210 | -5.112 | -5.516 | -5.737 | -4.152 | -3.736 | -4.356 | -4.922 |
| 3000 | -5.154 | -5.143 | -5.405 | -5.511 | -4.206 | -3.659 | -4.132 | -4.569 |
| 3100 | -5.102 | -5.171 | -5.300 | -5.299 | -4.257 | -3.584 | -3.923 | -4.239 |
| 3200 | -5.052 | -5.197 | -5.203 | -5.100 | -4.304 | -3.515 | -3.728 | -3.930 |
| 3300 | -5.006 | -5.221 | -5.111 | -4.914 | -4.349 | -3.451 | -3.544 | -3.639 |
| 3400 | -4.962 | -5.244 | -5.024 | -4.738 | -4.391 | -3.391 | -3.372 | -3.366 |
| 3500 | -4.920 | -5.265 | -4.942 | -4.572 | -4.430 | -3.334 | -3.210 | -3.108 |
| 3600 | -4.881 | -5.285 | -4.865 | -4.416 | -4.467 | -3.280 | -3.056 | -2.865 |
| 3700 | -4.843 | -5.304 | -4.791 | -4.267 | -4.503 | -3.230 | -2.912 | -2.636 |
| 3800 | -4.807 | -5.321 | -4.721 | -4.127 | -4.536 | -3.182 | -2.775 | -2.418 |
| 3900 | -4.773 | -5.338 | -4.655 | -3.994 | -4.568 | -3.137 | -2.646 | -2.212 |
| 4000 | -4.741 | -5.353 | -4.592 | -3.867 | -4.598 | -3.095 | -2.523 | -2.016 |

**Appendix B2**—(*continued*)

| T, K | C₂H₂ | C₂H₄ | HCO | CH₂O | N | CN | HCN | HNCO |
|------|------|------|-----|------|---|-----|-----|------|
| 600 | -16.687 | -7.652 | -1.169 | 8.868 | -38.081 | -32.567 | -9.957 | 7.166 |
| 700 | -13.882 | -7.114 | -.653 | 7.358 | -32.177 | -27.149 | -8.286 | 5.879 |
| 800 | -11.784 | -6.728 | -.271 | 6.214 | -27.744 | -23.088 | -7.036 | 4.911 |
| 900 | -10.155 | -6.438 | .021 | 5.317 | -24.292 | -19.932 | -6.065 | 4.155 |
| 1000 | -8.856 | -6.213 | .251 | 4.595 | -21.528 | -17.409 | -5.290 | 3.548 |
| 1100 | -7.795 | -6.034 | .436 | 4.000 | -19.265 | -15.347 | -4.658 | 3.051 |
| 1200 | -6.913 | -5.889 | .588 | 3.502 | -17.377 | -13.631 | -4.132 | 2.635 |
| 1300 | -6.168 | -5.766 | .715 | 3.079 | -15.778 | -12.180 | -3.688 | 2.282 |
| 1400 | -5.531 | -5.664 | .822 | 2.715 | -14.406 | -10.939 | -3.308 | 1.979 |
| 1500 | -4.979 | -5.575 | .913 | 2.398 | -13.217 | -9.864 | -2.979 | 1.716 |
| 1600 | -4.497 | -5.497 | .992 | 2.121 | -12.175 | -8.925 | -2.691 | 1.486 |
| 1700 | -4.072 | -5.430 | 1.060 | 1.876 | -11.256 | -8.097 | -2.438 | 1.282 |
| 1800 | -3.695 | -5.369 | 1.120 | 1.658 | -10.437 | -7.362 | -2.213 | 1.101 |
| 1900 | -3.358 | -5.316 | 1.173 | 1.462 | -9.705 | -6.705 | -2.013 | .938 |
| 2000 | -3.055 | -5.267 | 1.220 | 1.285 | -9.046 | -6.115 | -1.832 | .791 |
| 2100 | -2.782 | -5.223 | 1.261 | 1.125 | -8.449 | -5.582 | -1.669 | .659 |
| 2200 | -2.532 | -5.183 | 1.299 | .980 | -7.905 | -5.097 | -1.521 | .538 |
| 2300 | -2.306 | -5.146 | 1.332 | .846 | -7.409 | -4.655 | -1.385 | .427 |
| 2400 | -2.098 | -5.113 | 1.361 | .724 | -6.954 | -4.251 | -1.262 | .326 |
| 2500 | -1.906 | -5.081 | 1.389 | .612 | -6.535 | -3.879 | -1.148 | .233 |
| 2600 | -1.730 | -5.052 | 1.413 | .507 | -6.149 | -3.536 | -1.043 | .146 |
| 2700 | -1.566 | -5.025 | 1.435 | .411 | -5.790 | -3.218 | -.946 | .066 |
| 2800 | -1.415 | -5.000 | 1.455 | .321 | -5.457 | -2.924 | -.856 | -.008 |
| 2900 | -1.274 | -4.977 | 1.474 | .237 | -5.147 | -2.650 | -.772 | -.078 |
| 3000 | -1.142 | -4.955 | 1.490 | .159 | -4.858 | -2.394 | -.694 | -.143 |
| 3100 | -1.019 | -4.934 | 1.506 | .085 | -4.587 | -2.155 | -.621 | -.204 |
| 3200 | -.903 | -4.915 | 1.520 | .016 | -4.332 | -1.931 | -.552 | -.261 |
| 3300 | -.795 | -4.897 | 1.532 | -.049 | -4.093 | -1.720 | -.488 | -.315 |
| 3400 | -.693 | -4.880 | 1.544 | -.110 | -3.868 | -1.522 | -.428 | -.366 |
| 3500 | -.597 | -4.864 | 1.555 | -.168 | -3.656 | -1.336 | -.371 | -.414 |
| 3600 | -.506 | -4.848 | 1.565 | -.223 | -3.455 | -1.159 | -.317 | -.459 |
| 3700 | -.420 | -4.834 | 1.573 | -.276 | -3.265 | -.993 | -.267 | -.503 |
| 3800 | -.339 | -4.821 | 1.581 | -.325 | -3.086 | -.835 | -.219 | -.544 |
| 3900 | -.262 | -4.808 | 1.589 | -.372 | -2.915 | -.685 | -.174 | -.583 |
| 4000 | -.189 | -4.796 | 1.596 | -.417 | -2.752 | -.543 | -.130 | -.620 |

**Appendix B2**—(*continued*)

| T, K | HO₂ | HNO | F | Cl | BR | HF | HCl | HBR |
|------|------|--------|--------|--------|--------|--------|-------|-------|
| 600  | -4.134 | -10.990 | -3.814 | -7.710 | -5.641 | 24.077 | 8.530 | 5.036 |
| 700  | -3.903 | -9.779  | -2.810 | -6.182 | -4.431 | 20.677 | 7.368 | 4.374 |
| 800  | -3.732 | -8.874  | -2.053 | -5.031 | -3.522 | 18.125 | 6.494 | 3.876 |
| 900  | -3.603 | -8.172  | -1.462 | -4.133 | -2.814 | 16.137 | 5.812 | 3.486 |
| 1000 | -3.501 | -7.612  | -.988  | -3.413 | -2.245 | 14.544 | 5.265 | 3.173 |
| 1100 | -3.419 | -7.153  | -.599  | -2.822 | -1.799 | 13.240 | 4.816 | 2.917 |
| 1200 | -3.351 | -6.772  | -.273  | -2.328 | -1.389 | 12.152 | 4.442 | 2.702 |
| 1300 | -3.294 | -6.449  | .003   | -1.909 | -1.059 | 11.230 | 4.124 | 2.520 |
| 1400 | -3.246 | -6.172  | .240   | -1.549 | -.775  | 10.438 | 3.852 | 2.364 |
| 1500 | -3.205 | -5.932  | .447   | -1.236 | -.527  | 9.752  | 3.615 | 2.229 |
| 1600 | -3.168 | -5.722  | .627   | -.962  | -.311  | 9.191  | 3.408 | 2.110 |
| 1700 | -3.137 | -5.536  | .788   | -.720  | -.119  | 8.420  | 3.225 | 2.006 |
| 1800 | -3.109 | -5.371  | .930   | -.504  | .053   | 8.147  | 3.062 | 1.913 |
| 1900 | -3.084 | -5.223  | 1.058  | -.310  | .207   | 7.724  | 2.916 | 1.829 |
| 2000 | -3.061 | -5.090  | 1.173  | -.136  | .346   | 7.343  | 2.785 | 1.754 |
| 2100 | -3.041 | -4.970  | 1.277  | .022   | .472   | 6.998  | 2.666 | 1.686 |
| 2200 | -3.023 | -4.860  | 1.372  | .166   | .587   | 6.684  | 2.558 | 1.625 |
| 2300 | -3.007 | -4.760  | 1.459  | .298   | .692   | 6.396  | 2.459 | 1.568 |
| 2400 | -2.992 | -4.668  | 1.539  | .419   | .789   | 6.134  | 2.368 | 1.517 |
| 2500 | -2.978 | -4.584  | 1.613  | .530   | .879   | 5.892  | 2.285 | 1.469 |
| 2600 | -2.966 | -4.506  | 1.681  | .633   | .962   | 5.668  | 2.208 | 1.425 |
| 2700 | -2.954 | -4.434  | 1.744  | .729   | 1.039  | 5.460  | 2.136 | 1.384 |
| 2800 | -2.944 | -4.367  | 1.802  | .818   | 1.110  | 5.268  | 2.070 | 1.347 |
| 2900 | -2.935 | -4.305  | 1.857  | .900   | 1.178  | 5.088  | 2.008 | 1.311 |
| 3000 | -2.926 | -4.246  | 1.908  | .978   | 1.240  | 4.920  | 1.950 | 1.278 |
| 3100 | -2.918 | -4.192  | 1.956  | 1.050  | 1.299  | 4.763  | 1.896 | 1.248 |
| 3200 | -2.910 | -4.141  | 2.001  | 1.118  | 1.355  | 4.616  | 1.845 | 1.219 |
| 3300 | -2.903 | -4.093  | 2.043  | 1.182  | 1.407  | 4.478  | 1.798 | 1.192 |
| 3400 | -2.897 | -4.048  | 2.082  | 1.242  | 1.459  | 4.347  | 1.753 | 1.166 |
| 3500 | -2.891 | -4.006  | 2.120  | 1.299  | 1.503  | 4.224  | 1.710 | 1.142 |
| 3600 | -2.886 | -3.966  | 2.155  | 1.353  | 1.547  | 4.108  | 1.670 | 1.119 |
| 3700 | -2.881 | -3.928  | 2.189  | 1.404  | 1.589  | 3.998  | 1.632 | 1.098 |
| 3800 | -2.876 | -3.893  | 2.220  | 1.452  | 1.629  | 3.894  | 1.596 | 1.077 |
| 3900 | -2.872 | -3.859  | 2.251  | 1.498  | 1.666  | 3.795  | 1.562 | 1.058 |
| 4000 | -2.868 | -3.827  | 2.280  | 1.541  | 1.703  | 3.700  | 1.529 | 1.039 |

## Appendix B3

### Heat of formation of some common pure fuels and oxidizers at 25°C (298.15 K) (Ideal gas) (kJ/mole; 1 cal = 4.184 J)‡

| Substance | Formula | $\Delta H_f^\circ$ | Substance | Formula | $\Delta H_f^\circ$ |
|---|---|---|---|---|---|
| Methane | $CH_4$ | −74.85 | n-heptane | $C_7H_{16}$ | −187.78 |
| Ethane | $C_2H_6$ | −84.68 | Toluene | $C_7H_8$ | 50.00 |
| Ethylene | $C_2H_4$ | 52.30 | n-Octane | $C_8H_{18}$ | −208.45 |
| Acetylene | $C_2H_2$ | 226.73 | Ethylene oxide | $C_2H_4O$ | −52.63 |
| Propane | $C_3H_8$ | −103.85 | Propylene oxide | $C_3H_6O$ | −92.76 |
| Cyclopropane | $C_3H_6$ | 53.30 | Methyl alcohol | $CH_3OH$ | −201.17 |
| Propene | $C_3H_6$ | 20.42 | Ethyl alcohol | $C_2H_5OH$ | −234.81 |
| Allene | $C_3H_4$ | 192.13 | Acetone | $C_3H_6O$ | −217.57 |
| Propyne | $C_3H_4$ | 185.43 | Dimethyl ether | $C_2H_6O$ | −184.05 |
| n-butane | $C_4H_{10}$ | −126.15 | Diethyl ether | $C_4H_{10}O$ | −252.21 |
| iso-butane | $C_4H_{10}$ | −134.52 | Nitromethane | $CH_3NO_2$ | −74.73 |
| 1-2 butadiene | $C_4H_6$ | 162.21 | Methyl nitrate | $CH_3NO_3$ | −120.50 |
| 1-3 butadiene | $C_4H_6$ | 110.16 | Hydrazine | $N_2H_4$ | −95.19 |
| n-pentane | $C_5H_{12}$ | −146.44 | Hydrogen peroxide | $H_2O_2$ | −136.11 |
| iso-pentane | $C_5H_{12}$ | −154.47 | Cyanogen | $C_2N_2$ | 308.95 |
| n-hexane | $C_6H_{14}$ | −167.19 | Ozone | $O_3$ | 142.26 |
| Cyclohexane | $C_6H_{12}$ | −123.14 | Sulfur dioxide | $SO_2$ | −296.85 |
| Benzene | $C_6H_6$ | 82.93 | Sulfur trioxide | $SO_3$ | −395.26 |

‡ Taken from D. R. Stull, E. F. Westrum, and G. C. Sinke, *The Chemical Thermodynamics of Organic Compounds*, Wiley, New York (1969), with permission.

# APPENDIX C‡

## COMBUSTION PROPERTIES OF SOME SOLID AND LIQUID FUELS AND THE PROPERTIES OF DRY AIR AND HEAT OF VAPORIZATION OF WATER

### Appendix C1

#### Analysis of typical U.S. coals, as mined

| State | Rank‡ | ASH | % ultimate analysis | | | | | | High heating value MJ/kg |
| | | | $H_2O$ | C | $H_2$ | S | $O_2$ | $N_2$ | |
|---|---|---|---|---|---|---|---|---|---|
| RI | A | 18.9 | 13.3 | 64.2 | 0.4 | 0.3 | 2.7 | 0.2 | 21.66 |
| CO | B | 8.0 | 2.5 | 83.9 | 2.9 | 0.7 | 0.7 | 1.3 | 31.91 |
| NM | B | 8.9 | 2.9 | 82.3 | 2.6 | 0.8 | 1.3 | 1.2 | 31.02 |
| PA | C | 9.7 | 3.0 | 80.2 | 3.3 | 0.7 | 2.0 | 1.1 | 31.28 |
| VA | C | 19.6 | 3.1 | 70.5 | 3.2 | 0.6 | 2.2 | 0.8 | 27.56 |
| AR | D | 8.6 | 3.4 | 79.6 | 3.9 | 1.0 | 1.8 | 1.7 | 31.86 |
| OK | D | 8.7 | 2.6 | 80.1 | 4.0 | 1.0 | 1.9 | 1.7 | 32.09 |
| PA | E | 6.2 | 3.3 | 80.7 | 4.5 | 1.8 | 2.4 | 1.1 | 33.28 |
| VA | E | 7.2 | 3.1 | 80.1 | 4.7 | 1.0 | 2.4 | 1.5 | 32.63 |
| AL | F | 2.8 | 5.5 | 80.3 | 4.9 | 0.6 | 4.2 | 1.7 | 33.04 |
| CO | F | 11.7 | 1.4 | 73.4 | 5.1 | 0.6 | 6.5 | 1.3 | 30.72 |
| KY | F | 3.0 | 3.1 | 79.2 | 5.4 | 0.6 | 7.2 | 1.5 | 33.23 |
| IL | G | 8.4 | 8.0 | 68.7 | 4.5 | 1.2 | 7.6 | 1.6 | 28.21 |
| WY | G | 4.6 | 5.1 | 73.0 | 5.0 | 0.5 | 10.6 | 1.2 | 28.04 |
| IL | H | 8.6 | 12.1 | 62.8 | 4.6 | 4.3 | 6.6 | 1.0 | 26.70 |
| CO | I | 4.0 | 19.6 | 58.8 | 3.8 | 0.3 | 12.2 | 1.3 | 23.56 |
| ND | J | 6.2 | 34.8 | 42.4 | 2.8 | 0.7 | 12.4 | 0.7 | 16.77 |

‡ Rank key: A, meta-anthracite; B, anthracite; C, semianthracite; D, low-vol. bituminous; E, med.-vol. bituminous; F, high-vol. bituminous A; G, high-vol. bituminous B; H, high-vol. bituminous C; I, subbituminous; J, lignite.

‡ Unless otherwise indicated all the data in this appendix were taken with permission from *Combustion, Fossil Power Systems*, J. G. Singer, ed., Combustion Engineering Inc., Windsor, Conn. (1981).

## Appendix C2

### Typical analyses of dry wood

| | \multicolumn{6}{c}{% by weight} | HHV |
|---|---|---|---|---|---|---|---|
| | C | H$_2$ | S | O$_2$ | N$_2$ | Ash | MJ/kg |
| **Softwoods** | | | | | | | |
| Cedar, white | 48.80 | 6.37 | — | 44.46 | — | 0.37 | 19.53‡ |
| Fir, Douglas | 52.3 | 6.3 | — | 40.5 | 0.1 | 0.8 | 21.04 |
| Pine, pitch | 59.00 | 7.19 | — | 32.68 | — | 1.13 | 26.32‡ |
| white | 52.55 | 6.08 | — | 41.25 | — | 0.12 | 20.70‡ |
| yellow | 52.60 | 7.02 | — | 40.07 | — | 0.31 | 22.35‡ |
| **Hardwoods** | | | | | | | |
| Ash, white | 49.73 | 6.93 | — | 43.04 | — | 0.30 | 20.74‡ |
| Birch, white | 49.77 | 6.49 | — | 43.45 | — | 0.29 | 20.12‡ |
| Maple | 50.64 | 6.02 | — | 41.74 | 0.25 | 1.35 | 19.95 |
| Oak, black | 48.78 | 6.09 | — | 44.98 | — | 0.15 | 19.02‡ |
| red | 49.49 | 6.62 | — | 43.74 | — | 0.15 | 20.21‡ |
| white | 50.44 | 6.59 | — | 42.73 | — | 0.24 | 20.49‡ |

‡ Calculated from reported high heating value of kiln-dried wood assumed to contain 8-percent moisture.

## Appendix C3

### Typical analyses and properties of fuel oils

| Grade | No. 1 fuel oil | No. 2 fuel oil | No. 4 fuel oil | No. 5 fuel oil | No. 6 fuel oil |
|---|---|---|---|---|---|
| Type | Distillate (kerosene) | Distillate | Very light residual | Light residual | Residual |
| Color | Light | Amber | Black | Black | Black |
| Specific gravity 15/15°C | 0.8251 | 0.8654 | 0.9279 | 0.9529 | 0.9861 |
| kg/m$^3$ 15°C | 823.2 | 863.5 | 925.9 | 950.8 | 984.0 |
| Viscosity m$^2$/sec × 10$^6$; 38°C | 1.6 | 2.68 | 15.0 | 50.0 | 360.0 |
| Pour point, °C | Below zero | Below zero | −12.0 | −1.0 | 18.0 |
| Temp. for pumping, °C | Atmospheric | Atmospheric | −10 min. | 2 min. | 38.0 |
| Temp. for atomizing, °C | Atmospheric | Atmospheric | −4 min. | 54.0 | 93.0 |
| Carbon residue, % | Trace | Trace | 2.5 | 5.0 | 12.0 |
| Sulfur, % | 0.1 | 0.4–0.7 | 0.4–1.5 | 2.0 max. | 2.8 max. |
| Oxygen and nitrogen, % | 0.2 | 0.2 | 0.48 | 0.70 | 0.92 |
| Hydrogen, % | 13.2 | 12.7 | 11.9 | 11.7 | 10.5 |
| Carbon, % | 86.5 | 86.4 | 86.10 | 85.55 | 85.70 |
| Sediment and water, % | Trace | Trace | 0.5 max. | 1.0 max. | 2.0 max. |
| Ash, % | Trace | Trace | 0.02 | 0.05 | 0.08 |
| mJ/kg | 46.4 | 45.5 | 43.9 | 43.4 | 42.5 |

# Appendix C4

## Composition of dry air and Heat of vaporization of water

| Dry air‡ | |
|---|---|
| Species | Mole fraction |
| $N_2$ | 0.78084 |
| $O_2$ | 0.20946 |
| $CO_2$ | 0.00033 |
| Ar | 0.00934 |
| Ne | 0.00002 |
| He | 0.00001 |

Heat of vaporization of water
$(\Delta H_{vap})_{298.15} = 43.98$ kJ/mole

‡ Data taken with permission from the CRC Handbook of Chemistry and Physics, R. C. Weast and M. J. Astle, eds., 58th ed. CRC Press Inc., West Palm Beach, Fla. (1978).

# APPENDIX D

## Van der Waals and critical constants of gases‡

$a$   Pa m$^6$/mole$^2$
$b$   m$^3$/mole
$T_c$   critical temperature, K
$P_c$   critical pressure, Pa
$\rho_c$   critical vapor density, kg/m$^3$
$V_c$   critical volume, m$^3$/mole

| Gases | $a \times 10^7$ | $b \times 10^5$ | $T_c$ | $P_c \times 10^{-6}$ | $\rho_c \times 10^{-3}$ | $V_c \times 10^6$ |
|---|---|---|---|---|---|---|
| Argon | 1.493 | 3.231 | 151.2 | 4.86 | 0.533 | 75 |
| Acetylene | 4.873 | 5.154 | 309.5 | 6.24 | 0.2304 | 113 |
| Ammonia | 4.630 | 3.737 | 405.5 | 11.28 | 0.235 | 72.5 |
| Carbon dioxide | 3.978 | 4.280 | 304.2 | 7.39 | 0.486 | 94 |
| Carbon monoxide | 1.606 | 3.954 | 133.0 | 3.50 | 0.3012 | 93 |
| Chlorine | 7.161 | 5.621 | 417.0 | 7.71 | 0.572 | 124 |
| Ethylene | 4.973 | 5.750 | 283.1 | 5.12 | 0.226 | 124 |
| Hydrogen chloride | 4.051 | 4.085 | 324.6 | 8.26 | 0.419 | 87 |
| Hydrogen | 0.2715 | 2.668 | 33.3 | 1.30 | 0.03102 | 65 |
| Methane | 3.489 | 4.271 | 190.7 | 4.64 | 0.162 | 99 |
| Methyl chloride | 8.241 | 6.480 | 416.3 | 6.68 | 0.353 | 143 |
| Nitrogen | 1.490 | 3.864 | 126.2 | 3.39 | 0.311 | 90 |
| Oxygen | 1.503 | 3.187 | 154.4 | 5.04 | 0.432 | 74 |
| Sulfur dioxide | 7.472 | 5.678 | 430.7 | 7.88 | 0.525 | 122 |
| Water | 6.017 | 3.042 | 647.4 | 22.12 | 0.322 | 56 |

‡ With permission from O. A. Hougen, and K. M. Watson, *Chemical Process Principles*, Part II, *Thermodynamics*, 2d ed. Wiley (1947).

# SUBROUTINE WEIN AND CALCULATION OF SHOCK PROPERTIES IN AN IDEAL REACTIVE GAS WITH REAL GAS ENTHALPIES

# Appendix E1

## Subroutine WEIN‡§

A subroutine to calculate the equilibrium composition for an ideal gas CHNO$I$ system (where $I$ is an inert species such as argon).

At the call the following variables must be specified:

| | |
|---|---|
| T | the temperature, K ($600 < T < 5000$) |
| P | the pressure, pascals |
| RAO | the ratio of inert molecules to atomic oxygen |
| RCO | the carbon–oxygen atomic ratio |
| RHO | the hydrogen–oxygen atomic ratio |
| RNO | the nitrogen–oxygen atomic ratio |
| IA | = 0 if new guesses of species composition are to be used |
| | = 1 if values from previous calculations are to be used |

The subroutine returns with the mole fractions of the species as $X(1)$, $X(2)$, ..., $X(14)$ in the order $H_2$, $O_2$, $H_2O$, CO, $CO_2$, OH, H, O, $N_2$, N, NO, $NO_2$, $CH_4$, $I$ where $I$ is an inert molecule. Additionally, the counter JA contains the number of iterations required for convergence.

The subroutine has been tested and has been found to converge over the range

$$700 \leq T \leq 4700 \text{ K} \qquad : \qquad 25 \text{ kPa} \leq P \leq 1 \text{ MPa}$$
$$3.76 \leq RNO \leq 0.00001 \quad : \qquad 0.3 \leq \Phi \leq 2.0$$
$$10^{-2} \leq RHC \, (= RHO/RCO) \leq 10^{+2}$$

RAO can be zero but RCO, RHO, and RNO must have positive values when the routine is called. It is suggested that if the system does not contain C, H, or N as a component element the ratio for that missing element be set to $10^{-5}$.

Since this subroutine will sometimes not converge when $\Phi = 1.0$, $\Phi$ is set to 1.00001 whenever it is calculated to be 1.0 internally. The subroutine uses JANAF log $(Kp)_f$ values tabulated at 100 intervals. It linearly interpolates log $(Kp)_f$ vs. $1/T$ when $T$ is not an integral multiple of 100. The "guessed" species are $CO_2$, $H_2O$, and $O_2$ for lean mixtures and $CO_2$, $H_2O$, and $H_2$ for rich mixtures. Over a wide range of conditions convergence occurs with 3 to 5 interations. The average machine time for an iteration is about 300 $\mu$s on the CYBER 175.

---

‡ A deck of punched cards of this subroutine can be obtained from: Professor Roger A. Strehlow, 104 S. Mathews Avenue, 105 Transportation Building, University of Illinois, Urbana, Illinois 61801. The cost of punching and mailing is $8.00 domestic and US $27.00 foreign.

§ Based on F. J. Weinberg, *Proc. Roy. Soc.*, **241A**: 132–140 (1975).

```
        SUBROUTINE WEIN
        COMMON P,T,X(14),RAO,RCO,RHO,RNO,IA,JA
        DIMENSION AX1(45),AX2(45),AX3(45),AX4(45),AX5(45),
       *AX6(45),AX7(45),AX8(45),AX9(45),AX10(45)
C
C       AX1=LOG 10 KP H20
C
        DATA AX1/ 18.633,  15.583,  13.289,  11.498,  10.062,
       *           8.883,   7.899,   7.064,   6.347,   5.725,
       *           5.180,   4.699,   4.270,   3.886,   3.540,
       *           3.227,   2.942,   2.682,   2.443,   2.224,
       *           2.021,   1.833,   1.658,   1.495,   1.343,
       *           1.201,   1.067,    .942,    .824,    .712,
       *            .607,    .507,    .413,    .323,    .238,
       *            .157,    .079,    .005,   -.065,   -.133,
       *           -.197,   -.259,   -.319,   -.376,   -.430/
C
C       AX2=LOG 10 KP CO
C
        DATA AX2/ 14.318,  12.946,  11.914,  11.108,  10.459,
       *           9.926,   9.479,   9.099,   8.771,   8.485,
       *           8.234,   8.011,   7.811,   7.631,   7.469,
       *           7.321,   7.185,   7.061,   6.946,   6.840,
       *           6.741,   6.649,   6.563,   6.483,   6.407,
       *           6.336,   6.269,   6.206,   6.145,   6.088,
       *           6.034,   5.982,   5.933,   5.886,   5.841,
       *           5.798,   5.756,   5.717,   5.679,   5.642,
       *           5.607,   5.573,   5.540,   5.508,   5.477/
C
C       AX3=LOG 10 KP CO2
C
        DATA AX3/ 34.405,  29.506,  25.830,  22.970,  20.680,
       *          18.806,  17.243,  15.920,  14.785,  13.801,
       *          12.940,  12.180,  11.504,  10.898,  10.353,
       *           9.860,   9.411,   9.001,   8.625,   8.280,
       *           7.960,   7.664,   7.388,   7.132,   6.892,
       *           6.668,   6.458,   6.260,   6.074,   5.898,
       *           5.732,   5.574,   5.425,   5.283,   5.149,
       *           5.020,   4.898,   4.781,   4.670,   4.563,
       *           4.460,   4.362,   4.268,   4.178,   4.091/
C
C       AX4=LOG 10 KP OH
C
        DATA AX4/ -2.568,  -2.085,  -1.724,  -1.444,  -1.222,
       *          -1.041,   -.890,   -.764,   -.656,   -.563,
       *           -.482,   -.410,   -.347,   -.291,   -.240,
       *           -.195,   -.153,   -.116,   -.082,   -.050,
       *           -.021,    .005,    .030,    .053,    .074,
       *            .094,    .112,    .129,    .145,    .160,
       *            .174,    .188,    .200,    .212,    .223,
       *            .234,    .244,    .253,    .262,    .270,
       *            .278,    .286,    .293,    .300,    .307/
```

```
C
C     AX5=LOG 10 KP H
C

      DATA AX5/-16.336,-13.599,-11.539, -9.934, -8.646,
     *          -7.589, -6.707, -5.958, -5.315, -4.756,
     *          -4.266, -3.833, -3.448, -3.102, -2.790,
     *          -2.508, -2.251, -2.016, -1.800, -1.601,
     *          -1.417, -1.247, -1.089,  -.941,  -.803,
     *           -.674,  -.553,  -.439,  -.332,  -.231,
     *           -.135,  -.044,   .042,   .123,   .201,
     *            .274,   .345,   .412,   .476,   .537,
     *            .595,   .651,   .705,   .757,   .806/
C
C     AX6=LOG 10 KP O
C

      DATA AX6/-18.574,-15.449,-13.101,-11.272, -9.807,
     *          -8.606, -7.604, -6.755, -6.027, -5.395,
     *          -4.842, -4.353, -3.918, -3.529, -3.178,
     *          -2.860, -2.571, -2.307, -2.065, -1.842,
     *          -1.636, -1.446, -1.268, -1.103,  -.949,
     *           -.805,  -.670,  -.543,  -.423,  -.310,
     *           -.204,  -.103,  -.007,   .084,   .170,
     *            .252,   .330,   .404,   .475,   .543,
     *            .608,   .671,   .730,   .788,   .843/
C
C     AX7=LOG 10 KP NO
C

      DATA AX7/ -7.210, -6.086, -5.243, -4.587, -4.062,
     *          -3.633, -3.275, -2.972, -2.712, -2.487,
     *          -2.290, -2.116, -1.962, -1.823, -1.699,
     *          -1.586, -1.484, -1.391, -1.305, -1.227,
     *          -1.154, -1.087, -1.025,  -.967,  -.913,
     *           -.863,  -.815,  -.771,  -.729,  -.690,
     *           -.653,  -.618,  -.585,  -.554,  -.524,
     *           -.496,  -.470,  -.444,  -.420,  -.397,
     *           -.375,  -.354,  -.333,  -.314,  -.296/
C
C     AX8=LOG 10 KP CH4
C

      DATA AX8/  2.001,   .951,   .146,  -.493, -1.011,
     *          -1.440, -1.801, -2.107, -2.372, -2.602,
     *          -2.803, -2.981, -3.139, -3.281, -3.408,
     *          -3.523, -3.627, -3.722, -3.809, -3.889,
     *          -3.962, -4.030, -4.093, -4.152, -4.206,
     *          -4.257, -4.304, -4.349, -4.391, -4.430,
     *          -4.467, -4.503, -4.536, -4.568, -4.598,
     *          -4.626, -4.653, -4.679, -4.704, -4.727,
     *          -4.750, -4.772, -4.793, -4.813, -4.832/
```

```
C
C      AX9=LOG 10 KP NO2
C
       DATA AX9/  -6.111,  -5.714,  -5.417,  -5.185,  -5.000,
     *            -4.848,  -4.721,  -4.612,  -4.519,  -4.438,
     *            -4.367,  -4.304,  -4.248,  -4.198,  -4.152,
     *            -4.111,  -4.074,  -4.040,  -4.008,  -3.979,
     *            -3.953,  -3.928,  -3.905,  -3.885,  -3.864,
     *            -3.846,  -3.828,  -3.812,  -3.797,  -3.783,
     *            -3.770,  -3.757,  -3.746,  -3.734,  -3.724,
     *            -3.714,  -3.705,  -3.696,  -3.688,  -3.680,
     *            -3.673,  -3.666,  -3.659,  -3.652,  -3.646/
C
C      AX10=LOG 10 KP N
C
       DATA AX10/-38.081,-32.177,-27.744,-24.292,-21.528,
     *           -19.265,-17.377,-15.778,-14.406,-13.217,
     *           -12.175,-11.256,-10.437, -9.705, -9.046,
     *            -8.449, -7.905, -7.409, -6.954, -6.535,
     *            -6.149, -5.790, -5.457, -5.147, -4.858,
     *            -4.587, -4.332, -4.093, -3.868, -3.656,
     *            -3.455, -3.265, -3.086, -2.915, -2.752,
     *            -2.598, -2.450, -2.310, -2.176, -2.047,
     *            -1.924, -1.807, -1.694, -1.585, -1.481/
C
C      CHECK TEMPERATURE RANGE AND SET INITIAL VALUES
C
       IF(T.LE.600.)PRINT 100
       IF(T.LE.600.)T=601.
       IF(T.GE.5000.)PRINT 110,T
       IF(T.GE.5000)T=4999.
       PT=P/101325.
       RNC=RNO/RCO
       ROC=1./RCO
       RHC=RHO/RCO
       RAC=RAO/RCO
C
C      SET INITIAL GUESSES.
C          NOTE: RICH & LEAN MIXTURES TREATED DIFFERENTLY
C
       IF(IA.EQ.1)GOTO 2
       CONO=1./(RCO+RHO+RNO+RAO+1.)
       CONN=CONO*RNO
       CONA=CONO*RAO
       CONH=CONO*RHO
       CONC=CONO*RCO
       PHI=2.*RCO+RHO/2.
       IF(PHI.EQ.1.00000)PHI=1.00001
       RHTH2O=TANH(10.0**(0.002353*(5000.0-T)-4.0))
       RHTCO2=TANH(10.0**(-0.00126262626*(T-4417.0)-2.0))
       IF(T .LE. 2600.) RHTCO2=RHTH2O=1.0
       DIV=4.4375*PHI**2-8.875*PHI+7.0975
```

```
        PEAK=0.1347*PHI+0.00459
        QUAL=(1.0-PHI+1.0E-9)/ABS(1.0-PHI+1.0E-9)
        CO2INT=(1.0+QUAL)*PEAK/2.0 + (1.0-QUAL)*(0.28-PEAK)/2.0
        MULT=1.0
        IF( PHI .GT. 1.8 ) MULT=(2.0-PHI)/0.2
        PPCO2=((SQRT(PT*CO2INT*TANH(1.0/RHC))/DIV)*MULT+
       *(PT*0.2/((1.0/RHC)+RHC))*(1.0-MULT))*RHTCO2
        PPH2O=PT*0.2*(TANH(RHC/10.0))*RHTH2O
        IF(PHI.GE.1.0)GOTO 1
C
C       GUESS FOR LEAN MIXTURES
C
        CNT=CONO/2.+CONH/4.+CONN/2.+CONA
        PPO2=PT*(CONO/2.-CONC-CONH/4.)/CNT
        GOTO 3
      1 CONTINUE
C
C       GUESS FOR RICH MIXTURES
C
        PPH2=7.0*PT*SQRT(RHC)/T
        GOTO 3
C
C       GUESS USING LAST VALUES
C
      2 PPH2O=X(3)*PT
        PPCO2=X(5)*PT
        PPH2= X(1)*PT
        PPO2= X(2)*PT

      3 CONTINUE
C
C       CALCULATE EQUILIBRIUM CONSTANTS AT TEMPERATURE T
C
        M=T/100
        T1=M*100
        T2=T1+100.
        M=M-5
        AH2O=AX1(M)+(AX1(M+1)-AX1(M))*(T1-T)*T2/((T1-T2)*T)
        ACO =AX2(M)+(AX2(M+1)-AX2(M))*(T1-T)*T2/((T1-T2)*T)
        ACO2=AX3(M)+(AX3(M+1)-AX3(M))*(T1-T)*T2/((T1-T2)*T)
        AOH =AX4(M)+(AX4(M+1)-AX4(M))*(T1-T)*T2/((T1-T2)*T)
        AH  =AX5(M)+(AX5(M+1)-AX5(M))*(T1-T)*T2/((T1-T2)*T)
        AO  =AX6(M)+(AX6(M+1)-AX6(M))*(T1-T)*T2/((T1-T2)*T)
        ANO =AX7(M)+(AX7(M+1)-AX7(M))*(T1-T)*T2/((T1-T2)*T)
        ACH4=AX8(M)+(AX8(M+1)-AX8(M))*(T1-T)*T2/((T1-T2)*T)
        ANO2=AX9(M)+(AX9(M+1)-AX9(M))*(T1-T)*T2/((T1-T2)*T)
        AN  =AX10(M)+(AX10(M+1)-AX10(M))*(T1-T)*T2/((T1-T2)*T)
        BK1=10.0**(ACO-ACO2)
        BK2=10.0**(-AH2O)
        BK3=10.0**(AOH-AH2O)
        BK4=10.0**(AH)
        BK5=10.0**(AO)
```

```
      BK6=10.0**(ANO)
      BK7=10.0**(ACH4+2*AH2O-ACO2)
      BK8=10.0**(ANO2)
      BK9=10.0**(AN)
C
C     START ITERATION (LOOP 14)
C
      DELH2O=0.
      DELH2=0.
      DELO2=0.
      DELCO2=0.
      DO 14 I=1,50
      JA=I
      KA=0
C
C     RESET PARTIAL PRESSURES USING THE DELTAS
C
      PP1CO2=PPCO2
      PP1H2O=PPH2O
      PPH2O=PPH2O+DELH2O
      IF(PPH2O.GE.0.)GOTO 4
      PPH2O=PP1H2O/2.0
      KA=1
    4 PPCO2=PPCO2+DELCO2
      IF(PPCO2.GE.0.)GOTO 5
      PPCO2=PP1CO2/2.0
      KA=1
    5 IF(PHI.GE.1.0)GOTO 7
      PP1O2=PPO2
      PPO2=PPO2+DELO2
      IF(PPO2.GE.0.)GOTO 6
      PPO2=PP1O2/2.0
      KA=1
    6 PPH2=PPH2O*BK2/SQRT(PPO2)
      PPCH4=0.0
      GOTO 9
    7 CONTINUE
      PP1H2=PPH2
      PPH2=PPH2+DELH2
      IF(PPH2.GE.0.)GOTO 8
      PPH2=PP1H2/2.0
      KA=1
    8 PPO2=(PPH2O*BK2/PPH2)**2.0
      PPCH4=BK7*PPCO2*(PPH2**4)/(PPH2O*PPH2O)
    9 CONTINUE
      PPCO=PPCO2*BK1/SQRT(PPO2)
      PPOH=PPH2O*BK3/SQRT(PPH2)
      PPO=BK5*SQRT(PPO2)
      PPH=BK4*SQRT(PPH2)
      PPA=RAC*(PPCO2+PPCO+PPCH4)
      PPNO=BK6**2*PPO2*(SQRT(1.0+8.0*RNC*(PPCO2+PPCO+PPCH4)/
```

```
      *(PPO2*BK6**2))-1.0)/4.0
       PPN2=.5*RNC*(PPCO2+PPCO+PPCH4)-.5*PPNO
       PPN=BK9*PPN2**.5
       PPNO2=BK8*PPO2*PPN2**.5
       PA=PT-PPNO2-PPN
C
C     CALCULATE CONSTANTS FOR THE DETERMINANTS
C
       IF(PHI.GE.1.0)GOTO 10
       C1=2.0+(1.0/PPH2O)*(2.0*PPH2+.5*PPOH+.5*PPH)
       C2=(1.0/PPO2)*(-PPH2+.25*PPOH-.25*PPH+.5*RHC*PPCO)
       C3=-RHC*(1.0+PPCO/PPCO2)
       C4=2.0*PPH2O+2.0*PPH2+PPOH+PPH-RHC*(PPCO2+PPCO)
       C5=1.0+.5*PPOH/PPH2O
       C6=2.0+(-.5*PPCO+.25*PPOH+.5*PPO+.5*ROC*PPCO)/PPO2+
      *(PPNO/PPO2)*(2.0*PPN2-.5*RNC*PPCO)/(4.0*PPN2+PPNO)
       C7=2.0-ROC-ROC*(PPCO/PPCO2)+(PPCO/PPCO2)+(RNC*PPNO/PPCO2)*
      *(PPCO+PPCO2)/(4.0*PPN2+PPNO)
       C8=2.0*PPCO2+PPCO+PPH2O+2.0*PPO2+PPOH+PPO-ROC*(PPCO2+PPCO)+PPNO
       C9=1.0+(1.0/PPH2O)*(PPH2+.5*PPOH+.5*PPH)+RAO*(1.0+PPOH/PPH2O)
       C10=1.0+(-.5*PPCO-.5*PPH2+.25*PPOH-.25*PPH+.5*PPO
      *-(PPN2*(RNC*PPCO+PPNO)+PPNO*(2.0*PPN2-.5*RNC*PPCO)))/
      *(4.0*PPN2+PPNO))/PPO2-0.5*RAC*PPCO/PPO2
       C11=1.0+(PPCO+2.0*RNC*(PPN2*(PPCO2+PPCO)+.5*PPNO*
      *(PPCO2+PPCO))/(4.0*PPN2+PPNO))/PPCO2+RAC*(1.0+PPCO/PPCO2)
       C12=PPCO2+PPCO+PPH2O+PPO2+PPH2+PPOH+PPH+PPO+PPN2+PPNO+PPA-PA
       GOTO 11
   10 CONTINUE
       CA=4.0*PPN2+PPNO
       CA1=2.0*PPN2*RNC*(PPCO2+PPCO+PPCH4)/(PPCO2*CA)
       CA2=-2.0*PPN2*(RNC*(PPCO+2.0*PPCH4)+PPNO)/(PPH2O*CA)
       CA3=2.0*PPN2*(RNC*(PPCO+4.0*PPCH4)+PPNO)/(PPH2*CA)
       CB1=PPNO*RNC*(PPCO2+PPCO+PPCH4)/(PPCO2*CA)
       CB2=(PPNO/PPH2O)*(1.0-((RNC*(PPCO+2.0*PPCH4)+PPNO)/CA))
       CB3=(PPNO/PPH2)*((RNC*(PPCO+4.0*PPCH4)+PPNO)/CA-1.0)
       C1=2.0+(-0.5*PPOH+0.5*PPH+16.0*PPCH4-RHC*(PPCO+4.0*PPCH4))/PPH2
       C2=2.0+(PPOH-8.0*PPCH4+RHC*PPCO+2.0*RHC*PPCH4)/PPH2O
       C3=-RHC*(1.0+PPCO/PPCO2+PPCH4/PPCO2)+4.0*PPCH4/PPCO2
       C4=2.0*PPH2O+2.0*PPH2+PPOH+PPH+4.0*PPCH4-RHC*(PPCO2+PPCO+PPCH4)
       C5=CB3+(PPCO-4.0*PPO2-0.5*PPOH-PPO-ROC*(PPCO+4.0*PPCH4))/PPH2
       C6=CB2+1.0+(-PPCO+4.0*PPO2+PPOH+PPO+ROC*(PPCO+2.0*PPCH4))/PPH2O
       C7=CB1+2.0-ROC+(PPCO-ROC*(PPCO+PPCH4))/PPCO2
       C8=PPNO+2.0*PPCO2+PPCO+2.0*PPO2+PPOH+PPO+PPH2O-
      *ROC*(PPCO2+PPCO+PPCH4)
       C9=CB3+CA3+1.0+(PPCO-2.0*PPO2-0.5*PPOH+0.5*
      *PPH-PPO+4.0*PPCH4)/PPH2+(RAC/PPH2)*(PPCO+4.0*PPCH4)
       C10=CB2+CA2+1.0+(-PPCO+2.0*PPO2+PPOH+PPO-
      *2.0*PPCH4)/PPH2O-(RAC/PPH2O)*(PPCO+2.0*PPCH4)
       C11=CB1+CA1+1.0+(PPCO+PPCH4)/PPCO2+RAC*(1.0+(PPCO+PPCH4)/PPCO2)
       C12=PPCO2+PPCO+PPH2O+PPO2+PPH2+PPOH+PPH+PPO+PPN2+PPNO+PPCH4+PPA-PA
   11 CONTINUE
```

```
C
C      EVALUATE THE DETERMINANTS
C
       DET1=C2*(C7*C12-C11*C8)-C3*(C6*C12-C10*C8)+C4*(C6*C11-C7*C10)
       DET2=C1*(C7*C12-C11*C8)-C3*(C5*C12-C9 *C8)+C4*(C5*C11-C9*C7 )
       DET3=C1*(C6*C12-C10*C8)-C2*(C5*C12-C9 *C8)+C4*(C5*C10-C6*C9 )
       DET4=C1*(C6*C11-C10*C7)-C2*(C5*C11-C9 *C7)+C3*(C5*C10-C9*C6 )
C
C      CALCULATE THE DELTA"S
C
       IF(DET4.EQ.0.0) DET4=1.0E-100
       DELCO2=-DET3/DET4
       IF(PHI.GE.1.0)GOTO 12
       DELH2O=-DET1/DET4
       DELO2 = DET2/DET4
       IF(KA.EQ.1)GOTO 14
       IF(I.LT.3)GOTO 14
       IF(ABS(DELO2/PPO2)-1.E-6)13,13,14
    12 CONTINUE
       DELH2=-DET1/DET4
       DELH2O=DET2/DET4
       IF(KA.EQ.1)GOTO 14
       IF(I.LT.3)GOTO 14
       IF(ABS(DELH2/PPH2).GT.1.E-6)GOTO 14
    13 IF(ABS(DELH2O/PPH2O).GT.1.E-6)GOTO 14
       IF(ABS(DELCO2/PPCO2).GT.1.E-6)GOTO 14
       IF(ABS(C12/PA).LE.1.E-6)GOTO 15
    14 CONTINUE
       PRINT 120
    15 X(1)=PPH2/PT
       X(2)=PPO2/PT
       X(3)=PPH2O/PT
       X(4)=PPCO/PT
       X(5)=PPCO2/PT
       X(6)=PPOH/PT
       X(7)=PPH/PT
       X(8)=PPO/PT
       X(9)=PPN2/PT
       X(10)=PPN/PT
       X(11)=PPNO/PT
       X(12)=PPNO2/PT
       X(13)=PPCH4/PT
       X(14)=PPA/PT
       RETURN
   100 FORMAT(10X,"TEMPERATURE LESS THAN 600 DEGREES KELVIN,"/
      *15X,"BELOW VALID TEMPERATURE RANGE. T IS SET TO 601.")
   110 FORMAT(10X,"TEMPERATURE GREATER THAN 5000 DEGREES KELVIN,"/
      *F7.1,8X,"ABOVE VALID TEMPERATURE RANGE. T IS SET TO 4999.")
   120 FORMAT(/" NO CONVERGENCE, WEIN")
       END
```

## Appendix E2

### Shock properties in an ideal reactive gas with real gas enthalpies

The basic equations for the transition across a steady normal shock wave are

$$\rho_1 u_1 = \rho_2 u_2 \tag{E-1}$$

$$P_1 + \rho_1 u_1^2 = P_2 + \rho_2 u_2^2 \tag{E-2}$$

$$h_1 + \tfrac{1}{2}u_1^2 = h_2 + \tfrac{1}{2}u_2^2 \tag{E-3}$$

For an ideal gas with composition change we have

$$P_1 = \rho_1 \mathscr{R}_1 T_1 \qquad P_2 = \rho_2 \mathscr{R}_2 T_2 \tag{E-4}$$

and for real gas enthalpies we have

$$h_1 = h_1(T_1) \qquad \text{and} \qquad h_2 = h_2(T_2, P_2, \lambda_{j,2})$$

where the $\lambda_j$'s are reaction coordinates for the gases in state 2. Note that in general

$$\mathscr{R}_2 = \mathscr{R}_2(T_2, P_2, \lambda_{j,2}) \tag{E-5}$$

To solve these equations for a given state one and a value of $T_2$ we define the volume ratio $\varepsilon = \rho_1/\rho_2$ across the shock. Substitution of the state equations (E-4) into (E-2) yields

$$\rho_1 \mathscr{R}_1 T_1 + \rho_1 u_1^2 = \rho_2 \mathscr{R}_2 T_2 + \rho_2 u_2^2$$

or

$$\varepsilon(\mathscr{R}_1 T_1 + u_1^2) = \mathscr{R}_2 T_2 + u_2^2$$

therefore

$$\varepsilon \mathscr{R}_1 T_1 + \varepsilon u_1^2 = \mathscr{R}_2 T_2 + \varepsilon^2 u_1^2$$

or

$$\varepsilon \mathscr{R}_1 T_1 - \mathscr{R}_2 T_2 = \varepsilon(\varepsilon - 1)u_1^2 \tag{E-6}$$

The energy equation (E-3) can be written as

$$h_1 + \tfrac{1}{2}u_1^2 = h_2 + \tfrac{1}{2}\varepsilon^2 u_1^2$$

or

$$2(h_2 - h_1) = (1 - \varepsilon^2)u_1^2 \tag{E-7}$$

If we multiply Eq. (E-6) by $-(\varepsilon + 1)$ and Eq. (E-7) by $\varepsilon$ we obtain

$$-(\varepsilon + 1)(\varepsilon \mathscr{R}_1 T_1 - \mathscr{R}_2 T_2) = \varepsilon(1 - \varepsilon^2)u_1^2$$

and

$$2\varepsilon(h_2 - h_1) = \varepsilon(1 - \varepsilon^2)u_1^2$$

We therefore obtain a quadratic equation in $\varepsilon$

$$\varepsilon^2 + \left[\frac{(2h_2 - \mathscr{R}_2 T_2) - (2h_1 - \mathscr{R}_1 T_1)}{\mathscr{R}_1 T_1}\right]\varepsilon - \frac{\mathscr{R}_2 T_2}{\mathscr{R}_1 T_1} = 0 \tag{E-8}$$

which has the solution

$$\varepsilon = \frac{-\psi + \sqrt{\psi^2 + 4\mathscr{R}_2 T_2/\mathscr{R}_1 T_1}}{2} \tag{E-9}$$

where

$$\psi = \frac{(2h_2 - \mathscr{R}_2 T_2) - (2h_1 - \mathscr{R}_1 T_1)}{\mathscr{R}_1 T_1} \tag{E-10}$$

Equations (E-9) and (E-10) can be solved explicitly for unreactive shocks because in this case $h_2 = h_2(T_2)$. Subsequent substitution into Eqs. (E-1) to (E-4) will yield all the properties across the shock. For reactive shocks an explicit solution is not available and one must iterate until the assumed pressure satisfies both the momentum equation and the ideal gas law.

# APPENDIX F

## NOMENCLATURE FOR *COMBUSTION FUNDAMENTALS*

**Letters**

| | |
|---|---|
| $A$ | molar Helmholtz free energy (or work function), J/mole |
| $A$ | constant coefficient (Chap. 5) |
| $A$ | proportionality constant, dimensionless (Chap. 1) |
| $A$ | species A |
| $A$ | area, $m^2$ (Chaps. 3, 5, 10, 15) |
| $A$ | frequency factor of a general chemical reaction (units depend on order of reaction) |
| $A_i$ | frequency factor of $i$th elementary chemical reaction (same units as $k_i$) |
| $A_s$ | frequency factor for a surface reaction [units defined in Eq. (7-44)] |
| $A_i$ | atomic weight of $i$th element, kg/mole |
| AIT | Auto ignition temperature, K |
| $\mathscr{A}$ | phonon absorption rate, $J/m^3$-s (Chap. 5) |
| $\mathscr{A}$ | Avagadro's number, $6.02283 \times 10^{23}$ molecules/mole |
| $\mathscr{A}$ | dimensionless velocity of sound (Chap. 9) |
| $a$ | specific Helmholtz free energy, J/kg (Chap. 4) |
| $a$ | velocity of sound, m/s |
| $a$ | average distance, m (Chap. 3) |
| $a$ | Van der Waals constant, Pa $m^6/mole^2$ (Chap. 3) |
| $a_i$ | activity of the $i$th component in a solution |
| $a'$ | fluctuating mole fraction difference from mean, $x_A - \bar{x}_A$ |
| $B$ | constant coefficient (Chap. 5) |
| $B$ | proportionality constant, dimensionless (Chap. 1) |
| $B$ | species $B$ |
| $B$ | transfer coefficient for droplet evaporation, dimensionless |
| $B$ | constants for propellants and rocket engines (Chap. 15) |

| | |
|---|---|
| $B(T)$ | first virial coefficient, $m^3$/mole |
| $B^*$ | dimensionless first virial coefficient |
| $b_0$ | effective volume of one mole of molecules, $m^3$/mole |
| $b(T)$ | first virial coefficient, $Pa^{-1}$ |
| $C$ | chain carrier, reactive molecule |
| $C$ | capacity of a condenser, farads (Chap. 12) |
| $C$ | constant coefficient (either real or imaginary) |
| $C\dagger$ | collision integral ratio for thermal diffusion, dimensionless |
| $C^*$ | dimensionless second virial coefficient |
| $C_p$ | molar heat capacity of a gas at constant pressure, J/mole-K |
| $C_v$ | molar heat capacity of a gas at constant volume, J/mole-K |
| $C(T)$ | second virial coefficient, $m^6$/mole$^2$ |
| $c$ | molecular speed, m/s (Chaps. 1 and 3) |
| $c$ | specific heat capacity of a solid, J/kg-K (Chap. 2) |
| $c$ | number of component elements in system |
| $c$ | constant relating the diffusion coefficient to temperature, $m/s^2$-$K^2$ (Chap. 8) |
| $c$ | propagation velocity of an acoustic wave on a shock wave, m/s (Chap. 9) |
| $c_p$ | specific heat capacity of a gas at constant pressure, J/kg-K |
| $c_{pf}$ | translational heat capacity at constant pressure, J/kg-K (Chap. 3) |
| $c_v$ | specific heat of a gas at constant volume, J/kg-K |
| $c(T)$ | second virial coefficient, $Pa^{-2}$ |
| $D$ | binary diffusion coefficient, $m^2$/s |
| $D$ | induction zone distance in a detonation, m (Chap. 9) |
| $D$ | depth, m (Chap. 10) |
| $D'$ | multicomponent diffusion coefficient, $m^2$/s |
| $D^T$ | thermal diffusion coefficient, $kg$/m-s |
| Da | Damköhler number (see Table 5-1), dimensionless |
| $d$ | diameter, m |
| $d$ | quenching distance, mm |
| $d$ | mass flow eigenvalue ratio for a superadiabatic flame, $\mathcal{M}/\mathcal{M}_p$, dimensionless |
| $E$ | molar internal energy, J/mole |
| $E$ | spark energy, J (Chap. 12) |
| $E$ | acoustic energy, J (Chap. 14) |
| $E$ | a dimensionless constant (Chap. 7) |
| $E_c$ | critical energy for direct initiation of detonation, J (Chap. 13) |
| $\Delta E_f^\circ$ | molar internal energy of formation, J/mole |
| $^iE$ | symbol for $i$th element |
| $E^*$ | activation energy, J/mole |
| $\mathscr{E}$ | energy per unit volume, $J/m^3$ |
| $\mathscr{E}$ | phonon emission rate, $J/m^3$-s |
| $e$ | specific internal energy, J/kg |
| $e$ | excess air or oxygen per mole of fuel |
| $F$ | molar Gibbs free energy, J/mole |
| $F$ | fuel |
| Fr | Froude number (see Table 5-1), dimensionless |
| $\mathscr{F}$ | force, N (Chap. 3) |
| $\mathscr{F}$ | dimensionless heat flux (Chap. 8) |
| $f$ | specific Gibbs free energy, J/kg |
| $f$ | frequency, $s^{-1}$ |
| $f$ | constant relating thermal conductivity to temperature, $J/m\ s^2$-$K^2$ (Chap. 8) |
| $f(u, v, w)$ | molecular velocity distribution |
| $f(V)$ | molecular velocity distribution |
| $f(\ )$ | functional form |
| $\mathscr{f}$ | body force on $i$th specie, N |

| | |
|---|---|
| $f$ | fugacity, dimensionless |
| $G$ | momentum flux in a jet (Chap. 11) |
| $\mathcal{G}$ | total mass of mixture, kg |
| $g$ | body force due to gravity, N |
| $g_s$ | correlation function for scale of segregation |
| $g(u, v)$ | molecular velocity distribution in terms of the $x$ and $y$ components |
| $g(\ )$ | functional form |
| $(g)$ | gaseous state |
| $H$ | molar enthalpy, J/mole |
| $\Delta H_c$ | molar enthalpy (heat) of combustion, J/mole |
| $\Delta H_f^\circ$ | molar enthalpy (heat) of formation, J/mole |
| $\Delta H_r$ | molar enthalpy (heat) of reaction/mole of reaction, J/mole of reaction, as written |
| $\mathcal{H}$ | dimensionless heat transfer coefficient, $J/m^3$-s-K (Chap. 8) |
| $h$ | specific enthalpy, J/kg |
| $h$ | effective heat transfer coefficient, $J/m^2$-s-K (Chaps. 6, 8) |
| $h$ | flame height, m (Chap. 8) |
| $h$ | location of heat source (Chap. 14) |
| $h_D$ | mass transfer coefficient, $kg/m^2$-s (Chap. 7) |
| $h_s$ | sensible enthalpy, $J/m^3$ |
| $\Delta h_f^\circ$ | specific enthalpy (heat) of formation, J/kg |
| $h(u)$ | $x$ component of molecular velocity distribution |
| $h$ | Planck's constant, $6.6256 \times 10^{-34}$ J-s |
| $I$ | inert species |
| $I$ | Species I (a general species) |
| $I$ | intensity of turbulence, dimensionless |
| $I_s$ | intensity of segregation—mixing turbulent flows, dimensionless |
| $[I]$ | concentration of $i$th species, $moles/m^3$ |
| $J$ | molar flow rate, moles/s |
| $j$ | mass flux relative to mass velocity, $kg/m^2$-s |
| $K$ | Karlovitz number, dimensionless |
| $K.E.$ | kinetic energy J |
| $K_T$ | thermal diffusion ratio, dimensionless |
| $K_c$ | equilibrium constant expressed in terms of concentrations |
| $K_n$ | equilibrium constant expressed in terms of moles/kg |
| $K_p$ | equilibrium constant expressed in terms of dimensionless partial pressures, $\not{p}_i$ |
| $K_{pf}$ | equilibrium constant of formation expressed in terms of dimensionless partial pressures, $\not{p}_i$ |
| $\mathcal{K}$ | dimensionless heat transfer coefficient (Chap. 8) |
| $\mathcal{K}$ | rocket performance constant (Chap. 15) |
| $^n k_j$ | rate constant of $j$th elementary chemical reaction of order $n$ units: $n = 1$, $s^{-1}$ |
| | $\qquad n = 2$, $m^3$ moles$^{-1}$ $s^{-1}$ |
| | $\qquad n = 3$, $m^6$ moles$^{-2}$ $s^{-1}$ ($n$ usually omitted) |
| $k$ | coupling constant in acoustics, $s^{-1}$ |
| $k$ | droplet evaporation constant, $m^2/s$ (Chap. 7) |
| $k$ | Boltzman constant, $1.3805 \times 10^{-23}$ J/molecule K |
| $L$ | characteristic length, tube length, block length, wavelength, m |
| $L^*$ | characteristic rocket length $V/A$, m |
| $(L)$ | left-hand propagating wave |
| $L_s$ | linear scale of segregation |
| Le | Lewis number ($\kappa/c_p D\rho$), dimensionless |
| LEL | lean explosion limit |
| LFL | lean flammability limit |
| $\mathcal{L}$ | Schvab-Zel'dovich operator (Chap. 5) |

| | |
|---|---|
| $\mathscr{L}$ | dimensionless block length (Chap. 8) |
| $l$ | mean free path, m/collision |
| $l$ | direction cosine to the $x$ direction |
| $\ell$ | latent heat of evaporation, J/kg |
| $(\ell)$ | liquid state |
| $M$ | molecular weight, kg/mole |
| $M$ | Mach number ($\mathbf{V}/a$), dimensionless |
| $M$ | third body |
| $\dot{M}$ | mass flow, m (Chap. 7) |
| $\mathscr{M}$ | superadiabatic flame mass flow, kg/m$^2$-s (Chap. 8) |
| $\mathscr{M}_c$ | consumption of lean reactant at surface, kg/m$^2$-s |
| $\mathscr{M}_D$ | diffusive mass flow of lean reactant to surface, kg/m$^2$-s |
| $\mathscr{M}_p$ | planar mass flow, kg/m$^2$-s |
| $\mathscr{M}_s$ | cylindrical or spherical mass flow defined as $\pi \mathscr{M}_p$, kg/m$^2$-s |
| $m$ | mass of molecule, kg/molecule |
| $m$ | direction cosine to the $y$ direction |
| $\dot{m}$ | mass flow, kg/s |
| $mf$ | mole fraction, dimensionless |
| $N$ | molecules per unit volume, molecules/m$^3$ |
| $N_I$ | number of atoms of the $i$th element of mixture, moles/kg of system |
| $\mathbf{N}_i$ | net flux of species $i$, kg/m$^2$-s |
| $N_{I/J}$ | atomic ratio of the $I$th to the $J$th element in system, dimensionless |
| $\mathscr{N}$ | molar density, moles/m$^3$ |
| $\mathfrak{N}_i$ | number of atoms of the $i$th element per mass of system, moles/$\mathscr{G}$ kg |
| $n$ | reaction order |
| $n$ | $n$th term in series |
| $n$ | direction cosine to the $z$ direction |
| $n_i$ | number or moles of $i$th species per kg of mixture, moles/kg of system |
| $\Delta n_f$ | net moles of gas produced per mole of formation reaction, moles |
| $O$ | oxidizer |
| $P$ | pressure, Pa |
| $P(\ )$ | probability density function of the variable ( ), dimensionless |
| $P/d$ | pitch to diameter ratio (single spin detonation), dimensionless |
| $\mathscr{P}$ | pressure, dimensionless ($P/101,325.0$) |
| $\mathfrak{P}_i$ | fictitious partial pressure of the $i$th atom in a mixture for equilibrium calculations (Chap. 4) |
| Pe | Peclet number (see Table 5-1), dimensionless |
| Pe, $s$ | Peclet number for mass transfer (see Table 5-1), dimensionless |
| Pr | Prandtl number ($c_p \mu/\kappa$), dimensionless |
| P.E. | potential energy, J |
| $p$ | number of reactions occurring in system |
| $p_i$ | partial pressure of $i$th species, Pa |
| pdf | probability density function, dimensionless |
| $\not{p}_i$ | partial pressure of the $i$th species, dimensionless ($p_i/101,325.0$) |
| $Q$ | heat added to system, J |
| $Q$ | heat transferred across a plane, J (Chap. 3) |
| $q$ | effective heat of reaction per kilogram of mixture, J/kg |
| $q_{\mathrm{fu}}$ | heat of oxidation of fuel, J/kg |
| $\mathbf{q}$ | energy flux, J/m$^2$-s |
| $R$ | molar (or universal) gas constant, 8.31434 J/mole-K |
| $R$ | radius of curvature, m |
| $R_0$ | radius of a spherical or cylindrical object, m |
| $(R)$ | right-hand propagating wave |
| Re | Reynolds number ($LV/\nu$), dimensionless |

| | |
|---|---|
| Re( ) | real part of an imaginary function |
| $RR$ | reaction rate |
| $\mathscr{R}$ | specific gas constant, J/kg-K |
| $r$ | maximum number of reaction coordinates in a system $(= s - c)$ |
| $r$ | radial distance, m |
| $r$ | molecular separation distance, nm |
| $r$ | regression rate of propellant surface, m/s (Chap. 15) |
| $r_i$ | $i$th reaction in a system |
| $r_L$ | radius of a drop of liquid, m |
| $r_F$ | flame radius, m |
| $S$ | molar entropy, J/mole-K |
| $\Delta S_f^\circ$ | molar entropy of formation, J/mole-K |
| $[S]$ | total species concentration, moles/m$^3$ |
| $\mathscr{S}$ | swirl number (Chap. 11) |
| $S_u$ | normal burning velocity, m/s |
| $S_b$ | normal flow velocity downstream of a flame, m/s |
| Sc | Schmidt number $(\mu/\rho D)$, dimensionless |
| Sh | Sherwood number $(h_D d/\rho D)$, dimensionless |
| $s$ | specific entropy, J/kg-s |
| $s$ | number of species in system |
| $s$ | condensation $(\delta\rho/\rho_0$ or $\gamma\delta p/p_0)$, dimensionless |
| $s$ | surface area, m$^2$ |
| $s$ | coordinate attached to flame, m (Chap. 10) |
| $(s)$ | solid state |
| $T$ | temperature, K |
| $t$ | time, s |
| $U$ | uniform background velocity in $x$ direction only, m/s |
| UEL | upper explosion limit |
| UFL | upper flammability limit |
| $\mathscr{U}$ | dimensionless flow velocity (Chap. 9) |
| $u$ | $x$ component of velocity or radial velocity, m/s |
| $V$ | molar volume, m$^3$/mole |
| $V$ | vessel volume, m$^3$ (Chaps. 6, 15) |
| $V$ | volumetric flow ratio, m$^3$/s (Chap. 8) |
| $V$ | voltage, volts (Chaps. 8, 12) |
| $V^{\cdot}$ | rocket engine free volume, m$^3$ (Chap. 15) |
| $V_s$ | volumetric scale of segregation |
| $\mathbf{V}$ | velocity, m/s |
| $v$ | specific volume, m$^3$/kg |
| $v$ | $y$ component of velocity or tangential velocity, m/s |
| $v$ | Kolmogorov velocity scale of turbulence, m/s |
| $v_i$ | diffusion velocity of the $i$th species in the $x$ direction, m/s |
| $W$ | work performed on surroundings, J |
| $w$ | $z$ component of velocity, m/s |
| $w$ | velocity of triple-point trajectory, m/s (Chap. 8) |
| $X$ | halogen atom |
| $x$ | cartesian coordinate, m |
| $x$ | defined variable (Chap. 7) |
| $x_i$ | mole fraction of $i$th species, dimensionless |
| $Y_i$ | mass of $i$th element per kg of mixture, kg/kg of mixture |
| $y$ | cartesian coordinate, m |
| $y$ | pressure ratio, dimensionless |
| $y_i$ | mass of $i$th species per kg of mixture, kg/kg of mixture |

| $Z$ | compressibility factor, dimensionless |
|---|---|
| $Z$ | cell size in a detonation, m (Chaps. 9, 13) |
| $Z_{molecules}$ | collision number, molecules colliding/$m^3$-s |
| $Z_{wall}$ | molecular collisions with wall, molecule/$m^2$-s |
| $z$ | cartesian coordinate, m |
| $z$ | dimensionless mass flux (Chap. 8) |

**Greek letters**

| $\alpha$ | thermal diffusivity, $m^2/s$ |
|---|---|
| $\alpha$ | integration constant (Chap. 1) |
| $\alpha$ | reaction coordinate, mole of reaction/0.048 kg of mixture (Chap. 2) |
| $\alpha$ | degree of dissociation, dimensionless (Chap. 4) |
| $\alpha$ | chain cycle branching factor (Chap. 6) |
| $\alpha$ | angle between flame and approach flow (Chap. 8) |
| $\alpha$ | intersection angle of triple-point paths (Chap. 9) |
| $\alpha_i$ | thermal diffusion factor, dimensionless |
| $\alpha_{ij}$ | multicomponent thermal diffusion factor, dimensionless |
| $\beta$ | integration constant (Chaps. 1, 6) |
| $\beta$ | reaction coordinate, mass fraction of reaction (Chap. 5) |
| $\beta$ | reaction coordinate, moles of reaction/0.048 kg of mixture (Chap. 2) |
| $\beta$ | general scalar property, with subscript, dimensionless (Chap. 5) |
| $\beta$ | Schvab–Zel'dovich coupling variable, dimensionless (Chap. 5) |
| $\beta$ | $\mathcal{N} c_p u/\kappa$ in infinite activation energy asymptotics, $m^{-1}$ (Chap. 8) |
| $\beta$ | defined variable (Chap. 7) |
| $\beta$ | angle between triple-point trajectory and stretch orientation (Chap. 9) |
| $\beta$ | deflection angle of flow by an oblique flame |
| $\beta_f$ | frozen composition coefficient of thermal expansion, $K^{-1}$ |
| $\Gamma$ | net molecular flow of one species due to diffusion, molecules/$m^2$-s |
| $\Gamma$ | defined variable in asymptotic analysis (Chap. 8) |
| $\Gamma$ | Gruniesen coefficient (Chap. 15) |
| $\gamma$ | heat capacity ratio ($c_p/c_v$), dimensionless |
| $\Delta$ | effective reaction rate for nozzle flows, dimensionless (Chap. 5) |
| $\Delta$ | incremental value |
| $\Delta\Psi$ | potential (bond) energy change during a reactive collision, J/molecule |
| $\delta$ | incremental partial pressure, dimensionless (Chap. 4) |
| $\delta$ | incremental value |
| $\delta_{ij}$ | Kronecker delta (Chap. 5) |
| $\delta$ | branching factor [Eq. (6-13)] |
| $\delta$ | reciprocal dimensionless activation energy (Chap. 8) |
| $\delta$ | angle between flow direction and flame propagation direction (Chap. 10) |
| $\varepsilon$ | dimensionless activation energy relative to some reference temperature ($E^*/RT_{ref}$) |
| $\varepsilon$ | energy stored per molecule, J/molecule (Chap. 1) |
| $\varepsilon$ | volume ratio, dimensionless |
| $\dot{\varepsilon}$ | rate of energy supply to "equilibrium" turbulent regime, J/s |
| $\varepsilon$ | Lennard–Jones well depth, J/molecule |
| $\varepsilon_n$ | energy in the $n$th vibrational quantum state, J/molecule |
| $\varepsilon(t)$ | energy transferred to the equilibrium range of turbulence, J/s |
| $\varepsilon_r$ | rotational energy storage, J/molecule |
| $\varepsilon_t$ | translational energy storage, J/molecule |
| $\varepsilon_v$ | vibrational energy storage, J/molecule |
| $\zeta$ | molecule velocity distribution in spherical coordinates, (Chap. 1) |
| $\zeta$ | temperature-dependence of a third-order reaction, dimensionless |
| $\zeta$ | dimensionless distance (Chap. 7) |

| | |
|---|---|
| $\zeta_F$ | dimensionless flame height (Chap. 7) |
| $\eta$ | molecular velocity distribution in spherical coordinates (Chap. 1) |
| $\eta$ | Kolmogorov length scale of turbulence, m |
| $\eta_i$ | number of moles of species in mixture, moles/$\mathscr{G}$ kg of mixture (Chap. 4) |
| $\eta$ | transposed volume $[v_2/v_1 - (\gamma - 1)/(\gamma + 1)]$, dimensionless |
| $\eta$ | extent of reaction in well-stirred reactor or catalytic combustor (Chaps. 6, 7) |
| $\eta$ | dimensionless distance (Chap. 7) |
| $\eta_0$ | preheat zone thickness of a flame, m |
| $\eta(c)$ | molecular velocity distribution function, dimensionless (Chaps. 3 to 4) |
| $\Theta$ | expansion variable for inner region (Chap. 8) |
| $\hat{\Theta}$ | a modified expansion variable for the inner region, $\phi\Theta/(\phi - 1)$ |
| $\theta$ | angle in spherical coordinates, radians |
| $\theta$ | flow deflection angle or angle, radians |
| $\theta$ | dimensionless temperature, $T/T_0$ (Chap. 6) |
| $\theta_D$ | characteristic dissociation temperature, K |
| $\theta_v$ | characteristic vibrational temperature, K |
| $\kappa$ | thermal conductivity, J/m-s-K |
| $\Lambda$ | imaginary coefficient, s |
| $\Lambda_g, \Lambda_f$ | macroscale or integral scale of turbulence, m |
| $\Lambda$ | mass flow eigenvalue for flames ($i$ represents order of the reaction), dimensionless (Chap. 8) |
| $\lambda$ | reaction coordinate, moles of reaction/kg of mixture |
| $\lambda$ | constant coefficient in thermal explosion theory, K/s (Chap. 6) |
| $\lambda_g, \lambda_f$ | microscale or Taylor scale of turbulence, m |
| $\mu$ | coefficient of viscosity, kg/m-s |
| $\mu$ | Mach angle, radians |
| $\dot{\mu}$ | dimensionless mass flow (Chap. 7) |
| $\mu_i$ | chemical potential of the $i$th species, J/mole |
| $\mu A$ | current, microamperes |
| $v$ | kinematic viscosity, m$^2$/s |
| $v$ | stoichiometric coefficient for the oxidizer in an overall chemical reaction |
| $v_0$ | oscillator frequency, Hz |
| $v_{ij}$ | stoichiometric coefficient of $i$th species in $j$th chemical reaction, dimensionless integer |
| $\Xi(\varepsilon)$ | equilibrium distribution function for molecular energy |
| $\xi$ | transposed pressure ratio $[P_2/P_1 + (\gamma - 1)/(\gamma + 1)]$, dimensionless |
| $\xi$ | dimensionless block length (Chap. 8) |
| $\xi_0$ | dimensionless distance from back wall (Chap. 13) |
| $\xi$ | fugacity coefficient, dimensionless (Chap. 4) |
| $\xi_{ij}$ | coefficient for $i$th element in $j$th species to determine molecular weight, dimensionless integer |
| $\pi$ | 3.141592 (dimensionless) |
| $\pi$ | stress tensor, kg/m-s |
| $\rho$ | density, kg/m$^3$ |
| $\rho'$ | fluctuating density $(\rho - \bar{\rho})$, kg/m$^3$ |
| $\sigma$ | emissivity of a solid, dimensionless |
| $\sigma$ | dimensionless distance (Chap. 7) |
| $\sigma_i$ | effective density change due to the combined enthalpy and mass production for the $i$th species in a chemical reaction, kg/m$^3$ |
| $\sigma_0$ | molecular diameter, nm |
| $\tau$ | induction delay time |
| $\tau$ | dimensionless temperature (Chap. 8) |
| $\tau$ | viscous stress tensor, kg/m-s |
| $\tau_f$ | characteristic flow time, s |

| | |
|---|---|
| $\tau_r$ | characteristic reaction time, s |
| $\Phi$ | fuel equivalence ratio, dimensionless |
| $\phi$ | angle in spherical coordinates, radians |
| $\phi$ | dimensionless flame temperature (Chap. 8) |
| $\phi$ | extent of heat addition $q/q_{CJ}$ (Chap. 9) |
| $\phi$ | exit angle for a symmetric triple-point intersection (Chap. 9) |
| $\phi$ | quantum efficiency (Chap. 16) |
| $\phi_i$ | $i$th component of a flux (Chap. 14) |
| $\chi$ | stretched distance for asymptotic analysis, m |
| $\chi_{ij}$ | correction term for viscosity and thermal conductivity |
| $\Psi$ | Lennard–Jones potential, J/molecule |
| $\psi$ | oscillator frequency, radians/s |
| $\psi$ | explosion delay locus (Chap. 13) |
| $\Omega_{\mu/\kappa}$ | collision integral for viscosity and thermal conductivity, dimensionless |
| $\Omega_D$ | collision integral for diffusion, dimensionless |
| $\omega$ | rate of species production by a chemical reaction, kg/m$^3$-s |
| $\omega$ | wave number (1/wavelength), m$^{-1}$ |
| $\omega_i$ | thermal contribution to the molar Gibbs free energy of the $i$th species, J/mole (Chap. 4) |
| $\omega_0$ | oscillation frequency, s$^{-1}$ |

## Subscripts

| | |
|---|---|
| $A$ | species $A$ |
| $A$ | atomic species (Chap. 4) |
| $a$ | gasification temperature (Chap. 15) |
| air | air |
| amb | ambient |
| $B$ | species $B$ |
| $b$ | branching reactions |
| $b$ | blowout state (well-stirred reactors and catalytic combustors) |
| $b$ | burned state (flames) |
| CJ | Chapman–Jouguet conditions |
| $c$ | concentration |
| $c$ | critical state for a pure substance (Chap. 3) |
| $c$ | convergence length (Chap. 13) |
| calc | calculated |
| chem | due to chemical reaction |
| cr | critical conditions |
| cycle | for one cycle |
| $D$ | dissociation |
| $E$ | effective dissociation value |
| $e$ | exhaust |
| eq | equilibrium |
| $F$ | flame sheet |
| $f$ | forward chemical reaction |
| $f$ | final state |
| $f$ | frozen composition |
| fu | fuel |
| $g$ | gas |
| $I$ | integral scale of turbulence |
| $i$ | $i$th species |
| $i$ | initial state |
| $i$ | running index |
| ig | ignition |

| | |
|---|---|
| in | inert |
| inn | inner region of expansion (Chap. 8) |
| irr | irreversible |
| $j$ | running index |
| $K$ | Kolmogorov scale of turbulence |
| $L$ | liquid surface (of drop) |
| $L$ | lower flash point (Chap. 12) |
| $M$ | molecular species (Chap. 4) |
| $m$ | maximum value |
| meas | measured |
| mix | mixture |
| min | minimum value |
| mono | monatomic gas |
| $n$ | $n$th term |
| o | standard static conditions |
| out | outer region of expansion (Chap. 8) |
| ox | oxidizer |
| $p$ | constant pressure |
| $p$ | propellant |
| process | process |
| prod | products |
| $r$ | rotational (Chap. 1) |
| $r$ | reduced states relative to critical properties of a gas (Chap. 3) |
| $r$ | reverse chemical reaction |
| $r$ | of reaction |
| $r$ | universal concentration variable (Chap. 7) |
| reactants | reactants |
| rec | recombination |
| ref | reference state |
| rev | reversible |
| $rs$ | reflected shock |
| $s$ | constant entropy |
| $s$ | number of species |
| $s$ | solid |
| $s$ | time after shock passage (Chap. 13) |
| stoich | stoichiometric mixture |
| $T$ | temperature, constant temperature |
| $T$ | turbulent |
| $T$ | total number of molecules (Chap. 4) |
| $t$ | translational (Chap. 1) |
| $t$ | terminating reactions |
| throat | throat of nozzle |
| $U$ | upper flash point (Chap. 12) |
| $u$ | atomic content of carbon in a fuel, either moles/mole or moles/kg |
| $u$ | unburned state, flames |
| $us$ | surface temperature (Chap. 15) |
| $v$ | atomic content of hydrogen in a fuel, either moles/mole or moles/kg |
| $v$ | constant volume |
| v | vibrational (Chap. 1) |
| $w$ | atomic content of oxygen in a fuel, either moles/mole or moles/kg |
| $w$ | wall |
| $x$ | in the $x$ direction |
| $x$ | atomic content of nitrogen in a fuel, either moles/mole or moles/kg |

| $y$ | in the $y$ direction |
|---|---|
| $y$ | atomic content of sulfur in a fuel, either moles/mole or moles/kg |
| $z$ | in the $z$ direction |
| $\phi$ | in the transverse (rotational) direction |
| 0 | initial conditions, initial value or background value |
| 0 | round tube quenching distance (Chap. 12) |
| 1/2 | half-life of a species in a chemical reaction |
| 1 | state 1 |
| 1 | molecule 1 (Chap. 1) |
| 1 | reaction number one (Chap. 2) |
| 2 | state 2 |
| 2 | molecule 2 (Chap. 1) |
| 2 | reaction number two (Chap. 2) |
| 3 | reaction number three (Chap. 2) |
| 3 | state 3 |
| 4 | state 4 |
| 5 | state 5 |
| 6 | state 6 |
| $\infty$ | at infinity |
| $\parallel$ | traveling parallel to, also parallel-plate quenching distance (Chap. 12) |

**Superscripts**

| $(c)$ | due to thermal conduction |
|---|---|
| $(D)$ | due to concentration gradients |
| $(f)$ | due to body forces |
| $j$ | power for flow geometry, $j = 0, 1, 2$ for planar, cylindrical, or spherical coordinates |
| $\ell$ | concentration dependence for an inert in an overall Arrhenius expression |
| $m$ | concentration dependence for oxidizer in an overall Arrhenius expression |
| $n$ | concentration dependence for fuel in an overall Arrhenius expression, also general reaction order |
| $n$ | pressure exponent of rocket-propellant regression rate (Chap. 15) |
| $\circ$ | standard state conditions |
| $(p)$ | due to pressure gradients |
| $(R)$ | due to radiation |
| $(T)$ | due to temperature gradients |
| $-$ | average value |
| $\cdot$ | time derivative |
| $*$ | activated state of a molecule |
| $'$ | incorrect value, used in iteration technique (Chap. 4) |
| $'$ | modified swirl number or axial momentum flux (Chap. 11) |

**Math operators**

| cos | cosine |
|---|---|
| $D$ | total differential |
| $d$ | differential |
| $d$ | path process differential |
| exp, $e$ | exponential function |
| $\mathscr{L}$ | ordinary differential operation |
| log | logarithm to the base 10 |
| ln | logarithm to the base $e$ |
| sec | secant |
| sin | sine |

| | |
|---|---|
| tan | tangent |
| $\nabla$ | del — $(\partial/\partial x, \partial/\partial y, \partial/\partial z)$ |
| $\Delta$ | small change in property |
| $\delta$ | incremental change in property |
| $\prod$ | product sign |
| $\sum$ | summation sign |
| $\int$ | integral sign |
| $\sqrt{\phantom{x}}$ | square root |
| $\partial$ | partial differential |
| $\propto$ | is proportional to |
| [ ] | species concentration, moles/m$^3$ |
| $\rightarrow$ | approaches |
| $\vert_x$ | at the location $x$ |

# INDEX